IMMEDIATE EARLY GENES AND INDUCIBLE TRANSCRIPTION FACTORS IN MAPPING OF THE CENTRAL NERVOUS SYSTEM FUNCTION AND DYSFUNCTION

HANDBOOK OF CHEMICAL NEUROANATOMY

Series Editors: A. Björklund and T. Hökfelt

Volume 19

IMMEDIATE EARLY GENES AND INDUCIBLE TRANSCRIPTION FACTORS IN MAPPING OF THE CENTRAL NERVOUS SYSTEM FUNCTION AND DYSFUNCTION

Editors:

L. KACZMAREK
Department of Molecular and Cellular Neurobiology,
Nencki Institute, Pasteura 3, 02-093 Warsaw, Poland

H.A. ROBERTSON
Department of Pharmacology, Faculty of Medicine,
Dalhousie University, Halifax, NS B3H 4H7, Canada

2002

ELSEVIER

Amsterdam – Boston – London – New York – Oxford – Paris – San Diego
San Francisco – Singapore – Sydney – Tokyo

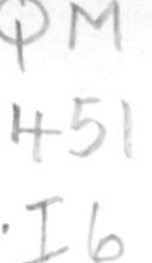

ELSEVIER SCIENCE B.V.
Sara Burgerhartstraat 25
P.O. Box 211, 1000 AE Amsterdam, The Netherlands

First edition 2002

Library of Congress Cataloging in Publication Data
A catalog record from the Library of Congress has been applied for.

ISBN: 0-444-50835-X (volume)
ISBN: 0-444-90340-2 (series)
ISSN: 0924-8196 (series)

⊗ The paper used in this publication meets the requirements of ANSI/NISO Z39.48-1992 (Permanence of Paper).
Printed in The Netherlands.

List of Contributors

H. BESTER
Département de Neurophysiologie in vivo
Sanofi-Synthélabo
31 Avenue P. Vaillant Couturier
92225 Bagneux Cedex
France

D. BILKEY
Department of Psychology
University of Otago
Dunedin
New Zealand

S. BRECHT
Institute of Pharmacology
University of Kiel
Hospitalstrasse 4
24105 Kiel
Germany

J. CABOCHE
Laboratoire de Neurochimie-Anatomie
Institut de Neurosciences
CNRS-UMR 7624
9 Quai Saint-Bernard
75005 Paris
France

A. CHAUDHURI
Department of Psychology
McGill University
1205 Dr. Penfield Avenue
Montreal, PQ H3A 1B1
Canada

M. DRAGUNOW
Department of Pharmacology
University of Auckland
Faculty of Medical and Health Sciences
Private Bag 92019
85 Park Avenue
Auckland
New Zealand

M.E. GUIDO
CIQUIBIC (CONICET)
Departamento de Química Biológica
Facultad de Ciencias Químicas
Universidad Nacional de Córdoba
5000 Cordoba
Argentina

T. HERDEGEN
Institute of Pharmacology
University of Kiel
Hospitalstrasse 4
24105 Kiel
Germany

S.P. HUNT
Department of Anatomy and
Developmental Biology
University College London
Medawar Building, Malet Place
London WC1E 6BT
UK

S.A. JOSSELYN
Departments of Neurobiology, Psychiatry,
Psychology and Brain Research Institute
Room 2357
UCLA
Box 951761
Los Angeles, CA 90095-1761
USA

L. KACZMAREK
Department of Molecular and Cellular
Neurobiology
Nencki Institute
Pasteura 3
02-093 Warsaw
Poland

B. KAMINSKA
Department of Cellular Biochemistry
Laboratory of Transcription Regulation
Nencki Institute of Experimental Biology
Pasteur Street 3
02-093 Warsaw
Poland

S. KIDA
Laboratory of Animal Molecular Biology
Department of Bioscience
Faculty of Applied Biosciences
Tokyo University of Agriculture
1-1-1 Sakuragaoka, Setagaya-ku
Tokyo, 156-8502
Japan

J. LEAH
Biomedical Sciences
Griffith University
Nathan Campus
Brisbane, QLD 4111
Australia

C.V. MELLO
Neurological Sciences Institute (NSI)
Oregon Health Sciences University (OHSU)
West Campus Building 1
505 NW 185th Avenue
Portland, OR 97006
USA

A. PEARSON
Department of Pharmacology
Faculty of Medical and Health Sciences
University of Auckland
Private Bag 92019
85 Park Avenue
Auckland
New Zealand

S. PEÑA DE ORTIZ
Department of Biology
University of Puerto Rico
P.O. Box 23360
San Juan, 00931-3360
Puerto Rico

R. PRZEWŁOCKI
Department of Molecular Neuropharmacology
Institute of Pharmacology
Polish Academy of Sciences
Smętna 12
31-343 Krakow
Poland

B. RUSAK
Departments of Psychiatry, Psychology and Pharmacology
Dalhousie University
Halifax, NS B3H 2E2
Canada

K. SEMBA
Departments of Anatomy & Neurobiology and Psychology
Dalhousie University
Halifax, NS B3H 4H7
Canada

A.J. SILVA
Department of Neurobiology, Psychiatry, Psychology and Brain Research Institute
Room 2357
UCLA Box 951761
695 Young Drive South, Box 951761
Los Angeles, CA 90095-1761
USA

P. VANHOUTTE
Laboratoire de Neurochimie-Anatomie
Institut des Neurosciences
CNRS-UMR 7624
9 Quai Saint-Bernard
75005 Paris
France

P.A. WILCE
Department of Biochemistry and Molecular Biology
University of Queensland
St. Lucia Campus
Brisbane, QLD 4072
Australia

S. ZANGENEHPOUR
Department of Psychology
McGill University
1205 Dr. Penfield Avenue
Montreal, PQ H3A 1B1
Canada

B. ZIÓŁKOWSKA
Department of Molecular
Neuropharmacology
Institute of Pharmacology
Polish Academy of Sciences
Smętna 12
31-343 Krakow
Poland

Preface

Just 15 years ago, the neuronal nucleus was most often regarded as a passive material supplier in the functioning of the nervous system. The discovery, reported in 1987 and 1988, of massive, rapid, and transient accumulation of *c-fos* mRNA and protein in response to a variety of stimuli delivered to the central nervous system (CNS) changed this view dramatically.

Presently, the nuclear events are believed to play a pivotal role in the CNS responses, especially in reorganizations of neural networks evoked by external stimuli — be it for good or for bad. The well-coordinated waves of specific gene expression are assumed to provide a foundation for such physiological and pathological reactions as, e.g. cortical plasticity, circadian rhythms, learning and memory, long-term pain responses, post-injury nerve regeneration as well as programmed cell death. The underlying common theme is stimulation of transcription factors, regulating the gene expression.

Furthermore, this rapid genomic response in stimulated cells has provided researchers of neuronal pathways and networks with a very useful tool to map the activated neurons in the CNS. The fact that extracellularly-regulated gene expression was often found to be very transient, has proven to be especially convenient, as it allowed for a good temporal resolution. Furthermore, as this phenomenon most often concerned the nuclear proteins, excellent spatial resolution facilitated their immunostaining-based detection.

For the virtue of this rapid, transient, and primary (i.e. not dependent on de novo protein synthesis) nature, the activated genes were called immediate-early (IEGs), by analogy to certain viral genes. Originally, genes coding for transcription factors were first to be identified in this category. Nowadays, this list is much longer, with genes encoding many other proteins, localized often to the cytoplasm, or even specifically to, e.g. dendrites.

Because of the number of studies in this field and the plethora of genes sharing similar regulatory features, etc., it was necessary to impose boundaries to present a comprehensive volume on this topic. Because of their proven importance to neuronal mapping and relatively advanced knowledge about them, this volume deals only with transcription factors (TFs). Furthermore, only the inducible transcription factors are extensively covered. However, this presentation is not limited to TFs encoded by IEGs. To understand better the IEG response it is mandatory to gain knowledge about their regulation by pre-existing TFs. Those are sometimes called constitutive transcription factors, but in fact they are also inducible, most often by phosphorylation by specific kinases. Notably, recent developments in obtaining of phospho-specific antibodies make them one more excellent mapping tool.

In the spirit of the series title "Handbook of Chemical Neuroanatomy" most of the volume deals with neuronal mapping of a variety of the CNS structures in a number of physiological and pathological responses. However, this analysis is preceded by a few chapters focused more on the technologies employed, to allow the reader to be introduced to the basis of the procedures, their virtues, and the caveats. Finally, the last few chapters present selected cases of more detailed functional analysis of individual transcription factors. This is meant as a reminder of what is the foremost biological significance of the molecules taken as the mapping tools.

Despite these 15 years of progress, marked by thousands of research papers published on IEGs/ITFs (inducible transcription factors) in the CNS, it is amazing how limited is our knowledge on their roles in physiology and pathology. Hence, this volume should not be viewed as a conclusion of the field, but rather as a background for further studies of which three major directions are probably the most promising. (1) Very careful analysis of spatio-temporal patterns on individual genes and proteins in specific phenomena. To date, the majority of the studies were mostly focused on obtaining a very general picture, documenting a gene expression component to a given physiological or pathological situation. (2) Identification of specific target genes. IEGs/ITFs have clearly defined molecular function, namely, regulation of gene expression. Finding their targets should tell us a lot about the function. So far, studies on this topic have concentrated mostly on searching for specific DNA responsive elements in gene promoters/enhancers and spatio-temporal correlations. More refined analysis, combining in vivo and in vitro investigation, including detailed analysis of DNA binding properties of ITFs and their responsive elements within the presumed target genes, should greatly advance our knowledge. (3) Application of interventive strategies allowing for forced expression or inhibition of the IEGs/ITFs. Lack of appropriate technologies allowing for modulation of gene function in the CNS in vivo has hampered progress in the field. Recent advances with transgenic animals, in which gene expression can be controlled, as well as development of inducible knockout mice should greatly promote the field.

This volume features chapters by many leaders in the field of IEG and transcription factors in nerve cells. It was very fortunate that they agreed to contribute to this book and we would like to take this opportunity to extend our gratitude to all of them for their excellent efforts.

LESZEK KACZMAREK
HAROLD A. ROBERTSON
Warsaw and Halifax, February 2002

Contents

CHAPTER I

Methods used in inducible transcription factor studies: focus on mRNA

BARBARA ZIÓŁKOWSKA AND RYSZARD PRZEWŁOCKI

1. INTRODUCTION

The term 'inducible transcription factors' (ITFs) refers to proteins which share two features: (1) they are products of immediate-early genes, and (2) they can specifically bind to regulatory DNA sequences and activate or repress transcription of target late response genes. ITFs include proteins belonging to the Fos family (c-Fos, Fra-1, Fra-2, FosB and its truncated form called ΔFosB), Jun family (c-Jun, JunB and JunD), as well as several other proteins, of which Krox-20 (also called Egr-2) and Krox-24 (also called NGFI-A, Zif268, Egr-1, Tis8, Zenk) are the most extensively studied. All these proteins are encoded by separate genes, except for ΔFosB, which is a splice variant of FosB (for any detailed information and literature references see the comprehensive review by Herdegen and Leah, 1998).

The Fos- and Jun-family proteins are structurally related to each other in that they possess a basic DNA-binding domain and a leucine zipper domain which enables dimerization with other leucine zipper-containing proteins. Jun proteins can form both homodimers and heterodimers with Fos-family proteins, other members of the Jun family, as well as several other (e.g. ATF) proteins. In contrast, Fos proteins do not form homodimers; they dimerize mainly with different Jun-family proteins. Different dimers of Fos and Jun proteins are collectively referred to as the activator protein-1 (AP-1) transcription factor. They all bind to the same DNA regulatory element which has the consensus sequence TGAC/GTCA, and is called the AP-1 binding site. The affinity of the binding and trans-activation potential of AP-1 depend on the dimer composition.

On the other hand, Krox-20 and Krox-24 belong to the zinc finger transcription factors. The two proteins show no homology to each other except for the zinc finger DNA-binding domain. They recognize similar DNA sequences called Krox response elements (KREs).

Expression of ITFs in the cells is highly variable. Basal levels of most ITFs are very low (c-Fos, FosB, c-Jun, JunB), while some others are expressed at relatively high basal levels (JunD and Krox-24). By definition, a common feature of all ITFs is their potential to be rapidly induced in the cells in response to activating stimuli. ITF induction was initially demonstrated in the cultured 3T3 fibroblasts, pheochromocytoma (PC12) and other cell lines in response to serum, phorbol esters, growth factors and cytokines, forskolin, membrane depolarization, and ultraviolet light (reviewed by Herdegen and Leah, 1998). These factors were shown to stimulate ITF gene transcription, which resulted in rapid and transient increases in the levels of the respective mRNA species, followed by the accumulation of ITF proteins. It was found

Handbook of Chemical Neuroanatomy Vol. 19: Immediate Early Genes and Inducible Transcription Factors in Mapping of the Central Nervous System Function and Dysfunction
L. Kaczmarek and H.A. Robertson, editors

that transcriptional activation of ITF genes is dependent solely on transcription factors which had already been present in the cells before stimulation, since, in all cases, the induction was not blocked by protein synthesis inhibitors.

Due to the inducibility of ITFs and their function as transcription factors, it was realized that these proteins may play a role of so-called 'third messengers' in the pathways which transduce extracellular information into intracellular signals and are responsible for the generation of cellular responses to the changes in the environment. It was proposed that ITFs may evoke (possibly long-lasting) changes in cellular phenotype by influencing expression of their target genes (Curran and Morgan, 1987; Curran and Franza, 1988). As observed by Goelet et al. (1986), this type of signaling at the genomic level could play, in particular, an important role in neurons which are known to undergo long-term plastic changes underlying the phenomena of sensitization, adaptation, learning and memory. The finding that stimulation of membrane receptors and voltage-sensitive calcium channels led to the induction of c-*fos*, the first known ITF gene, in neuron-like pheochromocytoma cells (Greenberg et al., 1986; Morgan and Curran, 1986) prompted a search for evidence that ITFs are expressed in the nervous system.

In 1987, c-Fos was shown to be present in the brain in vivo (Dragunow et al., 1987). Shortly thereafter, Morgan et al. (1987) demonstrated that both c-*fos* mRNA and c-Fos protein are induced in the mouse neocortex, hippocampus and some other limbic regions as a result of seizures elicited by pentylenetetrazole. At the same time, a similar observation was reported by Dragunow and Robertson (1987), while Hunt et al. (1987) found induction of c-Fos in the spinal cord dorsal horn neurons after noxious and innocuous stimulation of the skin and joints. Therefore, the stimulation of primary afferents led to induction of c-Fos in postsynaptic neurons of the spinal cord. Moreover, it was shown that electrical stimulation of the motor/sensory cortex dramatically elevated c-Fos expression in the target areas of cortical projection (mainly in the thalamus) (Sagar et al., 1988). The latter studies revealed that c-Fos could be induced in neurons in response to transsynaptic activation, and their authors proposed that c-Fos immunocytochemistry may serve as a method for mapping functional pathways in the central nervous system (CNS) (Morgan et al., 1987; Sagar et al., 1988).

In the period 1987–1989, other *fos*-family, *jun*-family and *krox* genes were cloned. They were found to be co-induced with c-*fos* in many experimental conditions and in many cell types, including neurons. However, it turned up that sets of ITFs induced by different stimuli and in different kinds of cells varied significantly. Time-courses of induction are also different for individual ITFs. Consequently, the ratios of levels of individual ITFs may change greatly with time after the onset of stimulation.

The combination of transcription factors that are present in the cell in a particular moment, and their relative abundance, are of great functional significance due to their binding to different DNA regulatory elements (in this case mainly AP-1 binding site versus KRE) and different transactivation potential of transcription factor proteins or different AP-1 dimers. While this is a subject of a separate scrutiny, much of the research concerning expression of ITFs in the CNS has left aside their physiological role and considered their induction only as a marker of neuronal activation. Thus, studying ITF induction has become a method to examine the CNS function. Such studies have been performed using numerous techniques designed for detection of particular mRNAs and proteins, which permitted to investigate the level of ITF gene expression, and to determine distribution of ITF mRNAs and proteins in the CNS both at the level of nervous structures and single nervous cells. In the present article, we discuss the application of ITF induction measurements as a method in neuroscience research, describe particular techniques which are used to study ITF expression in the CNS, especially in situ hybridization and other methods by which mRNA levels are assessed, and we compare the usefulness of these techniques for specific experimental purposes.

2. ITF INDUCTION AS A MARKER OF NEURONAL ACTIVATION

Various stimuli cause induction of ITFs in discrete populations of neurons in the CNS. As exemplified by the results of Sagar et al. (1988) which were mentioned previously, direct electrical stimulation of the brain structures or nerve fibers (including peripheral nerves) may result in ITF induction in those CNS regions that are innervated by the stimulated neurons.

A similar effect may be produced by sensory stimulation. Noxious (mechanical, chemical or thermal) stimulation of the skin, muscles and viscera, induces ITFs in the dorsal horn of the spinal cord and in some brain targets of the spinal nociceptive projection (Hunt et al., 1987; Bullitt, 1990; Wisden et al., 1990; Herdegen et al., 1991; Traub et al., 1992; Lanteri-Minet et al., 1995; Bon et al., 1996; cf. Herdegen and Leah, 1998, pp. 404–407). Non-noxious cutaneous stimulation induces c-Fos in the spinal cord in other laminae and to a much lesser extent than the noxious input (Hunt et al., 1987; Bullitt, 1990). Exposure to light after a period of darkness leads to induction of c-*fos*, *junB* and *krox-24* in the retina and several parts of the visual system, including the lateral geniculate nucleus, the superior colliculus, the suprachiasmatic nucleus and the visual cortex (Rea, 1989; Rusak et al., 1990; Sagar and Sharp, 1990; Chaudhuri et al., 1997; Kaczmarek and Chaudhuri, 1997; Herdegen and Leah, 1998, p. 402). Similarly, auditory stimuli induce c-*fos* in the consecutive parts of the brain auditory system: the cochlear nuclei, the lateral lemniscus, the inferior colliculus, the medial geniculate nucleus, and the primary auditory cortex (Friauf, 1992, 1995; Rouiller et al., 1992; Herdegen and Leah, 1998, pp. 402–403). Olfactory and gustatory stimuli also induce c-*fos* in the brain structures conveying sensory information of the given modality (in the olfactory bulb and in the nucleus of the solitary tract, respectively; Guthrie et al., 1993; Sallaz and Jourdan, 1993; Harrer and Travers, 1996).

Another extensively studied example of ITF induction produced by the activation of neuronal circuits is that occurring as a result of seizures in multiple brain regions, most notably in different parts of the hippocampal formation and of the cerebral cortex (Fig. 1; Morgan et al., 1987; White and Gall, 1987; Sonnenberg et al., 1989a,b; Gall et al., 1990; Wisden et al., 1990; Pennypacker et al., 1993; Kamińska et al., 1994; Przewłocki et al., 1995; Beer et al., 1998). ITFs are induced irrespective of the seizure-eliciting factor (chemical substance, lesion or electrical stimulation), and the spatial distribution of their expression changes in time: it seems to follow the spreading wave of seizure activity within the brain (Morgan et al., 1987; Jørgensen et al., 1989; Gall et al., 1990; Beer et al., 1998).

It should be emphasized that in many of the above-mentioned cases ITF induction takes place in neurons that are separated by at least one synapse from those directly affected by the stimulus. Hence, expression of ITF genes in these postsynaptic neurons must depend upon action of chemical neurotransmitters. Therefore, it is not surprising that ITF induction events can be elicited in the CNS by pharmacological agents which act at neurotransmitter receptors. For example, agonists of different types of glutamate receptors induce *fos-*, *jun-* and *krox*-family genes in neurons by triggering influx of calcium ions into the cells (Sonnenberg et al., 1989a; Lerea et al., 1992; Bading et al., 1995), while dopamine receptor agonists stimulate ITF expression apparently by elevating intracellular cyclic AMP levels, which is due to the activation of the D1 receptors (Robertson et al., 1991; Konradi et al., 1994; Cole et al., 1995; Gerfen et al., 1995). After systemic administration of such drugs to animals, the pattern of

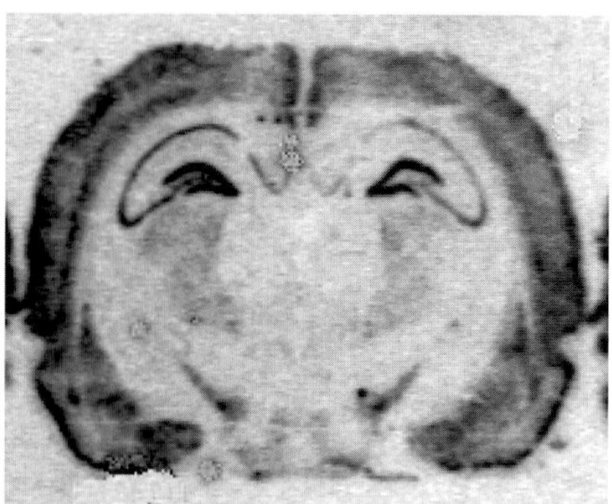

Fig. 1. In situ hybridization autoradiograms showing induction of c-*fos* mRNA in the rat brain elicited by kainic acid. The rats were injected with saline or kainic acid (12 mg/kg, i.p.) and decapitated 3 h later, during the kainic acid-produced status epilepticus. The brains were processed for in situ hybridization with a ^{35}S-labeled oligonucleotide probe for c-*fos*.

neurons expressing ITFs in response to them may be very complex. It reflects distribution of both the drug receptors within the CNS and sites excited by the drug indirectly, either by sequential activation of nerve cells in neuronal pathways or as a distant result of behavioral (e.g. motor) effects produced by the drug. For this reason, local injections of drugs into CNS structures are sometimes applied, as they may be more informative of the drug action within these structures.

2.1. VALIDITY OF ITF INDUCTION AS THE NEURONAL ACTIVITY MARKER

On the basis of the above-mentioned observations, it has become widely accepted that ITF induction reflects some kind of neuronal activation, although biochemical mechanisms underlying this type of activation seem to be very complex and are not satisfactorily understood. It is clear that the activation reflected by ITF induction cannot be equated with electrophysiological excitation of neurons. There exist several papers showing that direct depolarization of nerve cells by antidromic stimulation of their axons did not result in ITF expression in the cell bodies (cf. Herdegen and Leah, 1998, p. 451). There have also been reports showing no induction of ITFs after sensory stimulation in CNS regions which are known to be activated transsynaptically and to generate electrophysiological responses to the applied stimuli (Hunt et al., 1987; Bullitt, 1990; Sallaz and Jourdan, 1993; Kimpo and Doupe, 1997).

The state of neurons reflected by ITF induction has also been related to the metabolic activation depicted by uptake of 2-deoxyglucose (2-DG). Spatial patterns of ITF induction and 2-DG accumulation in the CNS are usually similar. However, there are cases when they differ; and the situations have been encountered in which an increase in glucose utilization in a given brain area is not accompanied by expression of ITF genes, and sometimes the opposite is observed (Sagar et al., 1988; Jørgensen et al., 1989; Sharp et al., 1993; Eells et al., 2000). These discrepancies may be partly due to the fact that ITF induction can be detected only in the cell body (as an increase in the level of the corresponding mRNA or protein), while 2-DG is taken up and accumulated mainly in neuronal processes. Even so, the opinion prevails that transcriptional activation of ITF genes takes place only in a subset of neurons which are activated metabolically (Sharp et al., 1993; Chaudhuri, 1997).

Therefore, whether a neuron in a certain CNS area will or will not express (at least some) ITF genes in response to a stimulus, is not only the issue of neuronal connections, but it is also dependent on some internal features of the neuron in question, as well as on ongoing activity of its numerous synaptic inputs. Crucial role in the cellular mechanisms leading to the activation of ITF transcription in neurons has been ascribed to calcium ions, which flow into the cells via the voltage-sensitive calcium channels (that are opened after cell depolarization) or via the cation channels of the glutamate NMDA receptors (Murphy et al., 1991; Lerea et al., 1992; Bading et al., 1993, 1995). Hence, it is likely that neurons which express calcium-binding proteins at high levels, and thereby have a high calcium-binding capacity, are less prone to transcriptional activation than other nerve cells (Herdegen and Leah, 1998, p. 450).

Induction of ITFs is also dependent on intensity and temporal pattern of stimulation. For example, high-frequency stimulation of the perforant path results in c-Fos and Krox-24 induction in the granule cells of the hippocampal dentate gyrus, which is not observed after low-frequency stimulation (Cole et al., 1989; Dragunow et al., 1989). Similarly, the induction of c-Fos in several limbic forebrain regions is elicited by electrical burst, but not by regular stimulation of the medial forebrain bundle (Chergui et al., 1996). This is probably due to cellular mechanisms which prevent excessive activation in response to stimuli of low or moderate amplitude (cf. Herdegen and Leah, 1998, p. 450).

All the above limitations must be taken into account when designing and interpreting the experiments in which ITF induction is treated as a neuronal activation marker. It is important to keep in mind that not all neurons which are activated in some way (electrophysiological or metabolic) will respond with induction of any ITF, and therefore no conclusions can be drawn from the mere *lack* of ITF induction in a CNS region under investigation. This is why the studies based solely on ITF expression cannot provide complete functional maps of neuronal circuits. On the contrary, *presence* of ITF induction may be very informative. First, induction of ITFs in some neurons indicates that these neurons are affected by the applied stimulus, provided that the experiment includes all the controls necessary to avoid false positive results (see Section 2.4). Second, ITF induction can be detected with cellular resolution, i.e. ITF gene products can be visualized within single neurons, which is not possible with the 2-DG method. Third, ITF detection can be combined with visualization of other proteins or mRNAs, as well as with retrograde labeling of nerve cells. This permits an exact identification of neuronal populations expressing ITFs on the basis of various biochemical markers (such as enzymes, receptors, neuropeptides) and of target areas of their projections. Fourth, levels of ITF expression can be measured in a quantitative or at least semi-quantitative manner (depending on the resolution), which enables a search for factors that modulate neuronal responses to ITF-inducing stimuli.

Therefore, despite all the shortcomings of approaches using induction of ITFs as an activation marker, much advantage can be taken from them in the studies of the CNS function.

2.2. EXAMINATION OF ITF INDUCTION AS A METHOD IN NEUROBIOLOGICAL AND NEUROPHARMACOLOGICAL STUDIES

In the 1990s, a plethora of studies were performed which used detection of ITFs for tracing activated neurons in the CNS and defining the degree of activation of CNS structures. The objectives of these studies were quite various and included:

- *mapping neuronal pathways activated by sensory input* (see previous sections);
- *studying functional anatomy of behavior* (e.g. neuronal pathways involved in sexual behavior were delineated; Biały and Kaczmarek, 1996; Veening and Coolen, 1998);
- *elucidating mechanisms of physiological/pathological processes* (the studies concerned e.g. pain, seizures, ischemia, neurodegeneration, development and ageing, learning and memory; cf Herdegen and Leah, 1998);
- *elucidating neurotransmitter interactions* (which was done by investigating the influence of pharmacologically characterized substances on the changes in ITF expression produced by other, endogenous or exogenous substances; e.g. Bhat and Baraban, 1993; Torres and Rivier, 1993; Boegman and Vincent, 1996; Konradi et al., 1996);
- *imaging CNS structures involved in action of drugs* (among the most extensively studied groups of drugs were neuroleptics and substances of abuse; Robertson et al., 1991, 1994; Nestler, 1993; Torres and Horowitz, 1996; Herdegen and Leah, 1998, pp. 428–430, 441–442; Erdtmann-Vourliotis et al., 1999b, 2000; Ryabinin et al., 2000);
- *drug classification and search for psychoactive substances having a desired profile of action* (these studies concerned atypical neuroleptics and antidepressant drugs whose mechanisms of action are not well understood, but which can be identified by their influence on ITF expression in specific brain regions; Robertson et al., 1994; Duncan et al., 1996).

2.3. c-Fos AND Krox-24: NEURONAL ACTIVITY MARKERS OF DIFFERENT TYPES

So far, c-Fos has been the most widely used among the ITFs as a neuronal activity marker. Although this has probably been dictated in part by historical reasons (c-Fos was the first ITF to be found in the CNS), c-Fos now seems a good choice being both reliable and convenient as a marker. In fact, cases in which induction of other ITFs is *not* accompanied by induction of c-Fos are very uncommon (an example is the selective, delayed and persistent expression of c-Jun taking place in axotomized neurons; Beer et al., 1998; cf. Herdegen and Leah, 1998, p. 407), while subsets of ITFs which are co-induced with c-Fos are quite various in different experimental paradigms. Other advantages of c-Fos as a marker include its low basal expression (owing to which induction of the gene is easily discernible) and the narrow time window at which the induction can be observed. The latter feature is the consequence of (1) the rapid onset of c-*fos* gene transcription after application of a stimulus, (2) the short half-life of c-*fos* mRNA and protein (of about 15 min and 1–2 h, respectively), and (3) the fact that transcription of c-*fos* gene is inhibited by c-Fos protein (cf. Herdegen and Leah, 1998, pp. 385–386). Usually, an increase in c-*fos* mRNA levels can be observed within minutes after the onset of stimulation, the mRNA level peaks between 20 and 60 min and falls back to the basal value by about 2 h. Expression of c-Fos protein is somewhat delayed, peaking at about 2–4 h and then declining to the control level for the next several hours (e.g. Morgan et al., 1987; Sonnenberg et al., 1989a,b; Dragunow et al., 1990; Nguyen et al., 1992; Moratalla et al., 1993; Sirinathsinghji et al., 1994; Bading et al., 1995; see also Chapter VIII by Kaczmarek). Taking these short and rather predictable time-courses, it is easy to design experiments on previously unexamined neuronal systems, where tedious multiple measurements over long time periods can be largely avoided.

Krox-24 is the second ITF marker used extensively in CNS studies, usually in addition to c-Fos. Krox-24 differs markedly from c-Fos as its basal expression is fairly high in many CNS regions (Gerfen et al., 1995; Kaczmarek and Chaudhuri, 1997; Herdegen and Leah, 1998, p. 400). In spite of this, Krox-24 is easily up-regulated in neurons by the same stimuli which induce c-Fos. However, the most interesting feature of Krox-24 is the fact that not only its up-regulation, but also is down-regulation can be studied easily. The latter is somehow precluded in the case of c-Fos because of its low basal expression which is barely detectable.

Decreases in Krox-24 expression have been observed in the cerebral cortex after sensory deprivation: visual deprivation reduced levels of Krox-24 in the primary visual cortex (cf. Kaczmarek and Chaudhuri, 1997), while clipping of rat's vibrissae decreased Krox-24 levels in the corresponding barrel cortex (Steiner and Gerfen, 1994). Expression of Krox-24 in the visual cortex was restored rapidly after visual stimulation (cf. Kaczmarek and Chaudhuri, 1997). These findings led to the concept that the level of expression of *krox-24* gene reflects ongoing synaptic activity, undergoing up-regulation or down-regulation in the conditions in which stimulation of a neuron is increased or decreased, respectively (Chaudhuri, 1997; Kaczmarek and Chaudhuri, 1997). This notion is supported by the observation that a lesion of nigrostriatal pathway, leading to a dopamine depletion in the striatum, results in an up-regulation of *krox-24* gene in the population of striatal projection neurons which are inhibited by dopamine (via D2 receptors) and a down-regulation of the gene in the neuronal population which is stimulated by dopamine (via D1 receptors) (Gerfen et al., 1995).

In conclusion, the two most commonly used ITF markers, c-*fos* and *krox-24*, owing to their different regulation characteristics, provide different pieces of information on the activity of neurons under study. While induction of c-*fos* gene reflects a response of neurons to a sudden and strong stimulation, expression of *krox-24* gene follows long-term bidirectional changes in neuronal activity.

2.4. SOURCES OF FALSE RESULTS IN IN VIVO ITF INDUCTION STUDIES

Expression of ITF genes in discrete brain regions is markedly affected by the stimuli which are either very common (many sensory stimuli) or inherent in numerous experimental paradigms. Stress, traumatic injury or ischemia of the nervous tissue, and nociceptive stimulation, all induce ITFs in the CSN, while they cannot be avoided in experiments which include injections, surgery, implantation of cannulae or electrodes into some CNS structures, etc. Many authors report induction of c-Fos in the brain or spinal cord after such basic procedures as intraperitoneal injection of saline, incision of the skin, injection of neutral vehicle into the hippocampus or fastening earclips in order to evoke an electroconvulsive shock (Hunt et al., 1987; Kaczmarek et al., 1988; Gubits et al., 1989; Nakajima et al., 1989; Johnston and Morris, 1994; Erdtmann-Vourliotis et al., 1999b).

Both basal and induced expression of ITFs can also be markedly influenced by anesthetic agents. Several anesthetics (e.g. urethane, pentobarbital, halothane, methoxyflurane) induce c-Fos in specific brain areas by themselves (Bullitt, 1990; Takayama et al., 1994; Telford et al., 1996; Ryabinin et al., 2000). On the other hand, in deeply anesthetized animals, ITF induction may be reduced or completely blocked. For instance, induction of c-Fos elicited by noxious stimuli in the spinal cord, thalamic nuclei and superior colliculus, is reduced in barbiturate or urethane anaesthesia (Bullitt, 1990; Williams et al., 1990; Telford et al., 1996). In addition, barbiturate anesthesia completely prevents ITF induction in the striatum by haloperidol (Ziółkowska and Przewłocki, unpublished observation) and cocaine, and that elicited in several brain areas by ethanol (Ryabinin et al., 2000).

In order to avoid false results in ITF induction studies, it is therefore important to carry out experiments in a stimulus-controlled environment and to control strictly all experimental procedures with respect to their influence on ITF expression. The latter is accomplished e.g. by using vehicle-injected, anesthetized-only, and sham-operated groups of animals. Although many problems can be overcome by appropriate control procedures, some ITF experiments cannot be performed under anesthesia.

3. METHODS USED IN ITF STUDIES

Goals of research in the field of ITFs are twofold. Some studies, such as those listed in Section 2.2, concern the conditions under which ITF genes are induced, intracellular mechanisms of these different kinds of induction, anatomical distribution of ITF induction produced in vivo by specific stimuli and/or factors which can modulate the induction. Such studies use methods which permit to detect the changes in the level of ITF gene expression. An overview of these methods is presented below, with a particular stress on the methods detecting mRNA in situ. In contrast, other studies address the issue of physiological role of ITFs, and, accordingly, apply methods which may help uncover functions of ITF proteins. Some of these methods are mentioned briefly at the end of the chapter.

Expression of ITF genes can be studied either at the level of mRNA or at the level of protein by the same methods which are used for any other genes. mRNA is detected mainly using hybridization techniques which rely on the propensity of single-stranded nucleic acids to form duplexes (i.e. hybridize) with (other) nucleic acid molecules bearing complementary sequence(s). To visualize the mRNA species under study, hybridization is carried out with complementary nucleic acid probes, which are labeled beforehand by incorporation of e.g. haptens or radioactive compounds. There exist also the methods of mRNA detection and quantification that are based on reverse-transcription (i.e. synthesis of DNA complementary to the RNA of interest) and amplification of complementary DNA; however, they are used only rarely for measurement of ITF expression. On the other hand, methods aimed at detection of proteins take advantage of their immunogenic properties and are based on the reaction of proteins with specific antibodies.

The choice of a method for detecting ITF expression depends on particular goals of an experiment. While some experiments require superior histological resolution, in others the measurement of ITF mRNA or protein levels is necessary. In all mapping or imaging studies cellular resolution of ITF product detection is of importance, but no quantification of results is needed. On the contrary, when modulation of ITF induction level is of interest, then the results must be quantified as exactly as possible, so that they can be compared between the samples. This is an important point because some methods which offer single-cell resolution have a relatively low potential to detect mRNAs or proteins in a quantitative manner.

3.1. ITF mRNA DETECTION

3.1.1. In situ hybridization

In situ hybridization (ISH) is a method by which mRNA can be detected in tissue sections, using specific nucleic acid probes, in a quantitative or at least semi-quantitative manner, and with single-cell resolution. Although there are many variants of this technique, the procedure always consists of four basic steps:

- tissue preparation (pretreatment),
- hybridization,
- post-hybridization washing,
- detection of hybridization signal.

Below, only a general description of the method and its variants is presented. Detailed ISH protocols, applicable for the nervous tissue, can be found in numerous publications (Boehringer Mannheim Biochemicals, 1992; Lewis et al., 1993; McGinty et al., 1993; Schäfer et al., 1993; Sirinathsinghji and Dunnett, 1993; Baldino et al., 1994; Petersen and McCrone, 1994; Wisden and Morris, 1994b; Wilkinson, 1995b; Scharif, 1996; Lynch and O'Mara, 1997b; Erdtmann-Vourliotis et al., 1999a). Exhaustive methodological considerations are also available (Schäfer et al., 1993; Tecott et al., 1994; Wilkinson, 1995a).

3.1.1.1. Tissue preparation (pretreatment)

Slices for ISH are prepared from fresh-frozen tissue which is cut on a cryotome (cryostat). Alternatively, the tissue can be fixed by perfusion of an animal with paraformaldehyde, and then sectioned using a vibratome (or a cryotome). In the latter case it is possible to combine ISH with immunocytochemical detection of specific proteins in the same sections. ISH can also be performed using slices of tissue embedded in paraffin, epoxy resins and other media. Usually, once cut, tissue sections are mounted and then processed on gelatin- or poly-L-lysine-coated slides. However, ISH can also be performed on free-floating sections (Lu and Haber, 1992; Trembleau et al., 1993; Meltzer et al., 1998).

Before hybridization, the slices are treated with several reagents in order to:

(1) *fix the tissue* (if it is not perfusion-fixed), so that its structure is preserved and that RNA does not leak out during hybridization and washing; paraformaldehyde and glutaraldehyde are the most commonly used fixatives;

(2) *make the tissue permeable*, so that the hybridization probe can reach its target RNA, which is embedded in the cell structural components; proteases, detergents and chloroform are used for this purpose;

(3) *neutralize positively charged amino groups in the tissue*, to which nucleic acid probes, being polyanions, could bind nonspecifically by electrostatic interactions; acetic anhydride is used to acetylate these groups.

3.1.1.2. Hybridization

During the hybridization reaction, tissue sections are incubated for 6–20 h with a labeled RNA or DNA probe complementary to the mRNA of interest. The reaction should be carried out under such conditions which favor formation of stable duplexes between the probe and the target mRNA, but prevent binding of the probe to unrelated nucleic acid sequences. Stability of duplexes depends primarily on the probe itself: on its length (longer probes form more stable hybrids than shorter ones), on its chemical nature (RNA : RNA hybrids are more stable than DNA : RNA hybrids) and on the degree of complementarity with the target (mismatched pairs of nucleotides destabilize hybrids). For a given probe, efficient and specific hybridization is attained by an appropriate choice of hybridization temperature (hybrids are disrupted at high temperatures) and the chemical content of the hybridization buffer (where the concentrations of such standard components as sodium ions and formamide are of particular importance for hybrid stability). Although the impact of these factors on hybridization kinetics and duplex stability is well characterized for nucleic acid hybridization in solution, interaction of a probe with its target mRNA which is embedded in a tissue is not quite predictable. Thus, conditions of hybridization with a given probe in situ sometimes need to be optimized empirically.

In addition to enabling probe association with the target mRNA, it is important to prevent its nonspecific binding to other tissue components. This is achieved by supplementation of the hybridization buffer with nucleic acids, proteins and polysaccharides (e.g. tRNA, salmon sperm DNA, albumin, Ficoll, polyvinyl pyrrolidone) which compete for the corresponding binding sites in the tissue.

3.1.1.3. Post-hybridization washing

After hybridization, tissue sections are washed several times, usually at increasing temperatures, with salts solutions containing sodium ions at decreasing concentrations. The aim of the washing is to remove the excess of unbound probe and to disrupt connections of the probe with nucleic acid sequences to which it is only partially complementary. Thus, selectivity of target mRNA detection is influenced not only by hybridization conditions, but also by washing conditions. In addition, if RNA probes are used, the digestion of unhybridized single-stranded probe by ribonuclease is often carried out during washing in order to reduce background binding.

3.1.1.4. Types of hybridization probes used in ISH experiments and methods of their labeling

Probes most commonly used in ISH experiments are either single-stranded oligodeoxynucleotides (oligonucleotide probes), complementary RNA (cRNA) or double-stranded complementary DNA (cDNA).

Oligonucleotide probes are usually 20–60 nucleotides long. They are obtained by chemical synthesis using a DNA synthesizer. If the device is not available, custom synthesis of oligonucleotides can be ordered in many biotechnological companies. Since oligonucleotide probes are short as compared to typical mRNA molecules, appropriate choice of the mRNA portion as a target for the probe is fundamental. Care must be taken to assure that the sequence complementary to the probe is unique to the mRNA of interest, i.e. that it (or its part) is not shared with other mRNAs; otherwise the probe will lack selectivity and hybridize with multiple transcripts. This is particularly important in the case of genes belonging to gene families, such as *fos* or *jun*, since these always display some homology to other family members. In practice, however, oligonucleotide probe design does not pose a problem in the case of ITFs. Selective probes to all *fos*, *jun* and *krox* mRNAs have already been designed, tested thoroughly for their specificity and used successfully in many ISH experiments, which relied largely on the same probe sequences (Wisden et al., 1990; Moratalla et al., 1993; Sirinathsinghji et al., 1994; Gerfen et al., 1995; Beer et al., 1998).

Oligonucleotide probes are labeled most frequently by 'tailing' reaction using terminal deoxynucleotidyltransferase (TdT) (Lewis et al., 1993; Sirinathsinghji and Dunnett, 1993; Baldino et al., 1994; Wisden and Morris, 1994a; Young, 1995; Erdtmann-Vourliotis et al., 1999a). TdT adds multiple labeled nucleotides (almost always dATP is used) at the 3′ end of the probe. Oligonucleotides can also be labeled using T4 polynucleotide kinase, but this enzyme adds only one labeled phosphate group at the 5′ end of the probe, generating probes with low specific activity. Specific activity (i.e. the amount of label built in one molecule or mole of the probe) of TdT-labeled probes is higher, but it is still low in comparison with cRNA and cDNA probes.

cRNA probes are obtained by antisense transcription of a cDNA fragment corresponding to the mRNA of interest. The cDNA is cloned into a plasmid vector in such a way that it is flanked by gene promoters of T3, T7 or SP6 bacteriophages (where there is a different promoter on each side of the cDNA insert). The cDNA is transcribed using an appropriate RNA polymerase (T3, T7 or SP6). To avoid inclusion of plasmid sequences into the probe, the plasmid is cut (linearized) beforehand with a restriction enzyme near the end of the cDNA insert. The probes are labeled during transcription by incorporation of the labeled nucleotides into the cRNA molecules (usually labeled UTP is used) (Lewis et al., 1993; Schäfer et al., 1993; Baldino et al., 1994; Wahle, 1994; Luo and Jackson, 1999).

The length of cRNA probes is uniform and equal to the length of the cDNA template (the usual size range is 200–2500 base pairs). The opinion prevails that the optimum length of cRNA probes for ISH is several hundred nucleotides. Longer probes (above 1000 nucleotides) may poorly penetrate tissue sections, and therefore long transcripts are sometimes partially degraded by alkaline hydrolysis if used for ISH (Lewis et al., 1993, p. 7; Wilkinson, 1995b, p. 22).

Double-stranded cDNA probes are obtained and labeled most frequently by the method of random primer extension (random primed labeling) (Boehringer Mannheim Biochemicals, 1992, p. 18; Luo and Jackson, 1999). A cDNA fragment corresponding to the mRNA of interest is excised from a vector in which it is cloned and is used as a template for DNA synthesis reaction catalyzed by the Klenow fragment of the bacterial DNA polymerase I. The synthesis is initiated by short random-sequence primers which, depending on their sequence, anneal to different sites of the template. The resulting probe is thus composed of DNA molecules of various lengths. They are labeled during synthesis due to inclusion of labeled nucleotides in the reaction mixture. The probes are denatured before hybridization in order to split them into single DNA strands.

Comparison of different types of probes and their usefulness for ITF mRNA detection. An ideal probe for ISH should be an easy to obtain and to handle, single-stranded, stable nucleic acid molecule which can be labeled to yield high specific activity. None of the above-mentioned types of probes possesses all these features.

cRNA probes can yield the best quality results because, due to their length and stability of RNA : RNA hybrids, they bind tightly to their target mRNAs. Moreover, they have high specific activity owing to the incorporation of many labeled nucleotides in the probe molecule. cRNA probes are thus the most sensitive among the different probe types. However, like other RNAs, they are liable to degradation by the ubiquitously present ribonucleases and therefore very unstable. In addition, the preparation of these probes is fairly complicated and requires access to molecular biological expertise and equipment.

The latter inconvenience concerns also the use of double-stranded cDNA probes. cDNA probes are more stable than cRNA probes and are characterized by high specific activity, but form less stable hybrids with mRNA. Another disadvantage of cDNA probes is that the sense strand of the probe competes with the target mRNA for binding of the antisense strand. These probes are therefore less sensitive than cRNA probes.

On the other hand, oligonucleotide probes are single-stranded, stable and easy to obtain, but suffer from lack of sensitivity due to their relatively low specific activity (as compared to cRNA and cDNA probes). The latter shortcoming is partly compensated for by the fact that oligonucleotides penetrate tissue slices better than longer probes.

Despite their low sensitivity, oligonucleotide probes are good enough to be used in most studies examining expression of ITF genes in the CNS, especially those which address the issue of *ITF gene induction*, because, when induced, ITF mRNAs are abundant in nerve cells, and therefore can be easily detected even with less sensitive methods. On the other hand, oligonucleotide probes may not be that suitable to study *basal ITF expression* (which is, anyway, assessed mainly by immunocytochemical detection of proteins). Should this be necessary, sensitivity of mRNA detection using oligonucleotide probes can be increased

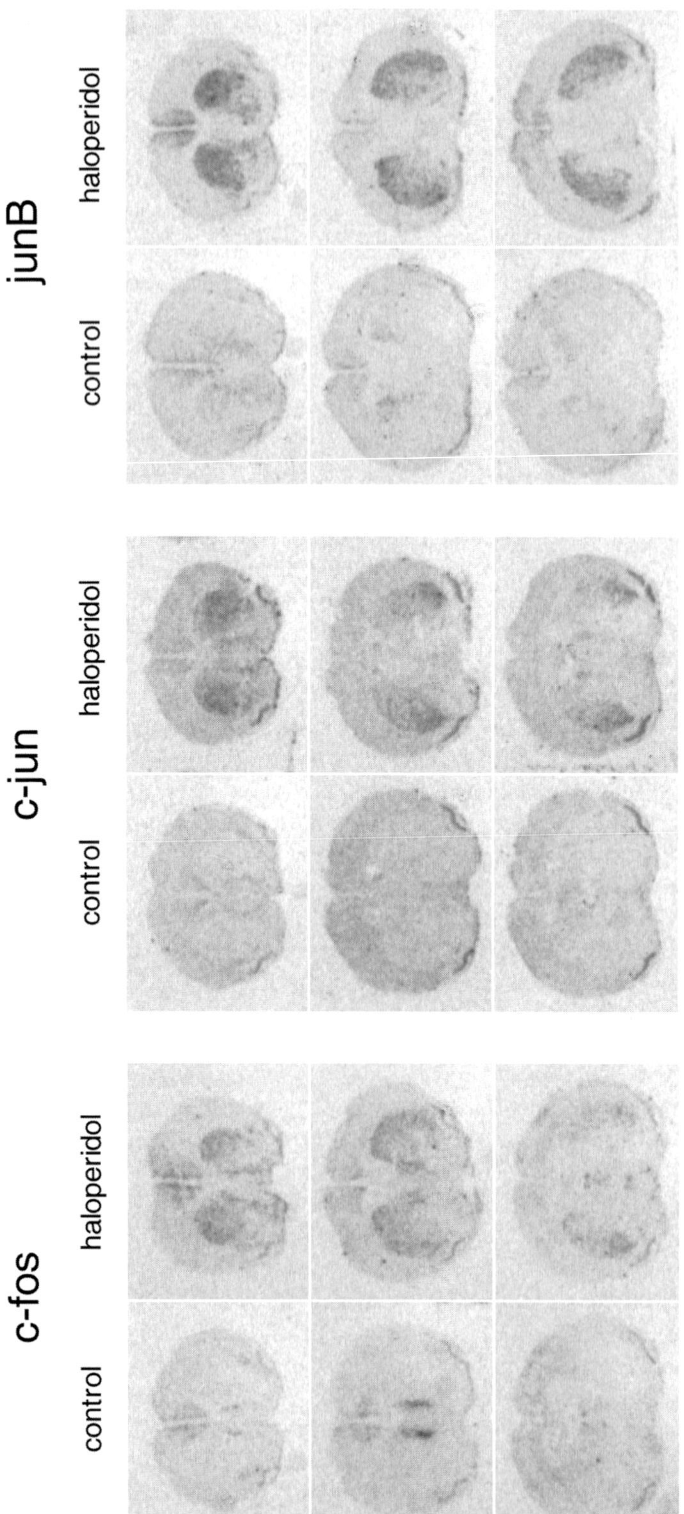

Fig. 2. In situ hybridization autoradiograms showing induction of *c-fos*, *c-jun* and *junB* mRNA in the mouse striatum elicited by haloperidol. The mice were injected with saline or haloperidol (1 mg/kg, i.p.) and killed 30 min later. The brains were processed for in situ hybridization with ^{35}S-labeled oligonucleotide probes. The three rows of coronal brain sections represent different levels of the striatum along its rostro-caudal axis.

by performing hybridization with a mixture of several probes targeted at different portions of the same mRNA (Trembleau and Bloom, 1995). While sequences of well-performing oligonucleotide probes for all ITF mRNAs can be found in the literature (Wisden et al., 1990; Moratalla et al., 1993; Sirinathsinghji et al., 1994; Gerfen et al., 1995; Beer et al., 1998), additional probes can be designed rationally according to the guidelines provided by Erdtmann-Vourliotis et al. (1999a).

The major advantages of using oligonucleotide probes (whenever their sensitivity is sufficient) rather than cRNA or cDNA probes rely on simplicity of the corresponding ISH protocols. ISH with oligonucleotide probes is an easy method; it is cheaper and less time-consuming than ISH with cRNA and cDNA probes; moreover, no cloning facilities are required. Excellent detailed oligonucleotide probe ISH protocols have been published by Wisden and Morris (1994a), as well as by Young (1995).

We have successfully used Young's protocol, and the radioactively labeled oligonucleotide probes described by Wisden et al. (1990) to detect induction of c-*fos*, c-*jun*, *junB* and *krox-24* mRNAs in the brains of rats treated with convulsants (Fig. 1; Przewłocki et al., 1995) and mice treated with classical neuroleptics (Figs. 2 and 4). Similar experiments were performed using cRNA probes to several ITF mRNAs, according to the protocol of Schäfer et al. (1993) (Ziółkowska and Höllt, 1993, and unpublished data). Although the latter probes were indeed more sensitive (shorter exposure times were required for signal detection), we observed some variability in tissue background and hybridization signal level between slides to which the same probe solution was applied. That problem arose probably due to differential degradation of the RNA probe during hybridization, despite our efforts to avoid contamination of slides with ribonucleases (by using nuclease-free reagents, autoclaving all solutions and/or treating them with the ribonuclease inhibitor diethylpyrocarbonate). This variability in signal intensity markedly complicated quantitative analysis of autoradiograms. In contrast, consistent quantitative results could be obtained using oligonucleotide probes, although longer film exposure times (2–3 weeks) were necessary for radioactive signal detection. Therefore, in our opinion, the use of cRNA probes may be cumbersome if quantification of hybridization signal is necessary, and the use of oligonucleotide probes seems much more convenient.

It is important to note, however, that our experience is confined to radioactively labeled probes. While oligonucleotides can be tailed, using TdT, with a string of radioactive nucleotides (typically 20–30), it is not possible to use probes containing so many non-radioactively labeled nucleotides (see Section 3.1.1.5), since detection of non-radioactive labels relies on their interaction with antibodies and/or other protein molecules, which occupy quite a lot of space. To enable efficient detection, steric hindrance of binding of detection system components must be prevented. To achieve this, non-radioactively labeled nucleotides in a probe must be separated from one another by stretches of non-labeled nucleotides. In practice, only 2–3 labeled nucleotides can be introduced into a 20–30 nucleotides long tail (Boehringer Mannheim Biochemicals, 1992, p. 21; Höltke et al., 1995). On the other hand, addition of much longer tails, which could contain more label, is not recommended because too long a tail would impede specific binding of the probe to its target mRNA.

Although non-radioactively tailed oligonucleotide probes have been used successfully to detect abundant transcripts (Cephalon Inc., 1992; Department of Cytochemistry and Cytometry, University of Leiden, 1992; Young, 1995), non-radioactive detection of ITF mRNAs was based so far largely on cRNA and cDNA probes (Chaudhuri et al., 1997; Beer et al., 1998; Meltzer et al., 1998; Luo and Jackson, 1999). Whether non-radioactive probe detection systems are sensitive enough to reliably detect ITF mRNAs using oligonucleotide probes remains therefore to be established. As proposed by Petersen and McCrone (1994), this can be done by hybridizing adjacent tissue sections with radioactively and non-radioactively labeled probes to the same mRNA and comparing distribution of the resulting signal.

3.1.1.5. Types of probe labels and systems of hybridization signal detection

Radioactive labels are introduced into probes in the form of nucleotides containing atoms of radioactive isotopes: ^{3}H, ^{32}P, ^{33}P or ^{35}S. They are detected after hybridization by exposure onto photographic emulsions. ^{35}S is used most commonly for probe labeling, since it offers good resolution of hybridization signal and has a reasonably short half-life (87 days), requiring exposure times of several days to several weeks. ^{33}P offers good resolution and requires even shorter exposure, but is more expensive. ^{3}H yields the best resolution, but requires very long exposure times (up to several months), while ^{32}P allows short exposure time, but due to the high energy of its radiation, offers poorer resolution than the other isotopes.

There are two options for detection of radioactive hybridization signal. For macroscopic assessment of signal intensity, tissue sections are exposed after hybridization to photographic films (Figs. 1, 2 and 4). The amount of probe bound to different portions of sections is reflected by local optical density in the resulting autoradiograms. Intensity of the signal can then be analyzed quantitatively using a densitometer or, more commonly, a computer-based image analysis system connected to a video camera which serves to digitize autoradiograms (Fig. 4). To make the analysis more accurate, radioactivity standards are usually exposed to films along with tissue sections.

For analysis of signal intensity at microscopic level, slides with tissue sections are coated after hybridization with liquid photographic emulsion, which is then dried, exposed and developed like a film. Tissue slices are counterstained, e.g. with cresyl violet, and hybridization signal can be observed in a light microscope as black grains of the emulsion over a single cell. The signal can be quantified by counting the number of grains per cell or the area covered by the grains.

Fluorescent labels are introduced into probes by incorporation of nucleotides conjugated with a fluorochrome molecule. The fluorochromes whose conjugates with nucleotides are available commercially include fluorescein, rhodamine and coumarin derivatives (Boehringer Mannheim Biochemicals, 1992). Upon excitation with light of an appropriate wavelength, the dyes fluoresce with green, red and blue light, respectively. After ISH, fluorochrome-labeled probes can be detected with single-cell resolution in tissue sections using a fluorescence microscope. Intensity of fluorescent light, which reflects the amount of hybridized probe, can be measured. Owing to multiplicity of fluorescent dyes, several probes labeled with different colors can be combined in one hybridization experiment, which permits simultaneous detection of several transcripts in the same cells. Alternatively, fluorochrome-labeled probes can be detected indirectly by using specific antibodies against the dye molecules (see next paragraph).

Hapten labels are introduced into the probes by incorporation of nucleotides conjugated with molecules (haptens) to which specific antibodies are available. The most commonly used hapten labels include digoxigenin, biotin and fluorescein. Hapten-labeled probes are detected in ISH tissue sections by means of the hapten-specific antibodies, which are conjugated with enzymes, fluorochromes or colloidal gold. The latter are reporter molecules which serve to visualize hybridization signal (reviewed in Boehringer Mannheim Biochemicals, 1992; Höltke et al., 1995).

Typical enzyme reporter molecules are peroxidase and alkaline phosphatase. They are used to catalyze reactions in which soluble substrates (e.g. diaminobenzidine and nitro blue tetrazolium/5-bromo-4-chloro-3-indolylphosphate [NBT/BCIP]) are transformed into precipitating colored products (brown and blue, respectively). In the case of biotin-labeled probes, the biotin-binding proteins avidin or streptavidin (conjugated with the same reporter enzymes) are often used instead of antibodies. Application of a detection system which includes the above enzyme reporters results in staining of the cells (or cell compartments) to which the labeled probe has bound with the colored reaction product. The stained cells can be observed under the light microscope, and the resolution of probe detection with this method is very high (in fact, probe distribution can be observed at subcellular level). However, chances to quantify the intensity of hybridization signal (and thereby to estimate the amount of the target mRNA in the region of interest) are rather poor. Although it is possible to measure optical density of color staining by the use of a microdensitometer or an image analysis system with a camera equipped with color filters, these measurements may not accurately reflect the amount of the probe bound to the tissue due to the complexity of the detection method, where the final signal is dependent on consecutive interactions of several molecules. Particularly inaccurate measures seem to derive from detection based on avidin–biotin rather than antibody–hapten interaction since the former reaction is not stoichiometric. Another problem is created by the fact that no external standards can be used in such experiments and therefore, if any intensity measurement is planned, color development must be monitored carefully to avoid saturation of the signal. In practice, quantitative analysis of the results obtained using hapten-labeled probes most commonly relies on counting the number of cells stained to any noticeable degree (above background) which are present in some predetermined area unit of the tissue slice. In the studies on expression of ITF genes quantification of this type may be quite reasonable and informative because the huge elevations of ITF mRNA levels taking place in the process of gene induction are reflected by dramatic increases in the number of cells containing detectable levels of ITF mRNAs.

The same fluorescent dyes as those which serve for direct probe labeling (see previous section), are used as fluorochrome reporters. After reaction of a fluorochrome-conjugated antibody with a hapten-labeled probe, the complex is detected, like directly labeled probes, by examination of tissue sections under a fluorescence microscope. Fluorescent signal intensity can be measured, similarly as for the directly labeled probes. However, one should take into account inaccuracy of such measurements which results from the same sources as in the case of enzyme reporter-based systems (i.e. multiplicity of the interacting molecules of the detection system and lack of stoichiometry). Alternatively, analysis based on counting of label-positive cells can be performed.

Finally, colloidal gold, serves as an antibody-conjugated reporter for electron microscopy, which permits mRNA localization at ultrastructural level. Since this is not a concern in ITF studies, the technique will not be discussed here. Information on ISH at the level discernible only by electron microscope can be found in the article by Binder (1995).

Enzyme labels are molecules of an active enzyme attached to a nucleic acid probe by a covalent bond. In practice, conjugates of oligonucleotide probes with alkaline phosphatase (AP) are used for ISH (Augood et al., 1994). The probes are detected after hybridization by means of a colored precipitate-producing reaction catalyzed by the tagging enzyme. Routinely, NBT/BCIP are used as the substrate for AP, being turned into the blue NBT formazan. Resolution of signal detection by AP-labeled probes is excellent, similar to that obtained with hapten-specific AP-conjugated antibodies. Moreover, since multiple detection steps are avoided in the case of probes labeled directly with AP, the amount of probe bound to the tissue can be estimated on the basis of staining intensity.

Comparison of hybridization signal detection systems and their application in ITF studies. The systems of in situ hybridization signal detection described above differ markedly in the resolution of the signal and possibility of quantitative measurements they offer, as well as in their sensitivity.

Quantitative measurements of signal intensity, reflecting the amount of the mRNA of interest, can be performed most reliably using radioactive probes and autoradiographic detection. In this case, the optical density of the autoradiogram or the number of irradiated photographic emulsion grains are directly related to the amount of radioactivity present in the tissue section (or its portion), and therefore to the amount of the probe bound to it. Even so, it is not an easy matter to calculate the absolute amount of the target mRNA present in a chosen area of the section on the basis of optical density measurements or grain counts. Although such absolute measurements are, in principle, possible (Schäfer et al., 1993), quite satisfactory data can be obtained in many experiments (including ITF studies) by relative quantification of hybridization results. This method relies on expressing signal intensity in some arbitrary optical density or radioactivity units, which is sufficient to detect relative differences in the target mRNA abundance in different CNS regions or between animal groups (Fig. 4). Moreover, if radioactivity standards are exposed with tissue sections, this approach also enables comparisons between different experiments.

Such a semi-quantitative type of analysis is commonly used in ITF studies. Optical density measurements of film autoradiograms are very useful whenever induction of ITF mRNAs takes place in well-defined patches of densely packed cells. Induction of c-*fos*, *fosB*, *fra-1*, *fra-2*, c-*jun* and *junB* mRNAs in the granule cell layer of the dentate gyrus and pyramidal cell layer of the hippocampus produced by seizure-eliciting treatments may serve as an example (Fig. 1; Gall et al., 1990; Przewłocki et al., 1995; Beer et al., 1998). Heterogenous distribution of ITF mRNA signal within large brain structures (e.g. in the striatum or in different layers of the cortex) can also be roughly assessed using that approach (Fig. 4).

Semi-quantitative analysis of hybridization signal at the cellular level is performed using liquid emulsion autoradiography by counting numbers of irradiated grains over individual cells or cell groups. Such measurements are useful in the cases if ITF mRNA induction takes place in sparsely scattered cells and/or if there is a need of defining the phenotype of ITF-expressing cells by staining them for other biochemical markers. For example, in the striatum, there are two predominant populations of medium-sized spiny GABAergic neurons which differ in their neuropeptide content, their neurotransmitter receptor complement and target areas of their projections, but which are largely interspersed among each other (Gerfen and Young, 1988; Albin et al., 1989; Gerfen, 1992). ITF induction is elicited by pharmacological agents differentially in these neuronal populations: while dopamine receptor D1 agonists and indirect dopamine agonists induce ITFs in the substance P- and dynorphin-containing neurons which project mainly to the substantia nigra pars reticulata, dopamine D2 receptor antagonists induce ITFs in the enkephalin-containing neurons projecting to the globus pallidus (Berretta et al., 1992; Robertson et al., 1992; Gerfen et al., 1995). Although in both cases ITF mRNA induction can be (and indeed is, e.g. Sirinathsinghji et al., 1994; Keefe and Adams, 1998; Steiner and Gerfen, 1999; Figs. 2 and 4) studied by film autoradiography, intensity of hybridization signal is underestimated due to its 'dilution' by signal-negative cells present in the areas of optical density scanning. This phenomenon may be overcome by quantification of hybridization signal at single-cell level using liquid emulsion autoradiography (e.g. Gerfen et al., 1995; Svenningsson et al., 2000). It must be noted, however, that such quantification is much more laborious than optical density measurements of film autoradiograms, and it is not a usual practice in ITF expression studies.

Quantitative potential of non-isotopic ISH is, in general, lower than that offered by radioactive methods. If probes labeled directly with fluorochromes or AP are used, then the amount of probe bound to the tissue sections is proportional to the fluorescence or color intensity, respectively, and therefore can be estimated (Luo and Jackson, 1999; Salehi et al., 1999). However, no external standards can be used with these labels, which constitutes a problem particularly in the case of AP-labeled probes, where there is a risk of color saturation during the detection reaction. Hapten-labeled probes detected by antibody- or avidin-based systems are of the lowest value for quantification purposes. The reasons for that have already been discussed.

Non-radioactive ISH methods are also considered to be less sensitive than the radioactive ones (Augood et al., 1994; Wilkinson, 1995a; Beer et al., 1998). Among them, the method using probes labeled directly with AP has been reported to be the most sensitive (Augood et al., 1994). In contrast, when fluorescent detection is applied, less intense hybridization signal is obtained with probes labeled directly with fluorochromes than with hapten-labeled probes which are detected using fluorochrome-conjugated antibodies (Department of Cytochemistry and Cytometry, University of Leiden, 1992).

On the other hand, a higher resolution of hybridization signal is obtained with non-radioactive than with radioactive ISH methods. Using colored or fluorescent probe detection, the typical location of mRNA in the cell body cytoplasm and sometimes also in fine processes is visualized, while the nucleus appears as a spot devoid of staining (Department of Cytochemistry and Cytometry, University of Leiden, 1992; Emson et al., 1994; Wahle, 1994; Meltzer et al., 1998). Although radioactive probes may just as well be detected at single-cell level by emulsion autoradiography, the resulting microscopic picture of signal distribution is rough and lacks any intracellular details: black emulsion grains are chaotically dispersed over the cell body (e.g. Moratalla et al., 1993; Emson et al., 1994; Johansson et al., 1994; Gerfen et al., 1995).

Due to its high resolution, non-radioactive ISH has been proposed to be the method of choice for neuronal activity mapping by ITF induction assessment at the mRNA level (Kaczmarek and Chaudhuri, 1997). As a matter of fact, however, it has little advantage over radioactive ISH followed by emulsion autoradiography when used for this purpose, since the same result is obtained with both methods, namely labeling of cell bodies of those neurons which are activated. While results of non-radioactive ISH are perhaps more elegant from the esthetic point of view, and less safety precautions need to be taken with non-radioactive methods, radioactive ISH followed by emulsion autoradiography seems to be the most balanced ISH variant, which offers both single-cell resolution and the possibility of quantitative measurement of the target mRNA abundance. Moreover, radioactive and non-radioactive ISH are often combined with each other, which allows simultaneous detection of multiple transcripts in the same neurons (see Section 3.3.4).

3.1.2. Northern blot

Northern blot is a method enabling quantitative and sequence-specific mRNA detection in RNA samples extracted from the tissue and size-fractionated by electrophoresis (for details of the technique see e.g. Sambrook et al., 1989; Farrell, 1993; Lynch and O'Mara, 1997c). Briefly, total RNA (including ribosomal, messenger, transfer and heterogenous nuclear RNA) is isolated from tissue samples by one of the few standard procedures, most commonly by the acid guanidinium thiocyanate–phenol–chloroform extraction (Chomczyński and Sacchi, 1987; Chomczyński, 1993). Concentration of RNA is then determined spectrophotometrically by measuring absorbance at the wavelength of 260 nm. RNA samples (usually equal in amount) are denatured, loaded on an agarose gel and subjected to electrophoresis whereby different RNA molecule pools present in the samples are separated into discrete bands according to their length. RNA is then transferred from the gel, by means of capillary forces, to a porous membrane made of nylon or nitrocellulose, and crosslinked to the membrane by baking or UV light exposure. Such a membrane containing size-fractionated RNA firmly bound to it is called a blot (since it is a 'copy' of the electrophoretic gel). Northern (i.e. RNA) blots are incubated with labeled nucleic acid probes, which results in hybridization of the probe to RNA band(s) bearing complementary sequence(s). Blots may be hybridized sequentially with several probes. cRNA and cDNA rather than oligonucleotide probes are used for blot hybridization. They are labeled by the same methods as probes for ISH, but the preferred label is the radioactive phosphorus isotope ^{32}P. In the last years, non-radioactive labels with chemiluminescent detection systems have also become quite popular (Boehringer Mannheim Biochemicals, 1995c). Hybridization signal is detected by film autoradiography of blots and its intensity is measured densitometrically (Fig. 3).

In principle, the method is quantitative since the optical density measurements are proportional to the amount of RNA detected by the probe (provided that optical density values fall into the linear range of the film). However, the method as a whole is not really precise because it consists of many steps at which quantitative errors may arise (extraction, concentration measurement, gel loading, RNA transfer). For that reason, differences between groups of equivalent samples that are processed independently can be reliably measured only if they exceed some 50 to 100%. The precision can be slightly improved by using internal standards, i.e. by measuring the signal of the mRNA of interest in relation to another RNA species whose levels should not differ between the samples. Transcripts of constitutively expressed (housekeeping) genes, e.g. cyclophilin, β-actin, α-tubulin, or ribosomal RNA most often serve this purpose (Fig. 3). The latter one is especially recommended, as rRNA species comprise dominating population of total RNA present in the cells. On the other hand, transcripts of each of the so-called housekeeping genes could (and it is well known that in fact they do) respond to some of the experimental treatments.

Even so, the difficulty of detecting small differences in RNA amount is not a problem in ITF studies where the changes between basal and induced mRNA levels range from at least severalfold to hundredfold. Whenever quantitative measurement of ITF mRNA levels is the main interest, Northern blot seems to be the method of choice and is indeed widely used. It is suitable not merely for detecting induction of ITF mRNAs, but also for studying factors which affect the magnitude of the induction to some extent. For example, dose-dependent effects of pharmacological agents which either elicit or reduce induction of some ITFs, as well as time-courses of mRNA induction can be assessed easily using Northern blot (e.g. Morgan et al., 1987; Sonnenberg et al., 1989b; Nguyen et al., 1992; Kamińska et al., 1994; Rogue and Malviya, 1994; Bading et al., 1995; Konradi et al., 1996). Beside its high quantitative value, the method has the advantages of being relatively straightforward and enabling measurement of levels of several mRNA species in the same samples (i.e. on the same blot). In addition, mRNA length of the bands detected by Northern hybridization probes can be estimated either on the basis of the position of 28S and 18S rRNA bands or, more exactly, using samples of multiple RNA size markers. It is thereby easy to notice nonspecific hybridization signal and distinguish it from that of the mRNA of interest (which is not the case in ISH). Similarly, when the probe hybridizes to multiple homologous transcripts, these can often be resolved and quantified separately using Northern blot, but not the ISH method. Therefore, the risk of misinterpreting hybridization results is much lower with Northern blot than with ISH.

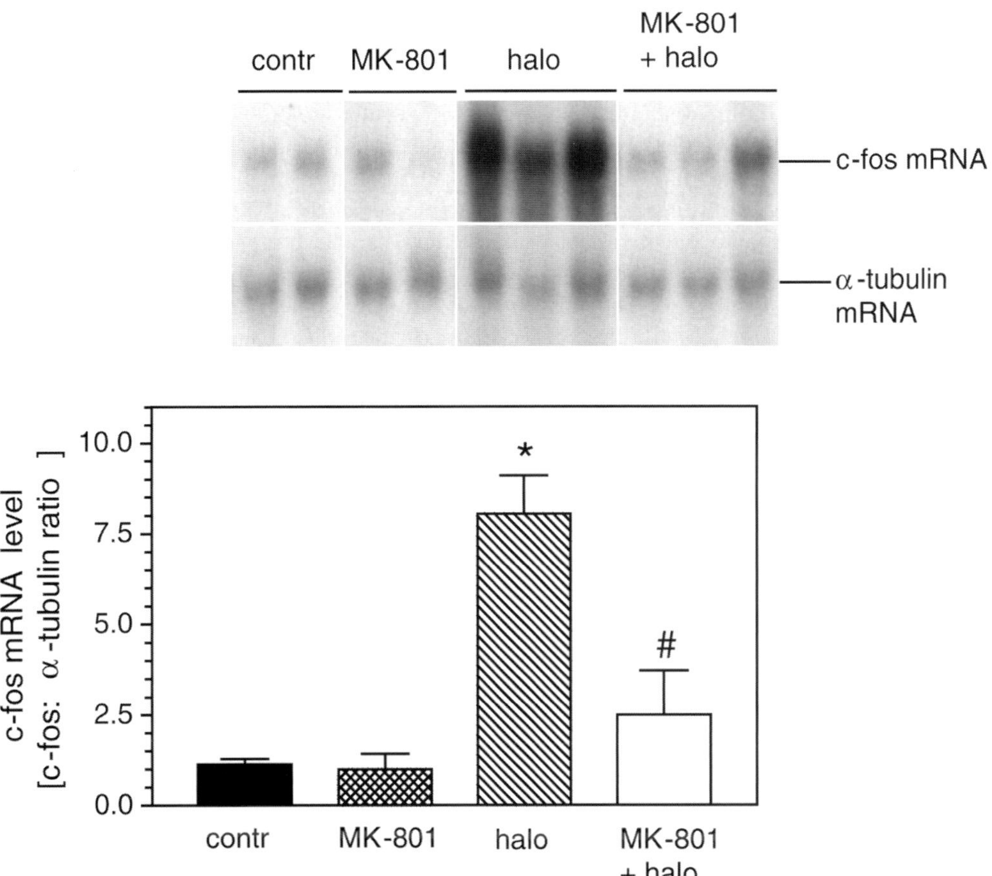

Fig. 3. A Northern blot analysis of the influence of the NMDA receptor antagonist MK-801 on the induction of c-*fos* mRNA elicited by haloperidol in the mouse striatum. The mice were injected i.p. with saline (contr), haloperidol (halo; 1 mg/kg), MK-801 (2 mg/kg) or both drugs. MK-801 was administered 30 min before haloperidol. The mice were killed 30 min after haloperidol or saline/60 min after MK-801 injection, striata were dissected and processed for Northern blot. The upper panel represents a typical autoradiogram of a blot hybridized with ^{32}P-labeled cRNA probes for c-*fos* and α-*tubulin* mRNAs (the latter served as internal standard to make a correction for unequal gel loading). Results of the densitometric analysis of the autoradiogram are given in the graph. The results are expressed as the ratio of c-*fos* to α-*tubulin* hybridization signal intensities. Statistical analysis: *, significantly different from 'contr'; #, significantly different from 'halo' by ANOVA followed by the Tukey test.

However, anatomical resolution of mRNA detection by Northern blot is very poor, because the method is based on RNA which is extracted from dissected and homogenized tissue. The minimum size of a nervous tissue sample which can be routinely analyzed by Northern blot is about 10 mg (which corresponds to the mass of e.g. one striatum or half a hippocampus of a mouse). Although smaller tissue samples can be identified and excised precisely, they do not contain a sufficient amount of total RNA to be used for blot preparation. In practice, one needs at least about 10 mg of total RNA: 5 or usually more micrograms per sample are loaded onto the gel, while another 1 or 2 mg are needed to measure RNA concentration in the sample.

3.1.3. Other methods

There are several other methods which are used to detect particular mRNAs and to measure their levels. Most of them require isolation of RNA from homogenized tissue samples and thus, similarly to Northern blot, they are characterized by poor anatomical resolution. Although each of these methods is superior to Northern blot in some aspect, e.g. being more sensitive or precise, permitting smaller tissue samples or enabling quantitative detection of many mRNA species in one experiment, these methods are not routinely used in ITF studies. This is because each of them has also major disadvantages (see below) which often overweigh their advantages in the case of ITF mRNA measurements, that are rather easy and do not require sophisticated techniques. First, ITF mRNAs, when induced, are so abundant that they do not require really sensitive methods to be detected. Second, the changes in ITF mRNA levels are so great that they do not require high precision to be assessed. Third, if ITF mRNA levels are to be estimated in very small samples, this can be done easily and with much higher resolution using semi-quantitative in situ hybridization. Fourth, the commonly studied ITFs are not really numerous and a reasonable proportion of their transcripts can be detected simultaneously or consecutively using a single Northern blot.

Nevertheless, although application of the methods listed below to studying expression of ITF genes is limited, they may be very useful indeed for some special purposes.

Ribonuclease protection assay (RPA) is a method in which RNA, extracted from the tissue, is hybridized in solution with a radiolabeled cRNA probe (Calzone et al., 1987; Sambrook et al., 1989a; Ambion, 1997; Einspanier and Plath, 1998). After hybridization, the samples are subjected to digestion with a ribonuclease specific for single-stranded RNA. In this step, RNA present in the sample is degraded except for those fragments of the target mRNA and the probe which form duplexes with each other. Samples are then separated by electrophoresis on polyacrylamide gels, and the gels are subjected to film autoradiography. Bands appearing on the resulting autoradiograms represent the target mRNA fragment protected from digestion by the probe, which should therefore be of the same length as the probe itself. However, if the probe is only partially complementary to the target, then the protected fragment(s) is (are) shorter. The method is quantitative and very sensitive, permitting the detection of rare transcripts, which cannot be readily detected using Northern blot. As already mentioned, high sensitivity is not needed for ITF mRNA detection, and since the RPA is more difficult to perform than Northern blot, requiring e.g. a better control over the probe synthesis and labeling, as well as more control samples, RPA is not in common use in ITF expression measurements.

However, there is one important exception, namely Δ*fosB*. Δ*fosB* mRNA is a splice variant of *fosB* mRNA, lacking 140 nucleotides in exon 4 of *fosB*. The deletion creates a stop codon, and the resulting ΔFosB protein lacks 101 C-terminal amino acids of FosB (Mumberg et al., 1991; Nakabeppu and Nathans, 1991). The difference in length between *fosB* and Δ*fosB* mRNAs is too small to be resolved by Northern blot, because the two bands migrate too close to each other on typical agarose gels (Bing et al., 1997). Still, the two mRNAs can be distinguished by RPA performed with a probe complementary to *fosB* mRNA which spans the splicing site (Chen et al., 1995). In RPA autoradiograms, the full-length *fosB* mRNA is then represented by a band corresponding to the probe length, while Δ*fosB* mRNA is represented by a shorter band corresponding to the probe fragment which is shared with Δ*fosB* mRNA. In this case, a fragment of the RNA probe hybridizing to Δ*fosB* mRNA is degraded because it is not protected from digestion by the exon 4 sequence which is absent from Δ*fosB* mRNA. The two bands representing *fosB* and Δ*fosB* mRNAs are readily separated on polyacrylamide gels.

Reverse transcription–polymerase chain reaction (RT-PCR) is a two-step method of selective mRNA detection in RNA solutions (for exhaustive descriptions of the technique see e.g. Kawasaki, 1990; Boehringer Mannheim Biochemicals, 1995b; Köhler et al., 1995). First, RNA extracted from the tissue, is subjected to reverse transcription into complementary DNA (cDNA) using the reverse transcriptase enzyme. In turn, a portion of cDNA corresponding to the mRNA of interest is selectively amplified, in an iterated DNA polymerase reaction (PCR), using pairs of sequence-specific primers. In order to detect the reaction product, the PCR reaction mixture is subjected to agarose gel electrophoresis and the gel is either stained directly with ethidium bromide or blotted onto a membrane which is then hybridized with a product-specific probe. There also exist a number of alternative methods of PCR product detection, which include high-performance liquid chromatography and enzyme-linked sorbent assays (Boehringer Mannheim Biochemicals, 1995a; Köhler et al., 1995).

Owing to the amplification step, the RT-PCR method is extremely sensitive, permitting detection of minute amounts of the target mRNA. Therefore, much smaller tissue samples can be used for RT-PCR than for Northern blot; in fact, the methods allowing to perform RT-PCR on RNA derived from a single cell have been described (Ziegler et al., 1992; Dixon et al., 2000). In general, RT-PCR is also much simpler, less time-consuming and much cheaper than Northern blot. However, being a perfect method for mRNA *detection*, RT-PCR is not as good for *quantitative measurements* of mRNA levels. On the one hand, amplification of the nucleic acid of interest is so extensive that even small differences between the samples in the amount of the original mRNA (arising either from experimental inaccuracy or from real differences in gene expression) may result in significant differences in the amount of the final PCR product. On the other hand, accumulation of the PCR product, which proceeds exponentially at the beginning of the reaction, reaches a plateau at some point. Should differences between original samples be inferred from the amount of PCR product measured in this plateau phase, they would be underestimated.

It is not to say, though, that quantitative measurements are not possible using RT-PCR. Quantitative or at least semi-quantitative data may be obtained by this method provided that optimum reaction conditions are chosen for each pair of PCR primers on the basis of control experiments in which the rate of PCR product synthesis is strictly monitored. More accurate measurements can be performed by co-amplification of external or internal DNA standards along with the cDNA of interest (Wang et al., 1989; Gilliland et al., 1990; Köhler et al., 1995). Nevertheless, taken the relatively high risk of experimental errors and complexity of more quantitative versions of the method, RT-PCR is used only rarely for typical studies on expression of ITFs (e.g. Giorgi et al., 1997) or other genes which can be assessed by less sensitive, but more accurate methods.

Again, however, Δ*fosB* mRNA is the exemption, and its levels can be measured most readily by RT-PCR. In this case, the used PCR primers recognize sequences located on both sides of the alternatively spliced exon of *fosB* mRNA (Bing et al., 1997). The PCR product corresponding to Δ*fosB* mRNA is about 140 base pairs shorter than that corresponding to the full-length *fosB* mRNA. The two products are separated by electrophoresis in high-concentration (about 2%) agarose gels.

Hybridization to DNA probe arrays (chips) is a method by which expression of thousands of genes can be compared simultaneously and quantitatively between the samples of extracted RNA (for details and overview see Lockhart and Winzeler, 2000). The method relies on the construction of arrays of localized, surface-bound cDNA or oligonucleotide probes. The arrays are available commercially, though at high prices, and with the current technologies it is possible to deposit up to several hundred thousand different probes per square centimeter of the support (high-density microchips). RNA, extracted from experimental samples, is labeled by tagging with biotin or by reverse transcription using fluorochrome-conjugated nucleotides, and then hybridized to the arrays. Hybridization signal is assessed by fluorescence intensity measurements.

The method is very expensive and serves for global estimation of gene expression (e.g. in disease states or under the influence of drugs), which has to be followed by a more detailed analysis of expression of particular chosen genes by other methods. As such, the use of DNA chips is much too advanced a method if expression of the mere ITFs is the main interest. However, less sophisticated, lower-density probe arrays may well be used for simultaneous detection of different ITF mRNAs and measurement of their levels in the extracted RNA samples. An example of such an extensive gene expression study has been presented by Soriano et al. (2000), who assessed expression of 750 genes, including multiple ITF genes, in several brain regions in a rat model of focal forebrain ischemia.

3.2. ITF PROTEIN DETECTION

ITF proteins can be detected either by the Western blot method or by immunocytochemistry. This subject is further covered in Chapter II by Dragunow and Bilkey. Briefly, in both methods, detection of ITF proteins is performed using specific antibodies which are either conjugated directly with fluorochrome molecules, permitting their visualization, or are recognized by secondary antibodies conjugated with enzymes (peroxidase or alkaline phosphatase) or fluorochromes. The systems of antibody visualization are largely the same as those used for the detection of hapten-labeled nucleic acid probes (see description of in situ hybridization probe labels), except that, in Western blot the substrates for alkaline phosphatase reactions are often used which are turned into chemiluminescent products that are detected by film autoradiography (Boehringer Mannheim Biochemicals, 1995c).

Immunocytochemistry permits detection of proteins in tissue slices, being the protein-based counterpart of in situ hybridization (McGinty et al., 1993; Wahle, 1994; Brahic and Ozden, 1995). Western blot, on the other hand, is the analogue of Northern blot, serving to detect proteins which are extracted from the homogenized tissue samples, size-fractionated by denaturing polyacrylamide gel electrophoresis and blotted (transferred) onto appropriate membranes (Sambrook et al., 1989b; Lynch and O'Mara, 1997c; Spector et al., 1998).

The quantitative potential of the two methods (i.e. the possibility of measuring levels of proteins in question) is lower than that of the corresponding mRNA detection techniques. While semi-quantitative analysis is possible in the case of Western blots, where intensity of protein bands can be measured by densitometry, immunocytochemistry is considered essentially as a qualitative rather than a quantitative method.

The main advantages of immunocytochemistry are its sensitivity and very high spatial resolution. Although the latter is similar to that obtained with non-radioactive ISH techniques, many more intracellular details may be revealed by staining proteins rather than mRNA. This is because mRNA is dispersed more or less homogeneously in the cell body cytoplasm, whereas proteins are distributed differentially in cellular compartments and organelles, and in neurons, fine processes can also be visualized by labeling their protein components. However, the latter is not the case with ITFs which are accumulated in the cell nucleus. When ITF immunocytochemistry is applied to the nervous tissue slices, it results in dotted labeling of nuclei of ITF-expressing neurons. Since the labeling is very distinct, the ITF-containing cells can be counted easily, which is the only way of expressing the results in numbers.

While in the case of mRNA hybridization, specificity of detection is defined by the unique nucleotide sequence of the target, it is much more difficult to ensure selectivity of protein detection by antibodies. In ITF studies, the antibodies were commonly used for several years which did not distinguish between different Fos-family members (Quinn et al., 1989; Young et al., 1991). Therefore, Fos-like proteins were detected collectively and could be distinguished only by Western blot (but not by immunocytochemistry) owing to differences in their molecular weight (e.g. Pennypacker et al., 1993). At present, that problem has already been overcome, since selective antibodies are now available to all Fos- and Jun-family members except for ΔFosB. The latter protein differs from FosB in lacking 101 amino acids at the C-terminus and its abundance can be studied using antibodies recognizing N-terminal portion of FosB in Western blot experiments, where ΔFosB and FosB are separated from each other by electrophoresis (Mumberg et al., 1991; Nakabeppu and Nathans, 1991; Chen et al., 1995, 1997; Hiroi et al., 1997). Antibodies raised against the C-terminus of FosB are also used; being selective for FosB, they do not bind to ΔFosB (Chen et al., 1995, 1997; Hiroi et al., 1997). ΔFosB protein studies are complicated further by the fact that it appears in Western blots as multiple bands, which are supposed to represent different post-translationally modified isoforms of the protein (Hope et al., 1994; Chen et al., 1995, 1997; Hiroi et al., 1997).

3.3. SUITABILITY OF THE METHODS USED TO STUDY ITF EXPRESSION FOR PARTICULAR EXPERIMENTAL PURPOSES

3.3.1. ITF mRNA versus ITF protein detection

Numerous studies have shown that the accumulation of ITF mRNAs, resulting from transcriptional activation of the respective genes, is followed by a huge increase in the ITF protein levels. Consequently, in very many studies concerning the CNS, ITF protein levels were treated as a measure of ITF gene induction. Such a practice is based on the presumption that ITF mRNAs must *always* be translated into proteins and that the rates of transcription and translation are proportional. However, this cannot be taken as a rule because in some cases elevation of ITF mRNA levels is not accompanied by an increase in the corresponding protein levels (which may be due to inefficient translation or rapid mRNA or protein degradation; Hisanaga et al., 1992; Kiessling et al., 1993; Worley et al., 1993) and therefore ITF gene induction may sometimes be overlooked or underestimated if examined only at the level of protein, and not mRNA. Nevertheless, the risk of such mistakes appears to be rather low.

While, in practice, the rises in mRNA and protein levels are used as equivalent indicators of ITF gene induction, it is necessary to adjust experimental paradigms to measurements of either ITF mRNA or protein. This is because time-courses of ITF mRNA and protein expression differ significantly. mRNA levels of the most commonly studied ITFs (c-*fos*, *fosB*, c-*jun*, *junB*, *krox-24*) rise immediately after application of the inducing stimulus, reach the maximum after 20–60 min and then decline very rapidly to reach the control values by 1.5–3 h. In contrast, ITF proteins accumulate with a delay, their levels peaking at several hours after the onset of stimulation (e.g. Morgan et al., 1987; Sonnenberg et al., 1989a,b; Dragunow et al., 1990; Nguyen et al., 1992; Moratalla et al., 1993; Rogue and Malviya, 1994; Sirinathsinghji et al., 1994; Bading et al., 1995). In order to obtain the most reliable data concerning ITF gene induction, the mRNAs and proteins should be detected at the time they peak. Thus, when ITF mRNA levels are to be measured, animals are sacrificed as early as 30–45 min after the stimulus or drug injection, while if protein detection is planned, animals are killed at 2–3 h.

Another difference between ITF mRNAs and proteins to be taken into account when tracing ITF expression at the cellular level, lies in their intracellular distribution. Whereas ITF mRNAs are present throughout the cell body cytoplasm, the corresponding proteins accumulate in the cell nucleus. The resulting lack of interference between staining of ITF proteins and cellular mRNAs may be considered an advantage when designing experiments in which ICC and ISH methods are combined (see Section 3.3.4).

3.3.2. Localizing ITF-expressing neurons

Typically, the objective of ITF expression studies is either to search the activated neurons within the CNS or to measure levels of ITF gene expression in some pre-selected CNS regions. More sophisticated studies adopt complex goals such as e.g. phenotypical definition of ITF-expressing neurons or simultaneous detection of multiple ITF gene products in the same cells. Most commonly, experiments are carried out using immunocytochemistry (ICC), ISH or, less frequently, Northern blot. However, the suitability of these methods and, in particular, of numerous ISH variants, to approach the above-mentioned goals, differs significantly.

The aim of mapping or imaging studies is to establish distribution of neurons in the CNS which are activated (in terms of expressing some ITFs) by a particular stimulus or a drug. Therefore, such experiments require techniques by which ITF gene products can be detected with single-cell resolution. In practice, ICC is almost always used for mapping neuronal pathways and imaging drug action. For example, the large majority of the results mentioned in Section 2 of this chapter, concerning pathways activated by sensory stimuli and targets of drug action, were obtained by immunocytochemical detection of c-Fos. In spite of its overwhelming popularity, ICC is not the only method appropriate for mapping and imaging purposes. The most serious alternative is ISH with hapten- or fluorochrome-labeled probes, since the resolution offered by the non-radioactive ISH techniques is not inferior to that of ICC. Another possibility is to use ISH with radioactively labeled probes, followed by coating of tissue sections with photographic emulsion. Despite their potential in tracing single ITF-expressing neurons, the above ISH variants are very uncommon in simple mapping or imaging experiments.

The strong preference for ICC may be due to the fact that the method is relatively straightforward, fast and inexpensive. Although many ISH procedures are just as simple, and can be performed successfully by inexperienced researchers, the experiments are somewhat complicated by the fact that they must be carried out in ribonuclease-free conditions in order to avoid degradation of target mRNA. This involves the use of molecular biology grade (i.e. more expensive) reagents, treating all solutions and containers with ribonuclease inhibitors and/or autoclaving them. Additional problems are inherent in work with radioactively labeled probes. First, safety precautions must be taken to prevent excessive exposure of experimenters to radioactivity. Second, radioactively labeled probes are short-lived due to radioactive decay and radiolysis. Third, harvesting hybridization results is delayed by the time necessary for emulsion exposure. Fourth, if liquid emulsion is used, then hybridization signal intensity cannot be monitored during exposure: if the sections are overexposed, there is no possibility to make a correction. On the other hand, in all techniques involving colored reaction products, color development can be controlled during reaction. In conclusion, radioactive ISH followed by emulsion coating may be too cumbersome for simple imaging experiments, but non-radioactive ISH techniques do not seem to be much more demanding than ICC and should be easy to be implemented, at least in the laboratories with some background in molecular biology.

3.3.3. Assessing levels of ITF gene expression

In contrast to drug action imaging experiments, in which the question is *where* in the CNS the investigated drug acts, many pharmacological studies aim to elucidate mechanisms of drug action by taking advantage of the changes in ITF gene expression produced by the substance(s) under investigation. Usually, in such experiments different receptor ligands or enzyme inhibitors are tested for their ability to influence the induction of ITFs elicited by the drug in question. This type of studies is exemplified by the extensive research on the mechanisms by which psychostimulants and neuroleptics affect ITF gene expression in the striatum/nucleus accumbens. It was found that the psychostimulants cocaine and amphetamine elicited induction of c-*fos*, c-*jun*, *junB* and *krox-24* in the striatal complex (Graybiel et al., 1990; Berretta et al., 1992; Cole et al., 1992; Moratalla et al., 1992, 1993; Nguyen et al., 1992; Konradi et al., 1996). These effects were demonstrated to depend largely on the stimulation of D1 receptors by dopamine, because they were markedly reduced by D1 receptor antagonists, and absent in D1 receptor-deficient mice (Graybiel et al., 1990; Berretta et al., 1992; Cole et al., 1992; Moratalla et al., 1992; Nguyen et al., 1992; Drago et al., 1996). However, D2 receptor antagonists were also shown to have some inhibitory influence on c-*fos* induction by cocaine and amphetamine in the dorsal and central parts of the striatum, suggesting a synergy between D1 and D2 receptors in mediating the effect of both psychostimulants (Ruskin and Marshall, 1994; Daunais and McGinty, 1996). On the other hand, the involvement of serotonin in the mechanism of ITF induction after cocaine administration was proposed on the basis of the findings that disruption of the serotonin system attenuated the cocaine-induced c-*fos* and *krox-24* expression and that selective serotonin reuptake inhibitors potentiated striatal induction of c-*fos* and *krox-24* elicited by selective blockers of the dopamine transporter (Bhat and Baraban, 1993; Humblot et al., 1998). In addition, glutamate was also shown to play a part in mediating the psychostimulant effects on ITF induction in the striatum, since the induction was attenuated by pretreatment with NMDA and AMPA receptor antagonists, as well as by lesion to the glutamatergic corticostriatal pathway (Cenci and Björklund, 1993; Torres and Rivier, 1993; Wang et al., 1994a,b; Vargo and Marshall, 1995; Konradi et al., 1996). Studies on the mechanisms of neuroleptic-elicited ITF induction in the striatum followed a similar scheme. Classic neuroleptics were shown to induce expression of several ITFs in the striatum due to the blockade of dopamine D2 receptors, since the neuroleptics' effects were mimicked by selective D2 antagonists, and reversed by D2 agonists (Dragunow et al., 1990; Nguyen et al., 1992; Robertson et al., 1992; Rogue and Malviya, 1994; Sirinathsinghji et al., 1994). However, the induction of c-*fos* was prevented to a great extent by the administration of either NMDA, adenosine A2A or muscarinic receptor antagonists, indicating that induction of ITFs was not elicited by the mere blockade of D2 receptors, but required simultaneous action of glutamate, adenosine and acetylcholine on their respective receptors (Figs. 2–4; Dragunow et al., 1990; Guo et al., 1992; Ziółkowska and Höllt, 1993; Boegman and Vincent, 1996; Keefe and Adams, 1998).

Owing to the above-mentioned and many similar studies, complex interactions of several neurotransmitters and their involvement in drug action can be watched in the brain structures, and actually even in single defined neurons. However, to address such issues, it is essential to use the methods which permit to measure the extent of ITF gene expression. In other words, one needs to answer the question *how much* ITF gene product has accumulated in the CNS region of interest and to compare the amount of the gene product in different animal groups. Two approaches have been widely used to quantitatively measure ITF gene expression. One of them relies on counting the number of neurons present in a defined brain area which express

c-fos

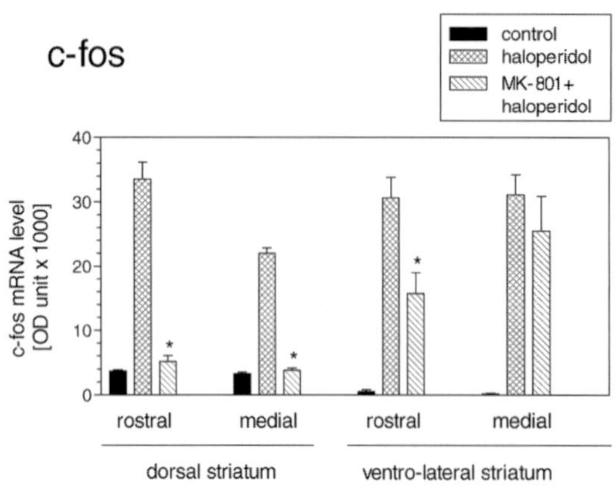

haloperidol MK-801 + haloperidol

c-jun

c-jun mRNA level [OD unit x 1000]
60
50
40
30
20
10
0
rostral medial rostral medial
dorsal striatum ventro-lateral striatum

haloperidol MK-801 + haloperidol

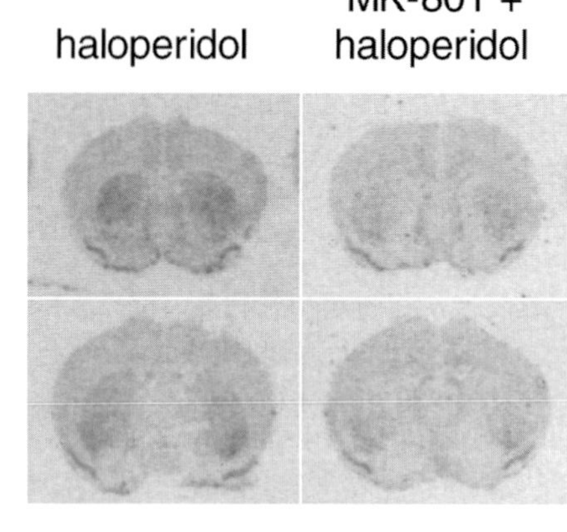

junB

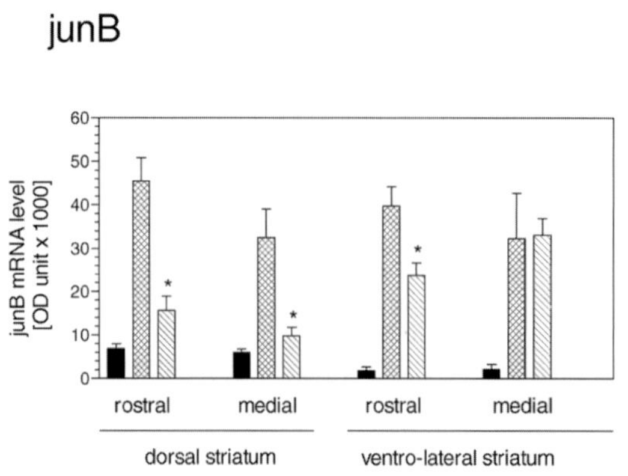

haloperidol MK-801 + haloperidol

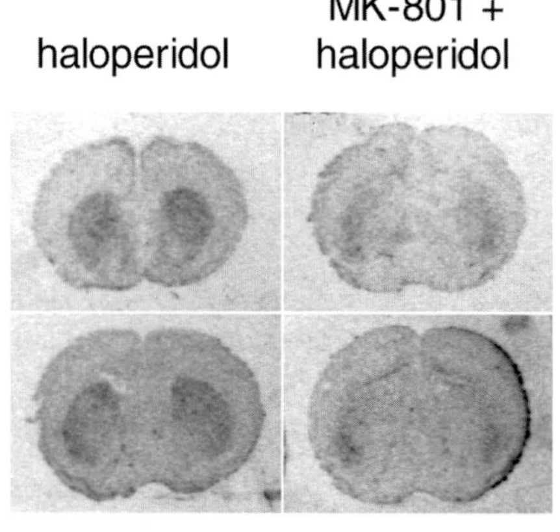

the ITF under study at any detectable level. This method has been used mainly to quantify results of ICC, but it can also be applied for any variant of single-cell-resolution ISH. The second approach is based on densitometric measurements of the detection signal intensity within the CNS region of interest, and it is used to quantify the results of radioactive ISH followed by film autoradiography, some non-radioactive ISH variants, Northern and Western blot. Counting of emulsion grains over emulsion-coated ISH sections also represents this type of analysis.

The first of the above-mentioned approaches (cell counting) takes advantage of the fact that basal expression of several ITF genes is so low that in most neurons their products are not detectable by ICC or ISH. When induced, levels of their respective mRNAs and proteins rise rapidly and dramatically, thus becoming easily detectable in many cells, which can then be counted as ITF-positive. This method of quantification has one, potentially serious, disadvantage: it is based solely on the detection threshold and it does not reflect concentrations of ITF gene products within individual cells. The latter may be important whenever the experimenter aims to determine the ability of one substance to reduce or potentiate ITF induction elicited by another substance or a given stimulus. Conceivably, if the investigated drug inhibits ITF induction only partially, e.g. by reducing levels of ITF protein or mRNA by 50% in each ITF-expressing neuron, the inhibition may not be reflected by a proportional decrease in the number of cells containing detectable amounts of the ITF gene products. This is because the inhibition is 'visible' only in those cells in which levels of ITF gene products have dropped below the detection threshold of the applied method. However, the above limitation does not preclude the cell-counting approach from being used in pharmacological experiments. In fact, inhibition of ITF gene induction was detected successfully in many studies using this quantification method (e.g. Graybiel et al., 1990; Guo et al., 1992; Torres and Rivier, 1993; Ruskin and Marshall, 1994; Vargo and Marshall, 1995; Boegman and Vincent, 1996). Nevertheless, it seems reasonable to keep in mind that quantitative data based on cell counting may be only a rough indicator of a change, and that negative results obtained by this method (i.e. lack of change) may not be reliable. Another problem connected with the cell-counting approach arises due to the fact that the method does not seem suitable for the studies on genes whose basal expression is high, including the widely used ITF marker *krox-24*.

In contrast to cell counting, quantification based on densitometric measurements or grain counts reflects concentrations of ITF gene products, mainly mRNAs, within the CNS region of interest or single neurons, thus overcoming the flaws of the former approach. It allows a more precise analysis of the influence of experimental treatment on the actual *level* of ITF gene expression, which is particularly useful for assessment of the combined effects of several drugs or stimuli in the same neurons, examination of dose-dependence of drug

←
Fig. 4. An in situ hybridization analysis of the influence of MK-801 on the induction of c-*fos*, c-*jun* and *junB* mRNA elicited by haloperidol in the mouse striatum. The mice were injected i.p. with saline, haloperidol (1 mg/kg) or haloperidol (1 mg/kg) and MK-801 (1 mg/kg). MK-801 was administered 30 min before haloperidol. The mice were killed 30 min after haloperidol or saline injection. The brains were processed for in situ hybridization with ^{35}S-labeled oligonucleotide probes. On the right side, autoradiograms are shown corresponding to the rostral and medial levels of the striatum. The graphs on the left side represent results of the densitometric analysis of the autoradiograms. The analysis was performed separately for two parts of the striatum in each section: dorsal and ventro-lateral. Note that expression of c-*fos*, c-*jun* and *junB* genes is regulated differentially in different parts of the striatum. Statistical analysis: $*$, significantly different from 'haloperidol' by ANOVA followed by the Tukey test.

effects and following time-courses of ITF gene induction. Radioactive ISH and Northern blot are often employed in such experiments, and quantitative data are obtained by densitometric measurements of film autoradiograms (e.g. Figs. 3 and 4; Morgan et al., 1987; Sonnenberg et al., 1989b; Kamińska et al., 1994; Rogue and Malviya, 1994; Sirinathsinghji et al., 1994; Wang et al., 1994a,b; Przewłocki et al., 1995; Konradi et al., 1996; Keefe and Adams, 1998). The most exact quantitative data can be obtained from Northern blot analysis. The method enables measurement of basal ITF mRNA levels, which is much more difficult using the less sensitive ISH. However, as discussed in Section 3.1.2, the application of Northern blot is limited to large CNS regions which can be dissected precisely. On the other hand, analysis of ISH autoradiograms permits delineation of small CNS structures measuring about 0.5 mm (e.g. hypothalamic nuclei). It is most exact in the cases where hybridization signal is distributed uniformly within densely packed patches of neurons, while the measurements over large areas where ITF-expressing neurons are sparse results in signal 'dilution' and, consequently, loss of precision. In the latter situations, more refined analyses of ITF mRNA levels within individual neurons can be performed by liquid emulsion coating of tissue slices and the subsequent grain counting. Quantitative fluorimetric measurements of c-*fos* and c-*jun* mRNA levels in single cells following ISH with fluorochrome-labeled probes have also been reported (Luo and Jackson, 1999; Salehi et al., 1999).

While optical density measurements appear to be more appropriate than cell counting for quantitative purposes, they require costly equipment such as an image analysis system or a (micro)densitometer. In contrast, only a light microscope is necessary for cell counting. However, in the latter case, analysis of the results is very time-consuming, since it involves inspection of numerous tissue slices under the microscope, and 'manual' counting of the stained neurons. Densitometric measurements of film autoradiograms are less tedious, allowing a larger number of sections to be analyzed in one experiment. Thus, radioactive ISH followed by film autoradiography and densitometry can be used as a high throughput method for semi-quantitative ITF gene expression analysis, but its utility and reliability is limited whenever ITF induction takes place in sparse neurons. ICC and ISH followed by cell counting enable only a rough estimation of the level of gene expression, and are more time-consuming, but offer a higher resolution of signal and require no special equipment. Finally, relatively high quantitative potential and high resolution are combined in the methods which are the most demanding from the technical point of view, such as radioactive ISH with subsequent liquid emulsion coating and emulsion grain counting, or fluorescent ISH followed by fluorescence intensity measurements in individual neurons.

3.3.4. Phenotypical definition of ITF-expressing neurons and detection of multiple ITF gene products

Many anatomically well-defined CNS regions are composed of several interspersed populations of neurons, some of which cannot be distinguished on the basis of their cellular morphology. When ITF induction takes place in such heterogenous regions, it cannot be circumscribed to any particular neuronal population, even if detected with single-cell resolution. In order to gain an insight into the phenotype of ITF-expressing neurons, it may be necessary to perform additional staining of the tissue for population-specific biochemical markers, e.g. neuropeptides, enzymes or neurotransmitter receptors. This is accomplished by using simultaneously multiple ISH probes, multiple antibodies in ICC and/or combining ISH with ICC in one experiment on the same tissue sections, which enables simultaneous detection of several mRNAs and/or proteins in the same cells. Of course, different labels and detection systems must be used for different probes and antibodies, so that their respective ISH and ICC reaction products can be distinguished from each other. For instance, detection based on peroxidase reaction with diaminobenzidine (giving brown precipitating product) can be combined with blue staining resulting from alkaline phosphatase reaction with NBT/BCIP, and with radioactive ISH followed by autoradiographic emulsion coating (which produces dotted black labeling of cells). Another possibility is to use different fluorochromes for probe or antibody labeling: ISH or ICC signals are then distinguished owing to different spectral properties of the fluorochromes (i.e. differences in the excitation wavelength and the color of the emitted light). It is important to note that combining ISH with ICC may require adjustments in both techniques (Emson et al., 1994; Wahle, 1994; Brahic and Ozden, 1995).

Typical experiments in which multiple methods were combined to identify phenotypes of ITF-expressing neurons are represented by the number of studies on ITF induction elicited in the striatum by dopamine D1 and D2 receptor ligands and indirect dopamine agonists. Previous co-expression studies indicated that the striatum contains two predominant, morphologically similar populations of GABAergic projection neurons, one of which expresses the neuropeptide enkephalin and the D2 (but not D1) receptors, while the other expresses the neuropeptides substance P and dynorphin, as well as the D1 (but not D2) receptors (Albin et al., 1989; Gerfen et al., 1990; Le Moine et al., 1990, 1991; Gerfen, 1992). By combination of ICC for c-Fos detection with radioactive ISH for the visualization of enkephalin mRNA, it was demonstrated that the induction of c-Fos elicited by the D2 receptor antagonist haloperidol takes place mainly in the population of enkephalinergic neurons. In contrast, c-Fos induced by the D1 receptor agonist SKF-38393 in the striatum of dopamine-depleted rats was found in neurons identified as enkephalin mRNA-negative (Robertson et al., 1992). Similarly, SKF-38393 was shown to induce *krox-24* mRNA selectively in enkephalin mRNA-negative striatal neurons (Gerfen et al., 1995). In that experiment, double ISH was performed to detect *krox-24* and enkephalin mRNA with ^{35}S- and digoxigenin-labeled probes, respectively (the latter detected by alkaline phosphatase reaction with NBT/BCIP). The selective activation of enkephalin-negative striatal neurons by dopamine agonists was further corroborated by the studies on effects of cocaine and amphetamine. Both drugs were shown to induce c-*fos* mRNA or c-Fos protein in enkephalin- and enkephalin mRNA-negative, D2 receptor-negative neurons in the striatum which, in contrast, did contain substance P and D1 receptor mRNAs (Berretta et al., 1992; Johansson et al., 1994; Badiani et al., 1999). The latter experiments were carried out using double staining ICC or double probe ISH. ICC to detect c-Fos and Met-enkephalin combined either immunogold-silver enhancement reaction with peroxidase/diaminobenzidine reaction or immunofluorescent labeling with fluorescein isothiocyanate and Texas Red (Berretta et al., 1992). Double probe ISH was either performed with each (radioactively labeled) probe independently on adjacent tissue sections, and the neurons were aligned after their staining with cresyl violet, or radioactively and non-radioactively labeled probes were applied simultaneously to the same sections (Johansson et al., 1994; Badiani et al., 1999). Thus, by combination of ISH and ICC techniques, D2 receptor antagonists and D1 receptor agonists were shown to induce c-*fos* in separate populations of striatal neurons, defined by their neuropeptide and dopamine receptor complement.

Additional information, concerning the target brain areas innervated by ITF-expressing neurons, can be obtained by the combination of ISH or ICC with retrograde labeling of nerve cells. In the latter technique, tract tracers such as fluorogold or horseradish peroxidase are injected into putative projection targets; the tracers are taken up by nerve fibers and moved by retrograde axonal transport to the corresponding cell bodies, which results in cell body staining (Lynch and O'Mara, 1997a). The method of retrograde labeling with fluorogold has been used, for example, together with radioactive ISH (followed by liquid emulsion autoradiography) and ICC (employing antibodies tagged with a fluorochrome or peroxidase) to identify target areas of different populations of striatal projection neurons, including those which respond with c-*fos* induction to treatment with the dopamine receptor ligands. It was demonstrated that a great majority of neurons expressing enkephalin and the D2 receptor project to the globus pallidus, while most neurons containing substance P, dynorphin and the D1 receptor innervate the substantia nigra pars reticulata (Gerfen and Young, 1988; Gerfen et al., 1990). In agreement with the finding that striatal enkephalinergic neurons respond with c-*fos* induction to D2 receptor antagonists, while in substance P- and dynorphin-containing neurons c-*fos* is induced by D1 receptor agonists, haloperidol-induced c-Fos protein was found mainly in the neurons retrogradely labeled with fluorogold injected into the globus pallidus (but not the substantia nigra), whereas SKF-38393 induced c-Fos in the unlabeled neurons from the globus pallidus (Robertson et al., 1992). Studies of this type, which combine several methods to characterize ITF-expressing neurons with respect to their biochemical properties (the content of neurotransmitters/neuropeptides, receptors, enzymes and other proteins), as well as their projection targets, permit to gain increasingly more insight into the complexities of the CNS functional anatomy and mechanisms of drug action.

By using multiple probe ISH or multiple antibody ICC, it is also possible to detect several ITF mRNAs or proteins within the same neurons, and thus to monitor their in vivo expression simultaneously at the cellular level. So far, such studies have been scarce. They employed mainly fluorescent probe–ISH, which in this case seems to be a good choice because of the availability of several appropriate fluorochrome labels and the possibility of hybridization signal quantification based on fluorescence intensity (Luo and Jackson, 1999; Salehi et al., 1999). On the other hand, the combination of ISH and ICC cannot be used to detect ITF mRNAs and proteins induced by the same stimulus, which is due to the fast and poorly overlapping time-courses of mRNA versus protein expression for most ITFs. However, a particular technique, called dual activity mapping, which combines ISH and ICC for one ITF (*krox-24*) has been introduced in order to discriminate between neurons responsive to *two* different stimuli (Chaudhuri et al., 1997, 2000). In this method, two different ITF-inducing stimuli are applied at about a 2 h interval, and tissue is taken for ISH and ICC at 20–30 min after the second stimulus. At that time, ITF mRNA induced by the first stimulus has already vanished, but the protein is present at high level. In contrast, the level of ITF mRNA induced by the second stimulus is nearly maximal, while the corresponding protein has not accumulated yet. Therefore, ITF protein synthesized in response to the first stimulus and ITF mRNA induced by the second stimulus can be measured concomitantly by ICC and ISH. The method has so far been used to study responsiveness of neocortical neurons to visual and acoustic stimuli (Chaudhuri et al., 1997, 2000).

4. CONCLUDING REMARKS

The numerous methods discussed in this chapter are used to examine expression of ITF genes in the CNS, which is considered mainly as a useful indicator of neuronal activity. However, no information concerning the function of ITFs as transcription factors within nerve cells can be drawn from experiments which rely on the detection and quantitative measurements of ITF mRNA or protein levels. Functional issues are addressed using a completely different set of techniques. Among them are electrophoretic mobility shift and supershift assays employed to study binding of ITFs to specific gene promoters or regulatory promoter elements, and several methods by which intracellular levels of specific ITF mRNAs can be changed purposely. The latter include e.g. gene knock-down techniques (where levels of a specific mRNA are decreased in chosen CNS regions or cultured cells using antisense oligonucleotides, triplex-forming oligonucleotides or ribozymes), gene knock-out (where a specific gene is inactivated in the whole animal) and overexpression of a gene (whose additional copy, cloned in an expression vector, is introduced into the cells). These methods are used to identify putative target genes regulated by ITFs, and to search for the role of ITFs in various CNS functions at the level of single neurons, neuronal circuits, as well as complex behaviors. Such functional approaches are of great importance because the roles ITFs play in nerve cells are still largely unknown in spite of the fact that induction of ITFs in the CNS appears to be a common phenomenon.

5. ABBREVIATIONS

2-DG	2-deoxyglucose
AP-1	activator protein-1
CNS	central nervous system
GABA	γ-amino-butyric acid
ICC	immunocytochemistry
ISH	in situ hybridization
ITF	inducible transcription factors
NMDA	*N*-methyl-D-aspartate

6. ACKNOWLEDGEMENTS

This work was supported in part by grant of the Polish Committee for Scientific Research (KBN) No. 6 P05A 107 20.

7. REFERENCES

Albin RL, Young AB, Penney JB (1989): The functional anatomy of basal ganglia disorders. *Trends Neurosci 12*:366–375.

Ambion (Ed) (1997): *RPA II. Ribonuclease Protection Assay Kit. Instruction Manual.* Ambion. (Internet version available at: http://www.ambion.com/techlib/prot/fm_1410.pdf)

Augood SJ, McGowan EM, Finsen BR, Heppelman B, Emson PC (1994): Non-radioactive in situ hybridization using alkaline phosphatase-labelled oligonucleotides. In: Wisden W, Morris BJ (Eds), *In Situ Hybridization Protocols for the Brain, Ch 8.* London: Academic Press, pp 81–97.

Badiani A, Oates MM, Day HE, Watson SJ, Akil H, Robinson TE (1999): Environmental modulation of amphetamine-induced c-*fos* expression in D1 versus D2 striatal neurons. *Behav Brain Res 103*:203–209.

Bading H, Ginty DD, Greenberg ME (1993): Regulation of gene expression in hippocampal neurons by distinct calcium signaling pathways. *Science 260*:181–186.

Bading H, Segal MM, Sucher NJ, Dudek H, Lipton SA, Greenberg ME (1995): *N*-methyl-D-aspartate receptors are critical for mediating the effect of glutamate on intracellular calcium concentration and immediate early gene expression in cultured hippocampal neurons. *Neuroscience 64*:653–664.

Baldino F Jr, Robbins E, Lewis ME (1994): Enzyme histochemical detection of neuronal mRNA. In: Eberwine JH, Valentino KL, Barchas JD (Eds), *In Situ Hybridization in Neurobiology, Ch 4.* New York: Oxford University Press, pp 63–77.

Beer J, Mielke K, Zipp M, Zimmermann M, Herdegen T (1998): Expression of c-*jun*, *junB*, c-*fos*, *fra-1* and *fra-2* mRNA in the rat brain following seizure activity and axotomy. *Brain Res 794*:255–266.

Berretta S, Robertson HA, Graybiel AM (1992): Dopamine and glutamate agonists stimulate neuron-specific expression of Fos-like protein in the striatum. *J Neurophysiol 68*:767–777.

Bhat RV, Baraban JM (1993): Activation of transcription factor genes in striatum by cocaine: role of both serotonin and dopamine systems. *J Pharmacol Exp Ther 267*:496–505.

Biały M, Kaczmarek L (1996): c-Fos expression as a tool to search for the neurobiological base of the sexual behaviour of males. *Acta Neurobiol Exp 56*:567–577.

Binder M (1995): In situ hybridization at the electron microscope level. In: Wilkinson DG (Ed), *In Situ Hybridization. A Practical Approach, Ch 8*. New York: IRL Press at Oxford University Press, pp 105–120.

Bing G, Wang W, Qi Q, Zhehui F, Hudson P, Jin L, Zhang W, Bing R, Hong J-S (1997): Long-term expression of Fos-related antigen and transient expression of ΔFosB associated with seizures in the rat hippocampus and striatum. *J Neurochem 68*:272–279.

Boegman RJ, Vincent SR (1996): Involvement of adenosine and glutamate receptors in the induction of c-fos in the striatum by haloperidol. *Synapse 22*:70–77.

Boehringer Mannheim Biochemicals (Ed) (1992): *Nonradioactive In Situ Hybridization Application Manual*. Boehringer Mannheim (and the updated Internet version of the manual at http://biochem.roche.com).

Boehringer Mannheim Biochemicals (Ed) (1995a): Use of nonradioactive labels in PCR. In: Boehringer Mannheim Biochemicals (Ed), *PCR Applications Manual, Ch 5*. Boehringer Mannheim, pp 55–78 (and the updated Internet version of the manual at http://biochem.roche.com).

Boehringer Mannheim Biochemicals (Ed) (1995b): Reverse transcription–PCR (RT-PCR). In: Boehringer Mannheim Biochemicals (Ed), *PCR Applications Manual, Ch 6*. Boehringer Mannheim, pp 79–94 (and the updated Internet version of the manual at http://biochem.roche.com).

Boehringer Mannheim Biochemicals (Ed) (1995c): Detection of DIG-labeled nucleic acids. In: Boehringer Mannheim Biochemicals (Ed), *The DIG System User's Guide for Filter Hybridization, Ch 9*. Boehringer Mannheim, pp 58–67 (and the updated Internet version of the manual at http://biochem.roche.com).

Bon K, Lanteri-Minet M, de Pommery J, Michiels JF, Menetrey D (1996): Cyclophosphamide cystitis as a model of visceral pain in rats. A survey of hindbrain structures involved in visceroception and nociception using the expression of c-Fos and Krox-24 proteins. *Exp Brain Res 108*:404–416.

Brahic M, Ozden S (1995): Simultaneous detection of cellular RNA and proteins. In: Wilkinson DG (Ed), *In Situ Hybridization. A Practical Approach, Ch 7*. New York: IRL Press at Oxford University Press, pp 85–104.

Bullitt E (1990): Expression of c-fos-like protein as a marker for neuronal activity following noxious stimulation in the rat. *J Comp Neurol 296*:517–530.

Calzone FJ, Britten RS, Davidson EH (1987): Mapping of gene transcripts by nuclease protection assays and cDNA primer extension. *Methods Enzymol 152*:611–632.

Cenci MA, Björklund A (1993): Transection of corticostriatal afferents reduces amphetamine- and apomorphine-induced striatal Fos expression and turning behaviour in unilaterally 6-hydroxydopamine-lesioned rats. *Eur J Neurosci 5*:1062–1070.

Cephalon Inc. (1992): Detection of neuronal mRNA of the rat pituitary pars intermedia with a digoxigenin-tailed oligonucleotide probe for proopiomelanocortin (POMC). In: Boehringer Mannheim Biochemicals (Ed), *Nonradioactive In situ Hybridization Application Manual*. Published by Boehringer Mannheim, pp 52–54.

Chaudhuri A (1997): Neural activity mapping with inducible transcription factors. *Neuroreport 8*:5–9.

Chaudhuri A, Nissanov J, Larocque S, Rioux L (1997): Dual activity maps in primate visual cortex produced by different temporal patterns of zif268 mRNA and protein expression. *Proc Natl Acad Sci USA 94*:2671–2675.

Chaudhuri A, Zangenehpour S, Rahbar-Dehgan F, Ye F (2000): Molecular maps of neural activity and quiescence. *Acta Neurobiol Exp 60*:403–410.

Chen J, Nye HE, Kelz MB, Hiroi N, Nakabeppu Y, Hope BT, Nestler EJ (1995): Regulation of ΔFosB and FosB-like proteins by electroconvulsive seizure and cocaine treatment. *Mol Pharmacol 48*:880–889.

Chen J, Kelz MB, Hope BT, Nakabeppu Y, Nestler EJ (1997): Chronic Fos-related antigens: stable variants of ΔFosB induced in brain by chronic treatments. *J Neurosci 17*:4933–4941.

Chergui K, Nomikos GG, Mathe JM, Gonon F, Svensson TH (1996): Burst stimulation of the medial forebrain bundle selectively increase Fos-like immunoreactivity in the limbic forebrain of the rat. *Neuroscience 72*:141–156.

Chomczyński P (1993): A reagent for the single-step simultaneous isolation of RNA, DNA and proteins from cell and tissue samples. *Biotechniques 15*:532–537.

Chomczyński P, Sacchi N (1987): Single-step method of RNA isolation by acid guanidinium thiocyanate–phenol–chloroform extraction. *Anal Biochem 162*:156–159.

Cole AJ, Saffen DW, Baraban JM, Worley PF (1989): Rapid increase of an immediate early gene messenger RNA in hippocampal neurons by synaptic NMDA receptor activation. *Nature 340*:474–476.

Cole AJ, Bhat RV, Patt C, Worley PF, Baraban JM (1992): D1 dopamine receptor activation of multiple transcription factor genes in rat striatum. *J Neurochem 58*:1420–1426.

Cole RL, Konradi C, Douglass J, Hyman SE (1995): Neuronal adaptations to amphetamine and dopamine: molecular mechanisms of prodynorphin gene regulation in rat striatum. *Neuron 14*:813–823.

Curran T, Franza Jr. BR (1988): Fos and Jun: the AP-1 connection. *Cell 55*:395–397.

Curran T, Morgan JI (1987): Memories of fos. *Bioessays 7*:255–258.

Daunais JB, McGinty JF (1996): The effects of D1 or D2 dopamine receptor blockade on zif/268 and pre-prodynorphin gene expression in rat forebrain following a short-term cocaine binge. *Mol Brain Res 35*:237–248.

Department of Cytochemistry and Cytometry, University of Leiden (1992): Detection of neuropeptide mRNAs in tissue sections using oligonucleotides tailed with fluorescein-12-dUTP or digoxigenin-dUTP. In: Boehringer Mannheim Biochemicals (Ed), *Nonradioactive In Situ Hybridization Application Manual*. Boehringer Mannheim, pp 54–55.

Dixon AK, Richardson PJ, Pinnock RD, Lee K (2000): Gene-expression analysis at the single-cell level. *Trends Pharmacol Sci 21*:65–70.

Drago J, Gerfen CR, Westphal H, Steiner H (1996): D1 dopamine receptor-deficient mouse: cocaine-induced regulation of immediate-early gene and substance P expression in the striatum. *Neuroscience 74*:813–823.

Dragunow M, Robertson HA (1987): Generalized seizures induce c-fos protein(s) in mammalian neurons. *Neurosci Lett 82*:157–161.

Dragunow M, Peterson MR, Robertson HA (1987): Presence of c-fos-like immunoreactivity in the adult rat brain. *Eur J Pharmacol 135*:113–114.

Dragunow M, Abraham WC, Goulding M, Mason SE, Robertson HA, Faull RL (1989): Long-term potentiation and the induction of c-*fos* mRNA and proteins in the dentate gyrus of unanesthetized rats. *Neurosci Lett 101*:274–280.

Dragunow M, Robertson GS, Faull RLM, Robertson HA, Jansen K (1990): D_2 dopamine receptor antagonists induce fos and related proteins in rat striatal neurons. *Neuroscience 37*:287–294.

Duncan GE, Knapp DJ, Johnson KB, Breese GR (1996): Functional classification of antidepressants based on antagonism of swim stress-induced fos-like immunoreactivity. *J Pharmacol Exp Ther 277*:1076–1089.

Eells JB, Clough RW, Miller JW, Jobe PC, Browning RA (2000): Fos expression and 2-deoxyglucose uptake following seizures in developing genetically epilepsy-prone rats. *Brain Res Bull 52*:379–389.

Einspanier R, Plath A (1998): Detecting mRNA by use of the ribonuclease protection assay (RPA). In: Rapley R, Walker JM (Eds), *Molecular Biomethods Handbook, Ch 5*. Totowa, NJ: Humana Press, pp 51–57.

Emson PC, Heppelmann B, Augood SJ (1994): Development of techniques to combine isotopic and nonisotopic in situ hybridization and immunocytochemistry for phenotypic characterization of individual neurons. In: Eberwine JH, Valentino KL, Barchas JD (Eds), *In Situ Hybridization in Neurobiology, Ch 3*. New York: Oxford University Press, pp 43–62.

Erdtmann-Vourliotis M, Mayer P, Riechert U, Händel M, Kriebitzsch J, Höllt V (1999a): Rational design of oligonucleotide probes to avoid optimization steps in in situ hybridization. *Brain Res Protoc 4*:82–91.

Erdtmann-Vourliotis M, Mayer P, Riechert U, Höllt V (1999b): Acute injection of drugs with low addictive potential (delta(9)-tetrahydrocannabinol, 3,4-methylenedioxymethamphetamine, lysergic acid diamide) causes a much higher c-*fos* expression in limbic brain areas than highly addicting drugs (cocaine and morphine). *Mol Brain Res 71*:313–324.

Erdtmann-Vourliotis M, Mayer P, Riechert U, Höllt V (2000): Prior experience of morphine application alters the c-*fos* response to MDMA ('ecstasy') and cocaine in the rat striatum. *Mol Brain Res 77*:55–64.

Farrell RE Jr (1993): *RNA Methodologies. A Laboratory Guide for Isolation and Characterization*. San Diego, CA: Academic Press.

Friauf E (1992): Tonotopic order in the adult and developing auditory system of the rat as shown by c-fos immunocytochemistry. *Eur J Neurosci 4*:798–812.

Friauf E (1995): C-fos immunocytochemical evidence for acoustic pathway mapping in rats. *Behav Brain Res 66*:217–224.

Gall C, Lauterborn J, Isackson P, White J (1990): Seizures, neuropeptide regulation, and mRNA expression in the hippocampus. *Prog Brain Res 83*:371–390.

Gerfen CR (1992): The neostriatal mosaic: multiple levels of compartmental organization. *Trends Neurosci 15*:133–139.

Gerfen CR, Young WS (1988): Distribution of striatonigral and striatopallidal peptidergic neurons in both patch and matrix compartments: an in situ hybridization histochemistry and fluorescent retrograde tracing study. *Brain Res 460*:161–167.

Gerfen CR, Engber TM, Mahan LC, Susel Z, Chase TN, Monsma Jr. FJ, Sibley DR (1990): D_1 and D_2 dopamine receptor-regulated gene expression of striatonigral and striatopallidal neurons. *Science 250*:1429–1432.

Gerfen CR, Keefe KA, Gauda EB (1995): D1 and D2 dopamine receptor function in the striatum: coactivation of D1- and D2-dopamine receptors on separate populations of neurons results in potentiated immediate early gene response in D1-containing neurons. *J Neurosci 15*:8167–8176.

Gilliland G, Perrin S, Bunn HF (1990): Competitive PCR for quantitation of mRNA. In: Innis MA, Gelfand DH, Sninsky JJ, White TJ (Eds), *PCR Protocols. A Guide to Methods and Applications, Ch 8*. San Diego, CA: Academic Press, pp 60–69.

Giorgi L, Fanfani E, Paternostro E, Trovati F, Falchetti A, Novelli GP (1997): Immediate-early genes expression in spinal cord as related to acute noxious stimulus. *Int J Clin Pharmacol Res 17*:59–61.

Goelet P, Castellucci VF, Schacher S, Kandel ER (1986): The long and the short of long-term memory — a molecular framework. *Nature 322*:419–422.

Graybiel AM, Moratalla R, Robertson HA (1990): Amphetamine and cocaine induce drug-specific activation of the c-*fos* gene in striosome–matrix compartments and limbic subdivisions of the striatum. *Proc Natl Acad Sci USA 87*:6912–6916.

Greenberg ME, Ziff EB, Greene LA (1986): Stimulation of neuronal acetylcholine receptors induces rapid gene transcription. *Science 234*:80–83.

Gubits RM, Smith TM, Fairhurst JL, Yu H (1989): Adrenergic receptors mediate changes in c-*fos* mRNA levels in brain. *Mol Brain Res 6*:39–45.

Guo N, Robertson GS, Fibiger HC (1992): Scopolamine attenuates haloperidol-induced c-*fos* expression in the striatum. *Brain Res 588*:164–167.

Guthrie KM, Anderson KJ, Leon M, Gall C (1993): Odor-induced increases in c-*fos* mRNA expression reveal an anatomical 'unit' for odor processing in olfactory bulb. *Proc Natl Acad Sci USA 90*:3329–3333.

Harrer MI, Travers SP (1996): Topographic organization of Fos-like immunoreactivity in the rostral nucleus of the solitary tract evoked by gustatory stimulation with sucrose and quinine. *Brain Res 711*:125–137.

Herdegen T, Leah JD (1998): Inducible and constitutive transcription factors in the mammalian nervous system: control of gene expression by Jun, Fos and Krox, and CREB/ATF proteins. *Brain Res Rev 28*:370–490.

Herdegen T, Kummer W, Fiallos-Estrada CE, Leah JD, Bravo R (1991): Specific temporal and spatial distribution of JUN, FOS and KROX-24 proteins in spinal neurons following noxious transsynaptic stimulation. *J Comp Neurol 313*:178–192.

Hiroi N, Brown JR, Haile CN, Ye H, Greenberg ME, Nestler EJ (1997): FosB mutant mice: Loss of chronic cocaine induction of Fos-related proteins and heightened sensitivity to cocaine's psychomotor and rewarding effects. *Proc Natl Acad Sci USA 94*:10397–10402.

Hisanaga K, Sagar SM, Sharp FR (1992): *N*-Methyl-D-aspartate antagonists block fos-like protein expression induced via multiple signaling pathways in cultured cortical neurons. *J Neurochem 58*:1836–1844.

Höltke HJ, Ankenbauer W, Mühlegger K, Rein R, Sagner G, Seibl R, Walter L (1995): The digoxigenin (DIG) system for non-radioactive labelling and detection of nucleic acids — an overview. *Cell Mol Biol 41*:883–905.

Hope BT, Nye HE, Kelz MB, Self DW, Iadarola MJ, Nakabeppu Y, Duman RS, Nestler EJ (1994): Induction of a long-lasting AP-1 complex composed of altered Fos-like proteins in brain by chronic cocaine and other chronic treatments. *Neuron 13*:1235–1244.

Humblot N, Thiriet N, Gobaille S, Aunis D, Zwiller J (1998): The serotonergic system modulates the cocaine-induced expression of the immediate early genes *egr-1* and c-*fos* in rat brain. *Ann NY Acad Sci 844*:7–20.

Hunt SP, Pini A, Evan G (1987): Induction of c-fos-like protein in spinal cord neurons following sensory stimulation. *Nature 328*:632–634.

Johansson B, Lindstrom K, Fredholm BB (1994): Differences in the regional and cellular localization of c-*fos* messenger RNA induced by amphetamine, cocaine and caffeine in the rat. *Neuroscience 59*:837–849.

Johnston HM, Morris BJ (1994): Induction of c-*fos* gene expression is not responsible for increased proenkephalin mRNA levels in the hippocampal dentate gyrus following NMDA stimulation. *Mol Brain Res 25*:147–150.

Jørgensen MB, Deckert J, Wright DC, Gehlert DR (1989): Delayed c-*fos* proto-oncogene expression in the rat hippocampus induced by transient global cerebral ischemia: an in situ hybridization study. *Brain Res 484*:393–398.

Kaczmarek L, Chaudhuri A (1997): Sensory regulation of immediate-early gene expression in mammalian visual cortex: implications for functional mapping and neural plasticity. *Brain Res Rev 23*:237–256.

Kaczmarek L, Siedlecki JA, Danysz W (1988): Proto-oncogene c-*fos* induction in rat hippocampus. *Brain Res 427*:183–186.

Kamińska B, Filipkowski RK, Zurkowska G, Lasoń W, Przewłocki R, Kaczmarek L (1994): Dynamic changes in the composition of the AP-1 transcription factor DNA-binding activity in rat brain following kainate-induced seizures and cell death. *Eur J Neurosci 6*:1558–1566.

Kawasak, ES (1990): Amplification of RNA. In: Innis MA, Gelfand DH, Sninsky JJ, White TJ (Eds), *PCR Protocols. A Guide to Methods and Applications, Ch 3*. San Diego, CA: Academic Press, pp 21–27.

Keefe KA, Adams AC (1998): Differential effects of *N*-methyl-D-aspartate receptor blockade on eticlopride-induced immediate early gene expression in the medial and lateral striatum. *J Pharmacol Exp Ther 287*:1076–1083.

Kiessling M, Stumm G, Xie Y, Herdegen T, Aguzzi A, Bravo R, Gass P (1993): Differential transcription and translation of immediate early genes in the gerbil hippocampus after transient global ischemia. *J Cereb Blood Flow Metab 13*:914–924.

Kimpo RR, Doupe AJ (1997): FOS is induced by singing in distinct neuronal populations in a motor network. *Neuron 18*:315–325.

Köhler T, Laßner D, Rost A-K, Thamm B, Pustowoit B, Remke H (1995): *Quantitation of mRNA by Polymerase Chain Reaction: Nonradioactine PCR Methods*. Berlin: Springer.

Konradi C, Cole RL, Heckers S, Hyman SE (1994): Amphetamine regulates gene expression in rat striatum via transcription factor CREB. *J Neurosci 14*:5623–5634.

Konradi C, Leveque J-C, Hyman SE (1996): Amphetamine and dopamine-induced immediate early gene expression in striatal neurons depends on postsynaptic NMDA receptors and calcium. *J Neurosci 16*:4231–4239.

Lanteri-Minet M, Bon K, de Pommery J, Michiels JF, Menetrey D (1995): Cyclophosphamide cystitis as a model of visceral pain in rats: model elaboration and spinal structures involved as revealed by the expression of c-Fos and Krox-24 proteins. *Exp Brain Res 105*:220–232.

Le Moine C, Normand E, Guitteny AF, Fouque B, Teoule R, Bloch B (1990): Dopamine receptor gene expression by enkephalin neurons in rat forebrain. *Proc Natl Acad Sci USA 87*:230–234.

Le Moine C, Normand E, Bloch B (1991): Phenotypical characterization of the rat striatal neurons expressing the D_1 dopamine receptor gene. *Proc Natl Acad Sci USA 88*:4205–4209.

Lerea LS, Butler LS, McNamara JO (1992): NMDA and non-NMDA receptor-mediated increase of c-*fos* mRNA in dentate gyrus neurons involves calcium influx via different routes. *J Neurosci 12*:2973–2981.

Lewis ME, Robbins E, Baldino F Jr (1993): In situ hybridization histochemistry with radioactive and non-radioactive cRNA and DNA probes. In: Sharif NA (Ed), *Molecular Imaging in Neuroscience, Ch 1*. New York: IRL Press at Oxford University Press, pp 1–22.

Lockhart DJ, Winzeler EA (2000): Genomics, gene expression and DNA arrays. *Nature 405*:827–836.

Lu W, Haber SN (1992): In situ hybridization histochemistry: a new method for processing material stored for several years. *Brain Res 578*:155–160.

Luo LG, Jackson IM (1999): Advantage of double labeled in situ hybridization for detecting the effects of glucocorticoids on the mRNAs of protooncogenes and neural peptides (TRH) in cultured hypothalamic neurons. *Brain Res Protoc 4*:201–208.

Lynch MA, O'Mara SM (Eds) (1997a): Neuroanatomical tracing based on cellular transport: horseradish peroxidase, phaseolus vulgaris-leucoagglutinin and biotinylated dextran amine. In: Lynch MA, O'Mara SM (Eds), *Neuroscience Labfax, Ch 3*. San Diego, CA: Academic Press, pp 33–46.

Lynch MA, O'Mara SM (Eds) (1997b): Receptor mRNA localization. In: Lynch MA, O'Mara SM (Eds), *Neuroscience Labfax, Ch 11*. San Diego, CA: Academic Press, pp 165–179.

Lynch MA, O'Mara SM (Eds) (1997c): Gel electrophoresis, Western, Southern and Northern blotting. In: Lynch MA, O'Mara SM (Eds), *Neuroscience Labfax, Ch 12*. San Diego, CA: Academic Press, pp 181–198.

McGinty J, Couce ME, Ways DK (1993): Comparison of the localization of protein kinase C subspecies by in situ hybridization and immunocytochemical methods. In: Sharif NA (Ed), *Molecular Imaging in Neuroscience, Ch 2*. New York: IRL Press at Oxford University Press, pp 23–42.

Meltzer JC, Sanders V, Grimm C. PC, Stern E, Rivier C, Lee S, Rennie SL, Gietz RD, Hole AK, Watson PH, Greenberg AH, Nance DM (1998): Production of digoxigenin-labelled RNA probes and the detection of cytokine mRNA in rat spleen and brain by in situ hybridization. *Brain Res Protoc 2*:339–351.

Moratalla R, Robertson HA, Graybiel AM (1992): Dynamic regulation of NGFI-A (zif268, egr1) gene expression in the striatum. *J Neurosci 12*:2609–2622.

Moratalla R, Vickers EA, Robertson HA, Cochran BH, Graybiel AM (1993): Coordinate expression of c-*fos* and *junB* is induced in the rat striatum by cocaine. *J Neurosci 13*:423–433.

Morgan JI, Curran T (1986): Role of ion flux in the control of c-*fos* expression. *Nature 322*:552–555.

Morgan JI, Cohen DR, Hempstead JL, Curran T (1987): Mapping patterns of c-*fos* expression in the central nervous system after seizure. *Science 237*:192–197.

Mumberg D, Lucibello FC, Schuermann M, Muller R (1991): Alternative splicing of *fosB* transcripts results in differentially expressed mRNAs encoding functionally antagonistic proteins. *Genes Dev 5*:1212–1223.

Murphy TH, Worley PF, Baraban JM (1991): L-type voltage-sensitive calcium channels mediate synaptic activation of immediate early genes. *Neuron 7*:625–635.

Nakabeppu Y, Nathans D (1991): A naturally occurring truncated form of FosB that inhibits Fos/Jun transcriptional activity. *Cell 64*:751–759.

Nakajima T, Daval JL, Gleiter CH, Deckert J, Post RM, Marangos PJ (1989): c-*fos* mRNA expression following electrical-induced seizure and acute nociceptive stress in mouse brain. *Epilepsy Res 4*:156–159.

Nestler EJ (1993): Cellular responses to chronic treatment with drugs of abuse. *Crit Rev Neurobiol 7*:23–39.

Nguyen TV, Kosofsky BE, Birnbaum R, Cohen BM, Hyman SE (1992): Differential expression of c-Fos and Zif268 in rat striatum after haloperidol, clozapine, and amphetamine. *Proc Natl Acad Sci USA 89*:4270–4274.

Pennypacker KR, Walczak D, Thai L, Fannin R, Mason E, Douglass J, Hong JS (1993): Kainate-induced changes in opioid peptide genes and AP-1 protein expression in the rat hippocampus. *J Neurochem 60*:204–211.

Petersen SL, McCrone S (1994): Characterization of the receptor complement of individual neurons using dual-label in situ hybridization histochemistry. In: Eberwine JH, Valentino KL, Barchas JD (Eds), *In Situ Hybridization in Neurobiology, Ch 5*. New York: Oxford University Press, pp 78–95.

Przewłocki R, Kamińska B, Łukasiuk K, Nowicka DZ, Przewłocka B, Kaczmarek L, Lasoń W (1995): Seizure-related changes in the regulation of opioid genes and transcription factors in the dentate gyrus of rat hippocampus. *Neuroscience 68*:73–81.

Quinn JP, Takimoto M, Iadarola MJ, Holbrook N, Levens D (1989): Distinct factors bind the AP-1 consensus sites in gibbon ape leukemia virus and simian virus 40 enhancers. *J Virol 63*:1737–1742.

Rea MA (1989): Light increases Fos-related protein immunoreactivity in the rat suprachiasmatic nuclei. *Brain Res Bull 23*:577–581.

Robertson GS, Vincent SR, Fibiger HC (1992): D1 and D2 dopamine receptors differentially regulate c-*fos* expression in striatonigral and striatopallidal neurons. *Neuroscience 49*:285–296.

Robertson GS, Matsumura H, Fibiger HC (1994): Induction patterns of Fos-like immunoreactivity in the forebrain as predictors of atypical antipsychotic activity. *J Pharmacol Exp Ther 271*:1058–1066.

Robertson HA, Paul ML, Moratalla R, Graybiel AM (1991): Expression of the immediate early gene c-*fos* in basal ganglia: induction by dopaminergic drugs. *Can J Neurol Sci 18 (3 Suppl.)*:380–383.

Rogue P, Malviya AN (1994): Neuroleptics differentially induce *jun* family genes in the rat striatum. *Neuroreport 5*:501–503.

Rouiller EM, Wan XS, Moret V, Liang F (1992): Mapping of c-*fos* expression elicited by pure tones stimulation in the auditory pathways of the rat, with emphasis on the cochlear nucleus. *Neurosci Lett 144*:19–24.

Rusak B, Robertson HA, Wisden W, Hunt SP (1990): Light pulses that shift rhythms induce gene expression in the suprachiasmatic nucleus. *Science 248*:1237–1240.

Ruskin DN, Marshall JF (1994): Amphetamine- and cocaine-induced fos in the rat striatum depends on D2 dopamine receptor activation. *Synapse 18*:233–240.

Ryabinin AE, Wang Y, Bachtell RK, Kinney AE, Grubb MC, Mark GP (2000): Cocaine- and alcohol-mediated expression of inducible transcription factors is blocked by pentobarbital anesthesia. *Brain Res 877*:251–261.

Sagar SM, Sharp FR (1990): Light induces a Fos-like nuclear antigen in retinal neurons. *Mol Brain Res 7*:17–21.

Sagar SM, Sharp FR, Curran T (1988): Expression of c-fos protein in brain: metabolic mapping at the cellular level. *Science 240*:1328–1331.

Salehi M, Barron M, Merry BJ, Goyns MH (1999): Fluorescence in situ hybridization analysis of the fos/jun ratio in the ageing brain. *Mech Ageing Dev 107*:61–71.

Sallaz M, Jourdan F (1993): C-fos expression and 2-deoxyglucose uptake in the olfactory bulb of odour-stimulated awake rats. *Neuroreport 4*:55–58.

Sambrook J, Fritsch EF, Maniatis T (1989): Extraction, purification, and analysis of messenger RNA from eukaryotic cells. In: Ford N, Nolan C, Ferguson M (Eds), *Molecular Cloning. A Laboratory Manual, 2nd ed, Vol 1, Ch 7*. New York: Cold Spring Harbor Laboratory Press, pp 7.1–7.87.

Sambrook J, Fritsch EF, Maniatis T (1989a): Mapping of RNA with ribonuclease and radiolabeled RNA probes. In: Ford N, Nolan C, Ferguson M (Eds), *Molecular Cloning. A Laboratory Manual, 2nd ed, Vol 1*. New York: Cold Spring Harbor Laboratory Press, pp 7.71–7.78.

Sambrook J, Fritsch EF, Maniatis T (1989b): Detection and analysis of proteins expressed from cloned genes. In: Ford N, Nolan C, Ferguson M (Eds), *Molecular Cloning. A Laboratory Manual, 2nd ed, Vol 3, Ch 18*. New York: Cold Spring Harbor Laboratory Press, pp 18.1–18.88.

Schäfer MK-H, Herman JP, Watson SJ (1993): In situ hybridization histochemistry. In: London ED (Ed), *Imaging Drug Action in the Brain, 1st ed, Ch 13*. Boca Raton, FL: CRC Press, pp 337–378.

Scharif NA (1996): Quantitative autoradiographic methods. In: Toga AW, Mazziotta JC (Eds), *Brain Mapping, 1st ed, Ch 5*. San Diego, CA: Academic Press, pp 115–144.

Sharp FR, Sagar SM, Swanson RA (1993): Metabolic mapping with cellular resolution: *c-fos* vs. 2-deoxyglucose. *Crit Rev Neurobiol 7*:205–228.

Sirinathsinghji DJS, Dunnett SB (1993): Imaging gene expression in neural grafts. In: Sharif NA (Ed), *Molecular Imaging in Neuroscience, Ch 3*. New York: IRL Press at Oxford University Press, pp 43–70.

Sirinathsinghji DJS, Schuligoi R, Heavens RP, Dixon A, Iversen SD, Hill RG (1994): Temporal changes in the messenger RNA levels of cellular immediate early genes and neurotransmitter/receptor genes in the rat neostriatum and substantia nigra after acute treatment with eticlopride, a dopamine D_2 receptor antagonist. *Neuroscience 62*:407–423.

Sonnenberg JL, Mitchelmore C, Macgregor-Leon PF, Hempstead J, Morgan JI, Curran T (1989a): Glutamate receptor agonists increase the expression of Fos, Fra, and AP-1 DNA binding activity in the mammalian brain. *J Neurosci Res 24*:72–80.

Sonnenberg JL, Rauscher III FJ, Morgan JI, Curran T (1989b): Regulation of proenkephalin by Fos and Jun. *Science 246*:1622–1625.

Soriano MA, Tessier M, Certa U, Gill R (2000): Parallel gene expression monitoring using of multiple transcripts with an animal model of focal ischemia. *J Cereb Blood Flow Metab 20*:1045–1055.

Spector LS, Goldman RD, Leinwand LA (1998): Immunoblotting and immunoblot affinity purification. In: Janssen K (Ed), *Cells. A Laboratory Manual, Vol 1, Ch 73*. Cold Spring Harbor: Cold Spring Harbor Laboratory Press, pp 73.1–73.12.

Steiner H, Gerfen CR (1994): Tactile sensory input regulates basal and apomorphine-induced immediate-early gene expression in rat barrel cortex. *J Comp Neurol 344*:297–304.

Steiner H, Gerfen CR (1999): Enkephalin regulates acute D2 dopamine receptor antagonist-induced immediate-early gene expression in striatal neurons. *Neuroscience 88*:795–810.

Svenningsson P, Fredholm BB, Bloch B, Le Moine C (2000): Co-stimulation of D1/D5 and D2 dopamine receptors leads to an increase in c-*fos* messenger RNA in cholinergic interneurons and a redistribution of c-*fos* messenger RNA in striatal projection neurons. *Neuroscience 98*:749–757.

Takayama K, Suzuki T, Miura M (1994): The comparison of effects of various anesthetics on expression of Fos protein in the rat brain. *Neurosci Lett 176*:59–62.

Tecott LH, Eberwine JH, Barchas JD, Valentiono KL (1994): Methodological considerations in the utilization of in situ hybridization. In: Eberwine JH, Valentino KL, Barchas JD (Eds), *In Situ Hybridization in Neurobiology, Ch 1*. New York: Oxford University Press, pp 3–23.

Telford S, Wang S, Redgrave P (1996): Analysis of nociceptive neurones in the rat superior colliculus using c-fos immunohistochemistry. *J Comp Neurol 375*:601–617.

Torres G, Horowitz JM (1996): Individual and combined effects of ethanol and cocaine on intracellular signals and gene expression. *Prog Neuropsychopharmacol Biol Psychiatry 20*:561–596.

Torres G, Rivier C (1993): Cocaine-induced expression of striatal c-fos in the rat is inhibited by NMDA receptor antagonists. *Brain Res Bull 30*:173–176.

Traub RJ, Herdegen T, Gebhard GF (1992): Differential expression of c-FOS and c-JUN in two regions of the rat spinal cord following noxious distension. *Neurosci Lett 160*:121–125.

Trembleau A, Bloom FE (1995): Enhanced sensitivity for light and electron microscopic in situ hybridization with multiple simultaneous non-radioactive oligodeoxynucleotide probes. *J Histochem Cytochem 43*:829–841.

Trembleau A, Roche D, Calas A (1993): Combination of non-radioactive in situ hybridization with immunohistochemistry: a new method allowing the simultaneous detection of two mRNAs and one antigen in the same brain tissue section. *J Histochem Cytochem 41*:489–498.

Vargo JM, Marshall JF (1995): Time-dependent changes in dopamine agonist-induced striatal Fos immunoreactivity are related to sensory neglect and its recovery after unilateral prefrontal cortex injury. *Synapse 20*:305–315.

Veening JG, Coolen LM (1998): Neural activation following sexual behavior in the male and female rat brain. *Behav Brain Res 92*:181–193.

Wahle P (1994): Combining non-radioactive in situ hybridization with immunohistological and anatomical techniques. In: Wisden W, Morris BJ (Eds), *In Situ Hybridization Protocols for the Brain, Ch 9*. London: Academic Press, pp 98–120.

Wang AM, Doyle MV, Mark DF (1989): Quantitation of mRNA by the polymerase chain reaction. *Proc Natl Acad Sci USA 86*:9717–9721.

Wang JQ, Daunais JB, McGinty JF (1994a): NMDA receptors mediate amphetamine-induced upregulation of zif/268 and preprodynorphin mRNA expression in rat striatum. *Synapse 18*:343–353.

Wang JQ, Daunais JB, McGinty JF (1994b): Role of kainate/AMPA receptors in induction of striatal zif/268 and preprodynorphin mRNA by a single injection of amphetamine. *Mol Brain Res 27*:118–126.

White JD, Gall CM (1987): Differential regulation of neuropeptide and proto-oncogene mRNA content in the hippocampus following recurrent seizures. *Mol Brain Res 3*:21–29.

Wilkinson DG (1995a): The theory and practice of in situ hybridization. In: Wilkinson DG (Ed), *In Situ Hybridization. A Practical Approach, Ch 1*. New York: IRL Press at Oxford University Press, pp 1–13.

Wilkinson DG (Ed) (1995b): *In Situ Hybridization. A Practical Approach*. New York: IRL Press at Oxford University Press.

Williams S, Evan GI, Hunt SP (1990): Changing patterns of c-fos induction in spinal neurons following thermal cutaneous stimulation in the rat. *Neuroscience 36*:73–81.

Wisden W, Morris BJ (1994a): In situ hybridization with synthetic oligonucleotide probes. In: Wisden W, Morris BJ (Eds), *In Situ Hybridization Protocols for the Brain, Ch 1*. London: Academic Press, pp 9–34.

Wisden W, Morris BJ (Eds) (1994b): *In Situ Hybridization Protocols for the Brain*. London: Academic Press.

Wisden W, Errington ML, Williams S, Dunnett SB, Waters C, Hitchcock D, Evan G, Bliss TV, Hunt SP (1990): Differential expression of immediate early genes in the hippocampus and spinal cord. *Neuron 4*:603–614.

Worley PF, Bhat RV, Baraban JM, Erickson CA, McNaughton BL, Barnes CA (1993): Thresholds for synaptic activation of transcription factors in hippocampus: correlation with long-term enhancement. *J Neurosci 13*:4776–4786.

Young ST, Porrino LJ, Iadarola MJ (1991): Cocaine induces striatal c-Fos-immunoreactive proteins via dopaminergic D1 receptors. *Proc Natl Acad Sci USA 88*:1291–1295.

Young WS III (1995): In situ hybridization with oligodeoxyribonucleotide probes. In: Wilkinson DG (Ed), *In Situ Hybridization. A Practical Approach, Ch 3*. New York: IRL Press at Oxford University Press, pp 33–44.

Ziegler BL, Lamping C, Thoma S, Thomas CA (1992): Single-cell cDNA-PCR: removal of contaminating genomic DNA from total RNA using immobilized DNase I. *Biotechniques 13*:726–729.

Ziółkowska B, Höllt V (1993): The NMDA receptor antagonist MK-801 markedly reduces the induction of c-*fos* gene by haloperidol in the mouse striatum. *Neurosci Lett 156*:39–42.

CHAPTER II

Neuroanatomical and functional mapping using activation of transcription factors

MIKE DRAGUNOW AND DAVID BILKEY

The ability to visualize the location of transcription factor expression and activation provides us with a powerful method for the functional mapping of brain pathways and regions. In this chapter we will briefly review, using specific examples, factors to consider when undertaking transcription factor mapping studies and what we can and can not reasonably infer from this work. We will not review the broad category of inducible transcription factors (ITFs) or constitutively expressed transcription factors (CTFs) and readers are referred to a number of recent reviews for more information on this topic (Dragunow and Preston, 1995; Hughes and Dragunow, 1995; Dragunow, 1996; Beckmann and Wilce, 1997; Hughes et al., 1999; Walton et al., 1999a,b).

Currently, there are two main transcription factor based methods for mapping brain areas activated by physiological, pharmacological and/or pathological stimulation: (1) immunocytochemistry (ICC) or in situ hybridization (ISH) to detect inducible transcription factors (ITFs) that are normally expressed at zero to very low levels in neurons; (2) ICC, using phospho-specific antisera, to detect 'activated' constitutively expressed transcription factors (CTFs) or ITFs in neurons.

Additionally, ICC or ISH can be used to detect a loss of CTFs (e.g. ATF-2, CREB) and ITFs (e.g. Krox 24) that are normally expressed at high levels in neurons under basal conditions. This loss can occur after various types of neuronal injury or in transgenic brains (e.g. the axotomy-induced loss of ATF-2 which is high in normal neurons). This approach allows for mapping by measuring a loss of signal in stressed/damaged/dead neurons, rather than an increase in signal expression as utilized in methods 1 and 2 above. In combination, these three approaches provide powerful tools for studying functional neuroanatomy and neuropathology. Perhaps even more importantly, they provide clues about the types of processes occurring in responding neurons and an insight into the functions of the transcription factors (TFs) that respond to the various stimuli.

The use of TF techniques for mapping purposes requires, however, attention to several factors. It is, for example, vital that such studies:

(1) evaluate a range of time-points after the stimulus, so that the evolving pattern of TF expression/activation may be fully evaluated. This is particularly necessary because it is possible that different brain regions will display different induction and decay kinetics;
(2) utilize a range of transcription factors, since there is differential responsiveness of TFs, even within the same family, to various stimuli. For example, manipulation of dopamine receptors and muscarinic receptors strongly induces Jun B, but has no effect on c-Jun

Handbook of Chemical Neuroanatomy Vol. 19: Immediate Early Genes and Inducible Transcription Factors in Mapping of the Central Nervous System Function and Dysfunction
L. Kaczmarek and H.A. Robertson, editors

expression in various neuronal populations (Hughes and Dragunow, 1994; MacGibbon et al., 1995), and dopamine-denervation induces Fos-related proteins, but not c-Fos in striatal neurons (Dragunow et al., 1991, 1995). Furthermore, although c-Fos is induced in glial and ependymal cells after brain injury, Fos B is not (Dragunow, 1990).

(3) wherever possible, utilize more than one antibody to any particular transcription factor. For example, we used a range of antisera to Jun to investigate its expression in post-mortem normal and Alzheimer's disease human brain material (MacGibbon et al., 1997). With one antibody we detected nuclear staining, but with a range of others only cytoplasmic staining was observed. Others have also observed differences in Jun staining using different antisera, with a number of commercial c-Jun antisera also cross-reacting with apoptosis-specific protein (ASP, see details in Terwel and Van De Berg, 2000). Thus, some investigators who have used only one Jun antibody in their assays may, depending upon which antibody they chose, inadvertently have been detecting ASP rather than Jun, thus invalidating their results. In our studies on Alzheimer's disease brain material, we found that 'Jun' immunostaining went up in CA1 pyramidal cells using the Oncogene Science PC07 Jun antiserum (MacGibbon et al., 1997). It has since been determined, however, that this antiserum also detects ASP, and indeed it may be ASP that is elevated in these neurons.

(4) utilize Western blotting to validate their antisera immuno results. Due to the issues outlined in point 3 above, validation using Western blotting is more than desirable, but indeed vital, to avoid specificity issues and provide definitive results.

With carefully constructed studies and attention to the factors outlined above, TFs can be used to attain information about brain function that may be difficult to gather through other methods. For example, ITFs can be used to discover novel regions that are activated by particular neurotransmitter receptor-specific compounds. For example, the D1 dopamine receptor agonist SKF 38393, administered i.p. to rats induces a strong expression of the Krox 24 transcription factor in dentate granule cells (Figs. 1 and 2). Although the question of whether the action is due to direct effects of D1 receptors on granule cells or to polysynaptic interactions is presently unclear, these results do suggest a possible anatomical (hippocampal granule cells) and biochemical (Krox 24-regulated genes) basis for the effects of dopamine on memory systems.

By mapping the activation and expression of TFs we may also gain information that provides important early clues to CNS function. For example, a number of years ago we showed that mechanical injury or hypoxia–ischemia leads to ITF expression in ependymal cells lining ventricles (Dragunow et al., 1990; Gunn et al., 1990; Dragunow and Hughes, 1993). In hindsight, it is possible that these findings may be related to proliferation and or differentiation of neural stem cells that are now known to emerge from the ventricular linings. Although these examples illustrate the potential power of the TF-based approach to mapping, experiments using ITFs need to be undertaken with precise control and understanding of the relevant experimental conditions.

Several years ago we demonstrated that the constitutive expression of Krox 24, which is high in forebrain regions, was completely abolished by sodium pentobarbital anaesthesia over a period of approximately 4 h (Richardson et al., 1992). Since the suppression of Krox 24 by anaesthetic is very strong, it is critical that this factor is monitored carefully when using ITFs in mapping studies. Unfortunately, the situation is not simply one of anaesthesia masking ITF induction. For example, we have found that in the presence of urethane anaesthesia, stimulation that elicits hippocampal theta activity induces a range of ITFs (Fos, JunB, FRAs) including Krox 24 mRNA and protein in dentate granule cells. This seems to be atropine-

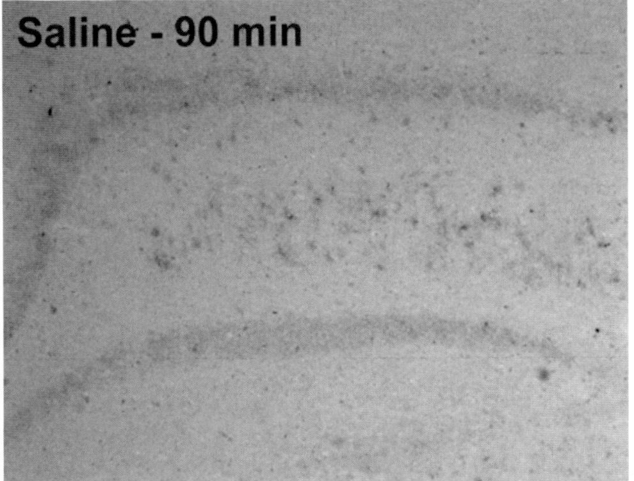

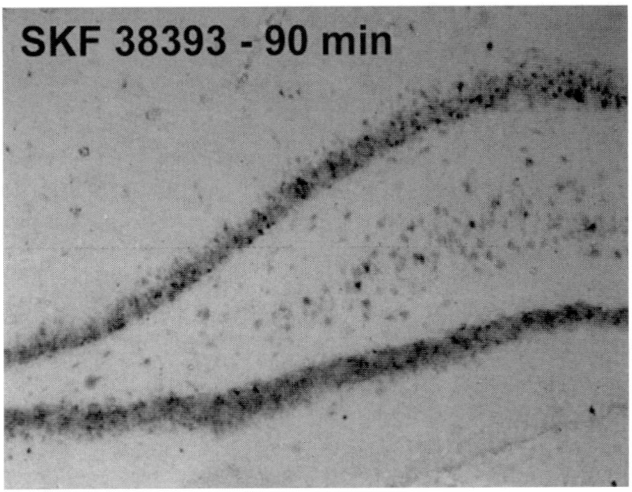

Fig. 1. Krox 24 expression in dentate granule cells 90 min after i.p. injection of saline or SKF 38393 (20 mg/kg). Note the strong induction of Krox 24 after SKF 38393 (a D1 dopamine receptor agonist, RBI, S-168) injection. These experiments were carried out under ethical approval from the University of Otago Animal Ethics Committee to DB. Rats were injected i.p. with the saline vehicle or SKF 38393, perfused 90 min later and coronal brain sections were immunostained for the detection of Krox 24 (Santa Cruz polyclonal antibody sc-189, 1 : 1000 dilution) using standard immunocytochemical methods as previously described (Hughes and Dragunow, 1994).

sensitive suggesting that it is cholinergically mediated theta activity. Without this anaesthesia, however, the induction of ITFs after theta-generating stimulation is far less reliable. The reason for this effect is presently unclear but may be related to the ability of urethane to augment theta activity. What this observation does show is that under some conditions anaesthesia can augment, rather than inhibit, ITF expression in neurons.

As with any assay technique, one must bear in mind that the sensitivity of ITF induction may not always be sufficient to allow for reliable detection of all brain activation under all conditions. Furthermore, there may be circumstances where high levels of brain activity may occur, but without correspondingly strong activation of ITFs (and vice versa). For example, in an exhaustive study of the effects of novelty and habituation on activation of a range of

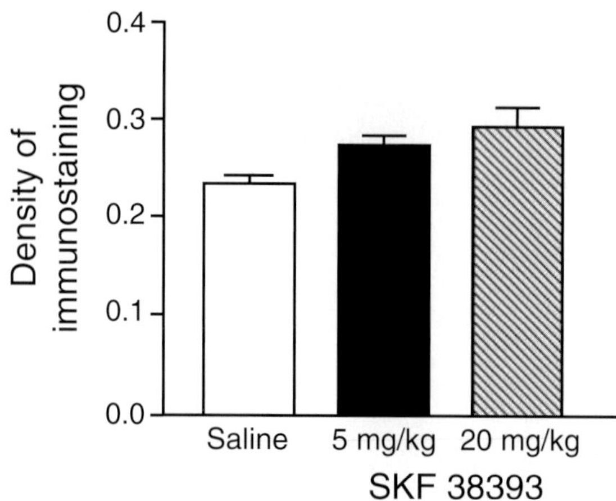

Fig. 2. Graph showing the dose-dependent induction of Krox 24 (measured as density of immunostaining over the dentate granule cell layer) after saline or 5 and 20 mg/kg SKF 38393 injection (i.p.).

ITFs (Fos, FRAs, JunB, Krox 24) we were unable to detect consistent changes in the above ITFs (data not shown) or in the activation of CREB (Walton et al., 1999a,b). Since these novel environments are likely to be of considerable ecological significance to the animal and have been associated with changes in behavior and brain state, the failure to find associated ITF changes indicates that there are clear limitations to the use of ITFs in mapping brain pathways.

ITFs and more recently CTFs have been very useful in elucidating the molecular basis of nerve cell apoptosis and survival. For example, in the hippocampus both ischemia and status epilepticus-induced nerve cell apoptosis occurs mainly in the pyramidal cells, leaving the dentate granule cells intact. We have previously used this selective vulnerability of hippocampal neuronal populations to determine the function of TFs in neuronal death and survival. In our initial studies we found that c-Jun is induced rapidly and transiently in granule cells but delayed and for prolonged periods in dying pyramidal cells (reviewed in Dragunow and Preston, 1995). In contrast, more recently, using phospho-specific antibodies, we have found that the Jun-dimerizing partner ATF-2 is phosphorylated only in dying pyramidal cells (Walton et al., 1998), whereas CREB is phosphorylated only in surviving granule cells (reviewed in Walton and Dragunow, 2000). Thus, activated ATF-2 may provide a useful tool for identifying neurons undergoing apoptosis, whereas activated CREB may provide a tool for identifying neurons expressing an active survival pathway, in addition to its well characterized role as a marker of neurons involved in memory consolidation (Viola et al., 2000).

As mentioned above, in contrast to ATF-2 and CREB which are selectively activated in dying and surviving hippocampal neurons, respectively, the ITF Jun is induced in both types of neurons albeit with different kinetics. Different phosphorylation sites on Jun may also determine different functions. For example, we have recently demonstrated that phosphorylation on serine-73 may be involved in the effects of c-Jun on neurite (and possibly axonal) sprouting (Dragunow et al., 2000), whereas others have shown that serine-63 phosphorylation of Jun might be involved in its ability to induce nerve cell death (Watson et al., 1998). Thus, phospho-specific antisera for Jun may be useful for dissecting out brain regions undergoing death or axonal regeneration (we await proof of this hypothesis).

In conclusion, ITFs and CTFs have proved to be powerful tools for dissecting out brain structure and function and they have provided great insights into neuronal functioning in studies of both the normal and pathological brain. While the potential benefits of such techniques are large, however, their use requires careful consideration and interpretation if the results of such studies are to be regarded as meaningful and valid. For the future, it is clear that more tools will become available as our knowledge of post-translational regulation of TFs increases. There is little doubt, therefore, that with careful use of these tools we will be able to further discern the complexities of neuronal activity and how these relate to brain function.

1. ABBREVIATIONS

ATF-2	activating transcription factor 2
CREB	cAMP and Calcium Responsive Element Binding protein
CTF	Constitutive Transcription Factor
ICC	immunocytochemistry
ISH	in situ hybridization histochemistry
ITF	inducible transcription factor
TF	transcription factor

2. ACKNOWLEDGEMENTS

Supported by a grant from the Marsden Fund and the Health Research Council of New Zealand. We are also indebted to our many students and collaborators, especially Professor Cliff Abraham, Dr Marshall Walton, Dr Ping Liu and Dr Ernest Sirimanne.

3. REFERENCES

Beckmann A, Wilce P (1997): Egr transcription factors in the nervous system. *Neurochem Int 31*:477–510.

Dragunow M (1990): Presence and induction of Fos B-like immunoreactivity in neural, but not non-neural, cells in adult rat brain. *Brain Res 533*:324–328.

Dragunow M (1996): A role for immediate-early genes in learning and memory. *Behav Genet 26*:293–299.

Dragunow M, Hughes P (1993): Differential expression of immediate-early proteins in non-nerve cells after focal brain injury. *Int J Dev Neurosci 11*:249–255.

Dragunow M, Preston K (1995): The role of inducible transcription factors in apoptotic nerve cell death. *Brain Res Rev 21*:1–28.

Dragunow M, Goulding M, Faull RLM, Ralph R, Mee E, Frith R (1990): Induction of c-fos mRNA and protein in neurons and glia after traumatic brain injury: pharmacological characterization. *Exp Neurol 107*:236–248.

Dragunow M, Leah J, Faull RLM (1991): Prolonged and selective induction of fos-related antigen(s) in striatal neurons after 6-hydroxydopamine lesions of the rat substantia nigra pars compacta. *Mol Brain Res 10*:355–358.

Dragunow M, Butterworth N, Waldvogel H, Faull RLM, Nicholson LFB (1995): Prolonged expression of FRPs, Jun B and TrkB in dopamine-denervated striatal neurons. *Mol Brain Res 30*:393–396.

Dragunow M, Xu R, Walton M, Woodgate A-M, Lawlor P, MacGibbon G, Young D, Gibbons H, Lipski J, Mouravlev A, Pearson A, During MJ (2000): c-Jun promotes neurite outgrowth and survival in PC12 cells. *Mol Brain Res 83*:20–33.

Gunn A, Dragunow M, Faull RLM, Gluckman P (1990): Hypoxia–ischaemia and seizures have opposed effects on neuronal c-fos protein levels in the infant rat. *Brain Res 531*:105–116.

Hughes P, Dragunow M (1994): Activation of pirenzepine-sensitive muscarinic receptors induces c-fos, jun B, and krox 24 but not c-jun gene expression in rat brain. *Mol Brain Res 24*:166–178.

Hughes P, Dragunow M (1995): Immediate-early genes and the control of neurotransmitter-regulated gene expression within the nervous system. *Pharmacol Rev 47 (1)*:133–178.

Hughes P, Alexi T, Walton M, Williams C, Dragunow M, Clark RG, Gluckman PD (1999): Activity and injury-dependent expression of inducible transcription factors, growth factors and apoptosis-related genes within the CNS. *Prog Neurobiol 57*:421–450.

MacGibbon G, Lawlor P, Hughes P, Young D, Dragunow M (1995): Differential expression of inducible transcription factors in basal ganglia neurons. *Mol Brain Res 34*:294–302.

MacGibbon G, Lawlor P, Walton M, Sirimanne E, Faull RLM, Mee E, Dragunow M (1997): Expression of Fos, Jun and Krox family proteins in Alzheimer's disease. *Exp Neurol 147*:316–332.

Richardson CL, Tate WP, Mason SE, Lawlor PA, Dragunow M, Abraham WC (1992): Correlation between the induction of an immediate-early gene, *zif 268*, and long-term potentiation in the dentate gyrus. *Brain Res 580*:147–154.

Terwel D, Van De Berg W (2000): c-Jun/AP-1 (N) directed antibodies cross-react with apoptosis-specific protein which marks an autophagic process during neuronal apoptosis. *Neuroscience 96*:445–446.

Viola H, Furman M, Izquierdo LA, Alonso M, Barros DM, de Souza MM, Izquierdo I, Medina J (2000): Phosphorylated cAMP response element-binding protein as a molecular marker of memory processing in rat hippocampus: effect of novelty. *J Neurosci 20*:RC112.

Walton M, Dragunow M (2000): Is CREB a key to neuronal survival?. *Trends Neurosci 23*:48–53.

Walton M, Woodgate A-M, Sirimanne E, Gluckman P, Dragunow M (1998): ATF-2 phosphorylation in apoptotic neuronal death. *Mol Brain Res 63*:198–204.

Walton M, Henderson C, Mason-Parker S, Abraham WC, Bilkey D, Dragunow M (1999a): Immediate-early gene transcription and synaptic modulation. *J Neurosci Res 58*:96–106.

Walton M, Connor B, Lawlor P, Young D, Sirimanne E, Cole G, Gluckman P, Dragunow M (1999b): Neuronal death and survival in two models of hypoxic–ischemic brain damage. *Brain Res Rev 29*:137–168.

Watson A, Eilers A, Lallemand D, Kyriakis J, Rubin L, Am J (1998): Phosphorylation of c-Jun is necessary for apoptosis induced by survival signal withdrawal in cerebellar granule neurons. *J Neurosci 18*:751–762.

CHAPTER III

Studies of the DNA binding activity of transcription factors in mapping brain function

BOZENA KAMINSKA

1. INTRODUCTION

The 'regulated' gene expression in eucaryotic cells depends on the binding of specific transcription factors to both the promoter and the enhancer regions of the particular gene. These factors can have either positive (activating) or a negative (repressing) effect on transcription. A great number of sequence-specific transcription factors (TFs) has been described until now; they differ from one another in the DNA sequence that they recognize and in the way that they are being regulated. Expression of some TFs is restricted to particular cell types, other are expressed only at certain stages of development, yet others are activated in response to a particular stimulus: neurotransmitter, hormone, electrical activity. Assembly of different combinations of these sequence-specific factors results in highly complex and unique patterns of gene expression.

In general, transcription factors have two distinct functional activities. A DNA binding activity referred to as its ability to bind the specific DNA sequence present in the regulatory element of genes is responsible for directing TFs to the vicinity of the promoter. A transactivating potential is the capability to stimulate gene expression through interactions with the general transcriptional machinery. The latter feature can be studied mostly in cells cultured in vitro using cell transfection and reporter gene assays.

2. RATIONALE BEHIND STUDYING DNA BINDING ACTIVITY OF TRANSCRIPTION FACTORS

Many transcription factors act as dimers composed of various members of the same or related families. For example, the transcription factor AP-1 (activator protein-1) was first defined as a DNA binding activity specific for positive regulatory element in SV40 early promoter, important for phorbol esters (TPA)-inducible gene expression (Angel et al., 1987). Further studies revealed that AP-1 DNA binding activity is not a single molecule but a dimer consisting of various members of Fos and Jun families (Angel and Karin, 1991; Karin et al., 1997). The Fos family contains five proteins (c-Fos, Fos B, ΔFosB, Fra-1, Fra-2) and Jun family includes three known members (c-Jun, JunB, JunD). Jun proteins, in contrast to Fos proteins, can form homo- and heterodimers. Although the AP-1 complex consists

Handbook of Chemical Neuroanatomy Vol. 19: Immediate Early Genes and Inducible Transcription Factors in Mapping of the Central Nervous System Function and Dysfunction
L. Kaczmarek and H.A. Robertson, editors

of dimers formed between two Jun or between Jun and Fos molecules, other proteins such as ubiquitously expressed ATF/CREB, can dimerize with Jun proteins at the AP-1 site (Benbrook and Jones, 1990; Hai and Curran, 1991).

Expression of Fos and Jun, at both mRNA and protein levels, is sensitive to a number of different stimuli. The increases, especially at the mRNA levels, are often large in magnitude but very transient. On the other hand, some AP-1 family members (often Fra-1, Fra-2, and JunD) are expressed under unstimulated conditions and in consequence the total increase in AP-1 DNA binding activity is less impressive than would be expected based on analysis of mRNA expression. Furthermore, Fos and Jun proteins differ substantially in their spatio-temporal patterns of expression in the brain (Herdegen et al., 1991a,b; Gass et al., 1992; Takeuchi et al., 1993). For example, following synaptic stimulation c-Jun is expressed together with c-Fos and JunB. However, after traumatic insults (transection of a nerve fiber) c-Jun and to lesser extent JunD are induced, but not other AP-1 proteins (Herdegen et al., 1997). In the sciatic-nerve transection experiments c-Jun is expressed in the absence of c-Fos in axotomized motor neurons, meanwhile in Schwann cells of the cut nerve, c-Fos is expressed persistently in the absence of c-Jun (Morgan and Curran, 1995). These results suggest that particular Fos and Jun proteins must have distinct partners in these two cell types and presumably they might serve entirely different biological functions (Dragunov and Preston, 1995).

The data concerning the expression of Fos/Jun proteins and mRNAs under pathological conditions, should be considered with a particular caution. First of all, an expression of *fos* and *jun* mRNA does not necessarily lead to protein synthesis. Kiessling et al. (1993) demonstrated that transient global ischemia induced TFs mRNA, but not protein, in neurons destined to die in the gerbil hippocampus, whereas both mRNA and protein were induced in surviving neurons. In situ hybridization visualizes a steady-state level of mRNA which reflects the balance between synthesis and degradation rate. Since the expression of genes coding for inducible TFs is a subject of negative regulation, an inhibition of protein synthesis can increase c-*fos* and c-*jun* mRNAs levels (as cycloheximide — an inhibitor of protein synthesis often does). Another complication stems from the fact that the rate of protein degradation may change under stress conditions and during apoptotic cell death. Under such circumstances, the stress-activated protein kinases c-Jun N-terminal kinase (JNK) and p38 MAPK phosphorylate transcription factors c-Jun, ATF2, Elk1, and p53, that results in the increase of protein stability and transcriptional activity, without up-regulation of mRNA expression (Minden et al., 1994; Gupta et al., 1995; Chihab et al., 1998).

The composition of the AP-1 complex depends upon the relative proportions of different proteins present in cells at a given time. Studying expression of exclusively one member of either Fos or Jun family provides a partial and insufficient information about participation of AP-1 transcription factor in the studied phenomena.

3. PROCEDURAL ASPECTS AND TECHNICAL CONSIDERATIONS

3.1. DNA BINDING ASSAYS FOR TRANSCRIPTION FACTORS IN NUCLEAR PROTEIN EXTRACTS

Once a promoter region is identified and some critical sequence elements are revealed, the activity of particular TF that recognizes this element can be measured in crude nuclear extracts. Such activity is then a candidate for regulating transcription of the gene of interest in vivo. The common method used to detect specific TFs–DNA interactions is the gel mobility

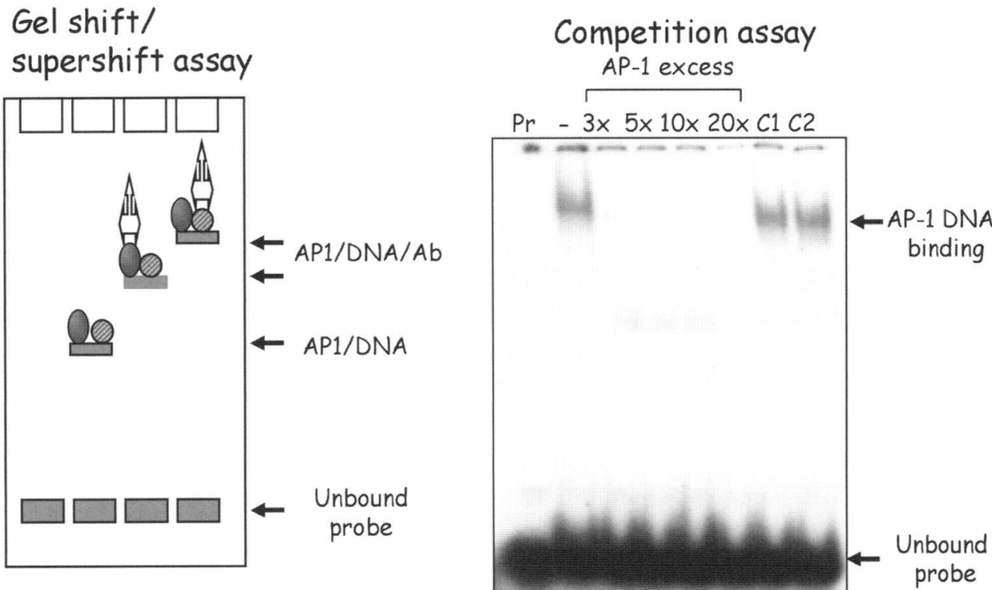

Fig. 1. Gel shift analysis of AP-1 DNA-binding activity in nuclear extracts from visual cortex of rats. Left panel: the scheme of gel shift/supershift assay. Double-stranded DNA fragments carrying specific binding sites are end-labeled with ^{32}P-dATP using T4 polynucleotide kinase. Typically 10–20 μg of nuclear proteins is incubated with 0.2–0.5 ng (30,000–40,000 Cerenkov's cpm) of end-labeled probe for 20–30 min at room temperature and protein–DNA complexes are separated from unbound probe by native polyacrylamide electrophoresis. Right panel: EMSA reaction and competition studies showing the specificity of binding to AP-1 consensus sequence. The first lane (*Pr*) shows 0.25 ng of end-labeled probe without extract; the second (–) shows the retardation of the AP-1 probe caused by nuclear proteins from visual cortex of rats exposed to light for 2 h. The retarded bands are specifically inhibited by increasing amounts of unlabeled AP-1 oligonucleotide. The inclusion of 15 ng of mutant AP-1 (C1) and unrelated SP-1 (C2) oligonucleotides, representing a 60-fold excess, failed to inhibit binding.

shift assay (also called EMSA — electrophoretic mobility shift assay, gel retardation assay). This assay was developed primarily by Freid and Crothers (1981) and Garner and Revzin (1981) for analyzing protein–DNA interactions.

In the gel mobility shift assay, the binding of a protein to a radiolabeled DNA fragment reduces the mobility of the DNA in a non-denaturing polyacrylamide gel electrophoresis, and thus reveals a complex that can be easily distinguished from the unbound (free) probe (Fig. 1, scheme and example). This method is relatively simple and sensitive. It provides a quantitative measure of a particular DNA binding activity and allows rapid, quantitative analysis of a large number of samples. An additional advantage of the EMSA is that complexes between a DNA probe and various proteins usually differ from each other in electrophoretic mobility so it is possible to distinguish between multiple proteins interacting with the same DNA binding site.

3.2. PREPARATION OF PROTEIN EXTRACTS AND ANALYSIS OF DNA BINDING ACTIVITIES

Since transcription factors are not abundant proteins, to generate sufficient quantities of protein of interest it is necessary to obtain 100–200 μg of nuclear protein extracts (protein concentration 10–20 μg/μl) from solid brain tissues or 20–50 μg of nuclear total extracts (2–5 μg/μl) from cell culture. Nuclear proteins must be extracted from fresh tissue (thawing frozen tissue will result in disruption of cell nuclei) to avoid protein degradation and immediately processed for purification of active nuclear extracts (Kaminska and Kaczmarek, 1993; Kaminska et al., 1994, 1995, 1999). Keeping the time period for preparation as short as possible is crucial for obtaining good quality extracts.

The tissue is manually pulverized with a Teflon pestle and suspended in 0.5 ml of buffer A: 10 mM Hepes [pH 7.9], 1.5 mM $MgCl_2$, 10 mM KCl, 1 mM DTT, and protease inhibitors: 1 mM PMSF, 10 mg/ml aprotinin, 10 mg/ml leupeptin, 1 mg/ml pepstatin A. After incubation for 15 min on ice, Nonidet NP-40 is added to final concentration of 1% (gentle treatment is necessary to preserve intact nuclei) before centrifugation at 10,000 rpm for 1 min at 4°C. The crude pellet is resuspended in freshly prepared buffer B: 20 mM Hepes, 0.8 M NaCl, 1.5 mM $MgCl_2$, 0.4 mM EDTA, 1 mM DTT, protease inhibitors as above (gentle treatment to avoid shearing DNA) and incubated for 15 min at 4°C. After centrifugation for 10 min at 12,000 rpm, the supernatant should be frozen at −70°C. The protein content is estimated and verified by Coomassie staining of SDS-PAGE gels.

3.3. PREPARATION OF THE DNA PROBE

Typically a synthetic oligonucleotide (20–50 bp) carrying specific binding site for TF is employed as a probe. The simplest way is to use the double-stranded, chemically synthesized oligonucleotides carrying the specific, consensus binding site for transcription factors. Such oligonucleotides are commercially available from various suppliers. We have compared the binding of nuclear protein extracts from the brain to various oligonucleotides carrying AP-1 and CRE (cyclic AMP responsive element) consensus binding sites surrounded by different flanking sequences, and found no difference in the pattern and intensity of binding of AP-1 and CREB transcription factors to the mentioned sequences. We have also compared the binding of nuclear proteins to the consensus CRE sequence and CRE binding site from c-*fos* promoter and we found no difference in the pattern and intensity of DNA binding. Elliott and Gall (2000) compared binding of nuclear proteins (isolated from hippocampi of control and lesioned rats) to the consensus AP-1 oligonucleotides and NGF-like AP-1 site-containing oligomers. The binding to the NGF-like AP-1 site-containing oligomers showed a similar pattern but was weaker than binding to the AP-1 consensus oligonucleotides.

Double-stranded DNA fragments carrying specific binding sites can be easily end-labeled with ^{32}P-dATP using T4 polynucleotide kinase. Alternatively the 3′-end of the DNA probe may be labeled using 5′-3′ polymerase activity of the Klenow fragment of DNA polymerase to fill in from a recessed 3′-end. Unincorporated nucleotides can be removed by ethanol precipitation or 'spun column' chromatography. For short oligonucleotide probes (<50 bp) removal of unincorporated nucleotides by ethanol precipitation gives poor recovery of labeled probe, so spun columns offer higher quantity of recovered labeled probe.

3.4. THE BINDING REACTION AND ELECTROPHORESIS OF PROTEIN–DNA COMPLEXES

Typically 10–20 μg of nuclear proteins is preincubated for 10 min at room temperature in binding buffer and subsequently incubated with 0.2–0.5 ng (30,000–40,000 Cerenkov's cpm) of end-labeled probe for 20–30 min at room temperature. A typical binding buffer consists of 10 mM Hepes, 25 mM KCl, 0.5 mM EDTA, 0.25 μg/ml of bovine serum albumin, 1 mM DTT, 20 μg/ml of poly d(I-C). Preincubation of proteins with binding buffer allows for nonspecific binding of some proteins to DNA. The addition of a non-specific competitor DNA to the binding buffer is important to select a single DNA binding protein from a vast number of proteins in nuclear extracts. Commonly used competitor DNA are synthetic copolymers, such as poly d(I-C) or poly d(A-T) and they provide excess of low affinity binding sites which absorb DNA binding proteins non-specifically. Heterologous DNA, such as *Escherichia coli* DNA, calf thymus or salmon sperm DNA can also be used but it contains binding sites which will reduce the amount of specific protein–DNA complexes.

A wide variety of parameters can influence protein–DNA interactions thus affecting results of the gel shift assay. Of particular importance are conditions of nuclear protein isolation and binding buffer: monovalent (Na^+, K^+) and divalent (Mg^{2+}) cations, pH, non-ionic detergents (e.g. Nonidet P-40), binding temperature and time of incubation, protein concentration, type and concentration of competitor DNA can influence protein–DNA interactions. Furthermore, gel composition and electrophoresis conditions can alter the mobility of complexes. Each parameter should be tittered for every type of transcription factor to be studied.

In the gel mobility shift assay, the specific binding of a protein to a radiolabeled DNA fragment reflects the amount of active transcription factor present in nuclear extracts. Densitometric measurements of the signal intensity on autoradiography films are used to quantify results of gel shift and compare the abundance of particular TFs in nuclear extracts under various conditions. It allows a more precise assessment of the influence of experimental treatments on the levels of TFs in studied brain structures (e.g. Kaminska et al., 1996; Yoneda et al., 1997, 1998).

3.5. OLIGONUCLEOTIDE COMPETITION STUDIES

To determine whether a DNA–protein binding is specific and a DNA binding activity corresponds to a particular transcription factor, it is necessary to perform competition studies. The specificity of a protein–DNA interactions can be measured easily by adding increasing amounts of unlabeled competitor DNA to a constant amount of the labeled DNA probe. In such an analysis, addition of the oligonucleotide fragment containing the consensus sequence for the particular factor (or high-affinity binding site for a protein) should effectively abolish the binding of protein to the labeled DNA in a dose-dependent manner (Fig. 1, right panel). Ideally, these studies should be conducted with both the wild-type consensus sequence and with the mutated sequence. The latter sequence should not compete for the protein binding and the DNA–protein complexes should be preserved.

3.6. SUPERSHIFT ASSAY

The most conclusive way of demonstrating that the DNA binding activity contains a certain transcription factor is to utilize antibodies that specifically recognize this factor in gel mobility ‘supershift assay’. This method is based on the assumption that if this factor is part of the DNA–protein complex, the inclusion of antibody against the protein will result in binding antibody to the protein in the DNA–protein complex and further retard its electrophoretic mobility. Obviously such a result could be obtained only if the antibody binding does not affect an interaction between transcription factor under study and DNA. Addition of antibodies produces either a supershift due to decreased electrophoretic mobility of the IgG containing complexes or disappearance of the DNA–protein complex due to disruption of the complex. The last situation occurs when an antibody is directed against DNA binding or dimerization domain of proteins acting as dimers. Fig. 2 shows results of the supershift analysis of AP-1 composition in the rat visual cortex after physiological, visual stimulation.

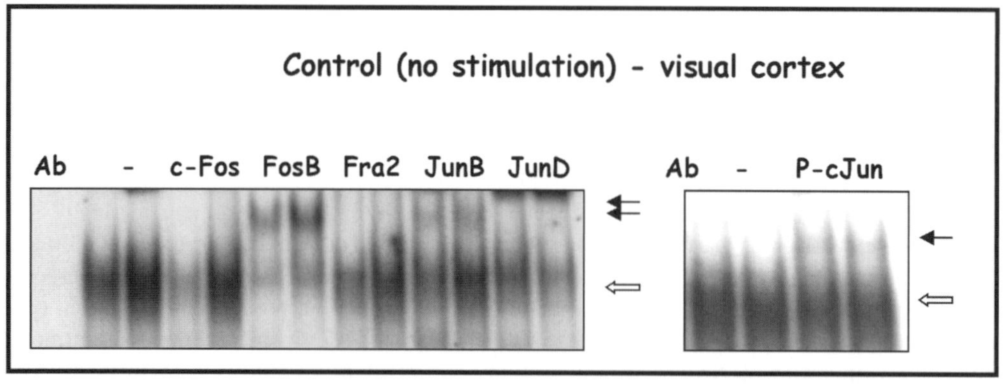

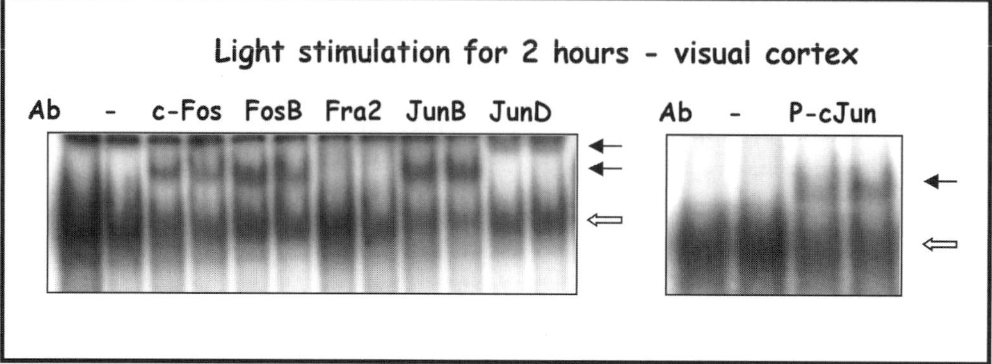

Fig. 2. The protein composition of the AP-1 transcription factor in visual cortex of rats. Ten micrograms of protein extract from visual cortices of naïve rats (control) or rats reared in darkness and exposed to light for 2 h were subjected to EMSA-supershift analysis. The results are shown for two animals for each condition. The designations at the top indicate which antibody was added to each sample. The supershifted bands, the positions of which are indicated with arrows, show the participation of specific proteins in the AP-1 complex; position of non-shifted protein–DNA complexes is indicated with white arrows. Addition of antibodies produces a supershift due to decreased electrophoretic mobility of the IgG containing complexes.

4. VISUALIZATION OF DNA BINDING ACTIVITIES OF TRANSCRIPTION FACTORS IN BRAIN STRUCTURES

Evaluation of DNA binding activities of transcription factors is not a good mapping tool since its use relies on homogenized tissue as the starting material and cellular-level resolution can not be achieved. Although Wang et al. (1992) demonstrated DNA binding activities of transcription factors in situ on a brain section, the specificity of DNA binding has not been sufficiently proved and the procedure could not apparently be reproduced. However, it is possible to evaluate DNA binding activities of transcription factors in various brain structures and it depends on the accuracy of brain regions dissection.

For instance, Kaminska et al. (1995) have demonstrated developmental changes in DNA binding activities of CREB, AP-1, and Octamer transcription factors in the barrel field of somatosensory cortex and visual cortex of 1, 3, 7, 14, 21 days old mice. In another study Tomoya et al. (1999) isolated nuclear extracts from various parts of the hippocampus in order to identify the exact region responsive to kainate stimulation. Mice were injected intraperitoneally with kainic acid (KA), followed by dissection of frozen coronal sections of the whole brain (with a thickness of 600 μm) and subsequent punching out of the pyramidal and granular cell layers in the hippocampus under the stereoscopic microscope. They found that KA administration induced marked increase of AP-1 DNA binding in the dentate gyrus of the granule cells at 2 h, with only small increase in the CA1 subfield of the pyramidal cells, and lack of early increase of AP-1 binding in the CA3 subfield. The time course of AP-1 DNA binding was also different in various hippocampal subfields, with persistent increase for 18 h in the DG neurons and gradual increase of AP-1 DNA binding activities in the CA1 and CA3 pyramidal neurons.

A similar analysis of extracts isolated from the precisely dissected parts of the hippocampus has been performed on the gerbils subjected to transient ischemia (Yoneda et al., 1994, 1998). In addition, the frontal cortex and caudate putamen were dissected out from the frontal portion of the brain, and the thalamus and hypothalamus were manually dissected from the middle part. Both basal and inducible levels of AP-1 DNA binding activity apparently differed in various brain structures (Yoneda et al., 1994); however, DNA binding activities of the constitutive transcription factor CREB varied also among the samples that may indicate differences in total nuclear proteins loading.

A similar approach was employed by Kashihara et al. (1996) who punched out the striatum and globus pallidus from the same coronal section of the brain sliced in the rostral to caudal direction. They studied the effects of agonists of D1 and D2 dopamine receptor on AP-1 DNA binding activity in the striatum and globus pallidus of rats with unilateral 6-OHDA (6-hydroxydopamine) lesions of the medial forebrain bundle (Kashihara et al., 1996).

5. CHANGES OF DNA BINDING ACTIVITIES OF AP-1 TRANSCRIPTION FACTORS IN THE BRAIN AFTER VARIOUS STIMULATION

Table 1 summarizes the vast majority of data concerning the kinetics and patterns of the activation of AP-1 DNA binding activity in the brain studied by gel shift techniques (EMSA). Depending on the stimulus and particular brain structure examined, two basic temporal patterns could be distinguished. The induction of AP-1 DNA binding can be a rapid and transient type, with the increase as early as 60 min after the onset of stimulation and the maximal level at 90 to 120 min. A second type of AP-1 activation represents a moderate but long-lasting (for days or weeks) increase in AP-1 DNA binding activity. In several cases the biphasic profile of AP-1 induction has been observed in the brain structures, with a first, transient wave of AP-1 induction at 2–6 h after stimulus and a second, long-lasting increase for several days (Hope et al., 1994a,b; Kaminska et al., 1994a,b; Elliott and Gall, 2000).

Hope et al. (1994a,b) demonstrated the presence of two specific AP-1 complexes detected by gel shift in the frontal cortex and the hippocampus of rats following acute and chronic electroconvulsive shock (ECS). The AP-1 complex induced following chronic ECS treatment is formed by the lower, and faster migrating band of the doublet. The AP-1 complex induced following chronic ECS treatment is represented by a single, slowly migrating band. The difference in migration rates of protein–DNA complexes for acutely and chronically induced AP-1 bands raised the possibility that the composition of the two complexes was different.

6. ALTERATIONS IN THE COMPOSITION OF AP-1 DNA BINDING ACTIVITY AND ITS IMPACT ON TRANSACTIVATING POTENTIAL

The molecular composition of the AP-1 complex is heterogeneous. Both the Fos and Jun families of nuclear phosphoproteins may participate in the formation of a homo-

TABLE 1. *Stimuli evoking AP-1 DNA binding activity in the brain*

Type of stimulation	AP-1 activation	Type of activation in brain structures	References
Kainate-induced seizures	+++	Prolonged (2–24 h) in rat brain	Sonnenberg et al., 1989a,b; Feng et al., 1997
Kainate-induced seizures	+++	Prolonged (4 h and 1–14 days) in the CA1, CA3, and dentate gyrus of the hipocampus	Pennypacker et al., 1993
Kainate-induced seizures	+++	Biphasic (2–6 h, 48–72 h) in the hippocampus, the sensory and enthorinal cortex of rats	Kaminska et al., 1994
Kainate-induced seizures	+++	Prolonged (2 h to 7 days) in the rat hippocampus	Kasof et al., 1995
Kainate-induced seizures	+++	Prolonged (2–24 h) in murine brain	Azuma et al., 1996
PTZ-induced seizures	+++	Rapid, transient (2–6 h) in the rat hippocampus, the sensory and enthorinal cortex	Kaminska and Kaczmarek, 1993; Lukasiuk and Kaczmarek, 1994
Intracerebroventricular injection of NMDA	++	Rapid, transient (2–6 h) in murine striatum, cerebral cortex, hypothalamus	Yoneda and Ogita, 1994; Yoneda et al., 1998
Transient forebrain ischemia	++	Rapid, at 3–6 h in the dentate gyrus, CA1 and CA3 subfields of the gerbil hippocampus	Yoneda et al., 1994, 1998
Transient forebrain ischemia	++	Rapid, at 1–3 h in the gerbil hippocampus	Domanska-Janik et al., 1999; Kapinya et al., 2000
Acute electroconvulsive seizure (ECS) treatment in rats	+++	Rapid, transient (2–4 h) in the dorsal parietal cortex of rats	Hope et al., 1994a
Chronic electroconvulsive seizure (ECS) treatment in rats	++	Prolonged (1–7 days) in the dorsal parietal and frontal cortex, the hippocampus of rats	Hope et al., 1994a
Kindling stimulation in amygdala	++	Rapid, transient (2–4 h) in the rat hippocampus	Kashihara et al., 1997
Cortical impact injury	+++	Rapid (1–3 h), prolonged (24 h) in the rat cerebral cortex	Yang et al., 1994
Hilus-lesion-induced seizures	++	Biphasic: rapid (4 h) and prolonged (10–24 h) in the dentate gyrus of the rat hippocampus	Elliott and Gall, 2000
Chronic morphine treatment	++	Prolonged (4–7 days) in the rat nucleus accumbens, the cortex and hippocampus.	Hope et al., 1992, 1994b
Acute morphine treatment (rats)	++	Rapid, transient (1 h) in the rat nucleus accumbens	Hope et al., 1992, 1994b
Intraperitoneal injection of Δ9-tetrahydrocannabinol	++	Rapid, transient (2–6 h) in the nucleus accumbens, the cingulate cortex and caudate putamen of the rats	Porcella et al., 1998
Single injection of caffein	++	Rapid and transient (4 h) of the striatum of the rats	Svenningsson et al., 1995
Hypertonic saline injection (hyperosmolarity)	++	Rapid and transient (2 h) in the rat hypothalamic supraoptic and paraventricular nuclei, but not striatum	Ying et al., 1996
D1 and D2 dopamine receptor agonists in 6-OHDA lesioned rats	++	Rapid and transient (2–6 h) of the striatum and the globus pallidus of the rats	Doucet et al., 1996; Huang and Walters, 1996; Kashihara et al., 1996;

TABLE 1. *(continued)*

Type of stimulation	AP-1 activation	Type of activation in brain structures	References
Levodopa treatment	++	Prolonged in the dorsolateral parts of the striatum	Kashihara et al., 1995
Radiation	++	In the rat cerebral cortex	Raju et al., 2000
Trimethyltin neurotoxicity	++	Prolonged in the rat hippocampus and cerebellum	Xiao et al., 1999
Estrogen injection	++	Rapid (1 h), long lasting (up to 72 h) in the pituitary of ovariectomized rats	Zhu and Pfaff, 1998
The induction of long-term potentiation	++	Rapid and transient in the rat dentate gyrus	Williams et al., 2000
Intraperitoneal injection of neuroleptics	++	Rapid (at 2 h) in murine brain	Ozaki et al., 1998
Physiological stimuli			
Brief exposure to novelty	+	Rapid, transient (2 h) in murine brain	Kinney and Routtenberg, 1993
Visual stimulation after period of darkness	+	Rapid, transient (2–6 h) in the rat visual, but not frontal cortex	Kaminska et al., 1996, 1999
Passive avoidance training	+	Rapid, transient (2–6 h) in rat hippocampus, sensory and enthorinal cortex	Lukasiuk et al., 1999

+++ = strong activation, ++ = moderate activation, + = weak activation.

or heterodimeric complex that binds to the DNA consensus sequence TGAC/GTCA and modulates expression of the target genes (Hirai et al., 1989; Angel and Karin, 1991; Morgan and Curran, 1995). Although all combinations of Fos–Jun or Jun–Jun dimers bind the consensus AP-1 target element, functional assays revealed some differences in the ability to bind unique AP-1 sites in various promoters, stability of binding, and transcriptional activation. For example, binding of heterodimers of Fos–Jun to DNA is more stable than binding of homodimers (Halazonetis et al., 1988; Abate et al., 1991). In vitro studies using recombinant proteins revealed that a stability of binding AP-1 complexes to DNA differs depending on the presence of particular proteins. The stability of binding for complexes containing c-Jun is higher than those containing JunD or JunB; heterodimers consisting of Fos B bind to DNA more stably than heterodimers made of Fra-1 or c-Fos (Ryseck and Bravo, 1991). Also transactivating potentials of various AP-1 components differ significantly showing for example opposing activities of c-Fos and Fra-2 (Yashioka et al., 1995; Rutberg et al., 1997) or c-Jun and JunB (Deng and Karin, 1993; Kobierski et al., 1993).

The composition of AP-1 DNA binding complexes can be probed by addition of antibodies specific for each Fos and Jun protein. Using supershift analysis many studies demonstrated that the composition of the AP-1 complex differs in various physiological situations and even closely related members of the same family might contribute to quite distinct biological phenomena. Table 2 summarizes some data concerning the composition of the AP-1 complex under physiological and pathological conditions. To facilitate comparisons only supershift data employing similar antibodies were considered. The proteins more abundant in AP-1 complex are written in bold type. Although the data are not complete, one can see a tendency that in unstimulated brain AP-1 is predominantly composed of **FosB** and **JunD**, while stimulation of neuronal activity is associated with the appearance of AP-1 complexes containing **c-Fos**,

TABLE 2. *Composition of AP-1 DNA binding activity under various conditions determined by supershift analysis with similar antibodies (Santa Cruz Biotechnology)*

Extract source and biological phenomena	Jun proteins	Fos proteins	References
Rat visual cortex:			
naïve	**JunD**	**FosB**	Kaminska et al., 1996, 1999
light deprived	**JunD**	**FosB**	
light stimulation for 2 h	**P-cJun**, **JunB**, JunD	**c-Fos**, FosB	
light stimulation for 6 h	**JunB**, **JunD**	**FosB**	
Rat hippocampus after seizures induced by kainate:			
early phase – 2 h	**P-cJun**, **JunB**	**c-Fos**, FosB	Kaminska et al., 1994, 1997
late phase – 6 h	**JunB**	c-Fos, **Fra-2**, **FosB**	
neurodegeneration – 72 h	**JunD**, **P-cJun**	**FosB**	
Rat hippocampus after seizures induced by pentylenotrazole (at 2 h posttreatment)	**JunB**	**c-Fos**, FosB	Lukasiuk and Kaczmarek, 1994
Rat suprachiasmatic nucleus:			
constant darkness	**JunD**	**Fra-2**	Francois-Bellan et al., 1999
2 h after light stimulation	**JunB**	**c-Fos**, Fra-2, FosB	
Rat visual cortex after learning session:			
not trained	**JunD**, c-Jun, JunB	**FosB**, Fra-2, c-Fos	Lukasiuk et al., 1999
2 h after training	**JunB**, c-Jun, JunD	**c-Fos**, FosB,	
Chronic electroconvulsive shock	**FosB** (**ΔFosB**)	**JunD**	Hiroi et al., 1998

c-Jun, **JunB**. A growing body of evidence suggests that neurodegeneration resulting from programmed cells death is associated with an accumulation of heterodimers composed of **FosB** and **Jun** proteins or **Jun** homodimers, with particular abundance of phosphorylated **c-Jun** and **JunD** (Kaminska et al., 1994; for review see Herdegen and Leah, 1998, and Mielke et al., 2000).

Prolonged AP-1 activity observed under conditions of chronic neuronal stimulation is associated with the presence of chronic **Fra** that appear to be a truncated form of FosB (**ΔFosB**) (Hope et al., 1994a,b; Hiroi et al., 1998). Competition studies revealed similar affinities of acute and chronic AP-1 complexes for the AP-1 sites derived from various promoters: HMT AP-1 (the human metallothionein gene), TH AP-1 (tyrosine hydroxylase gene), the prodynorphin non-canonical AP-1 (Hope et al., 1994a,b). However, the chronic AP-1 complexes have 2-fold higher affinity for the AP-1 site from the Fos CARE oligonucleotide (a CRE and AP-1-like site in the c-*fos* gene) (Hope et al., 1994a,b; Nestler et al., 1999). This suggests that AP-1 complexes with specific composition and/or various phosphorylation status may have distinct properties and effect on AP-1-dependent gene expression. We demonstrated that apoptosis of rat glioma cells occurs in association with persistent activation of JNK, that results in the stabilization and activation of c-Jun and ATF-2 followed by activation of AP-1 DNA binding activity. AP-1 complexes accumulating in apoptotic cells have altered transactivating potentials since the collagenase TRE/AP-1-dependent promoter is downregulated while the Fas Ligand gene promoter carrying AP-1 site becomes activated in apoptotic cells (Pyrzynska et al., 2000).

The functional consequences of changes in subunit composition of AP-1 complex are poorly understood, although it is clear from a great variety of in vitro and cell culture studies that different Fos and Jun proteins are distinct in terms of their ability to affect transcriptional control.

7. ABBREVIATIONS

AP-1	activatory protein 1, transcription factor
ATF	activating transcription factor
cAMP	cyclic adenosinomonophosphate
CRE	Ca^{2+}/cAMP responsive element, gene regulatory sequence
CREB	CRE binding proteins, transcription factors
DTT	dithiotreithol
ECS	electroconvulsive shock
EMSA	electrophoretic mobility shift assay
FRA	Fos-related antigens
JNK	c-Jun N-terminal kinase
KA	kainic acid
MAP	kinases: mitogen activated protein kinases
NGF	nerve growth factor
NMDA	receptors: *N*-methyl-D-aspartate receptors
6-OHDA	6-hydroxydopamine
PMSF	phenylmethylsulfonyl fluoride
SRE	serum responsive element, gene regulatory sequence
TF	transcription factor
TPA	tetradecyl phorbol 13-acetate
TRE	TPA responsive element
HMT	human metallothionein
TH	tyrosine hydroxylase

8. ACKNOWLEDGEMENTS

Supported by a grant 6 P04A024 18 from the State Committee for Scientific Research.

9. REFERENCES

Abate C, Luk D, Curran T (1991): Transcriptional regulation by Fos and Jun in vitro: interaction among multiple activator and regulatory domains. *Mol Cell Biol 11*:3624–3632.

Angel P, Karin M (1991): The role of Jun, Fos and the AP-1 complex in cell-proliferation and transformation. *Biochim Biophys Acta 1072*:129–157.

Angel P, Imagawa M, Chiu R, Stein B, Imbra RJ, Rahmsdorf HJ, Jonat C, Herrlich P, Karin M (1987): Phorbol ester-inducible genes contain a common cis element recognized by a TPA modulated trans-acting factor. *Cell 49*:729–739.

Azuma Y, Ogita K, Yoneda Y (1996): Binding of double stranded oligonucleotide probes for particular transcription factors with leucine-zipper motifs in discrete brain structures of mice with acquired and inherent spontaneous seizures. *Neurochem Int 29*:323–333.

Benbrook DM, Jones NC (1990): Heterodimer formation between CREB and JUN proteins. *Oncogene 5*:295–302.

Chihab R, Ferry C, Koziel V, Monin P, Daval JL (1998): Sequential activation of activator protein-1-related transcription factors and JNK protein kinases may contribute to apoptotic death induced by transient hypoxia in developing brain neurons. *Mol Brain Res 63*:105–120.

Dash PK, Moore AN, Dixon CE (1995): Spatial memory deficits, increased phosphorylation of the transcription factor CREB, and induction of the AP-1 complex following experimental brain injury. *J Neurosci 15*:2030–2039.

Deng T, Karin M (1993): JunB differs from c-Jun in its DNA-binding and dimerization domains, and represses c-Jun by formation of inactive heterodimers. *Genes Dev 7*:479–490.

Domanska-Janik K, Bong P, Bronisz-Kowalczyk A, Zajac H, Zablocka B (1999): AP1 transcriptional factor activation and its relation to apoptosis of hippocampal CA1 pyramidal neurons after transient ischemia in gerbils. *J Neurosci Res 57*:840.

Doucet JP, Nakabeppu Y, Bedard PJ, Hope BT, Nestler EJ, Jasmin BJ, Chen JS, Iadarola MJ, St-Jean M, Wigle N, Blanchet P, Grondin R, Robertson GS (1996): Chronic alterations in dopaminergic neurotransmission produce a persistent elevation of deltaFosB-like protein(s) in both the rodent and primate striatum. *Eur J Neurosci 8*:365–381.

Dragunow M, Preston K (1995): The role of inducible transcription factors in apoptotic nerve cell death. *Brain Res Rev 21*:1–28.

Elliott RC, Gall CM (2000): Changes in activating protein 1 (AP-1) composition correspond with the biphasic profile of nerve growth factor mRNA expression in rat hippocampus after hilus lesion-induced seizures. *J Neurosci 20*:2142–2149.

Feng Z, Zhang W, Hudson P, Bing G, Feng W, Hong JS (1997): Characterization of the long-lasting activator protein-1 complex induced by kainic acid treatment. *Brain Res 770*:53–59.

Francois-Bellan AM, Deprez P, Becquet D (1999): Light-induced variations in AP-1 binding activity and composition in the rat suprachiasmatic nucleus. *J Neurochem 72*:841–847.

Freid M, Crothers D (1981): Equilibria and kinetics of lac repressor–operator interactions by polyacrylamide gel electrophoresis. *Nucleic Acids Res 9*:6505–6525.

Garner M, Revzin A (1981): A gel electrophoresis method for quantifying the binding of proteins to specific DNA regions: application to components of the *Escherichia coli* lactose operon regulatory system. *Nucleic Acids Res 9*:3047–3060.

Gass P, Herdegen T, Bravo R, Kiessling M (1992): Induction of immediate early gene encoded proteins in the rat hippocampus after bicuculline-induced seizures: differential expression of KROX-24, FOS and JUN proteins. *Neuroscience 48*:315–324.

Gupta S, Campbell D, Dérijard B, Davis RJ (1995): Transcription factor ATF2 regulation by the JNK signal transduction pathway. *Science 267*:389–393.

Hai T, Curran T (1991): Cross-family dimerization of transcription factors Fos/Jun and ATF/CREB alters DNA binding specificity. *Proc Natl Acad Sci USA 88*:3720–3724.

Halazonetis TD, Georgopoulos K, Greenberg ME, Leder P (1988): c-Jun dimerizes with itself and with c-Fos, forming complexes of different DNA binding affinities. *Cell 55*:917–924.

Herdegen T, Leah JD (1998): Inducible and constitutive transcription factors in the mammalian nervous system: control of gene expression by Jun, Fos and Krox, and CREB/ATF proteins. *Brain Res Rev 28*:370–490.

Herdegen T, Kovary K, Leah J, Bravo R (1991a): Specific temporal and spatial distribution of JUN, FOS, and KROX-24 proteins in spinal neurons following noxious transsynaptic stimulation. *J Comp Neurol 313*:178–191.

Herdegen T, Leah JD, Manisali A, Bravo R, Zimmermann M (1991b): c-JUN-like immunoreactivity in the CNS of the adult rat: basal and transynaptically induced expression of an immediate-early gene. *Neuroscience 41*:643–654.

Herdegen T, Skene P, Bahr M (1997): The c-Jun transcription factor–bipotential mediator of neuronal death, survival and regeneration. *Trends Neurosci 20*:227–231.

Hirai SI, Ryseck RP, Mechta F, Bravo R, Yaniv M (1989): Characterization of junD: a new member of the jun proto-oncogene family. *EMBO J 8*:1433–1439.

Hiroi N, Marek GJ, Brown JR, Ye H, Saudou F, Vaidya VA, Duman RS, Greenberg ME, Nestler EJ (1998): Essential role of the fosB gene in molecular, cellular, and behavioral actions of chronic electroconvulsive seizures. *J Neurosci 18*:6952–6962.

Hope B, Kosofsky B, Hyman SE, Nestler EJ (1992): Regulation of immediate early gene expression and AP-1 binding in the rat nucleus accumbens by chronic cocaine. *Proc Natl Acad Sci USA 89*:5764–5768.

Hope BT, Nye HE, Kelz MB, Self DW, Iadarola MJ, Nakabeppu Y, Duman RS, Nestler EJ (1994a): Induction of a long-lasting AP-1 complex composed of altered Fos-like proteins in brain by chronic cocaine and other chronic treatments. *Neuron 13*:1235–1244.

Hope BT, Kelz MB, Duman RS, Nestler EJ (1994b): Chronic electroconvulsive seizure (ECS) treatment results in expression of a long-lasting AP-1 complex in brain with altered composition and characteristics. *J Neurosci 14*:4318–4328.

Huang KX, Walters JR (1996): Dopaminergic regulation of AP-1 transcription factor DNA binding activity in rat striatum. *Neuroscience 75*:757–775.

Lukasiuk K, Kaczmarek L (1994): AP-1 and CRE DNA binding activities in rat brain following pentylenetetrazole induced seizures. *Brain Res 643*:227–233.

Kaminska B, Kaczmarek L (1993): Robust induction of AP-1 transcription factor DNA binding activity in the hippocampus of aged rats. *Neurosci Lett 153*:189–191.

Kaminska B, Filipkowski RK, Zurkowska G, Lason W, Przewlocki R, Kaczmarek L (1994): Dynamic changes in the composition of the AP-1 transcription factor DNA-binding activity in rat brain following kainate-induced seizures and cell death. *Eur J Neurosci 6*:1558–1566.

Kaminska B, Mosieniak G, Gierdalski M, Kossut M, Kaczmarek L (1995): Elevated AP-1 transcription factor DNA binding activity at the onset of functional plasticity during development of rat sensory cortical areas. *Mol Brain Res 33*:295–304.

Kaminska B, Kaczmarek L, Chaudhuri A (1996): Visual stimulation regulates the expression of transcription factors and modulates the composition of AP-1 in visual cortex. *J Neurosci 15*:3968–3978.

Kaminska B, Filipkowski RK, Biedermann IW, Konopka D, Nowicka D, Hetman M, Dabrowski M, Gorecki DC, Lukasiuk K, Szklarczyk AW, Kaczmarek L (1997) Kainate-evoked modulation of gene expression in rat brain. *Acta Biochim Pol 44*:781–789. (Review.)

Kaminska B, Kaczmarek L, Zangenehpour S, Chaudhuri A (1999): Rapid phosphorylation of Elk-1 transcription factor and activation of MAP kinase signal transduction pathways in response to visual stimulation. *Mol Cell Neurosci 13*:405–414.

Kapinya K, Penzel R, Sommer C, Kiessling M (2000): Temporary changes of the AP-1 transcription factor binding activity in the gerbil hippocampus after transient global ischemia, and ischemic tolerance induction. *Brain Res 872*:282–293.

Karin M, Liu Z, Zandi E (1997): AP-1 function and regulation. *Curr Opin Cell Biol 9*:240–246.

Kashihara K, Ishihara T, Akiyama K, Kuroda S, Morimasa T, Shomori T (1995): Levodopa induces AP-1 and CREB DNA-binding activities in the rat striatum. *Psychiatry Clin Neurosci 49*:291–294.

Kashihara K, Akiyama K, Ishihara T, Shiro Y, Shohmori T (1996): Synergistic effect of D1 and D2 dopamine receptors on AP-1 DNA-binding activity in the striatum and globus pallidus of the rat with a unilateral 6-OHDA lesion of the medical forebrain bundle. *Life Sci 59*:1683–1693.

Kashihara K, Sato K, Akiyama K, Okada S, Ishihara T, Hayabara T, Shomori T (1997): Temporal pattern of AP-1 DNA-binding activity in the rat hippocampus following a kindled seizure. *Neuroscience 80*:753–761.

Kasof GM, Mandelzys A, Maika SD, Hammer RE, Curran T, Morgan JI (1995): Kainic acid-induced neuronal death is associated with DNA damage and a unique immediate-early gene response in c-fos-lacZ transgenic rats. *J Neurosci 15*:4238–4249.

Kiessling M, Stumm G, Xie Y, Herdegen T, Aguzzi A, Bravo R, Gass P (1993): Differential transcription and translation of immediate early genes in the gerbil hippocampus after transient global ischemia. *J Cereb Blood Flow Metab 13*:914–924.

Kinney W, Routtenberg A (1993): Brief exposure to a novel environment enhances binding of hippocampal transcription factors to their DNA recognition elements. *Brain Res Mol Brain Res 20*:147–152.

Kobierski LA, Chu HM, Tan Y, Comb MJ (1991): cAMP-dependent regulation of proenkephalin by JunD and JunB: positive and negative effects of AP-1 proteins. *Proc Natl Acad Sci USA 88*:10222–10226.

Lukasiuk K, Kaczmarek L (1994): AP-1 and CRE DNA binding activities in rat brain following pentylenetetrazole induced seizures. *Brain Res 643*:227–233.

Lukasiuk K, Savonenko A, Nikolaev E, Rydz M, Kaczmarek L (1999): Defensive conditioning-related increase in AP-1 transcription factor in the rat cortex. *Mol Brain Res 67*:64–73.

Mielke K, Herdegen T (2000): JNK and p38 stresskinases – degenerative effectors of signal-transduction-cascades in the nervous system. *Prog Neurobiol 61*:45–60.

Minden A, Lin A, Smeal T, Derijard B, Cobb M, Davis R, Karin M (1994): c-Jun N-terminal phosphorylation correlates with activation of the JNK subgroup but not the ERK subgroup of mitogen-activated protein kinases. *Mol Cell Biol 14*:6683–6688.

Morgan JI, Curran T (1995): Immediate-early genes: ten years on. *Trends Neurosci 18*:66–67.

Nestler EJ, Kelz MB, Chen J (1999): DeltaFosB: a molecular mediator of long-term neural and behavioral plasticity. *Brain Res 835*:10–17.

Ozaki T, Katsumoto E, Mui K, Furutsuka D, Yamagami S (1998): Distribution of Fos- and Jun-related proteins and activator protein-1 composite factors in mouse brain induced by neuroleptics. *Neuroscience 84*:1187–1196

Pandey SC, Xu T, Zhang D (1999): Regulation of AP-1 gene transcription factor binding activity in the rat brain during nicotine dependence. *Neurosci Lett 264*:21–24.

Pennypacker KR, Walczak D, Thai L, Fannin R, Mason E, Douglass J, Hong JS (1993): Kainate-induced changes in opioid peptide genes and AP-1 protein expression in the rat hippocampus. *J Neurochem 60*:204–211.

Porcella A, Gessa GL, Pani L (1998): Delta9-tetrahydrocannabinol increases sequence-specific AP-1 DNA-binding activity and Fos-related antigens in the rat brain. *Eur J Neurosci 10*:1743–1751.

Pyrzynska B, Mosieniak G, Kaminska B (2000): Changes of the trans-activating potential of AP-1 transcription factor during cyclosporin A-induced apoptosis of glioma cells are mediated by phosphorylation and alterations of AP-1 composition. *J Neurochem 74*:42–51.

Raju U, Gumin GJ, Tofilon PJ (2000): Radiation-induced transcription factor activation in the rat cerebral cortex. *Int J Radiat Biol 76*:1045–1053.

Rutberg SE, Saez E, Lo S, Jang SI, Markova N, Spiegelman BM, Yuspa SH (1997): Opposing activities of c-Fos and Fra-2 on AP-1 regulated transcriptional activity in mouse keratinocytes induced to differentiate by calcium and phorbol esters. *Oncogene 5*:1337–1346.

Ryseck RP, Bravo R (1991): c-JUN, JUN B, and JUN D differ in their binding affinities to AP-1 and CRE consensus sequences: effect of FOS proteins. *Oncogene 6*:533–542.

Sonnenberg JL, Mitchelmore C, Macgregor-Leon PF, Hempstead J, Morgan JI, Curran T (1989a): Glutamate receptor agonists increase the expression of Fos, Fra, and AP-1 DNA binding activity in the mammalian brain. *J Neurosci Res 24*:72–80.

Sonnenberg JL, Macgregor-Leon PF, Curran T, Morgan JI (1989b): Dynamic alterations occur in the levels and composition of transcription factor AP-1 complexes after seizure. *Neuron 3*:359–365.

Svenningsson P, Strom A, Johansson B, Fredholm BB (1995): Increased expression of c-jun, junB, AP-1, and preproenkephalin mRNA in rat striatum following a single injection of caffeine. *J Neurosci 15*:3583–3593.

Takeuchi J, Shannon W, Aronin N, Schwartz WJ (1993): Compositional changes of AP-1 DNA-binding proteins are regulated by light in a mammalian circadian clock. *Neuron 11*:825–836.

Wang XB, Watanabe Y, Osugi T, Ikemoto M, Hirata M, Miki N (1992): In situ DNA-protein binding: a novel method for detecting DNA-binding activity of transcription factor in brain. *Neurosci Lett 146*:25–28.

Williams JM, Beckmann AM, Mason-Parker SE, Abraham WC, Wilce PA, Tate WP (2000): Sequential increase in Egr-1 and AP-1 DNA binding activity in the dentate gyrus following the induction of long-term potentiation. *Mol Brain Res 77*:258–266.

Xiao Y, Harry GJ, Pennypacker KR (1999): Expression of AP-1 transcription factors in rat hippocampus and cerebellum after trimethyltin neurotoxicity. *Neurotoxicology 20*:761–766.

Yang K, Mu XS, Xue JJ, Whitson J, Salminen A, Dixon CE, Liu PK, Hayes RL (1994): Increased expression of c-fos mRNA and AP-1 transcription factors after cortical impact injury in rats. *Brain Res 664*:141–147.

Ying Z, Reisman D, Buggy J (1996): AP-1 DNA binding activity induced by hyperosmolality in the rat hypothalamic supraoptic and paraventricular nuclei. *Mol Brain Res 39*:109–116.

Yoneda Y, Ogita K (1994): Rapid and selective enhancement of DNA binding activity of the transcription factor AP1 by systemic administration of *N*-methyl-D-aspartate in murine hippocampus. *Neurochem Int 25*:263–271.

Yoneda Y, Ogita K, Inoue K, Mitani A, Zhang L, Masuda S, Higashihara M, Kataoka K (1994): Rapid potentiation of DNA binding activities of particular transcription factors with leucine-zipper motifs in discrete brain structures of the gerbil with transient forebrain ischemia. *Brain Res 667*:54–66.

Yoneda Y, Azuma Y, Inoue K, Ogita K, Mitani A, Zhang L, Masuda S, Higashihara M, Kataoka K (1997): Positive correlation between prolonged potentiation of binding of double-stranded oligonucleotide probe for the transcription factor AP1 and resistance to transient forebrain ischemia in gerbil hippocampus. *Neuroscience 79*:1023–1037.

Yoneda Y, Kuramoto N, Azuma Y, Ogita K, Mitani A, Zhang L, Yanase H, Masuda S, Kataoka K (1998): Possible involvement of activator protein-1 DNA binding in mechanisms underlying ischemic tolerance in the CA1 subfield of gerbil hippocampus. *Neuroscience 86*:79–97.

Yoshioka K, Deng T, Cavigelli M, Karin M (1995): Antitumor promotion by phenolic antioxidants: inhibition of AP-1 activity through induction of Fra expression. *Proc Natl Acad Sci USA 92*:4972–4976.

Zhu YS, Pfaff DW (1998): Differential regulation of AP-1 DNA binding activity in rat hypothalamus and pituitary by estrogen. *Mol Brain Res 55*:115–125.

CHAPTER IV

Immediate-early gene (IEG) expression mapping of vocal communication areas in the avian brain

CLAUDIO V. MELLO

1. INTRODUCTION

Songbirds are among the very few animals that, besides humans, evolved vocal learning (Nottebohm, 1972; Brenowitz, 1997). This trait consists in the ability to develop vocalizations through imitation of a model (the father's or a tutor's song) rather than solely by instinct (Thorpe, 1958; Marler and Peters, 1977). Learned vocal signals play central roles in various aspects of vocal communication in songbirds, such as individual recognition, territorial defense and mate selection (Catchpole and Slater, 1995; Kroodsma and Miller, 1996). The existence of several brain nuclei and projections dedicated to vocal production and learning, referred to as the song control system, reflects the importance of learned vocalizations in songbirds.

The purpose of the present chapter is to describe how the analysis of immediate-early gene (IEG) expression has been used to map vocal communication areas in the brain of songbirds and other avian groups. As explained elsewhere in the present volume, IEGs are genes that are rapidly and transiently induced in neuronal cells when these undergo depolarization (Morgan and Curran, 1989; Sheng and Greenberg, 1990; for a recent review, see Clayton, 2000). While some IEGs code for proteins that exert direct effects on neuronal cell function, several others encode transcription factors thought to modulate neuronal function indirectly by triggering gene regulatory cascades (Goelet et al., 1986; Morgan and Curran, 1989). Although the function of the IEG response is not yet fully understood, analysis of induced IEG expression can be used to identify patterns of brain activation associated with specific stimuli or behaviors (Hunt et al., 1987; Morgan et al., 1987; Rusak et al., 1990; Worley et al., 1991; Mello et al., 1992; Herrera and Robertson, 1996; Chaudhuri, 1997; Jarvis and Nottebohm, 1997; Tischmeyer and Grimm, 1999). The present discussion focuses on how this approach has furthered our understanding of the organization of the song system and of brain mechanisms underlying vocal communication and learning in songbirds. Emphasis will be on *zenk*, the IEG for which most data in birds are currently available.

1.1. THE SONG CONTROL SYSTEM OF SONGBIRDS

The brain system that controls song production and learning in songbirds has been most extensively studied in zebra finches (*Taenyopygia guttata*) and canaries (*Serinus canaria*). Both species are relatively easy to breed in captivity, but they differ considerably in the

Handbook of Chemical Neuroanatomy Vol. 19: Immediate Early Genes and Inducible Transcription Factors in Mapping of the Central Nervous System Function and Dysfunction
L. Kaczmarek and H.A. Robertson, editors

songs produced and their modes of vocal learning (Catchpole and Slater, 1995; Kroodsma and Miller, 1996; Brenowitz et al., 1997a). Zebra finches require exposure to an adequate song model during an early critical period in order to develop normal adult song (Immelmann, 1969; Eales, 1985), which is characterized by a broad range of harmonic frequencies and a strong rhythmic component (Sossinka and Bohner, 1980; Williams and Staples, 1992). Canaries, in contrast, have rich repertoires, sing with a more tonal-like quality (Güttinger, 1985), and retain the ability to change their song seasonally during adulthood (Nottebohm and Nottebohm, 1978).

The brain nuclei and projections that form the song control system were identified and characterized primarily based on the effects of localized brain lesions on song production and/or learning, and on studies of the connectivity of vocal control areas utilizing neuronal tract-tracing. Several of the brain areas thus defined have distinct cytoarchitectonic features and are readily visualized as discrete nuclei upon Nissl staining or, in some cases, inspection of unstained sections (Nottebohm et al., 1982). The song control nuclei have also been analyzed for the presence of markers such as neurotransmitters and respective receptors, neuromodulators, steroid hormone receptors, calcium-binding proteins, and others (for example, see Ball, 1994; Brenowitz, 1997; Clayton, 1997). It is noteworthy that the song system of several species presents a marked degree of sexual dimorphism, being often highly developed in males and underdeveloped or absent in females (Nottebohm and Arnold, 1976; Arnold, 1997). Furthermore, the song nuclei of birds that exhibit seasonal singing and breeding undergo marked morphological changes on a seasonal basis (Nottebohm, 1981; for reviews, see Brenowitz, 1997; Tramontin and Brenowitz, 2000).

As diagrammed in Fig. 1, the song system can be thought of as composed of two distinct pathways (Nottebohm et al., 1982; Bottjer et al., 1989; Vicario, 1991, 1993; Vates and Nottebohm, 1995; Foster et al., 1997; Margoliash, 1997; Vates et al., 1997; for reviews, see Wild, 1997; Wild et al., 1997b, 2000; Foster and Bottjer, 1998; Bottjer et al., 2000) [1]. The posterior or direct vocal motor pathway consists of sequential projections from nucleus HVC, meant here as a proper name (see Brenowitz et al., 1997a), to the robust nucleus of the archistriatum (RA) and from RA to motor neurons in the tracheosyringeal portion of the hypoglossal nucleus (nXIIts) in the medulla, which on their turn innervate the musculature of the vocal organ (syrinx). An indirect projection from RA to nXIIts also exists, through the dorsomedial nucleus (DM) of the midbrain's intercollicular complex (ICo). RA and DM also project to respiratory centers in the medulla. The other major subdivision of the song system, the anterior forebrain pathway, consists basically of a loop: the lateral magnocellular nucleus of the neostriatum (lMAN) projects to area X of the paleostriatum, which in turn projects to the dorsolateral nucleus of the thalamus (DLM), which then projects back to lMAN. This loop receives input from the direct motor pathway via the HVC to area X projection, and sends input back to that pathway via the lMAN to RA projection. Other projections have been described, the major one consisting of a projection of RA to the dorsomedial thalamic nucleus (DMP), which projects to the medial division of the magnocellular nucleus of the anterior neostriatum (mMAN), which then sends a major projection to HVC. HVC, a nodal nucleus in the song system, also receives direct projections from nucleus interfacialis (NIf) in the telencephalon, and from thalamic nucleus uvaeformis (Uva); the latter also projects indirectly

[1] The current avian neuroanatomical nomenclature is based on the early assumption that cortical structures were lacking in birds, and that the avian telencephalon corresponded to basal ganglia (striatal) structures of mammals. This view is now considered incorrect; while the paleostriatum is currently thought to be the avian equivalent of the mammalian striatum (Reiner et al., 1984, 1998), other telencephalic subdivisions — hyperstriatum, neostriatum and archistriatum — are considered to represent cortico-like structures, as they contain neurons that appear to correspond to cells present in specific layers of the mammalian cortex (Karten, 1991). There is as yet no consensus for how to best modify the prevailing nomenclature.

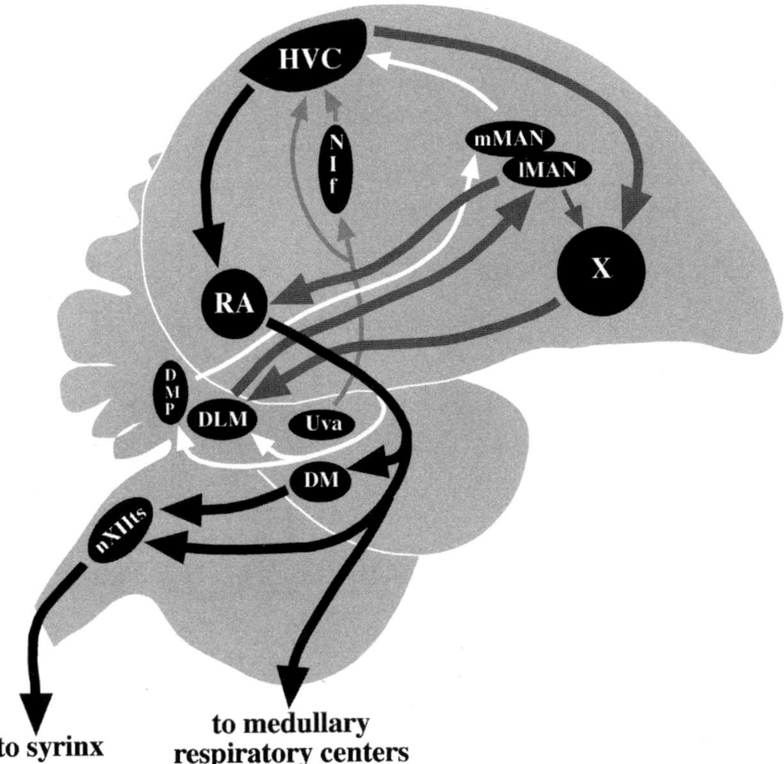

Fig. 1. Schematic representation of the song control system of the songbird brain. Projections of the direct motor pathway for song production are represented by thick black lines; projections of the anterior forebrain pathway are in thick dark-gray lines; re-entrant loops are in thin white lines; and the Uva/NIf pathway is in thin light-gray lines. For details and references, see text; for abbreviations, see list.

to HVC through NIf. The projections above are unilateral, with the exception of the DMP to mMAN projection. The latter, together with a more recently described bilateral ascending projection from DM to Uva (Striedter and Vu, 1998), provides a possible pathway for the interhemispheric integration required for coordinated singing behavior.

Localized lesions have shown that HVC and RA are necessary for the normal production of song (Nottebohm et al., 1976). Electrophysiological recordings demonstrate that these nuclei fire sequentially in association with song production, preceded by the activation of NIf (McCasland, 1987; Yu and Margoliash, 1996). In conjunction with the anatomical data, these findings attest to the active role of the direct motor pathway in the control of vocal production. Lesions to nuclei of the anterior forebrain pathway in juveniles have profound disruptive effects on song learning (Bottjer et al., 1984; Sohrabji et al., 1990; Scharff and Nottebohm, 1991). These nuclei are thought not to be necessary for song production, as their lesion in adults results in no apparent change in the song produced (see, however, Section 5.1 below).

1.2. THE ROLE OF AUDITORY INFORMATION AND AUDITORY PATHWAYS

Vocal communication signals need to be auditorily processed, perceptually integrated, and memorized by the recipient (Hauser and Konishi, 1999). These processes are essential for several aspects of songbird behavior (Catchpole and Slater, 1995; Kroodsma and Miller, 1996; Searcy and Nowicki, 1999). For example, song perception and discrimination by adult males provide a basis for the establishment and maintenance of breeding territories (for instance, see Brooks and Falls, 1975a,b; Searcy et al., 1981; for discussion, see Stoddard, 1996). In females, the ability to discriminate the songs produced by different males is required for the expression of song-based preferences and mate choice (for discussions, see Ratcliffe and Otter, 1996; Searcy and Yasukawa, 1996). In juveniles, the acquisition of an auditory memory or model of the song(s) heard is a necessary (and often first) step in the song learning process (Konishi, 1965; Marler and Peters, 1977). Later on, during the imitation phase of vocal learning, the bird's own attempts at singing influence the development and function of the song system through auditory feedback (for a recent review, see Nottebohm, 1999). For the latter to occur, the song system needs to have access to auditory information.

In spite of the central importance of auditory and perceptual aspects of vocal communication, the exact brain areas involved in song auditory processing, and the pathway(s) through which auditory information enters the song system, are still matters of debate. Auditory information ascends the avian brainstem along a pathway consisting of a series of nuclei (cochlear, lemniscal, midbrain and thalamic) comparable to the stations in the central auditory pathways of mammals, and arguably representing a conserved system in vertebrates (Butler and Hodos, 1996). It then gains access to cortico-like structures through projections from the thalamic nucleus ovoidalis to field L in the telencephalon (Karten, 1967, 1968; Kelley and Nottebohm, 1979; Brauth et al., 1987). An alternative pathway from the lemniscal nuclei to the basalis complex in the anterior forebrain was described in budgerigars (Brauth et al., 1987). Several telencephalic projections from field L have been described, representing possible ways of conveying auditory information to higher-order auditory areas (Kelley and Nottebohm, 1979; Fortune and Margoliash, 1995; Vates et al., 1996). Of particular interest, HVC is thought to be a major entry site of auditory information into the song system, and to represent a central node for sensory–vocal integration. Field L projects to the shelf area underneath nucleus HVC and to the cup area next to RA, but direct projections into HVC and RA were initially not observed (Kelley and Nottebohm, 1979). Recent studies, partly motivated by the findings of IEG mapping analysis (see below), indicate the existence of alternative pathways through which auditory information could potentially enter HVC and thereby gain access to the song system (Fortune and Margoliash, 1995; Vates et al., 1996; Mello et al., 1998b).

In addition to their established role in song production, song control nuclei present evoked responses to auditory stimuli (Katz and Gurney, 1981; Margoliash, 1983; Williams and Nottebohm, 1985; Vicario and Yohay, 1993). Such responses can be recorded at all levels of the direct motor pathway for song production, down to nXIIts motor neurons and syringeal muscles, as well as within the nuclei of the anterior pathway. These auditory responses show high selectivity towards conspecific song stimuli, particularly the song produced by the bird being recorded, i.e. the bird's own song, or BOS (Margoliash, 1986; Doupe and Konishi, 1991; Margoliash and Fortune, 1992; Volman, 1996). These selective patterns of response are dependent on experience, developing as a result of continued exposure to song (Doupe, 1997; Solis and Doupe, 1999, 2000). Although the role of these responses is not entirely clear, they attest to the fact that auditory activity, under certain circumstances, can gain access to the song system. Along a distinct line, lesions targeted at song nuclei in males have been reported to affect the birds' performance in operant tasks involving song auditory discrimination (Scharff et al., 1998), while lesions targeted to song nuclei in females appear to affect the birds' ability to recognize and properly respond to conspecific song stimuli (Brenowitz, 1991). Such findings have been invoked as evidence for a role of the song control nuclei in various aspects of auditory processing and perception, and in the acquisition of a song auditory template during the course of song learning (but see MacDougall-Shackleton et al., 1998; and discussion in Mello et al., 1998b).

1.3. THE *zenk* GENE

zenk is the avian homologue of (and acronym for) *zif-268*, *egr-1*, *ngfi-a* and *krox-24* (Milbrandt, 1987; Christy et al., 1988; Lemaire et al., 1988; Sukhatme et al., 1988; Mello, 1993). This IEG encodes a zinc-finger transcription factor expressed in the nervous system and various other tissues, and whose DNA-binding motif is present in the promoter of a large number of genes (Christy and Nathans, 1989; Pavletich and Pabo, 1991; Swirnoff and Milbrandt, 1995). *zenk* expression in neuronal cells is highly responsive to membrane depolarization, whereas expression in glial cells has not been reported (Bartel et al., 1989; Cole et al., 1990). Its low expression during rest and high responsiveness to depolarization render *zenk* a highly sensitive marker for neuronal activation (Worley et al., 1991; Mello et al., 1992; Chaudhuri, 1997). Most brain regions contain large numbers of neuronal cells that express *zenk* in response to depolarization. This was clearly demonstrated both in rodents and songbirds by administration of the GABAergic antagonist pentylenetetrazole (metrazole), which results in widespread depolarization and associated *zenk* expression in the brain (Saffen et al., 1988; Mello and Clayton, 1995). *zenk* is not a universal marker of neuronal activation, though, as some neurons express it only under certain circumstances, and other neuronal cells lack a *zenk* induction response to depolarization (Mello and Clayton, 1994, 1995; Mello et al., 1995). The latter, however, are restricted to specific brain locations (see Section 4.0). It should therefore be emphasized that the vast majority of neuronal cells, when activated, can be identified through *zenk* expression analysis.

zenk induction represents an early regulatory event in a genomic cascade that potentially leads to long-lasting changes in the properties of neuronal cells (Goelet et al., 1986; Sheng and Greenberg, 1990; Mello, 1998; Clayton, 2000). Consistent with this notion, *zenk* induction has been linked to hippocampal long-term potentiation, or LTP, an intensively studied model of activity-dependent synaptic plasticity (Cole et al., 1989). Among several activity-dependent IEGs, *zenk* induction profile provides the tightest correlation with that of LTP (Wisden et al., 1990; Abraham et al., 1993; Worley et al., 1993; Roberts et al., 1996). *zenk* activation has also been demonstrated in the cortex of rodents in association with morphological changes triggered by the animals' exposure to a novel enriched environment (Wallace et al., 1995). It seems therefore likely that *zenk* expression analysis reveals brain activation patterns associated with neuronal plasticity. Indeed, a recent study utilizing a targeted disruption of the *zenk* (zif-268) gene in a mouse strain provides more direct evidence linking *zenk* to the maintenance of late LTP and the expression of long-term memories (Jones et al., 2001). For the purposes of most of the present chapter, though, *zenk* will be treated primarily as a marker for subjacent neuronal activation; exceptions to this will be noted when relevant.

The *zenk* gene was cloned in birds by low-stringency screening of a canary brain cDNA library utilizing probes derived from *egr-1*. *zenk*'s predicted protein sequence presents an overall identity of 70% to its mammalian counterparts (Mello, 1993). The zinc-finger DNA-binding domain is 100% conserved between canary and mammals, indicating that the same DNA motif is recognized (Christy and Nathans, 1989; Mello, 1993) and that the same or very similar sets of genes are likely to be targets. One significant difference is that a serine/glycine-rich domain of about 30 residues present in mammals upstream of the zinc-finger sequence region is absent in the canary *zenk* sequence (Mello, 1993). The fact that this domain is also missing in the more recently cloned zebra fish homologue (Lanfear et al., 1991) suggests that the latter is the more ancestral condition. The *zenk* sequence downstream of the DNA-binding domain was compared by PCR methods in several avian species and is very conserved, particularly in the coding region and in presumably regulatory domains in the 3′ untranslated region (Long and Salbaum, 1998).

2. MAPPING *zenk* EXPRESSION IN THE BRAIN

In this section, some factors that are critical for mapping behavior-related IEG expression are discussed. The comments are based on songbird experiments utilizing *zenk*, but the issues should be of general relevance.

2.1. METHODOLOGY

2.1.1. General considerations

Due to the rapid and transient nature of their induction, analysis of the expression of IEGs is particularly useful for the detection of event-triggered or behaviorally triggered regional brain activation. In that regard, it is essential to consider the induction kinetics of IEGs to ensure that their expression at a given time actually reflects the events under study. It is equally important to monitor spontaneous behavior(s) and to minimize the influence of interactions with environmental factors and other individuals before starting the stimulation or behavioral paradigm. A useful strategy is to perform the experiments after a long quiescent or inactive period to minimize basal IEG expression levels. In the case of auditory and vocal control pathways in songbirds, a procedure routinely used has been to keep the animals in acoustic/social isolation for 12–24 h before starting an experiment. This takes advantage of the very low IEG expression in the brain at the end of the night period (Mello et al., 1992; Ribeiro et al., 1999; Cirelli and Tononi, 2000). It also ensures that diurnal behaviors such as singing or hearing songs have not been expressed for a considerable period, thus minimizing the associated IEG expression and optimizing the detection of stimulus-triggered expression in vocal communication areas.

Two different methodologies have been utilized for IEG expression analysis of brain activation in birds: in situ hybridization and immunocytochemistry (ICC). Detailed protocols have been described elsewhere (Mello et al., 1992, 1997; Jarvis and Nottebohm, 1997; Mello and Ribeiro, 1998). Below are highlighted some key methodological issues that need to be considered when choosing and applying these techniques.

2.1.2. In situ hybridization

The changes in *zenk* mRNA levels that occur in response to neuronal depolarization are very rapid. In songbirds, a significant rise in cytoplasmic *zenk* mRNA levels can be detected within minutes after stimulus onset (Mello and Clayton, 1994; Kruse et al., 2000). This induction peaks around 30 min, decreasing back to basal levels by 1 h. Thus, measuring *zenk* mRNA levels at a given time is optimal for assessing the effect of events occurring during a relatively narrow window before sacrifice.

The use of complementary (antisense) RNA strands, or riboprobes, yields high sensitivity for the in situ hybridization detection of *zenk* mRNA on thin cryostat sections of fresh-frozen brains (Clayton et al., 1988; Mello et al., 1992). Riboprobes are generated by in vitro transcription utilizing the cloned canary cDNA as template. Radioactive probe labeling followed by X-ray film densitometry or emulsion autoradiography is the preferred strategy for quantitative analysis (Mello et al., 1995); utilization of an image analysis program such as NIH's Image has proven very useful for grain counting (Jarvis and Nottebohm, 1997). Tissue counterstaining with Nissl (cresyl violet) after emulsion is used routinely for adequate identification of cell types based on morphology. The use of ^{35}S for labeling offers the advantages of a relatively long isotope half-life and adequate cellular resolution, but requires the probes to be kept under reducing conditions at all steps to prevent background due to cross-linking of oxidized probe to endogenous tissue components. ^{33}P has higher emission energy, requiring shorter exposure times without significant loss of resolution. It is less volatile than ^{35}S, posing less of a hazard, does not require reducing conditions, and is currently the isotope of choice.

Riboprobes can also be labeled and detected with a non-radioactive method, most frequently digoxigenin(dig)-labeled nucleotide. Dig-UTP is incorporated in the in vitro transcription reaction and can be detected utilizing an anti-digoxigenin antibody and a suitable detection system for histochemistry. Non-radioactive labeling provides better cellular resolution than the radioactive method, and is more convenient for double-labeling experiments. Non-radioactive in situ hybridization of brain sections from animals injected with fluorescently labeled tract-tracers has been useful for determining the connectivity of neurons expressing *zenk* or other specific markers (Jarvis et al., 1998; Denisenko-Nehrbass et al., 2000).

Two different hybridization protocols have been utilized (Clayton et al., 1988; Mello and Clayton, 1994; Mello et al., 1997). In the first one, both hybridization and subsequent washes are performed at high stringency (high temperature, low salt concentration, no RNase treatment). Hybridization and washing conditions resulting in specific *zenk* signal were initially defined on Northern blots (Mello and Clayton, 1994) and then applied for detection of *zenk* mRNA in frozen brain sections of canaries and zebra finches. With small adjustments, these conditions have been used for studies in a variety of bird species, including other songbirds, suboscines, parrots, hummingbirds, doves and chicken. This protocol circumvents the need for RNase treatment, preventing degradation of the Nissl substance and the consequent difficulty in identifying cell types based on morphology. The alternative is a more standard protocol where RNase digestion is used to eliminate unhybridized probe from the tissue. This approach is particularly useful for probes that show high levels of background, as canary c-jun (Nastiuk et al., 1994), but is not optimal for morphological analysis.

2.1.3. Immunocytochemistry (ICC)

Brain expression of the ZENK protein can be detected using ICC (Mello and Ribeiro, 1998). ZENK's carboxyterminal peptide (19 aa residues) shares about 70% identity with the corresponding peptides from mammalian homologues. A commercial antibody (Santa Cruz) raised against this peptide from the mouse *zenk* homologue (*egr-1*) recognizes a specific band of about 62.5 kDa, as predicted by the canary *zenk* sequence, on Western blots of the soluble fraction from songbird brain homogenates. On brain sections, this antibody detects a nuclear antigen; the staining is abolished by pre-absorption with the immunizing peptide (Mello and Ribeiro, 1998).

Some methodological factors need to be considered here. Optimal ZENK ICC results are obtained with the use of fresh-prepared fixative for perfusion, followed by cryoprotection and frozen sectioning of the brain. Either floating sections or thin (15–20 μm) sections mounted onto adhesive glass slides work well. Vector's ABC detection system with nickel enhancement of DAB (3,3′-diaminobenzidine) deposition is routinely utilized. The ABC reagent, based on avidin/biocytin binding, can result in considerable background in songbird brain tissue, presumably due to recognition of endogenous tissue components (Johnson et al., 2000). This artifact allows for the visualization of tissue and nuclear boundaries without counterstaining, but it can also mask specific labeling if excessive. This effect can be minimized with avidin/biotin blocking (Vector). Inadequately high titers of the anti-ZENK primary antibody also result in nonspecific staining, most prominently vascular (endothelial) labeling, which can be effectively eliminated by optimizing the serum dilution. Cytoplasmic localization of ZENK protein has been described in other systems (Day et al., 1990) but not examined in songbirds.

As expected, expression of ZENK protein is more protracted than of ZENK mRNA, and persists for several hours after stimulus onset (Mello and Ribeiro, 1998). In consequence, ZENK protein levels at a given time reflect brain activation during a long preceding period (hours). Conversely, a long period without stimulation is necessary for ZENK protein expression to return to basal levels; such a period should therefore be included when planning a ZENK protein mapping experiment. Overnight isolation from other individuals attains this goal satisfactorily in songbirds (Mello and Ribeiro, 1998).

When mapping large brain areas in many animals, or for comparisons across several experimental groups, it is advisable to utilize an automated system for the detection and mapping of IEG expression. Although commercial programs are available, a high-throughput system to analyze ZENK ICC experiments (Cecchi et al., 1999) was specifically developed for some of the experiments described below. High-resolution images of ICC sections are first captured with a CCD camera. The images are then stitched together and subjected to an algorithm for the recognition of labeled cell nuclei based on a set of morphological criteria or filters. Recognized cells are then separated based on labeling intensity (optical density, OD, in the tissue due to DAB/nickel deposition) into low, middle and high bins of expression. Finally, density maps are generated where information about cell position and labeling intensities are combined and color-coded. OD values are nonlinear with the amount of DAB/nickel precipitate present in the reacted sections; they are assumed to vary monotonically with the amount of ZENK protein and to give an indirect indication of the degree of neuronal activation.

2.2. *zenk* EXPRESSION IN CONTROL BIRDS

Little *zenk* expression is normally detected throughout the brain of unstimulated and quiet control songbirds after overnight acoustic/social isolation, particularly in auditory and vocal control brain regions (Mello et al., 1992, 1995; Mello and Ribeiro, 1998). This contributes to a high sensitivity of *zenk* expression analysis in the detection of activated neuronal populations by auditory stimulation or vocal behavior. Nonetheless, a definite percentage (3–5%) of neuronal cells in auditory brain regions of quiet unstimulated controls shows high *zenk* expression (Mello et al., 1995). These could represent cells that respond to the sounds generated by the birds while moving or feeding in their isolation boxes. *zenk*-expressing cells in controls are more numerous in lateral portions of the hyperstriatum and neostriatum (Mello and Ribeiro, 1998), which are known to contain visual processing areas (Benowitz and Karten, 1976; Karten and Shimizu, 1989). Some of this *zenk* expression may therefore be related to the visual stimulation that occurs when lights are turned on at the end of the night period. The number of these cells is higher by ICC than in situ, likely reflecting the longer half-life of ZENK protein.

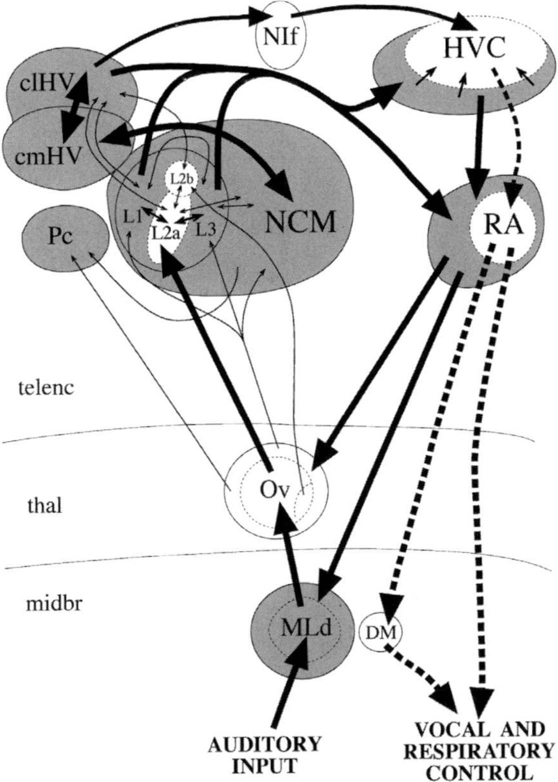

Fig. 2. Schematic representation of the auditory pathways and associated *zenk* expression patterns in the songbird brain. Areas that show *zenk* induction in response to song presentation are represented in gray; areas that lack a song-induced *zenk* response are represented in white. The area adjacent to nucleus HVC corresponds to the shelf, the area adjacent to nucleus RA corresponds to the cup. Thick solid arrows represent the major ascending and descending projections of the auditory system. Thin solid arrows represent either minor projections, or projections that have yet to be confirmed with both anterograde and retrograde tract-tracers. Thick dashed lines represent projections of the direct motor pathway for song production (compare with Fig. 1). For details and references, see text. Abbreviations: midbr, midbrain; telenc, telencephalon; thal, thalamus; for others, see list.

3. *zenk* INDUCTION BY SONG PRESENTATION

In contrast to quiet unstimulated controls, *zenk* induction occurs in various brain regions when songbirds hear stimuli such as playbacks of tape-recorded conspecific song (Mello et al., 1992; Mello and Clayton, 1994). *zenk* expression is most marked in the caudomedial telencephalon and occurs both in males hearing song without producing a vocal response (referred to as 'hearing only' birds), and in females, who usually do not sing and whose song control nuclei are rudimentary or lacking. *zenk* induction in the same areas also occurs in singing males but is abolished by deafening (Jarvis and Nottebohm, 1997; see Section 5.0). As discussed below, areas showing *zenk* induction in response to song playbacks are intimately associated with central auditory pathways. The conclusion is that the *zenk* induction response observed in 'hearing only' birds is related to the act of hearing song. As such, it is likely modulated by perceptual, attentional and/or motivation factors associated with hearing. These auditory areas, as shown diagrammatically in Fig. 2, precede the song system and are discussed in detail below; the results were obtained with a combination of mRNA and protein mapping.

The vast majority of *zenk*-labeled cells in song-stimulated birds (as well as in controls) have clear neuronal morphology (large soma, light-staining nuclei with prominent nucleoli, presence of Nissl substance) (Mello et al., 1992; Mello and Clayton, 1995). In contrast, cells with clear glial properties (small soma, small dark-staining nuclei, scant cytoplasm) are unlabeled. Thus, *zenk* expression in response to birdsong occurs primarily in neuronal cells, conforming to observations made in other systems (Worley et al., 1991). One should note, however, that a small percentage of *zenk*-labeled cells are small and their identity difficult to determine based on Nissl staining alone. Although some of these cells could potentially be non-neuronal cells, they most likely represent small-sized neurons (as shown by double-labeling with a neuronal tract-tracer, as in Jarvis et al., 1998).

3.1. MLd

The dorsal portion of the lateral mesencephalic nucleus (MLd) is the midbrain station of the avian ascending auditory pathways, and is considered the avian homologue of the mammalian inferior colliculus (Butler and Hodos, 1996). It receives major projections from the cochlear nuclei and a descending projection from the auditory telencephalon, and sends a prominent projection onto the thalamic auditory nucleus ovoidalis (Karten, 1967, 1968; Mello et al., 1998b). MLd is the first nucleus along the ascending auditory pathway where *zenk* induction by song presentation has been described (Mello and Clayton, 1994), the preceding stations having yet to be examined. *zenk* expression in MLd accords well with the cytoarchitectonic boundaries of this nucleus, as defined by Nissl staining. It is most prominent in the MLd core, which is the main midbrain termination site of ascending auditory fibers and also the source of the projection to the auditory thalamus. Electrophysiological studies of MLd in songbirds are lacking.

3.2. Pc

The paleostriatum is a complex and large structure in the avian telencephalon. A substantial portion of this structure is currently considered equivalent to the mammalian basal ganglia (Karten and Dubbeldam, 1973; Reiner et al., 1984, 1998). Several reports suggest that the caudal portion of the paleostriatum (Pc) receives inputs from auditory stations, particularly from the thalamic nucleus ovoidalis (Bonke et al., 1979a; Kelley and Nottebohm, 1979; Durand et al., 1992; Wild et al., 1993; Vates et al., 1996). This has been difficult to confirm, due to the presence of abundant fibers of passage in this brain region (Mello, 1993). Marked *zenk* induction occurs in Pc in association with song presentation, most notably in the PA (paleostriatum augmentatum) portion (Mello and Clayton, 1994; Mello and Ribeiro, 1998). This is consistent with the notion that Pc may represent an important auditory processing center in the telencephalon. Interestingly, these connectivity studies also suggest that Pc is a subcortical center, possibly comparable to a component of the mammalian amygdala, an intriguing possibility to be further explored.

3.3. FIELD L

Field L is the primary auditory thalamo-recipient zone of the avian telencephalon (Karten, 1968; Kelley and Nottebohm, 1979; Brauth et al., 1987). It is a complex region, composed of several subdivisions that can be distinguished based on cytoarchitectonics, connectivity and physiology (Bonke et al., 1979a; Scheich et al., 1979; Saini and Leppelsack, 1981;

Heil and Scheich, 1991a,b; Fortune and Margoliash, 1992; Wild et al., 1993; Vates et al., 1996; Gehr et al., 1999). Early studies considered field L to be the avian equivalent of the mammalian primary auditory cortex. This view is inaccurate, as significant components of the auditory cortex such as descending auditory projections are lacking in field L. A more recent interpretation is that field L contains neuronal populations comparable to those in some, but not all layers of the auditory cortex (Wild et al., 1993).

L2a, the core subdivision of field L, consists of a thin fiber-rich sheet of tissue that extends dorsally from the caudodorsal tip of the paleostriatum towards the ventricle and nucleus HVC (Fortune and Margoliash, 1992). L2a is characterized by small-sized neurons packed at high density in regular rows. It is the main recipient of projections from the auditory nucleus ovoidalis, and projects exclusively to field L subdivisions L1 and L3 (Karten, 1968; Kelley and Nottebohm, 1979; Brauth et al., 1987; Wild et al., 1993; Vates et al., 1996). Data from several avian species including zebra finches show that L2a has a high frequency of spontaneous discharges, and that auditory responses in L2a are of short latency, are organized tonotopically, have low selectivity for complex stimuli, and do not habituate to repeated stimulation (Bonke et al., 1979b; Scheich et al., 1979; Müller and Leppelsack, 1985; Heil and Scheich, 1991a, 1992; Chew et al., 1995; Stripling et al., 1997). Interestingly, L2a neurons have been shown to contain high levels of calcium-binding proteins, which may confer some degree of protection against transmitter-related cytotoxicity (Braun et al., 1991). L2b, a field L subdivision contiguous with L2a, is the main telencephalic target of a distinct subdivision of the nucleus ovoidalis and projects to a subset of L2a targets (Vates et al., 1996); it may represent a functional specialization within field L2.

Subdivisions L1 and L3 flank L2a rostrally and caudally, respectively (Fortune and Margoliash, 1992). They have larger neuronal cells than L2a, and constitute the main targets of fibers from L2a (Wild et al., 1993; Vates et al., 1996). They originate the projections from field L to other regions of the telencephalon, including the caudal portion of the hyperstriatum ventrale (CHV), the caudomedial neostriatum (NCM), as well as the shelf and cup regions adjacent to song nuclei HVC and RA, respectively (Kelley and Nottebohm, 1979; Vates et al., 1996). Direct projections from L1 and L3 to HVC have been suggested but not conclusively established (Fortune and Margoliash, 1995). Contralateral projections have been described in the pigeon (Wild et al., 1993). Recorded electrophysiological responses in L1 and L3 are of longer latency and show more selectivity towards complex auditory stimuli than those of L2a (Müller and Leppelsack, 1985; Heil and Scheich, 1991a).

Based on the studies above and other comparative analysis, the thalamo-recipient cells in L2a appear to correspond to the granular cell layer of the mammalian auditory cortex, whereas L1 and L3, which receive inputs from L2a and originate only intratelencephalic projections, would represent higher-order processing areas and correspond to neuronal populations in cortical supragranular layers (Karten and Shimizu, 1989; Vates et al., 1996). *zenk* expression in response to song presentation occurs in the neuronal cells present in subdivisions L1 and L3, but not in L2a (Mello and Clayton, 1994; see Section 4.0). It is not known if the *zenk*-labeled cells in L1 and L3 correspond to projection neurons or local interneurons.

3.4. NCM

The caudomedial neostriatum (NCM; see Fig. 3) is the brain area that presents the most robust *zenk* expression in response to song playbacks (Mello et al., 1992; Mello and Clayton, 1994) and where the *zenk* response to auditory stimulation has been characterized in most detail.

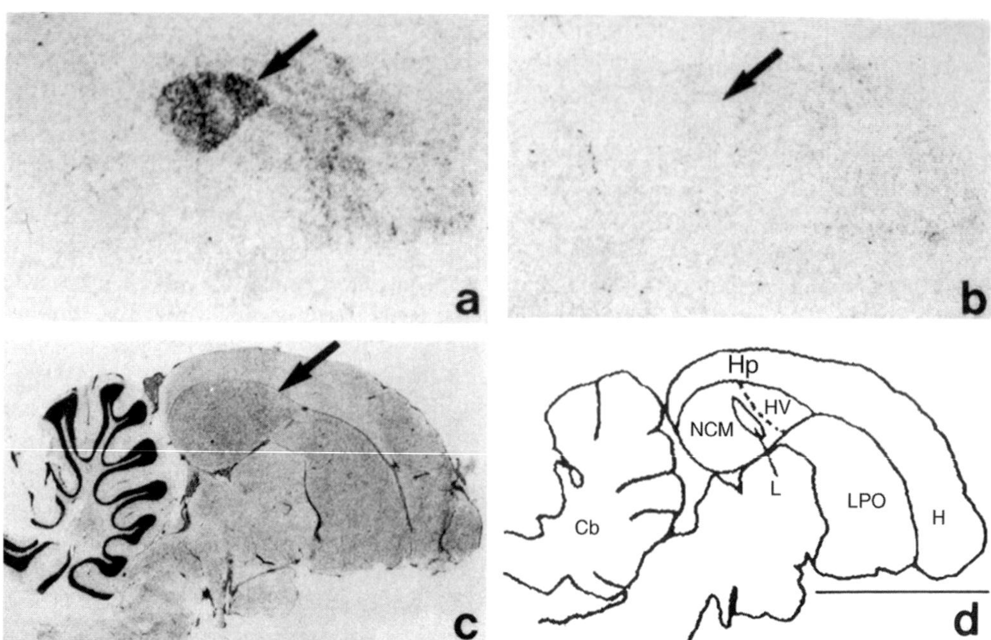

Fig. 3. Exposure to song induces *zenk* gene expression in the brain of songbirds (from Mello et al., 1992). Shown are in situ hybridization (X-ray film) autoradiograms of medial parasagittal sections hybridized with a radioactively labeled *zenk* riboprobe. (*a*) Adult male zebra finch exposed for 45 min to recorded conspecific song. (*b*) Unstimulated control. (*c*) Cresyl violet-stained section whose autoradiogram is shown in a. (*d*) Camera lucida drawing of the section shown in c. The arrows point to the caudomedial telencephalon where areas of high *zenk* expression are located, including NCM and the caudomedial HV. Bar size, 4 mm; for abbreviations, see list.

3.4.1. Definition and organization

NCM is the telencephalic area caudal to the lamina hyperstriatalis (LH) and immediately adjacent (medially and caudally) to field L (Mello and Clayton, 1994). It extends about 1.5 mm into the telencephalon from the lateral wall of the lateral ventricle. NCM is most easily visualized on parasagittal sections, starting medially as a small circular area completely encircled by the lateral ventricle. Laterally, it elongates until connecting with more rostral portions of the neostriatum. At this level, its boundaries are the ventricular zone dorsally, caudally and ventrally, and field L and LH rostrally. There are no obvious cytoarchitectonic boundaries laterally, even though a clear *zenk* expression boundary can be seen in frontal sections. The rostral portion of NCM is located at the same rostrocaudal level as field L3 laterally, and these two structures are possibly contiguous. NCM is not a discrete nucleus, but rather a large and heterogeneous expanse of the neostriatum, likely containing anatomical and/or functional subdomains (see below). NCM contains a diversity of neuronal cell types based on Golgi, but their functional significance is unknown (Saini and Leppelsack, 1981).

3.4.2. Connectivity

NCM receives dense projections from fields L1 and L3 and a lesser projection from the auditory thalamus, namely from cells located in the ovoidalis (Ov) shell (Mello, 1993; Vates et al., 1996). The projections from L1 and L3 provide evidence for anatomical subdomains within NCM, as their terminations are most prominent in the rostral NCM and much sparser or absent in its caudal portion; further studies are necessary to establish this point unambiguously. Another important source of input is the medial portion of the caudal hyperstriatum ventrale (cmHV), with which NCM is reciprocally connected (Vates et al., 1996). A possible projection from Pc remains to be confirmed (Mello, 1993). NCM also receives a significant noradrenergic projection, as shown by ICC staining for noradrenergic fibers (Mello et al., 1998a).

Detailed studies of the internal anatomical organization of NCM are lacking. Nonetheless, small injections of tracers such as biocytin reveals that NCM neurons have dense and extensively branching dendritic arborizations, and that NCM subdomains are highly interconnected by local projections (Mello, 1993), consistent with the scarce data from Golgi staining (Saini and Leppelsack, 1981). Although still preliminary, the emerging picture is of a densely interconnected network. The main known projection from NCM is to the cmHV, which is located rostrally to NCM and field L (Vates et al., 1996); this projection seems to be topographically organized. Another potential projection is to Pc (Mello, 1993). Interestingly, the medial portion of NCM, which shows the most robust *zenk* induction, connects with more lateral NCM areas that are contiguous with the shelf region under nucleus HVC. Whether the shelf receives a direct projection from the lateral NCM is not known. Such a pathway could provide a rather direct entry for auditory information processed in NCM into the song system.

3.4.3. Kinetics of *zenk* induction

The rise in *zenk* mRNA levels after song presentation has a rapid time course, being detectable within a few minutes after stimulus onset, peaking 30 min thereafter, and decreasing back to unstimulated levels by 1 h (Mello and Clayton, 1994; Kruse et al., 2000). *zenk* mRNA levels at peak expression are proportional to the amount of stimulation during the preceding period (Mello and Clayton, 1994). As little as one song presentation is enough to trigger a detectable rise in *zenk* mRNA levels in NCM, and stimulation beyond 10 min appears to result primarily in increased variability of *zenk* expression levels among different birds (Kruse et al., 2000). The time course of ZENK protein expression, determined by assessing the density of ZENK-labeled cells in an ICC assay, is protracted relative to that of *zenk* mRNA (Mello and Ribeiro, 1998). A rise in ZENK protein can also be detected as early as a few minutes after the start of song stimulation, but peaks between 1 and 2 h and decreases to unstimulated levels by 6 h after stimulus onset.

3.4.4. Selectivity of the *zenk* response

zenk induction in NCM is highest for same species (conspecific) song, followed by song of another species (heterospecific), the response being even lower or absent for synthetic tones (Mello et al., 1992). These data were obtained by densitometric analysis of X-ray film autoradiograms and represent an overall integration of the responses by units throughout NCM. No evidence has yet been obtained for sex differences in the *zenk* response to song in NCM (Mello, 1993; Duffy et al., 1999). Interestingly, a recent study in starlings, where song-based female preferences have been studied in detail (Gentner and Hulse, 2000), has revealed that the *zenk* response in NCM of females is highest for their preferred conspecific stimuli, i.e. longer songs (Gentner et al., 2001). This differential effect over shorter songs is observed even when compensating for total stimulus duration. Overall, these observations indicate that NCM is somehow tuned to auditory stimuli that are of higher behavioral relevance to the recipient bird.

3.4.5. Habituation of the *zenk* response

The *zenk* response to a particular conspecific song cannot be sustained, but rather decreases upon repeated presentations of that song (Mello et al., 1995). This phenomenon has been observed for both *zenk* mRNA and protein and has been referred to as 'habituation' of the *zenk* response, to borrow a term from the behavioral literature. This habituation can last 24 h or longer, depending on the presentation protocol. It is song-specific, as presentation of a novel song to animals habituated with a first song results in *zenk* re-induction in NCM (Mello et al., 1995). This finding provides further evidence that NCM neurons participate in song discrimination. As a very large percentage of neuronal cells in NCM show a *zenk* response to any particular conspecific song, it is very likely that many NCM neurons show a *zenk* induction response to more than one song, the response to each song being independent of that to other songs (Mello et al., 1995). These observations are consistent with the notion of ensemble coding for song representation, according to which the auditory representation of different songs involves the recruitment of distinct but partially overlapping cell groups from the total neuronal population.

3.4.6. Electrophysiological studies

Studies in starlings that preceded the finding of gene regulation by song had indicated that a relatively large portion of the caudal telencephalon, possibly including the lateral NCM, has electrophysiological responses to auditory stimulation, with many units showing some selectivity for complex stimuli (Müller and Leppelsack, 1985). Recordings targeted more specifically to areas that show the *zenk* response to song demonstrate that auditory responses are indeed prevalent throughout NCM (Chew et al., 1995, 1996a,b; Stripling et al., 1997; Ang, 2001), confirming that auditory processing is a central function of this brain area. Auditory responses in NCM are not as brisk and have longer latencies than those recorded in the adjacent field L, and apparently lack the degree of selectivity for specific song elements that has been reported for nuclei within the song control pathway. These observations are consistent with the notion that NCM occupies an intermediate position between the primary auditory telencephalic area and higher-order centers.

The auditory responses within NCM decrease quickly upon repeated song presentations (Chew et al., 1995, 1996a; Stripling et al., 1997). This phenomenon is best visualized by plotting the integrated amplitude of the responses in NCM to repeated presentations of a given song as a percentage of the response to the first presentation (Fig. 4). This electrophysiological 'habituation' is song-specific and can last for hours to days, depending on the presentation protocol. The habituation rate, a linear regression derivative of the habituation curve, provides a useful index of whether a particular song is 'remembered' by NCM neurons. While the response curves to familiar and remembered songs are characterized by no habituation or low habituation rates, the responses to novel or 'forgotten' songs have high rates of habituation. The electrophysiological habituation may help explain at least partially, the habituation of the *zenk* response to song. It should be noted, however, that electrophysiological responses are not abolished; rather, the response amplitude drops down to about 40% of the initial response. Once below a certain threshold, these responses appear to be no longer capable of eliciting a gene expression response.

It is unclear whether and how this decreased firing of NCM neurons to a particular song contributes to the bird's ability to remember that song. A possibility is that through habituation, a large population of units that are not tuned to the specific features of a particular

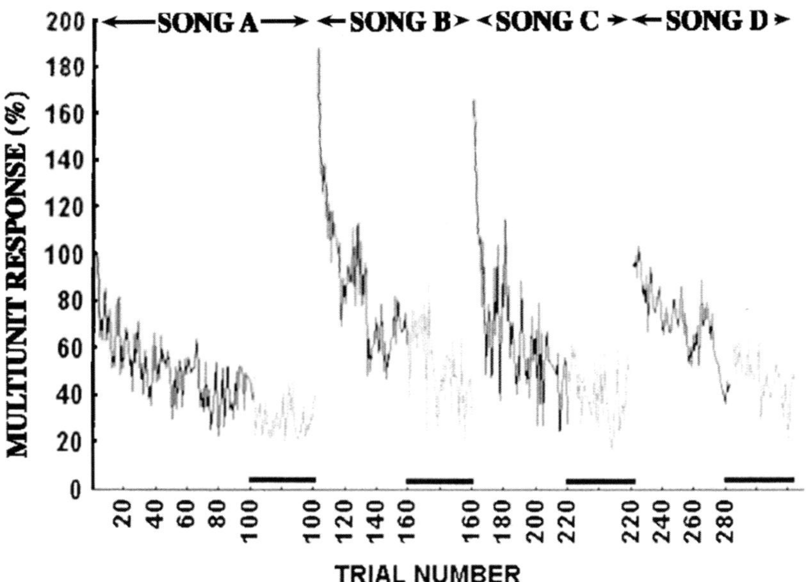

Fig. 4. Electrophysiological habituation of evoked auditory responses to song in NCM (from Chew et al., 1995). Shown are multi-unit activity responses to four different songs presented sequentially, recorded at a single site. The amplitudes of the evoked responses have been normalized to the initial response to song A. Notice how different songs elicited different initial responses, but that they all habituated upon repeated stimulus presentations (dark lines). Immediately after their presentation, songs A–D were again presented sequentially on trials 281–440. Notice how the response levels achieved during the first presentation period were retained during the second presentation period (light-gray lines, dark bars on x axis), even after training with other songs. This song-specific habituation appears to represent an electrophysiological memory trace for the songs the bird has recently heard.

song gets silenced. In consequence, the responses of a hypothetical small population of non-habituating units more highly tuned to that particular song might be highlighted. Most single units that have been analyzed in NCM so far show a habituation response (Chew et al., 1995) and other explanations are possible, but relatively little data are available (see also Ang, 2001).

3.4.7. Long-term habituation and gene expression

A possible function of song-induced gene expression is suggested by the studies on electrophysiological habituation. To test for the relationship between gene expression and auditory responses of NCM neurons, inhibitors of RNA and protein synthesis were injected at various times before and after exposure to song (Chew et al., 1995). The results show that (a) the auditory responses and the immediate habituation of NCM neurons to song are not affected by the blockers and are therefore independent of gene expression; the macromolecular assemblies present before stimulation are presumably sufficient for those processes, and (b) the long-term maintenance of the habituation of NCM neurons is blocked by the inhibitors during discrete periods after song presentation, and therefore depends on song-induced gene expression (Fig. 5). Such properties are comparable to those described for hippocampal LTP (Nguyen et al., 1994), and are consistent with the idea that the gene expression triggered by neuronal activation is necessary for long-term neuronal modification (Goelet et al., 1986; Morgan and Curran, 1989; Sheng and Greenberg, 1990) and the establishment of long-lasting

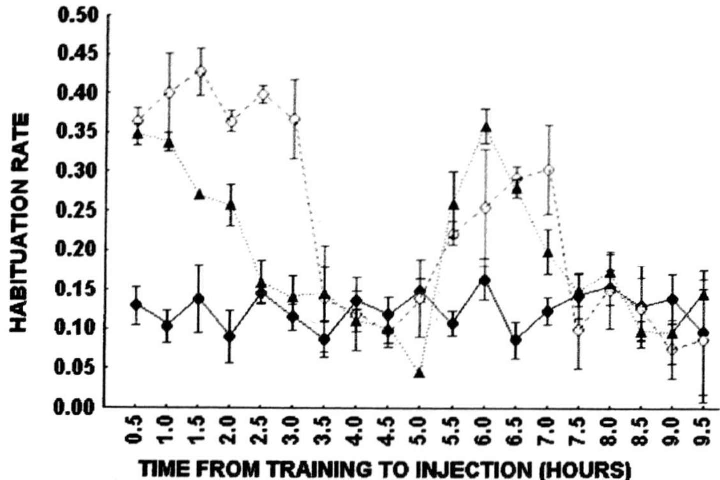

Fig. 5. Long-term habituation of evoked auditory responses to song in NCM is dependent on gene expression (from Chew et al., 1995). Plotted are the habituation rates (linear derivatives of the habituation curves, as in Fig. 4) ± SEM as a function of the interval from onset of stimulation to the local injection of cycloheximide (CYC; solid triangles), actinomycin (ACT, open diamonds) or saline (solid diamonds) in NCM. All habituation rates presented here were recorded 10 h after habituation to repeated song presentation, a time when songs are normally well remembered (low habituation rates). Notice that CYC and ACT were effective in blocking long-term habituation (recorded rates different from rates of control saline injections) at 0.5–3.0 and 5.5–7.0 h and at 0.5–2.0 and 5.5–6.5 h, respectively.

memories (Davis and Squire, 1984). Indeed, recent evidence shows that the expression of activity-dependent genes such as *zenk* is required for the maintenance of LTP and of spatial memories (Guzowski et al., 2000; Jones et al., 2001). The time course of song-induced *zenk* expression in NCM is compatible with its participation as an early regulator in genomic events leading to long-term habituation to song (Mello, 1998; Ribeiro and Mello, 2000). Testing that possibility directly requires a methodology to selectively block *zenk* expression in NCM during specific time windows. Alternatively, insights can be gained by identifying which among potential *zenk* targets are regulated by song.

3.4.8. Syllabic representation

In order to study the functional organization of NCM, we explored the fact that the song of canaries can be decomposed into phrases that are themselves repetitions of single-syllable types (Nottebohm and Nottebohm, 1978; Güttinger, 1985; Ribeiro et al., 1998). These syllables are simpler in their physical properties (for instance, the syllable type 'whistle' consists of pure tone-like vocalizations) and represent more discrete stimuli than complete songs. The resulting ZENK expression patterns in NCM were analyzed as global maps of the underlying neuronal activation utilizing a mapping system where both the position of labeled cells and their labeling intensities were computed (Cecchi et al., 1999).

As shown in Fig. 6, presentation of each syllable type normally present in the song of canaries of the Waterslager strain (whistles, combinations — stacks and sequences — and fast-frequency modulations) results in a distinct pattern of ZENK activation (Ribeiro et al., 1998). These patterns are more discrete than the patterns associated with whole songs or combinations of phrases. For instance, presentation of whistles results in clusters of

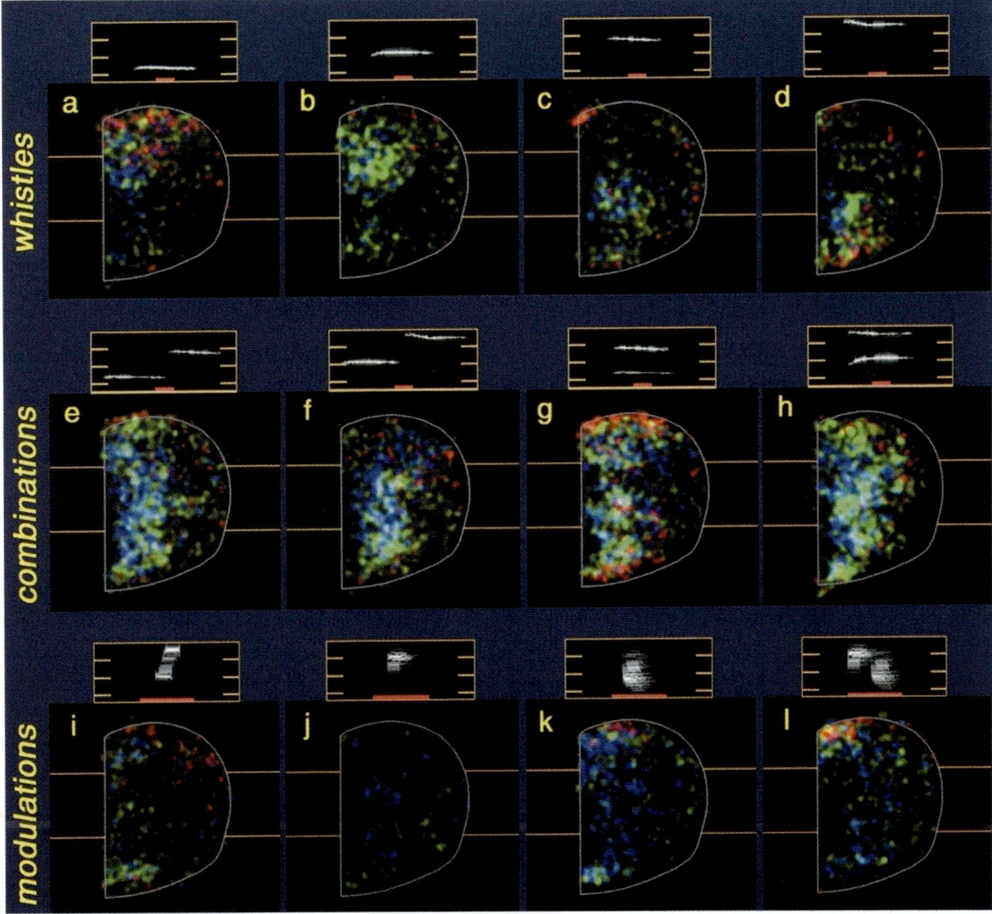

Fig. 6. Syllabic auditory representation in the canary brain (from Ribeiro et al., 1998). Panels show maps of *zenk* expression in NCM resulting from presentation of three families of song syllables, with the corresponding sonograms shown above each map. Each syllable family resulted in a family of *zenk* maps with distinct properties. Notice that some of the complex syllables are combinations of two individual whistles. For frequency scales in sonograms (1–4 kHz) and color keys in maps, see fig. 1 in Ribeiro et al. (1998). Red scale bars represent 100 ms.

ZENK-expressing cells located in the rostral NCM whose relative position in the dorsoventral axis changes as a function of whistle frequency. Importantly, patterns resulting from natural whistles are very different (more clustered spatially and in terms of cell labeling intensities) than those resulting from artificial whistles of comparable frequencies. This indicates that the organization of the rostral NCM according to the whistle frequency is not just tonotopic, but also depends on other features present in the natural vocalizations. Whistle combinations result in patterns that are distinct from the sum of the patterns resulting from the individual component whistles. Syllables characterized by fast-frequency modulations result in yet another distinct pattern, which is more distributed and shows little activation of the rostral NCM.

Some of the patterns above can be discriminated with the help of an index of clustering, a measure of the inhomogeneity of the spatial distribution of ZENK-labeled cells, indicating the importance of the spatial organization of the response. Further quantification with principal component analysis (a multi-dimensional linear regression analysis used in this study for comparisons among maps) shows that the patterns associated with different classes of syllables can be clearly distinguished based on information about the precise location and labeling intensities of ZENK-expressing neurons in NCM. In conclusion, the activation patterns in NCM resulting from different syllables provide a basis for an auditory syllabic representation in the songbird brain (Ribeiro et al., 1998). The data also show that discrete NCM subdomains can participate in the representation of various different syllables. Taken together, these results suggest that the auditory encoding of distinct syllables is provided by the recruitment of neuronal ensembles from a large population of available units.

3.4.9. Juveniles

To investigate a possible involvement of auditory-induced *zenk* expression in song learning, it is important to analyze *zenk* regulation during the song learning period. Consistent with what has been described in the rodent cortex, *zenk* expression throughout the songbird brain, including auditory portions of the neostriatum and hyperstriatum, is higher in juveniles than adults (Mello, 1993; Jin and Clayton, 1997). No significant *zenk* induction is observed upon presentation of conspecific song playbacks at 20 days of age (Jin and Clayton, 1997), which is before the beginning of the normal critical period for song learning. Basal *zenk* expression then gradually declines, until reaching the low levels characteristic of adults. *zenk* induction by song presentation is detectable by day 30 (Jin and Clayton, 1997), which is within the initial period for song learning, and increases thereafter.

These observations suggest that significant changes occur in the organization of the auditory forebrain and in song-induced *zenk* expression patterns in NCM in association with the song learning process. The interpretation of this correlation is not trivial. It has not been determined, for example, whether significant vocal learning results from the protocols that have been utilized for *zenk* expression analysis. It seems unlikely that the amount of song exposure that is sufficient to trigger a *zenk* induction response in zebra finches results in detectable song learning. Actually, the amount of song exposure that is optimal for vocal learning has yet to be determined. Recent evidence indicates that excessive exposure can, under certain circumstances, be detrimental (Tchernichovski et al., 1999). In addition, vocal learning in many species, including zebra finches, is influenced by several other parameters such as social context of song presentation (for instance, presence of a live tutor, female, or other siblings), the bird's arousal and motivation state, hormonal state, and others (see reviews in Brenowitz et al., 1997b).

Another study worth mentioning here investigated *zenk* expression in the NCM of adult males in response to the songs the birds had been exposed to as juveniles during the song learning period (Bolhuis et al., 2000). The authors found that *zenk* expression levels correlate with how much the birds learned from the songs presented, more specifically with the number of song elements copied from the tutors' songs. This finding is consistent with the notion that NCM is somehow involved in the formation and/or storage of song auditory memories, and that it may then impact song learning.

3.4.10. Neurochemical organization of NCM

The discussion here is restricted to studies addressing mechanisms that may participate in the gene expression response to song.

3.4.10.1. Norepinephrine

The noradrenergic system is required for IEG expression in the mammalian cortex (Cirelli et al., 1996). To test whether that also applies to the *zenk* response in NCM, an immunocytochemical study for a noradrenergic marker was conducted (Mello et al., 1998a). The results show that NCM receives a rich noradrenergic innervation, with terminal arborizations coming in close contact with *zenk*-labeled cells. In addition, song-induced *zenk* expression is blocked by systemic administration of noradrenergic antagonists (C.V. Mello and S. Ribeiro, unpublished observations), consistent with the notion that the noradrenergic system is required. These preliminary findings provide a plausible explanation for the habituation of the *zenk* response to song. Repeated presentations of the same stimulus would lead to decreased firing of the locus coeruleus, a nucleus that is the major source of noradrenergic projections to the brain and whose activity is highly dependent on stimulus novelty. This would in its turn result in decreased norepinephrine release and decreased *zenk* expression in NCM.

3.4.10.2. GABA

Systemic administration of the GABAergic antagonist metrazole leads to widespread neuronal depolarization and *zenk* expression throughout the brain, including NCM (Mello and Clayton, 1995). To investigate a possible relationship between the presence of this inhibitory neurotransmitter and song-induced *zenk* expression, the distribution of GAD, a GABAergic cell marker, was studied. Surprisingly, preliminary results indicate that the density of GABAergic cells in NCM is very low (S. Ribeiro and C.V. Mello, unpublished observations). In contrast, the density of these cells is high in field L, a major source of input to NCM. This suggests that a possible influence of GABA on *zenk* expression in NCM would be an indirect one.

3.4.10.3. Glutamate

In other systems induced IEGs expression is dependent on glutamatergic transmission (Cole et al., 1989; Bading et al., 1993; Xia et al., 1996). Glutamate also appears to be the main excitatory neurotransmitter in the avian brain, but the participation of glutamate receptors in song-induced *zenk* expression has not yet been investigated. Preliminary evidence indicates that the expression of several specific glutamate receptor subunits (NMDA, AMPA and metabotropic) are expressed in unique patterns in the caudomedial telencephalon, what could help explain the differential expression of *zenk* in NCM and different field L subfields (C.V. Mello and M. Hagiwar, unpublished observations).

3.4.10.4. Aromatase

It is highly intriguing that expression of aromatase, an enzyme that converts testosterone into its active metabolite estrogen, is expressed at very high levels in NCM (Schlinger, 1997; Saldanha et al., 2000). Also intriguing is the fact that there is no obvious overlap between brain areas of aromatase expression and those expressing estrogen receptors. It is tempting to hypothesize some link between the genomic response to song presentation and the brain's response to sex steroids. For instance, a locally generated hormone such as estrogen could be a modulator of the *zenk* response to song in NCM, possibly through non-genomic mechanisms. This action could then potentially affect the bird's ability to form song auditory memories.

3.5. CHV

Pronounced *zenk* induction by song also occurs in the caudal hyperstriatum ventrale (CHV), the portion of the HV that lies immediately rostral to field L (Mello and Clayton, 1994). On parasagittal sections, this area has a wedge shape medially, and is more elongated laterally. Tract-tracing studies in non-songbird avian species established that the CHV is reciprocally connected with field L, suggesting a role in the processing of auditory signals (Bonke et al., 1979a; Brauth and McHale, 1988; Wild et al., 1993). This is consistent with electrophysiological studies, which also established that responses in the CHV have a longer latency than those in field L and are tonotopically organized (Müller and Leppelsack, 1985; Heil and Scheich, 1991a). In addition, the CHV was reported to contain a high density of units showing some selectivity for complex auditory stimuli, indicating that it represents a higher-order auditory station than field L (Müller and Leppelsack, 1985).

The finding that *zenk* expression is induced by song presentation triggered renewed interest in this area. Tract-tracing studies in zebra finches have established several important points (Vates et al., 1996). CHV can be divided into medial and lateral subdomains, respectively cmHV and clHV. cmHV is reciprocally connected with NCM and clHV, and clHV is reciprocally connected with several field L subdivisions. In addition, clHV sends projections to the shelf region and to NIf. The latter projection is of particular importance, as it provides an anatomical substrate for a direct entry of auditory information into the song control system. Electrophysiological recordings of CHV in songbirds are still lacking. Indirect observations obtained during the study on the organization of canary NCM (see 3.4.8) suggest that auditory responses in CHV also have an organization based on frequencies. Importantly, lesions to CHV result in a disruption of normal song preferences in female zebra finches (MacDougall-Shackleton et al., 1998), providing evidence for a participation of this brain area in auditory processing and discrimination. By inference, this finding argues against a direct participation of song nucleus HVC in auditory discrimination in female songbirds (see Brenowitz, 1991).

3.6. SHELF AND CUP

zenk is also induced by song presentation in regions adjacent to two song control nuclei, the shelf under HVC and the cup rostroventral to RA (Mello and Clayton, 1994). For both structures, the Nissl delineation of the respective song nuclei provides a clear-cut boundary for the area of *zenk* expression (Fig. 7). Initial tract-tracing studies defined the shelf and cup regions as major termination zones of projections from field L and failed to reveal direct projections from field L to song nuclei HVC and RA (Kelley and Nottebohm, 1979). The intermediate location of the shelf and cup between auditory processing areas and the song control system suggested that they might constitute a site of entry for auditory information into the song system, and thereby play a role in sensorimotor integration (Nottebohm, 1996).

The finding that the shelf and cup express *zenk* in response to song presentation rekindled interest in these areas. Recent tract-tracing studies (Fortune and Margoliash, 1995; Vates et al., 1996; Mello et al., 1998b) have established and/or confirmed several points regarding their connectivity. (1) The projection from field L to both the shelf and cup originate from fields L1 and L3. (2) The shelf receives additional inputs from the clHV. (3) Neurons in the shelf send fibers into HVC, where they branch out extensively and terminate. (4) Dendrites of HVC neurons located close to its ventral border reach out into the shelf area, where they may receive terminations from field L projections. (5) No conclusive evidence has been uncovered for a direct projection from field L subfields to HVC. (6) The shelf projects in

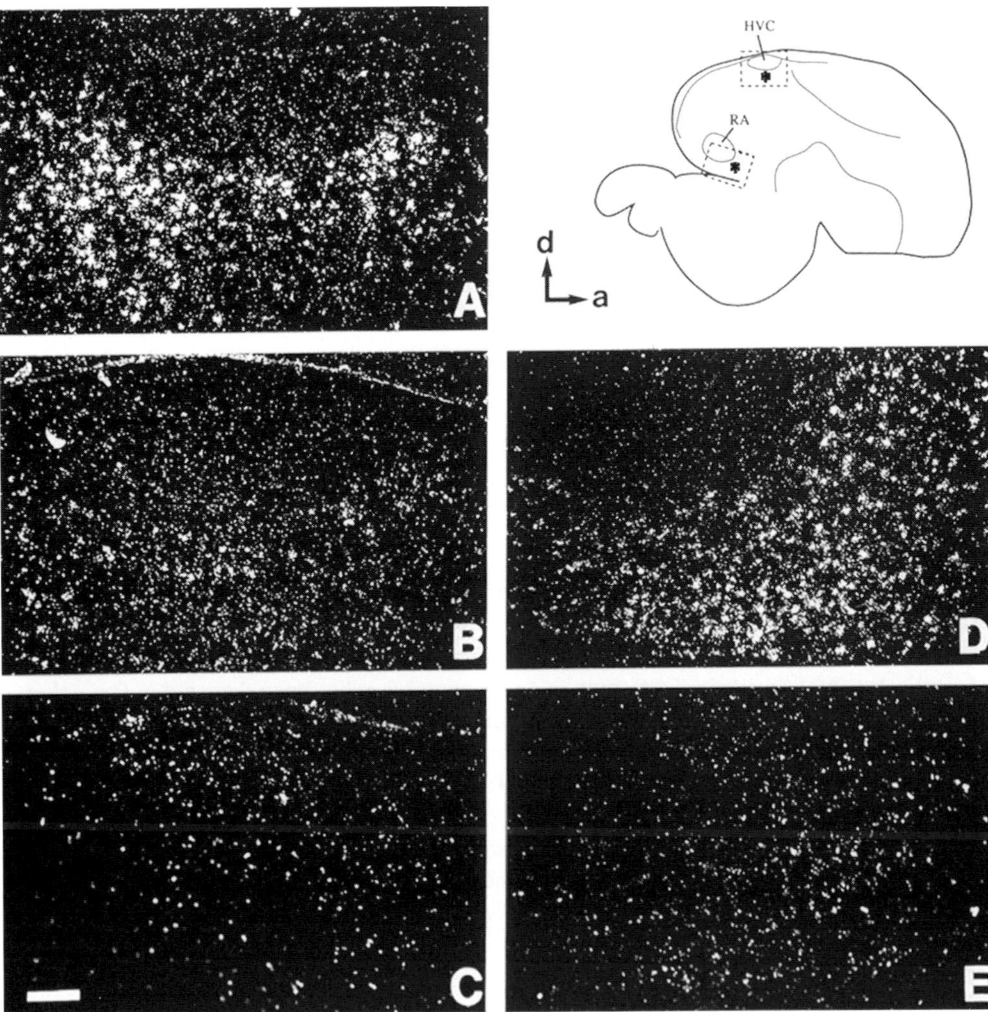

Fig. 7. Song presentation elicits *zenk* gene expression in the shelf and cup regions. Shown are X-ray film autoradiograms of parasagittal brain sections hybridized with a radioactively labeled *zenk* riboprobe. Sections are from a bird stimulated with playbacks of conspecific song (*A* and *D*) or from a quiet unstimulated control (*B* and *E*). The areas shown in the panels correspond to the dashed rectangles in the diagram on the upper right (*A* and *B*, HVC and shelf region; *D* and *E*, RA and cup region). Notice that song presentation triggers *zenk* expression in the shelf and cup regions (indicated by asterisks in the diagram) but not within HVC or RA. Panel *C* depicts a section hybridized with a sense control riboprobe. In the diagram, d is dorsal and a is anterior.

a spatially organized manner to the RA cup. (7) The latter region is complex, containing different subdomains; RA shell is possibly a more appropriate name for the RA surroundings. (8) The RA shell/cup originates a descending projection that terminates onto nuclei of the ascending auditory pathway; similar projections have been described in the pigeon (Wild et al., 1993), which lacks an RA, and are thought to correspond to descending projections that originate in infragranular layers of the mammalian auditory cortex. (9) There is no evidence for a direct projection from the cup into RA. In summary, the shelf and cup regions are integral components of the central auditory pathways, likely corresponding to the Nd and Aivm auditory regions of non-oscine avian species (Wild et al., 1993; Mello et al., 1998).

The shelf potentially provides auditory input to the song system; the cup is in a unique position to influence the flow of auditory information along the ascending auditory pathway. Further studies are needed to evaluate the functional role of these areas in auditory processing, sensorimotor integration and vocal learning.

4. *zenk* INDUCTION BY SONG PRESENTATION: NEGATIVE REGIONS

In contrast to the areas above, some other nuclei where song-induced *zenk* expression might be expected do not show such a response. These are discussed below.

4.1. AUDITORY THALAMUS

The ovoidalis (Ov) is a conspicuous oval-shaped nucleus in the dorsal thalamus. Connectivity studies in songbirds and other avian groups indicate that Ov is the main thalamic auditory nucleus (Karten, 1967, 1968; Kelley and Nottebohm, 1979; Brauth et al., 1987; Wild et al., 1993; Vates et al., 1996; Mello et al., 1998b). Ov receives a major projection from MLd in the midbrain, as well as a descending projection from the telencephalon (originating in the RA cup in songbirds). It is the major source of auditory input to the telencephalon through a projection to field L. Electrophysiological studies confirm that Ov is an auditory processing nucleus, and that it is highly activated by auditory stimuli such as birdsong (Bankes and Margoliash, 1993). Intriguingly, though, no *zenk* expression has been detected in Ov, either in unstimulated controls or in song-stimulated birds (Mello and Clayton, 1994). This provides a clear example of a dissociation or uncoupling between electrophysiological activation and *zenk* gene induction. It is possible that IEGs other than *zenk* are induced by auditory stimulation in Ov. However, no song-induced expression of at least one other IEG, c-jun, has been detected (Nastiuk et al., 1994), suggesting that the uncoupling of neuronal activation from IEG expression may be a general phenomenon in Ov.

What could be the functional meaning of this uncoupling? The link between *zenk* induction and neuronal plasticity suggests that it would prevent the activation of neuronal cells in Ov to be followed by long-lasting changes in their response properties. This would therefore provide a means for preserving or stabilizing auditory representations at the level of the thalamus. The exact mechanisms for uncoupling neuronal activation and *zenk* gene expression in the thalamus are unclear, but could include processes such as downregulation of membrane receptors or components of signal transduction pathways that are necessary for the IEG response, upregulation of calcium-buffering systems, or direct repression at the promoter level.

A few scattered *zenk*-positive cells are observed in the thalamic region that surrounds Ov, but their number does not change as a result of song stimulation (Mello and Clayton, 1994). This region represents a significant termination site of descending fibers from the auditory telencephalon (Mello et al., 1998b). It coincides with the Ov shell region that in ring-doves receives a projection from the MLd surroundings and projects to the hypothalamus (Durand et al., 1992). This structure has been postulated as relaying auditory information from the midbrain onto endocrine regulatory centers. Such projections have not been investigated in songbirds.

4.2. L2a

zenk expression after song stimulation is also conspicuously absent in L2a (Mello and Clayton, 1994). This has been consistently observed using a variety of natural and synthetic auditory stimuli (Ribeiro et al., 1998). The absence of *zenk* expression in L2a contrasts sharply with the high expression in the surrounding areas that receive projections from L2a. Interestingly, this contrast serves as a convenient marker for identifying L2a boundaries in sections reacted for ZENK ICC, without Nissl counterstaining (Mello and Ribeiro, 1998).

L2a is the main telencephalic termination zone of projections from the auditory thalamus, and is therefore a necessary site of entry for most of the auditory information reaching the telencephalon (Karten, 1968; Bonke et al., 1979a; Kelley and Nottebohm, 1979; Brauth et al., 1987; Wild et al., 1993; Vates et al., 1996). L2a is highly activated by auditory stimuli, as shown by electrophysiological and metabolic studies [Scheich, 1979 #966; Theurich, 1984 #991; Müller, 1985 #850; Müller, 1985 #992; Heil, 1991 #971; Heil, 1991 #972 (see Gehr et al., 1999)]. Similarly to Ov, L2a constitutes a clear example of an area where electrophysiological activation and the *zenk* induction response are dissociated. Here, again, although other IEGs might be induced by song presentation, the present evidence indicates that neuronal depolarization is dissociated from IEG induction.

As argued above for the auditory thalamus, the lack of *zenk* expression in the primary auditory telencephalic area may result in the stabilization of auditory representations at this level of the pathway. According to this view, L2a would serve primarily as a relay of auditory information into higher-order processing areas in the telencephalon, but this remains to be determined. An interesting observation in this regard comes from the studies utilizing metrazole (Mello and Clayton, 1995), a GABAergic antagonist that causes widespread neuronal activation. In contrast to the resulting *zenk* induction throughout the telencephalon, the primary termination zones for thalamic sensory projections lack the *zenk* response: L2a (auditory), the ectostriatum (visual), and nucleus basalis (somatosensory). This effect does not correlate with the known GABA or GABA receptor distributions (Müller, 1988; Grisham and Arnold, 1994). These observations are consistent with the notion that IEG expression has been generally uncoupled from neuronal depolarization at primary thalamo-recipient zones.

It is interesting to note that calcium-buffering proteins such as parvalbumin and calbindin also have high expression in L2a as compared to the rest of the brain (Braun et al., 1991). As calcium entry is a necessary step in the signaling cascade that leads from neuronal depolarization to IEG induction (Ghosh et al., 1994), preventing a rise in its intracellular concentration might provide an effective means of downregulating *zenk* induction in specific neuronal cells.

Neurons in field L2a can be considered functionally equivalent to neuronal cells in the granular layer of the mammalian auditory cortex that receive projections from the auditory thalamus. In case the differential pattern of *zenk* expression seen in field L is also characteristic of the mammalian auditory cortex, a possibility to be tested, one would predict that auditory stimulation of the appropriate kind would result in *zenk* induction in various cortical layers but not in the granular layer. Interestingly, *zenk* expression in response to visual stimulation occurs in several layers of the primary visual cortex, including the granular layer, but some specific subdivisions of the latter are negative for *zenk* (Kaplan et al., 1996; Okuno et al., 1997) Studies with further molecular markers will be necessary to establish a possible identity between *zenk*-negative neuronal cells in the avian and mammalian brains.

4.3. SONG CONTROL NUCLEI

zenk expression in hearing-only animals has not been observed in any of the nuclei that constitute the song control system (Mello and Clayton, 1994; Mello and Ribeiro, 1998). This finding was quite surprising at first, since many studies had demonstrated the existence of evoked electrophysiological responses to auditory stimuli including nuclei in the direct motor and anterior forebrain pathways for song control (Katz and Gurney, 1981; Margoliash, 1983, 1997; Williams and Nottebohm, 1985; Doupe and Konishi, 1991; Vicario and Yohay, 1993; Volman, 1996). The lack of a *zenk* induction response indicated that a dissociation of gene expression from electrophysiological activation also occurred within the song system. It is important to note, however, that while *zenk* expression studies were conducted in awake, freely behaving animals, most electrophysiological recordings of auditory responses were performed in anesthetized animals. Mapping *zenk* induction under conditions that matched those used for electrophysiology was not feasible, as the *zenk* expression response to song is blocked by pharmacological agents used for anesthesia, including nembutal, xylazine, ketamine, and urethane (C.V. Mello, unpublished observations).

More recent studies have been performed in awake birds utilizing chronically implanted electrodes. Although this area is still controversial, auditory responses in the nuclei of the direct motor pathway (HVC and RA) are much weaker, if not absent, in awake as compared to anesthetized animals (Dave et al., 1998; Schmidt and Konishi, 1998). Interestingly, robust auditory responses in song nuclei can be recorded during the night period, when birds are asleep. These responses resemble those obtained in anesthetized preparations (Dave et al., 1998), and are likely under the control of modulatory (e.g. noradrenergic or cholinergic) systems that regulate sleep/wakefulness states. The significance of these observations is not yet clear, but they appear to be inconsistent with the notion that auditory responses in song control nuclei play a central role in song perception in awake animals. They resonate closely, however, with the data from *zenk* expression analysis and with recent lesion experiments indicating that auditory processing areas that precede the song control system play a critical role in normal song perception and discrimination (see Section 3.0).

5. SINGING-RELATED *zenk* EXPRESSION

5.1. SINGING-INDUCED *zenk* EXPRESSION IN SONG CONTROL NUCLEI

As illustrated in Fig. 8, a systematic comparison of *zenk* expression patterns in adult songbirds singing before sacrifice ('hearing and singing' birds) with birds that heard conspecific song playbacks but did not sing, or birds prevented from singing in response to the playback ('hearing only' birds), reveals that *zenk* is induced in song control nuclei in association with singing behavior (Jarvis and Nottebohm, 1997; Jin and Clayton, 1997; Jarvis et al., 1998; Mello and Ribeiro, 1998). *zenk* expression follows singing activity, and in all nuclei examined its levels correlate positively with the number of song bouts produced during the period preceding sacrifice. As song control nuclei have evoked responses to song playbacks (at least in anesthetized animals; see above), it is also important to compare intact singing males ('hearing and singing') with deafened singers ('singing only'). As expected, *zenk* expression in 'singing only' birds is greatly reduced or abolished in auditory areas, including NCM, CHV, field L, and the shelf and cup regions. In contrast, these same birds show clear *zenk* expression in song control nuclei. Importantly, deafened males tend to sing less than intact birds, but *zenk* expression levels in these animals are similar to those in intact males that sang a comparable number of song bouts. In addition, *zenk* expression in song control nuclei is also present in males muted by sectioning the left hypoglossal nerve, and that attempted to sing a comparable number of times as intact males. These observations demonstrate that *zenk* expression in song control nuclei is associated with the motor act of singing and not with auditory feedback.

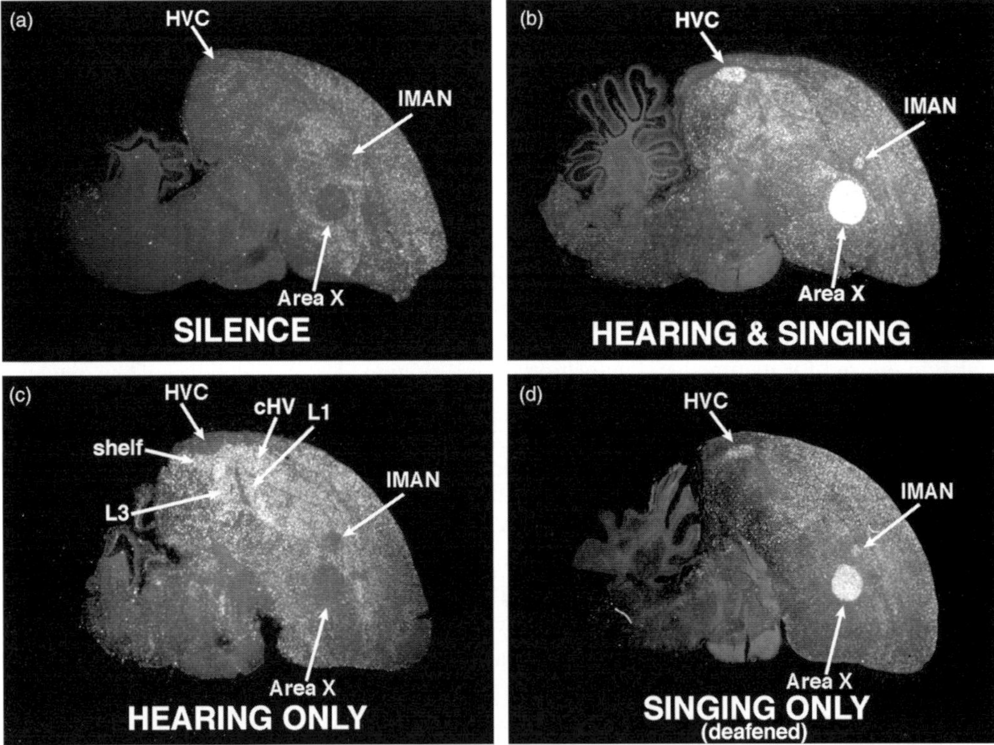

Fig. 8. Auditory- and vocalizing-induced *zenk* gene expression in the brain of male zebra finches (from Jarvis and Nottebohm, 1997). Shown are dark field views of cresyl violet-stained parasagittal sections hybridized with a radioactively labeled *zenk* riboprobe; sections are from representative birds in four experimental conditions. Notice that expression in song control nuclei only occurs in birds that were singing preceding sacrifice. Notice also that deafening abolishes singing-associated *zenk* induction in auditory regions but not in song control nuclei. For further details, see text; for abbreviations, see list.

One of the nuclei showing singing-induced *zenk* expression (DM) is located in the midbrain and is part of the intercollicular complex. DM is a conserved nucleus thought to be involved in vocal and respiratory control in all avian species examined to date (Wild, 1994, 1997). It projects to nXIIts and to respiratory centers in the medulla (Wild et al., 1997b), and its electrical stimulation results in vocalizations (de Lanerolle and Andrew, 1975; Vicario and Simpson, 1995). In songbirds, DM receives direct projections from the telencephalon, more specifically from the dorsal-most portion of RA. Such a projection is thought to be generally absent in avian species lacking vocal learning (Wild et al., 1993; Wild, 1994).

Seven other nuclei showing singing-induced *zenk* expression are in the telencephalon (HVC, RA, area X, MAN, Av, NIf and HVo). Of these, six have been previously detected with other techniques and linked to song behavior. Nuclei HVC and RA are necessary for song production (Nottebohm et al., 1976). They constitute the direct pathway for vocal control (Nottebohm et al., 1982) and participate actively in the motor control of song (Yu and Margoliash, 1996). NIf projects directly to HVC and its electrophysiological activation during singing precedes that of HVC (Nottebohm et al., 1982; McCasland, 1987) Nucleus Av receives a direct projection from HVC (Nottebohm et al., 1982), but its functional role has not been previously investigated. MAN and area X are part of the anterior forebrain pathway

(Okuhata and Saito, 1987; Bottjer et al., 1989) and are necessary for song learning (Bottjer et al., 1984; Sohrabji et al., 1990; Scharff and Nottebohm, 1991). For the latter three nuclei, the data from *zenk* expression analysis represent a first functional demonstration of their activation in association with singing behavior.

The other telencephalic nucleus showing singing-induced *zenk* expression is in the anterior hyperstriatum ventrale, immediately adjacent to lMAN. This location is very similar to that of HVo, a nucleus that is part of vocal control pathways in the telencephalon of budgerigars (Brauth et al., 1997; Durand et al., 1997). HVo was initially thought to be absent in songbirds, this being cited as a basic difference in the organization of vocal control pathways in songbirds and parrots (Striedter, 1994). The *zenk* mapping approach has now provided evidence for the existence of an HVo-like structure in songbirds (Jarvis et al., 1998), albeit smaller than in budgerigars, and for its functional activation during singing. The connectivity of this nucleus in songbirds remains to be established.

It is quite intriguing that singing-induced *zenk* expression occurs in MAN and area X. While lesions to these nuclei disrupt song learning when performed in juveniles, no effects on song are observed when the lesions are performed in adults (Bottjer et al., 1984; Sohrabji et al., 1990; Scharff and Nottebohm, 1991). These observations are consistent with a role for the anterior forebrain pathway in song learning, and argue against a direct participation of area X and MAN in the motor control of song production. Contrary to that notion, however, the *zenk* mapping experiments provide clear evidence for the activation of area X and MAN during singing, arguing for an active participation of the anterior forebrain pathway in the control of song behavior. In fact, area X shows the most marked *zenk* induction in association with singing (Jarvis and Nottebohm, 1997). Electrophysiological recordings with chronically implanted electrodes have confirmed the finding that nuclei of the anterior forebrain pathway are activated during singing behavior (Hessler and Doupe, 1999).

5.2. CONTEXT-DEPENDENT *zenk* EXPRESSION

While there is as yet no clear explanation for the apparent paradox between lesion and *zenk* expression studies, some clues as to the role of the anterior forebrain pathway in singing behavior come from the observation that *zenk* expression in area X of male zebra finches is context-dependent. It occurs in association with singing in the presence of other males or in a solo context (undirected singing), but not during female-directed singing (Jarvis et al., 1998). Some differences exist between female-directed and undirected song, the former being slightly shorter and containing more introductory notes (Sossinka and Bohner, 1980). Such differences, however, are rather minor and seem insufficient to explain the strong contrast in gene expression in the two situations above.

It is of interest to note that female-directed singing in zebra finches is accompanied by a complex series of movements and gestures equivalent to a courtship dance (Sossinka and Bohner, 1980; Jarvis et al., 1998). Area X, where the context effect on *zenk* expression is most dramatic, is located in a portion of the avian paleostriatum thought to correspond to the caudate-putamen in the mammalian basal ganglia (Reiner et al., 1998). This structure in mammals has been implicated, among other things, in the learning and maintenance of sequential motor actions that depend on sensorimotor integration (Lidsky et al., 1985; Aldridge and Berridge, 1998). It is therefore possible that area X is involved in coordinating singing with other behaviors that give contextual relevance to the song being produced. In this regard, it would be interesting to determine whether area X lesions in adult males can disrupt the birds' ability to coordinate singing with associated behaviors. An alternative but not excluding explanation for the context effect on *zenk* expression would be a difference in the birds' attention or arousal in the different contexts. As discussed above for auditory brain areas, catecholaminergic systems are intimately linked with the bird's arousal state, and could be key determinants of *zenk* induction in song control nuclei in association with singing behavior.

Yet another factor to be considered is that *zenk* mapping has revealed the existence of a medial extension of area X (Jarvis et al., 1998). The sparing of such an area could help explain the lack of an effect on song when area X was lesioned in adult birds. Based on *zenk* expression mapping, several song nuclei appear to be composed of medial and lateral portions, *zenk* expression in the former showing a more marked dependence on the behavioral context of singing (Jarvis et al., 1998). It remains to be determined whether such subdomains also differ in their connectivity and electrophysiological properties. The existence of functional subdomains and context-dependent *zenk* expression has also been detected in RA, *zenk* expression being lower or absent in association with female-directed singing (Jarvis et al., 1998). The latter is particularly intriguing, as activation of this nucleus is thought to be generally required for the motor act of singing irrespective of the context. As discussed above for some auditory areas, it seems likely that neuronal activation has been uncoupled from *zenk* expression in specific domains within RA.

5.3. JUVENILES

zenk induction in association with singing behavior has also been studied within the song control system of juvenile zebra finches, in search for possible direct correlations with the song learning process. The age of the birds studied ranged from 35 to 65 days, which represents the normal critical period for learning. Basal *zenk* expression in song nuclei of juveniles is very low or absent, but marked induction occurs in association with singing behavior (Jarvis and Nottebohm, 1997; Jin and Clayton, 1997). For several nuclei, including HVC, area X, MAN (both medial and lateral subdivisions), NIf and DM, this induction follows the same pattern observed during adulthood, and is proportional to the amount of singing. In contrast, an interesting developmental effect has been observed in RA (Jin and Clayton, 1997). In the anterior 2/3 of this nucleus, singing-induced *zenk* expression is higher in juveniles than in adults, whereas such an effect is not seen in the posterior 1/3. This finding led to the suggestion that singing-induced *zenk* expression levels in RA correlate with the degree of plasticity in the song being produced, which is higher during the learning phase (sub-song and plastic song). In another study, regulation of both *zenk* mRNA and protein in association with singing was investigated in juveniles. Interestingly, singing-induced ZENK protein expression in RA (as assessed by the number of immunoreactive cells) was found to be dissociated from *zenk* mRNA levels, being lower at an early age and increasing as song matures (Whitney et al., 2000). This suggests that post-transcriptional gene regulatory mechanisms may play a role in the song development process.

In interpreting the results in juveniles, though, one needs to consider that singing-induced *zenk* expression in RA is context-dependent, occurring as long as the birds are not singing female-directed song (Jarvis and Nottebohm, 1997). It should be noted in this regard that the amount of female-directed singing seems to increase with age. Furthermore, the observed *zenk* expression levels in RA at a given age are lowest in juvenile birds that presented most female-directed singing. Therefore, a significant portion of the age effect seen in RA may be related to the mode of singing rather than the degree of song plasticity. At any rate, these developmental studies provide functional support to the existence of subdomains within RA

(Vicario, 1991). Differences in the definition of such subdomains — anterior and posterior (Jin and Clayton, 1997) versus dorsal and ventral (Jarvis and Nottebohm, 1997) — likely result from differences in the exact alignments of parasagittal brain sections.

6. *zenk* EXPRESSION IN WILD BIRDS

The experiments discussed so far were performed in a laboratory context. A study conducted with song sparrows provided an opportunity to verify whether the phenomenon of song-induced brain gene expression also occurs in a natural setting (Jarvis et al., 1997). Song sparrows establish well-delimited territories early in the spring, and defend these territories and females from competing males during the breeding season (Searcy et al., 1981; Wingfield, 1985). By challenging a defending male with playbacks of tape-recorded song of an unfamiliar conspecific, one can promptly and reliably elicit robust singing behavior. By comparing *zenk* brain expression patterns in challenged birds versus unstimulated quiet controls it was possible to detect significant *zenk* induction in all the auditory processing areas and song control nuclei discussed in the preceding sections (Jarvis et al., 1997). Thus, singing and auditory-induced *zenk* expression occurs in wild birds, with patterns that are essentially the same as those observed in captivity.

It is thus possible to utilize IEG brain expression analysis in a natural setting to identify brain areas involved in complex behaviors that are not readily reproduced in the laboratory. In the field, however, it is harder to control all variables that affect neuronal activation and the associated brain gene expression. Some factors that were critical for the experiments with song sparrows are also likely to affect other field experiments. (1) Due to the territorial nature of their behavior during the breeding period, song sparrows tend to spend long periods at specific locations (such as conspicuous tree branches) from where they sing most often. This greatly facilitated locating and monitoring different individuals. (2) Generally, *zenk* brain levels were higher and the patterns more variable in the field than in the lab. To minimize basal *zenk* expression, it was important to utilize only birds that could be monitored for a long period preceding the experiment, and that did not engage in significant interactions with other birds during that period. Fortunately, birds in the field tend to concentrate most of their singing activity at dawn and dusk, and are generally quieter during mid-day. Obtaining controls and performing the challenge playbacks during this quiet mid-day period efficiently minimized basal *zenk* expression and optimized the detection of song-induced expression. (3) It is critical to capture the animals at a given time after stimulation or after they express a behavior of interest. The fact that song sparrows confine their activities to the area around the source of the playback greatly facilitated their quick capture at the end of the challenge period. (4) Assessment of IEG expression in field experiments is best done with in situ hybridization, as only the events that occur during the hour immediately preceding sacrifice will influence the results. Protein levels are affected by events that occur during a much broader time period, and are more likely to be affected by variables extraneous to the experimental paradigm. In summary, a detailed observation of the behavior of interest, a careful definition of the stimulation/behavioral paradigm, and a correct choice of the method of detection are all essential for mapping brain gene expression in wild animals.

7. COMPARATIVE APPROACH TO MAPPING AVIAN VOCAL CONTROL SYSTEMS

As discussed in the preceding sections, *zenk* expression analysis can be used to generate high-resolution global maps of brain areas involved in perceptual and motor aspects of vocal communication in songbirds. The present section deals with *zenk* mapping studies of vocal communication systems in other avian groups. The major question motivating this line of research has been an evolutionary one. Only three extant avian orders (parrots, hummingbirds and songbirds) are known to have developed vocal learning (Nottebohm, 1972; Brenowitz, 1997). These orders are separated by several orders (represented by birds such as suboscines, pigeons, cranes, and owls) that do not present the trait. It is therefore thought that vocal learning orders evolved vocal learning and associated brain structures independently. An alternative possibility is that vocal learning and related brain nuclei were present in a common ancestor to avian vocal learners and subsequently lost in the intervening orders.

Avian vocalizations are generally dependent on brainstem nuclei that control the vocal organ, the syrinx, in coordination with the respiratory function (Wild, 1997). Such nuclei, including DM and nXIIts, are highly conserved and are thought to be present irrespective of whether a particular species evolved vocal learning. In contrast, telencephalic nuclei that project to and control the activity of vocal and respiratory brainstem centers are present in vocal learners (songbirds and parrots), but appear to be absent in species that lack vocal learning (Brenowitz, 1997). It is therefore currently thought that the presence of telencephalic vocal control nuclei is a signature of the brain of vocal learners.

The contribution of *zenk* mapping to the comparative analysis of avian vocal communication systems is discussed below. One immediate goal of these experiments has been to define and compare the brain systems for vocal communication in all avian vocal learning orders. The idea is to determine the common features among these systems, as those may reflect general requirements for vocal learning. In contrast, features present in only one specific order or species are more likely to reflect specific adaptations and/or survival strategies. A second goal is to directly test for the presence of telencephalic vocal nuclei in vocal non-learners.

7.1. PARROTS

zenk expression was analyzed in the budgerigar, a small Australian parrot, to test whether IEG expression associated with vocal communication also occurs in another vocal learning order (Jarvis and Mello, 2000). Budgerigars have been the subject of intensive research regarding brain mechanisms for vocal production and learning in parrots. Several telencephalic vocal control nuclei have been identified based primarily on tract-tracing studies (Paton et al., 1981; Striedter, 1994; Brauth et al., 1997; Durand et al., 1997). These nuclei are in many regards similar to those described in songbirds, their projections converging onto brainstem nuclei that control the syrinx. As in songbirds, these telencephalic nuclei also segregate into an anterior and a posterior cluster, although significant differences seem to exist in their exact connectivity patterns (Striedter, 1994; Durand et al., 1997).

By systematically comparing 'hearing only' (birds exposed to playbacks of budgerigar's warble song) and 'quiet control' groups, *zenk* mapping revealed activation of the midbrain's MLd, as well as the presence of several auditorily activated areas in the caudomedial telencephalon (Jarvis and Mello, 2000). As in songbirds, some of the latter areas surround a central core (L2a) that is negative for *zenk*. Interestingly, no evidence was found for the activation of anterior telencephalic structures including nucleus basalis and some of its targets. This conflicts

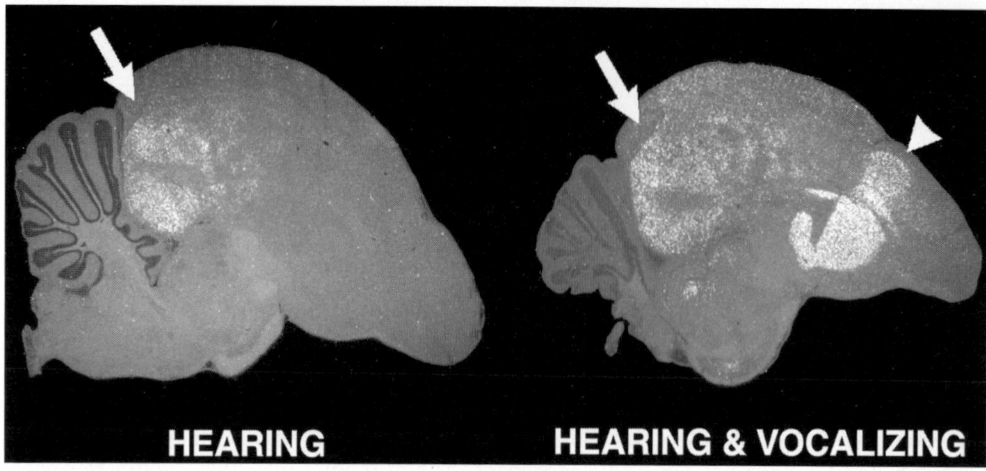

Fig. 9. Expression of *zenk* in the brain of a parrot (budgerigar) in association with vocal communication (from Jarvis and Mello, 2000). Shown are dark field views of cresyl violet-stained parasagittal sections hybridized with a radioactively labeled *zenk* riboprobe (expression is seen as white emulsion precipitate). Hearing species specific vocalizations induces *zenk* expression in caudal parts of the telencephalon, including NCM and CHV, but not field L2 (left panel, arrow; compare with Figs. 3 and 8C). Vocalizing in response to these sounds induces *zenk* expression in discrete telencephalic nuclei, including HVo, NAo and LPOm in the anterior forebrain pathway (right panel, arrowhead), in addition to expression in the caudal telencephalon (right panel, arrow). Quiet unstimulated controls have very low basal expression levels throughout the brain (not shown).

with studies indicating that such structures receive auditory projections and play a role in auditory processing and/or vocal learning and performance (Hall et al., 1993, 1994; Brauth et al., 1994; Striedter, 1994; Farabaugh and Wild, 1997; Wild et al., 1997a). A possible interpretation is that the activation of these structures in the anterior telencephalon has been uncoupled from *zenk* expression. Further studies including the recording of auditory responses in awake animals will be necessary to establish this point (for further discussion, see Jarvis and Mello, 2000).

By comparing the 'hearing and vocalizing' and 'hearing only' groups, two brainstem (midbrain's DM and thalamic DMm) as well as seven telencephalic (HVo, NAo, LPOm, NLc, AAc, lAN and lAHV) vocally activated nuclei were identified. Among the telencephalic structures, the former five nuclei constitute the known vocal control pathways in budgerigars, based on previous connectivity studies. The *zenk* mapping analysis provided a first functional demonstration of the activation of these nuclei in association with vocal production (Fig. 9). The latter two nuclei had not been previously linked to vocal control. *zenk* mapping has also revealed unsuspected features for some of the known vocal nuclei. The most striking example is LPOm, which based on *zenk* expression contains several complex extensions. The observed differences between *zenk* mapping and tract-tracing results may reflect the incompleteness of the latter approach, or alternatively it may represent evidence for the existence of functional/anatomical subdomains within the vocal nuclei of parrots. LPOm in budgerigars, as area X in songbirds, represents a basal ganglia component of the telencephalic vocal control system, whereas the other telencephalic structures represent cortico-like elements.

7.2. HUMMINGBIRDS

The third avian group known to have developed vocal learning is hummingbirds (Nottebohm, 1972; Baptista and Schuchmann, 1990; Brenowitz, 1997). Until recently, however, no data were available on brain mechanisms related to song production or learning in this order. By applying essentially the same approach as for songbirds and parrots, i.e. comparison of 'hearing and vocalizing', 'hearing only' and 'quiet control' groups, it was possible to rapidly identify and map vocal control centers in the brain of two tropical hummingbird species, the sombre hummingbird and the rufous-breasted hermit (Jarvis et al., 2000). Once such structures were defined by *zenk* mapping, they could be unambiguously identified in the brain of two other species (Jarvis et al., 2000), the four species examined covering the two existing hummingbird lineages.

Auditorily activated regions in hummingbirds consist of seven brain structures: one mesencephalic, one thalamic and five telencephalic. The auditory telencephalic areas are at very similar locations as the ones in songbirds and parrots. In particular, structures comparable to NCM, Pc and CHV are located caudomedially and surround a core area (presumably L2a) that is negative for *zenk*. The identified vocally activated regions consist of one mesencephalic (DM) and seven telencephalic (VAH, VAN, VAP, VMN, VMH, VLN and VA) structures. The latter can be divided into anterior, posterior–medial and posterior–lateral clusters. The vocally activated areas bear remarkable resemblance both in cytoarchitectonic features and in relative brain location to the vocal control nuclei of songbirds and parrots. For instance, VAP in hummingbirds, as area X in songbirds and LPOm in parrots, is located in the avian equivalent to a portion of the mammalian basal ganglia, providing evidence that such structures are also involved in vocal control in hummingbirds. The other vocally activated telencephalic structures, in contrast, appear to correspond to cortico-like structures also present in songbirds and parrots (see Fig. 10 for further detail). *zenk* expression levels in all vocally activated brain nuclei are proportional to the number of song bouts produced by the birds during the vocalizing period (Jarvis et al., 2000).

The data above provide a complete map of *zenk* expression associated with vocal communication in hummingbirds, and a first anatomical and functional demonstration of vocal control nuclei in this avian order. They also illustrate the power of utilizing the IEG expression approach to map brain systems involved in particular behaviors, particularly when no previous knowledge is available about such systems. It is worthwhile pointing out that this was accomplished with wild animals, without disturbing their behavior, and that a single study allowed the mapping of the entire brain, at cellular resolution. The results provide a starting point for exploring in detail the connectivity and electrophysiology of the vocal control system in hummingbirds. A recent study utilizing neuroanatomical tract-tracing methods has recently provided some initial connectivity data on vocal control pathways of another hummingbird species, *Calypte anna* (Gahr, 2000); how the nuclei thus identified relate to *zenk* expression patterns remains to be explored.

7.3. VOCAL NON-LEARNERS

Previous studies using Nissl-staining and tract-tracing methods have failed to demonstrate the existence of telencephalic vocal control areas in avian vocal non-learning species such as chicken (Kuenzel and Masson, 1988), pigeon (Karten and Hodos, 1967; Wild et al., 1997b), the eastern phoebe, a suboscine (Kroodsma and Konishi, 1991) and others (Gahr, 2000). *zenk* expression mapping during vocal communication likely represents a more sensitive test for whether telencephalic areas participate in vocal control in a given species, but such data are presently lacking in vocal non-learners. The possible detection of such areas would represent a significant challenge to the notion that telencephalic vocal control centers are present only in vocal learners (Brenowitz, 1997). Of interest, ZENK protein expression analysis with ICC has

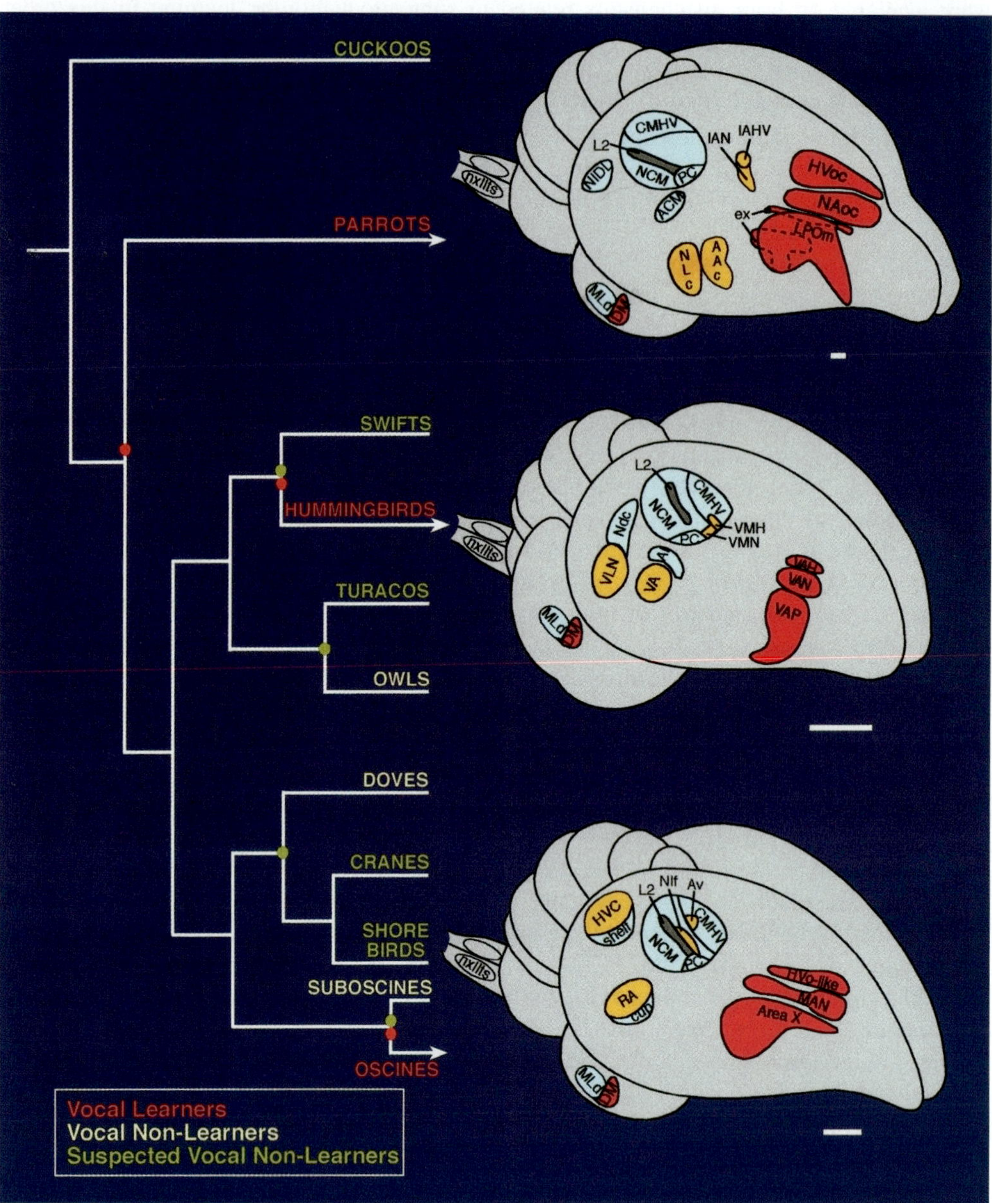

Fig. 10. Comparison of hearing- and vocalizing-induced *zenk* expression maps across vocal learning avian orders (from Jarvis et al., 2000). The tree diagram on the left represents phylogenetic relationships across extant avian orders; one common species name per order is represented. Nodes of possible independent gain of vocal learning are indicated by red dots, nodes of possible independent loss of vocal learning are indicated by green dots. The panels on the right represent semi-3D renditions of the brains of vocal learners. Regions of hearing-induced *zenk* expression are depicted in blue, regions of vocalizing-induced expression are depicted in red (similar positions across orders) or yellow (different positions across orders). Scale bar, 1 mm; for abbreviations, see list.

been utilized for the mapping of brain areas associated with sexual behavior in the Japanese quail (Ball et al., 1997). That study revealed activation of the nucleus of the stria terminalis, the intercollicular nucleus, and portions of the hyperstriatum ventrale in the telencephalon in the context of copulatory behavior, and activation of the latter areas alone in the context of appetitive (proximity) behavior. Vocalizations associated with copulatory and appetitive behaviors may have played some role in determining activation of the hyperstriatum ventrale, but that possibility was not directly examined.

7.4. CONCLUSION

The comparative *zenk* expression analysis described above has revealed several striking similarities in the organization of brain systems for vocal communication in independent avian vocal learning orders. Several of the features revealed (particularly the presence of seven telencephalic vocal control nuclei, their division into anterior and posterior clusters, and the participation of cortico-like and basal ganglia elements) are common across the three orders and may represent basic requirements for a vocal control system capable of vocal learning (Fig. 10). Whether these are general features of vocal learning systems in vertebrates (i.e. comparable structures are also present in mammalian vocal learning orders), and whether they are uniquely associated with vocal learning (i.e. comparable structures are absent in avian non-learners) are questions that remain to be answered.

8. OTHER GENES

8.1. c-fos

ICC has been utilized to study whether song activates expression of c-fos, another activity-dependent IEG that has been extensively studied in mammalian systems (Morgan and Curran, 1991; Herrera and Robertson, 1996); probes for in situ analysis of c-fos expression in songbirds are not yet available. No significant c-fos induction is observed in song control nuclei upon auditory stimulation (Kimpo and Doupe, 1997), which is in line with the *zenk* findings discussed in Section 4.3. Interestingly, c-fos induction is also not observed in NCM upon song presentation, suggesting that different classes of genes are differentially regulated by exposure to song (but see Bolhuis et al., 2000). In contrast, c-fos was shown to be strongly activated in song control nuclei HVC and RA in association with singing behavior (Kimpo and Doupe, 1997). As is the case for *zenk*, song-induced c-fos expression in song nuclei is independent of auditory feedback. The effect of singing context on c-fos expression has not been examined.

8.2. c-jun

Together with c-fos, c-jun comprises the activator protein-1 (AP-1) transcriptional activity, a known regulator of specific gene expression in the brain (Sonnenberg et al., 1989). The canary c-jun homologue was cloned and shown to be strongly activated throughout the brain by depolarization (metrazole) and weakly so in NCM by song presentation (Nastiuk et al., 1994). The detected song induction of c-jun was of the order of 1.5–1.7 fold, which characterizes this gene as much less sensitive than *zenk* for the mapping of song-activated brain areas. If confirmed, the apparent absence of c-fos induction in NCM would provide an interesting example of separate regulation of the two components of the AP-1 transcriptional activity. It remains to be tested whether c-jun is regulated in song control nuclei during song production.

8.3. OTHERS

Direct screening efforts involving as differential cDNA library screening and differential display have been utilized in an effort to identify other song-regulated genes (Mello et al., 1997). Preliminary evidence indicates the existence of early song-inducible genes whose expression patterns differ significantly from that of *zenk* (C. Mello and R. Malcher, unpublished observations); their identity has not yet been determined. In order to reveal more completely the genomic cascade triggered by song, it will be important to focus further efforts on the detection of late song-regulated genes, some of which may be direct targets of early transcriptional regulators such as *zenk*. It is also of relevance to mention here that BDNF, a protein with established functions both as a neurotrophin and as a regulator of synaptic transmission, is induced in NCM by song presentation and in song control nuclei in association with singing behavior (Li et al., 2000). The time course of BDNF regulation by song is protracted relative to *zenk*, consistent with the notion that BDNF represents a late effector song-regulated gene. The functional consequences of BDNF regulation by song remain to be determined.

9. SUMMARY AND PERSPECTIVES

The utilization of *zenk* expression mapping has revealed novel and unsuspected aspects of the functional organization of brain systems for vocal communication in birds. Analysis of *zenk* expression in response to presentation of song auditory stimuli has revealed the existence of a set of brain areas distinct from the song control system that are activated in association with hearing song. These areas form a system of interconnected nuclei whose organization resembles in many respects that of auditory cortical structures in mammals. Large areas of the songbird forebrain that precede the song control system are therefore likely involved in song auditory processing and perception, and possibly the formation of song auditory memories. Some of these stations project to song control nuclei, providing potential sources of auditory input to the song system. Information processed in these auditory brain regions can also potentially be fed to brain areas that subserve non-vocal aspects of the bird's response to song. A detailed quantitative and topographical analysis of the *zenk* expression patterns induced by song presentation in one auditory processing area (NCM) has revealed a possible basis for a syllabic representation in the songbird brain. The absence of induced gene expression in song control nuclei in association with song presentation argues against a direct participation of this system in perceptual aspects of vocal communication in songbirds.

Analysis of *zenk* expression to map activation within the song control system in association with singing behavior has also generated new sets of intriguing questions and possibilities regarding mechanisms controlling vocal production and learning. Of particular interest, this methodology has provided strong evidence for the active participation of the anterior forebrain pathway in the act of singing, as well as for the influence of the context of singing on brain activation patterns. A systematic application of *zenk* expression analysis in a comparative context has helped identify which features of the avian vocal control system are likely to be critical for vocal learning. Finally, the experiments with freely ranging animals demonstrate the feasibility of mapping brain pathways associated with complex behaviors that are of evolutionary relevance and difficult to reproduce out of a natural context.

10. ABBREVIATIONS

A	archistriatum
AAc	central nucleus of the anterior archistriatum
AC	anterior commissure
ACM	caudomedial neostriatum
Ai	intermediate archistriatum
Area X	area X of the paleostriatum
Av	nucleus avalanche
Cb	cerebellum
CHV	caudal hyperstriatum ventrale
clHV and cmHV	caudolateral and caudomedial hyperstriatum ventrale
DLM	medial nucleus of the dorsolateral thalamus
DM	dorsomedial nucleus of ICo
DMm	magnocellular nucleus of the dorsomedial thalamus
DMP	dorsomedial nucleus of the thalamus
ex	extensions of LPOm, H, hyperstriatum
Hp	hippocampus
HV	hyperstriatum ventrale
HVo	complex including the oval nucleus of the hyperstriatum ventrale and surrounds
ICo	intercollicular nucleus of the mesencephalon
(field) L	primary telencephalic auditory area
L1, L2a, L2b and L3	field L subdivisions
lAHV	lateral nucleus of the anterior hyperstriatum ventrale
lAN	lateral nucleus of the anterior neostriatum
LH	lamina hyperstriatalis
LPO	parolfactory lobe
LPOm	magnocellular nucleus of the parolfactory lobe
l and mMAN	lateral and medial subdivisions of the magnocellular nucleus of the anterior neostriatum
MLd	dorsal part of the lateral mesencephalic nucleus;
NAoc	complex including the oval nucleus of the anterior neostriatum and surrounds
NCM	caudomedial neostriatum
Ndc	dorsocaudal neostriatum
NIDL	neostriatum intermedium pars dorsolateralis
NIf	nucleus interfacialis
NLc	central nucleus of the lateral neostriatum
nXIIts	tracheosyringeal subdivision of the hypoglossal nucleus
Ov	nucleus ovoidalis of the thalamus
Pc	caudal paleostriatum
RA	robust nucleus of the archistriatum
Uva	nucleus uvaeformis
VA	vocal nucleus of the archistriatum
VAH	vocal nucleus of the anterior hyperstriatum ventrale
VAN	vocal nucleus of the anterior neostriatum
VAP	vocal nucleus of the anterior paleostriatum

VLN	vocal nucleus of the lateral neostriatum
VMH	vocal nucleus of the medial hyperstriatum ventrale
VMN	vocal nucleus of the medial neostriatum

11. REFERENCES

Abraham WC, Mason SE, Demmer J, Williams JM, Richardson CL, Tate WP, Lawlor PA, Dragunow M (1993): Correlations between immediate early gene induction and the persistence of long-term potentiation. *Neuroscience 56*:717–727.

Aldridge JW, Berridge KC (1998): Coding of serial order by neostriatal neurons: a 'natural action' approach to movement sequence. *J Neurosci 18*:2777–2787.

Ang CW-Y (2001): *Emerging Auditory Selectivity in the Caudomedial Neostriatum of the Zebra Finch Songbird.* PhD Thesis. New York: The Rockefeller University.

Arnold AP (1997): Sexual differentiation of the zebra finch song system: positive evidence, negative evidence, null hypotheses, and a paradigm shift. *J Neurobiol 33*:572–584.

Bading H, Ginty DD, Greenberg ME (1993): Regulation of gene expression in hippocampal neurons by distinct calcium signaling pathways. *Science 260*:181–186.

Ball GF (1994): Neurochemical specializations associated with vocal learning and production in songbirds and budgerigars. *Brain, Behav Evol 44*:234–246.

Ball GF, Tlemcani O, Balthazart J (1997): Induction of the Zenk protein after sexual interactions in male Japanese quail. *Neuroreport 8*:2965–2970.

Bankes SC, Margoliash D (1993): Parametric modeling of the temporal dynamics of neuronal responses using connectionist architectures. *J Neurophysiol 69*:980–991.

Baptista LF, Schuchmann KL (1990): Song learning in the anna hummingbird (*Calypte anna*). *Ethology 84*:15–26.

Bartel DP, Sheng M, Lau LF, Greenberg ME (1989): Growth factors and membrane depolarization activate distinct programs of early response gene expression: dissociation of fos and jun induction. *Genes Dev 3*:304–313.

Benowitz LI, Karten HJ (1976): Organization of the tectofugal visual pathway in the pigeon: a retrograde transport study. *J Comp Neurol 167*:503–520.

Bolhuis JJ, Zijlstra GG, den Boer-Visser AM, Van Der Zee EA (2000): Localized neuronal activation in the zebra finch brain is related to the strength of song learning. *Proc Natl Acad Sci USA 97*:2282–2285.

Bonke BA, Bonke D, Scheich H (1979a): Connectivity of the auditory forebrain nuclei in the guinea fowl (*Numida meleagris*). *Cell Tissue Res 200*:101–121.

Bonke BA, Scheich H, Langner G (1979b): Responsiveness of units in the auditory neostriatum of the guinea fowl (*Numida meleagris*) to species-specific calls and synthetic stimuli. I. Tonotopy and functional zones of field L. *J Comp Physiol 132*:243–255.

Bottjer SW, Miesner EA, Arnold AP (1984): Forebrain lesions disrupt development but not maintenance of song in passerine birds. *Science 224*:901–903.

Bottjer SW, Halsema KA, Brown SA, Miesner EA (1989): Axonal connections of a forebrain nucleus involved with vocal learning in zebra finches. *J Comp Neurol 279*:312–326.

Bottjer SW, Brady JD, Cribbs B (2000): Connections of a motor cortical region in zebra finches: relation to pathways for vocal learning. *J Comp Neurol 420*:244–260.

Braun K, Scheich H, Heizmann CW, Hunziker W (1991): Parvalbumin and calbindin-D28K immunoreactivity as developmental markers of auditory and vocal motor nuclei of the zebra finch. *Neuroscience 40*:853–869.

Brauth SE, McHale CM (1988): Auditory pathways in the budgerigar. II. Intratelencephalic pathways. *Brain, Behav Evol 32*:193–207.

Brauth SE, McHale CM, Brasher CA, Dooling RJ (1987): Auditory pathways in the budgerigar. I. Thalamo-telencephalic projections. *Brain, Behav Evol 30*:174–199.

Brauth SE, Heaton JT, Durand SE, Liang W, Hall WS (1994): Functional anatomy of forebrain auditory pathways in the budgerigar (*Melopsittacus undulatus*). *Brain, Behav Evol 44*:210–233.

Brauth SE, Heaton JT, Shea SD, Durand SE, Hall WS (1997): Functional anatomy of forebrain vocal control pathways in the budgerigar (*Melopsittacus undulatus*). *Ann NY Acad Sci 807*:368–385.

Brenowitz EA (1991): Altered perception of species-specific song by female birds after lesions of a forebrain nucleus. *Science 251*:303–305.

Brenowitz EA (1997): Comparative approaches to the avian song system. *J Neurobiol 33*:517–531.

Brenowitz EA, Margoliash D, Nordeen KW (1997a): An introduction to birdsong and the avian song system. *J Neurobiol 33*:495–500.

Brenowitz EA, Margoliash D, Nordeen KW (1997b): The neurobiology of birdsong. *J Neurobiol 33*: special issue.

Brooks RJ, Falls JB (1975a): Individual recognition by song in white-throated sparrows. I. Discrimination of songs of neighbours and strangers. *Can J Zool 53*:879–888.

Brooks RJ, Falls JB (1975b): Individual recognition by song in white-throated sparrows. III. Song features used in individual recognition. *Can J Zool 53*:1749–1761.

Butler AB, Hodos W (1996): *Comparative Vertebrate Neuroanatomy: Evolution and Adaptation*. New York, NY: Wiley–Liss.

Catchpole CK, Slater PJB (1995): *Bird Song: Biological Themes and Variations*. Cambridge: Cambridge University Press.

Cecchi GA, Ribeiro S, Mello CV, Magnasco MO (1999): An automated system for the mapping and quantitative analysis of immunocytochemistry of an inducible nuclear protein. *J Neurosci Methods 87*:147–158.

Chaudhuri A (1997): Neural activity mapping with inducible transcription factors. *Neuroreport 8*:iii–vii.

Chew SJ, Mello C, Nottebohm F, Jarvis E, Vicario DS (1995): Decrements in auditory responses to a repeated conspecific song are long-lasting and require two periods of protein synthesis in the songbird forebrain. *Proc Natl Acad Sci USA 92*:3406–3410.

Chew SJ, Vicario DS, Nottebohm F (1996a): A large-capacity memory system that recognizes the calls and songs of individual birds. *Proc Natl Acad Sci USA 93*:1950–1955.

Chew SJ, Vicario DS, Nottebohm F (1996b): Quantal duration of auditory memories. *Science 274*:1909–1914.

Christy B, Nathans D (1989): DNA binding site of the growth factor-inducible protein Zif268. *Proc Natl Acad Sci USA 86*:8737–8741.

Christy BA, Lau LF, Nathans D (1988): A gene activated in mouse 3T3 cells by serum growth factors encodes a protein with 'zinc finger' sequences. *Proc Natl Acad Sci USA 85*:7857–7861.

Cirelli C, Tononi G (2000): Differential expression of plasticity-related genes in waking and sleep and their regulation by the noradrenergic system. *J Neurosci 20*:9187–9194.

Cirelli C, Pompeiano M, Tononi G (1996): Neuronal gene expression in the waking state: a role for the locus coeruleus. *Science 274*:1211–1215.

Clayton DF (1997): Role of gene regulation in song circuit development and song learning. *J Neurobiol 33*:549–571.

Clayton DF (2000): The genomic action potential. *Neurobiol Learn Mem 74*:185–216.

Clayton DF, Huecas ME, Sinclair-Thompson EY, Nastiuk KL, Nottebohm F (1988): Probes for rare mRNAs reveal distributed cell subsets in canary brain. *Neuron 1*:249–261.

Cole AJ, Saffen DW, Baraban JM, Worley PF (1989): Rapid increase of an immediate early gene messenger RNA in hippocampal neurons by synaptic NMDA receptor activation. *Nature 340*:474–476.

Cole AJ, Abu-Shakra S, Saffen DW, Baraban JM, Worley PF (1990): Rapid rise in transcription factor mRNAs in rat brain after electroshock-induced seizures. *J Neurochem 55*:1920–1927.

Dave AS, Yu AC, Margoliash D (1998): Behavioral state modulation of auditory activity in a vocal motor system. *Science 282*:2250–2254.

Davis HP, Squire LR (1984): Protein synthesis and memory: a review. *Psychol Bull 96*:518–559.

Day ML, Fahrner TJ, Aykent S, Milbrandt J (1990): The zinc finger protein NGFI-A exists in both nuclear and cytoplasmic forms in nerve growth factor-stimulated PC12 cells. *J Biol Chem 265*:15253–15260.

De Lanerolle N, Andrew RJ (1975): Midbrain structures controlling vocalization in the domestic chick. *Brain, Behav Evol 10*:354–376.

Denisenko-Nehrbass NI, Jarvis E, Scharff C, Nottebohm F, Mello CV (2000): Site-specific retinoic acid production in the brain of adult songbirds. *Neuron 27*:359–370.

Doupe AJ (1997): Song- and order-selective neurons in the songbird anterior forebrain and their emergence during vocal development. *J Neurosci 17*:1147–1167.

Doupe AJ, Konishi M (1991): Song-selective auditory circuits in the vocal control system of the zebra finch. *Proc Natl Acad Sci USA 88*:11339–11343.

Duffy DL, Bentley GE, Ball GF (1999): Does sex or photoperiodic condition influence ZENK induction in response to song in European starlings?. *Brain Res 844*:78–82.

Durand SE, Tepper JM, Cheng MF (1992): The shell region of the nucleus ovoidalis: a subdivision of the avian auditory thalamus. *J Comp Neurol 323*:495–518.

Durand SE, Heaton JT, Amateau SK, Brauth SE (1997): Vocal control pathways through the anterior forebrain of a parrot (*Melopsittacus undulatus*). *J Comp Neurol 377*:179–206.

Eales LA (1985): Song learning in zebra finches: some effects of song model availability on what is learnt and when. *Anim Behav 33*:1293–1300.

Farabaugh SM, Wild JM (1997): Reciprocal connections between primary and secondary auditory pathways in the telencephalon of the budgerigar (*Melopsittacus undulatus*). *Brain Res 747*:18–25.

Fortune ES, Margoliash D (1992): Cytoarchitectonic organization and morphology of cells of the field L complex in male zebra finches (*Taeniopygia guttata*). *J Comp Neurol 325*:388–404.

Fortune ES, Margoliash D (1995): Parallel pathways and convergence onto HVc and adjacent neostriatum of adult zebra finches (*Taeniopygia guttata*). *J Comp Neurol 360*:413–441.

Foster EF, Bottjer SW (1998): Axonal connections of the high vocal center and surrounding cortical regions in juvenile and adult male zebra finches. *J Comp Neurol 397*:118–138.

Foster EF, Mehta RP, Bottjer SW (1997): Axonal connections of the medial magnocellular nucleus of the anterior neostriatum in zebra finches. *J Comp Neurol 382*:364–381.

Gahr M (2000): Neural song control system of hummingbirds: comparison to swifts, vocal learning (Songbirds) and nonlearning (Suboscines) passerines, and vocal learning (Budgerigars) and nonlearning (Dove, owl, gull, quail, chicken) nonpasserines. *J Comp Neurol 426*:182–196.

Gehr DD, Capsius B, Grabner P, Gahr M, Leppelsack HJ (1999): Functional organisation of the field-L-complex of adult male zebra finches. *Neuroreport 10*:375–380.

Gentner TQ, Hulse SH (2000): Female european starling preference and choice for variation in conspecific male song. *Anim Behav 59*:443–458.

Gentner TQ, Hulse SH, Duffy D, Ball GF (2001): Response biases in auditory forebrain regions of female songbirds following exposure to sexually relevant variation in male song. *J Neurobiol 46*:48–58.

Ghosh A, Ginty DD, Bading H, Greenberg ME (1994): Calcium regulation of gene expression in neuronal cells. *J Neurobiol 25*:294–303.

Goelet P, Castellucci VF, Schacher S, Kandel ER (1986): The long and the short of long-term memory — a molecular framework. *Nature 322*:419–422.

Grisham W, Arnold AP (1994): Distribution of GABA-like immunoreactivity in the song system of the zebra finch. *Brain Res 651*:115–122.

Güttinger HR (1985): Consequences of domestication on the song structures in the canary. *Behaviour 94*:254–278.

Guzowski JF, Lyford GL, Stevenson GD, Houston FP, McGaugh JL, Worley PF, Barnes CA (2000): Inhibition of activity-dependent arc protein expression in the rat hippocampus impairs the maintenance of long-term potentiation and the consolidation of long-term memory. *J Neurosci 20*:3993–4001.

Hall WS, Cohen PL, Brauth SE (1993): Auditory projections to the anterior telencephalon in the budgerigar (*Melopsittacus undulatus*). *Brain, Behav Evol 41*:97–116.

Hall WS, Brauth SE, Heaton JT (1994): Comparison of the effects of lesions in nucleus basalis and field 'L' on vocal learning and performance in the budgerigar (*Melopsittacus undulatus*). *Brain, Behav Evol 44*:133–148.

Hauser MD, Konishi M (Eds) (1999): *The Design of Animal Communication*. Cambridge, MA: MIT Press.

Heil P, Scheich H (1991a): Functional organization of the avian auditory cortex analogue. II. Topographic distribution of latency. *Brain Res 539*:121–125.

Heil P, Scheich H (1991b): Functional organization of the avian auditory cortex analogue. I. Topographic representation of isointensity bandwidth. *Brain Res 539*:110–120.

Heil P, Scheich H (1992): Spatial representation of frequency-modulated signals in the tonotopically organized auditory cortex analogue of the chick. *J Comp Neurol 322*:548–565.

Herrera DG, Robertson HA (1996): Activation of c-fos in the brain. *Prog Neurobiol 50*:83–107.

Hessler NA, Doupe AJ (1999): Social context modulates singing-related neural activity in the songbird forebrain. *Nat Neurosci 2*:209–211.

Hunt SP, Pini A, Evan G (1987): Induction of c-fos-like protein in spinal cord neurons following sensory stimulation. *Nature 328*:632–634.

Immelmann K (1969): Song development in the zebra finch and other estrilid finches. In: Hinde RA (Ed), *Bird Vocalizations*. Cambridge: Cambridge University Press, pp 61–74.

Jarvis ED, Mello CV (2000): Molecular mapping of brain areas involved in parrot vocal communication. *J Comp Neurol 419*:1–31.

Jarvis, E.D., Nottebohm, F. (1997): Motor-driven gene expression. *Proc Natl Acad Sci USA* 94:4097–4102.

Jarvis ED, Schwabl H, Ribeiro S, Mello CV (1997): Brain gene regulation by territorial singing behavior in freely ranging songbirds. *Neuroreport 8*:2073–2077.

Jarvis ED, Scharff C, Grossman MR, Ramos JA, Nottebohm F (1998): For whom the bird sings: context-dependent gene expression. *Neuron 21*:775–788.

Jarvis ED, Ribeiro S, da Silva ML, Ventura D, Vielliard J, Mello CV (2000): Behaviourally driven gene expression reveals song nuclei in hummingbird brain. *Nature 406*:628–632.

Jin H, Clayton DF (1997): Localized changes in immediate-early gene regulation during sensory and motor learning in zebra finches. *Neuron 19*:1049–1059.

Johnson F, Norstrom E, Soderstrom K (2000): Increased expression of endogenous biotin, but not BDNF, in telencephalic song regions during zebra finch vocal learning. *Dev Brain Res 120*:113–123.

Jones MW, Errington ML, French PJ, Fine A, Bliss TV, Garel S, Charnay P, Bozon B, Laroche S, Davis S (2001): A requirement for the immediate early gene Zif268 in the expression of late LTP and long-term memories. *Nature Neuroscience 4*:289–296.

Kaplan IV, Guo Y, Mower GD (1996): Immediate early gene expression in cat visual cortex during and after the critical period: differences between EGR-1 and Fos proteins. *Mol Brain Res 36*:12–22.

Karten HJ (1967): The organization of the ascending auditory pathway in the pigeon (*Columba livia*). I. Diencephalic projections of the inferior colliculus (nucleus mesencephalic lateralis, pars dorsalis). *Brain Res 6*:409–427.

Karten HJ (1968): The ascending auditory pathway in the pigeon (*Columba livia*). II. Telencephalic projections of the nucleus ovoidalis thalami. *Brain Res 11*:134–153.

Karten HJ (1991): Homology and evolutionary origins of the 'neocortex'. *Brain, Behav Evol 38*:264–272.

Karten HJ, Dubbeldam JL (1973): The organization and projections of the paleostriatal complex in the pigeon (*Columba livia*). *J Comp Neurol 148*:61–89.

Karten HJ, Hodos W (1967): *A Stereotaxic Atlas of the Brain of the Pigeon (Columba livia)*. Baltimore, MD: Johns Hopkins Press.

Karten HJ, Shimizu T (1989): The origins of neocortex: connections and lamination as distinct events in evolution. *J Cogn Neurosci 1*:291–301.

Katz LC, Gurney ME (1981): Auditory responses in the zebra finch's motor system for song. *Brain Res 221*:192–197.

Kelley DB, Nottebohm F (1979): Projections of a telencephalic auditory nucleus-field L-in the canary. *J Comp Neurol 183*:455–469.

Kimpo RR, Doupe AJ (1997): FOS is induced by singing in distinct neuronal populations in a motor network. *Neuron 18*:315–325.

Konishi M (1965): The role of auditory feedback in the control of vocalization in the white-crowned sparrow. *Z Tierpsychol 22*:770–783.

Kroodsma DE, Konishi M (1991): A suboscine bird (eastern phoebe, *Sayornis phoebe*) develops normal song without auditory feedback. *Anim Behav 42*:477–487.

Kroodsma DE, Miller EH (1996): *Ecology and Evolution of Acoustic Communication in Birds*. Ithaca, NY: Cornell University Press.

Kruse AA, Stripling R, Clayton DF (2000): Minimal experience required for immediate-early gene induction in zebra finch neostriatum. *Neurobiol Learn Mem 74*:179–184.

Kuenzel WJ, Masson M (1988): *A Stereotaxic Atlas of the Brain of the Chick (Gallus domesticus)*. Baltimore, MD: Johns Hopkins University Press.

Lanfear J, Jowett T, Holland PW (1991): Cloning of fish zinc-finger genes related to KROX-20 and KROX-24. *Biochem Biophys Res Commun 179*:1220–1224.

Lemaire P, Revelant O, Bravo R, Charnay P (1988): Two mouse genes encoding potential transcription factors with identical DNA-binding domains are activated by growth factors in cultured cells. *Proc Natl Acad Sci USA 85*:4691–4695.

Li XC, Jarvis ED, Alvarez-Borda B, Lim DA, Nottebohm F (2000): A relationship between behavior, neurotrophin expression, and new neuron survival. *Proc Natl Acad Sci USA 97*:8584–8589.

Lidsky TI, Manetto C, Schneider JS (1985): A consideration of sensory factors involved in motor functions of the basal ganglia. *Brain Res 356*:133–146.

Long KD, Salbaum JM (1998): Evolutionary conservation of the immediate-early gene ZENK. *Mol Biol Evol 15*:284–292.

MacDougall-Shackleton SA, Hulse SH, Ball GF (1998): Neural bases of song preferences in female zebra finches (*Taeniopygia guttata*). *Neuroreport 9*:3047–3052.

Margoliash D (1983): Acoustic parameters underlying the responses of song-specific neurons in the white-crowned sparrow. *J Neurosci 3*:1039–1057.

Margoliash D (1986): Preference for autogenous song by auditory neurons in a song system nucleus of the white-crowned sparrow. *J Neurosci 6*:1643–1661.

Margoliash D (1997): Functional organization of forebrain pathways for song production and perception. *J Neurobiol 33*:671–693.

Margoliash D, Fortune ES (1992): Temporal and harmonic combination-sensitive neurons in the zebra finch's HVc. *J Neurosci 12*:4309–4326.

Marler P, Peters S (1977): Selective vocal learning in a sparrow. *Science 198*:519–521.

McCasland JS (1987): Neuronal control of bird song production. *J Neurosci 7*:23–39.

Mello C, Nottebohm F, Clayton D (1995): Repeated exposure to one song leads to a rapid and persistent decline in an immediate early gene's response to that song in zebra finch telencephalon. *J Neurosci 15*:6919–6925.

Mello CV (1993): *Analysis of Immediate Early Gene Expression in the Songbird Brain Following Song Presentation*. PhD Thesis. New York: The Rockefeller University.

Mello CV (1998): Auditory experience, gene regulation and auditory memories in songbirds. *J Braz Assoc Adv Sci 50*:189–196.

Mello CV, Clayton DF (1994): Song-induced ZENK gene expression in auditory pathways of songbird brain and its relation to the song control system. *J Neurosci 14*:6652–6666.

Mello CV, Clayton DF (1995): Differential induction of the ZENK gene in the avian forebrain and song control circuit after metrazole-induced depolarization. *J Neurobiol 26*:145–161.

Mello CV, Ribeiro S (1998): ZENK protein regulation by song in the brain of songbirds. *J Comp Neurol 393*:426–438.

Mello CV, Vicario DS, Clayton DF (1992): Song presentation induces gene expression in the songbird forebrain. *Proc Natl Acad Sci USA 89*:6818–6822.

Mello CV, Jarvis ED, Denisenko N, Rivas M (1997): Isolation of song-regulated genes in the brain of songbirds. *Methods Mol Biol 85*:205–217.

Mello CV, Pinaud R, Ribeiro S (1998a): Noradrenergic system of the zebra finch brain: immunocytochemical study of dopamine-beta-hydroxylase. *J Comp Neurol 400*:207–228.

Mello CV, Vates GE, Okuhata S, Nottebohm F (1998b): Descending auditory pathways in the adult male zebra finch (*Taeniopygia guttata*). *J Comp Neurol 395*:137–160.

Milbrandt J (1987): A nerve growth factor-induced gene encodes a possible transcriptional regulatory factor. *Science 238*:797–799.

Morgan JI, Curran T (1989): Stimulus–transcription coupling in neurons: role of cellular immediate-early genes. *Trends Neurosci 12*:459–462.

Morgan JI, Curran T (1991): Stimulus–transcription coupling in the nervous system: involvement of the inducible proto-oncogenes fos and jun. *Annu Rev Neurosci 14*:421–451.

Morgan JI, Cohen DR, Hempstead JL, Curran T (1987): Mapping patterns of c-fos expression in the central nervous system after seizure. *Science 237*:192–197.

Müller CM (1988): Distribution of GABAergic perikarya and terminals in the centers of the higher auditory pathway of the chicken. *Cell Tissue Res 252*:99–106.

Müller CM, Leppelsack HJ (1985): Feature extraction and tonotopic organization in the avian auditory forebrain. *Exp Brain Res 59*:587–599.

Nastiuk KL, Mello CV, George JM, Clayton DF (1994): Immediate-early gene responses in the avian song control system: cloning and expression analysis of the canary c-jun cDNA. *Mol Brain Res 27*:299–309.

Nguyen PV, Abel T, Kandel ER (1994): Requirement of a critical period of transcription for induction of a late phase of LTP. *Science 265*:1104–1107.

Nottebohm F (1972): The origins of vocal learning. *Am Nat 106*:116–140.

Nottebohm F (1981): A brain for all seasons: cyclical anatomical changes in song control nuclei of the canary brain. *Science 214*:1368–1370.

Nottebohm F (1996): The King Solomon Lectures in Neuroethology. A white canary on Mount Acropolis. *J Comp Physiol [A] 179*:149–156.

Nottebohm F (1999): The anatomy and timing of vocal learning in birds. In: Hauser MD, Konishi M (Eds), *The Design of Animal Communication*. Cambridge, MA: MIT Press, pp 63–110.

Nottebohm F, Arnold AP (1976): Sexual dimorphism in vocal control areas of the songbird brain. *Science 194*:211–213.

Nottebohm F, Nottebohm ME (1978): Relationship between song repertoire and age in the canary, *Serinus canaria*. *Z Tierpsychol 46*:298–305.

Nottebohm F, Stokes TM, Leonard CM (1976): Central control of song in the canary, *Serinus canarius*. *J Comp Neurol 165*:457–486.

Nottebohm F, Kelley DB, Paton JA (1982): Connections of vocal control nuclei in the canary telencephalon. *J Comp Neurol 207*:344–357.

Okuhata S, Saito N (1987): Synaptic connections of thalamo-cerebral vocal nuclei of the canary. *Brain Res Bull 18*:35–44.

Okuno H, Kanou S, Tokuyama W, Li YX, Miyashita Y (1997): Layer-specific differential regulation of transcription factors Zif268 and Jun-D in visual cortex V1 and V2 of macaque monkeys. *Neuroscience 81*:653–666.

Paton JA, Manogue KR, Nottebohm F (1981): Bilateral organization of the vocal control pathway in the budgerigar, *Melopsittacus undulatus*. *J Neurosci 1*:1279–1288.

Pavletich NP, Pabo CO (1991): Zinc finger-DNA recognition: crystal structure of a Zif268-DNA complex at 2.1 A. *Science 252*:809–817.

Ratcliffe L, Otter K (1996): Sex differences in song recognition. In: Kroodsma DE, Miller EH (Eds), *Ecology and Evolution of Acoustic Communication in Birds*. Ithaca, NY: Cornell University Press, pp 340–355.

Reiner A, Davis BM, Brecha NC, Karten HJ (1984): The distribution of enkephalin-like immunoreactivity in the telencephalon of the adult and developing domestic chicken. *J Comp Neurol 228*:245–262.

Reiner A, Medina L, Veenman CL (1998): Structural and functional evolution of the basal ganglia in vertebrates. *Brain Res Rev 28*:235–285.

Ribeiro S, Mello CV (2000): Gene expression and synaptic plasticity in the auditory forebrain of songbirds. *Learn Mem 7*:235–243.

Ribeiro S, Cecchi GA, Magnasco MO, Mello CV (1998): Toward a song code: evidence for a syllabic representation in the canary brain. *Neuron 21*:359–371.

Ribeiro S, Goyal V, Mello CV, Pavlides C (1999): Brain gene expression during REM sleep depends on prior waking experience. *Learn Mem 6*:500–508.

Roberts LA, Higgins MJ, O'Shaughnessy CT, Stone TW, Morris BJ (1996): Changes in hippocampal gene expression associated with the induction of long-term potentiation. *Mol Brain Res 42*:123–127.

Rusak B, Robertson HA, Wisden W, Hunt SP (1990): Light pulses that shift rhythms induce gene expression in the suprachiasmatic nucleus. *Science 248*:1237–1240.

Saffen DW, Cole AJ, Worley PF, Christy BA, Ryder K, Baraban JM (1988): Convulsant-induced increase in transcription factor messenger RNAs in rat brain. *Proc Natl Acad Sci USA 85*:7795–7799.

Saini KD, Leppelsack HJ (1981): Cell types of the auditory caudomedial neostriatum of the starling (*Sturnus vulgaris*). *J Comp Neurol 198*:209–229.

Saldanha CJ, Tuerk MJ, Kim YH, Fernandes AO, Arnold AP, Schlinger BA (2000): Distribution and regulation of telencephalic aromatase expression in the zebra finch revealed with a specific antibody. *J Comp Neurol 423*:619–630.

Scharff C, Nottebohm F (1991): A comparative study of the behavioral deficits following lesions of various parts of the zebra finch song system: implications for vocal learning. *J Neurosci 11*:2896–2913.

Scharff C, Nottebohm F, Cynx J (1998): Conspecific and heterospecific song discrimination in male zebra finches with lesions in the anterior forebrain pathway. *J Neurobiol 36*:81–90.

Scheich H, Bonke BA, Bonke D, Langner G (1979): Functional organization of some auditory nuclei in the guinea fowl demonstrated by the 2-deoxyglucose technique. *Cell Tissue Res 204*:17–27.

Schlinger BA (1997): The activity and expression of aromatase in songbirds. *Brain Res Bull 44*:359–364.

Schmidt MF, Konishi M (1998): Gating of auditory responses in the vocal control system of awake songbirds. *Nat Neurosci 1*:513–518.

Searcy WA, Nowicki S (1999): Functions of song variation in song sparrows. In: Hauser MD, Konishi M (Eds), *The Design of Animal Communication* Cambridge, MA: MIT Press, pp 577–595.

Searcy WA, Yasukawa K (1996): Song and female choice. In: Kroodsma DE, Miller EH (Eds), *Ecology and Evolution of Acoustic Communication in Birds*. Ithaca, NY: Cornell University Press, pp 455–473.

Searcy WA, McArthur D, Peters S, Marler P (1981): Response of male song and swamp sparrows to neighbour, stranger and self songs. *Behaviour 77*:152–163.

Sheng M, Greenberg ME (1990): The regulation and function of c-fos and other immediate early genes in the nervous system. *Neuron 4*:477–485.

Sohrabji F, Nordeen EJ, Nordeen KW (1990): Selective impairment of song learning following lesions of a forebrain nucleus in the juvenile zebra finch. *Behav Neural Biol 53*:51–63.

Solis MM, Doupe AJ (1999): Contributions of tutor and bird's own song experience to neural selectivity in the songbird anterior forebrain. *J Neurosci 19*:4559–4584.

Solis MM, Doupe AJ (2000): Compromised neural selectivity for song in birds with impaired sensorimotor learning. *Neuron 25*:109–121.

Sonnenberg JL, Rauscher FJ, Morgan JI, Curran T (1989): Regulation of proenkephalin by Fos and Jun. *Science 246*:1622–1625.

Sossinka R, Bohner J (1980): Song types in the zebra finch (*Poephila guttata castanotis*). *Z Tierpsychol 53*:123–132.

Stoddard PK (1996): Vocal recognition of neighbors by territorial passerines. In: Kroodsma DE, Miller EH (Eds), *Ecology and Evolution of Acoustic Communication in Birds*. Ithaca, NY: Cornell University Press, pp 356–374.

Striedter GF (1994): The vocal control pathways in budgerigars differ from those in songbirds. *J Comp Neurol 343*:35–56.

Striedter GF, Vu ET (1998): Bilateral feedback projections to the forebrain in the premotor network for singing in zebra finches. *J Neurobiol 34*:27–40.

Stripling R, Volman SF, Clayton DF (1997): Response modulation in the zebra finch neostriatum: relationship to nuclear gene regulation. *J Neurosci 17*:3883–3893.

Sukhatme VP, Cao XM, Chang LC, Tsai-Morris CH, Stamenkovich D, Ferreira PC, Cohen DR, Edwards SA, Shows TB, Curran T et al. (1988): A zinc finger-encoding gene coregulated with c-fos during growth and differentiation, and after cellular depolarization. *Cell 53*:37–43.

Swirnoff AH, Milbrandt J (1995): DNA-binding specificity of NGFI-A and related zinc finger transcription factors. *Mol Cell Biol 15*:2275–2287.

Tchernichovski O, Lints T, Mitra PP, Nottebohm F (1999): Vocal imitation in zebra finches is inversely related to model abundance. *Proc Natl Acad Sci USA 96*:12901–12904.

Thorpe WH (1958): The learning of song patterns by birds with special reference to the song of the chaffinch, *Fringilla coelebs*. *Ibis 100*:535–570.

Tischmeyer W, Grimm R (1999): Activation of immediate early genes and memory formation. *Cell Mol Life Sci 55*:564–574.

Tramontin AD, Brenowitz EA (2000): Seasonal plasticity in the adult brain. *Trends Neurosci 23*:251–258.

Vates GE, Nottebohm F (1995): Feedback circuitry within a song-learning pathway. *Proc Natl Acad Sci USA 92*:5139–5143.

Vates GE, Broome BM, Mello CV, Nottebohm F (1996): Auditory pathways of caudal telencephalon and their relation to the song system of adult male zebra finches. *J Comp Neurol 366*:613–642.

Vates GE, Vicario DS, Nottebohm F (1997): Reafferent thalamo-'cortical' loops in the song system of oscine songbirds. *J Comp Neurol 380*:275–290.

Vicario DS (1991): Organization of the zebra finch song control system: II. Functional organization of outputs from nucleus Robustus archistriatalis. *J Comp Neurol 309*:486–494.

Vicario DS (1993): A new brain stem pathway for vocal control in the zebra finch song system. *Neuroreport 4*:983–986.

Vicario DS, Simpson HB (1995): Electrical stimulation in forebrain nuclei elicits learned vocal patterns in songbirds. *J Neurophysiol 73*:2602–2607.

Vicario DS, Yohay KH (1993): Song-selective auditory input to a forebrain vocal control nucleus in the zebra finch. *J Neurobiol 24*:488–505.

Volman SF (1996): Quantitative assessment of song-selectivity in the zebra finch 'high vocal center'. *J Comp Physiol [A] 178*:849–862.

Wallace CS, Withers GS, Weiler IJ, George JM, Clayton DF, Greenough WT (1995): Correspondence between sites of NGFI-A induction and sites of morphological plasticity following exposure to environmental complexity. *Mol Brain Res 32*:211–220.

Whitney O, Soderstrom K, Johnson F (2000): Post-transcriptional regulation of zenk expression associated with zebra finch vocal development [In Process Citation]. *Mol Brain Res 80*:279–290.

Wild JM (1994): The auditory–vocal–respiratory axis in birds. *Brain, Behav Evol 44*:192–209.

Wild JM (1997): Neural pathways for the control of birdsong production. *J Neurobiol 33*:653–670.

Wild JM, Karten HJ, Frost BJ (1993): Connections of the auditory forebrain in the pigeon (*Columba livia*). *J Comp Neurol 337*:32–62.

Wild JM, Reinke H, Farabaugh SM (1997a): A non-thalamic pathway contributes to a whole body map in the brain of the budgerigar. *Brain Res 755*:137–141.

Wild JM, Li D, Eagleton C (1997b): Projections of the dorsomedial nucleus of the intercollicular complex (DM) in relation to respiratory–vocal nuclei in the brainstem of pigeon (*Columba livia*) and zebra finch (*Taeniopygia guttata*). *J Comp Neurol 377*:392–413.

Wild JM, Williams MN, Suthers RA (2000): Neural pathways for bilateral vocal control in songbirds. *J Comp Neurol 423*:413–426.

Williams H, Nottebohm F (1985): Auditory responses in avian vocal motor neurons: a motor theory for song perception in birds. *Science 229*:279–282.

Williams H, Staples K (1992): Syllable chunking in zebra finch (*Taeniopygia guttata*) song. *J Comp Psychol 106*:278–286.

Wingfield JC (1985): Short-term changes in plasma levels of hormones during establishment and defense of a breeding territory in male song sparrows, *Melospiza melodia*. *Horm Behav 19*:174–187.

Wisden W, Errington ML, Williams S, Dunnett SB, Waters C, Hitchcock D, Evan G, Bliss TV, Hunt SP (1990): Differential expression of immediate early genes in the hippocampus and spinal cord. *Neuron 4*:603–614.

Worley PF, Christy BA, Nakabeppu Y, Bhat RV, Cole AJ, Baraban JM (1991): Constitutive expression of zif268 in neocortex is regulated by synaptic activity. *Proc Natl Acad Sci USA 88*:5106–5110.

Worley PF, Bhat RV, Baraban JM, Erickson CA, McNaughton BL, Barnes CA (1993): Thresholds for synaptic activation of transcription factors in hippocampus: correlation with long-term enhancement. *J Neurosci 13*:4776–4786.

Xia Z, Dudek H, Miranti CK, Greenberg ME (1996): Calcium influx via the NMDA receptor induces immediate early gene transcription by a MAP kinase/ERK-dependent mechanism. *J Neurosci 16*:5425–5436.

Yu AC, Margoliash D (1996): Temporal hierarchical control of singing in birds. *Science 273*:1871–1875.

CHAPTER V

Molecular activity maps of sensory function

AVI CHAUDHURI AND SHAHIN ZANGENEHPOUR

1. INTRODUCTION

The immediate-early genes (IEGs) c-*fos* and *zif268* have become popular neurobiological tools for mapping functional activity in the brain. We now know that the expression of these genes is strongly, though not exclusively, linked to synaptic stimulation and that their products may be involved in key aspects of normal cellular function. Although much progress has been made toward understanding the intracellular processes that guide the expression of these genes, the precise physiological roles of the proteins encoded by them remain largely unknown. Many members of the IEG family are activated shortly after neuronal stimulation and without the requirement for de novo protein synthesis. The products of many of these genes have been observed in a wide range of tissues and under a variety of stimulation conditions (recently reviewed in Herrera and Robertson, 1996; Kaczmarek and Chaudhuri, 1997; Herdegen and Leah, 1998). The rapid accumulation of IEG products in activated neurons, combined with histological methods that offer detection at the cellular level, are key features that have led to their wide use in visualizing activated neurons.

Our objective in this chapter is to discuss much of the extant literature on how molecular mapping strategies have been used to address fundamental questions of sensory function in the mammalian nervous system. Throughout this chapter, we will use the standard convention of denoting genes and their mRNA products in italics (e.g., c-*fos*, *zif268*) and the proteins encoded by them with a single capital letter (e.g., c-Fos, Zif268). It should also be noted that a varied nomenclature is used to refer to *zif268*, reflecting the independent identification of the same gene from different laboratories. These include *egr-1*, NGFI-A, Krox-24, and a concatenation of these four names from their first letter — ZENK. We will refer to this gene by the identity used by authors of a particular study when it is being considered.

1.1. CHARACTERISTICS OF INDUCIBLE TRANSCRIPTION FACTORS

The various members of the IEG family are known to encode proteins with diverse cellular functions. In the brain, IEGs that are linked to neural activity and that have mapping applications generally encode proteins that serve as transcription factors, i.e., complexes that bind to the promoter regions of certain genes and then either activate or repress their expression. These so-called inducible transcription factors (ITFs) are distinguished from other proteins that normally reside in the nucleus and which also serve as regulators of ongoing transcriptional activity.

Handbook of Chemical Neuroanatomy Vol. 19: Immediate Early Genes and Inducible Transcription Factors in Mapping of the Central Nervous System Function and Dysfunction
L. Kaczmarek and H.A. Robertson, editors

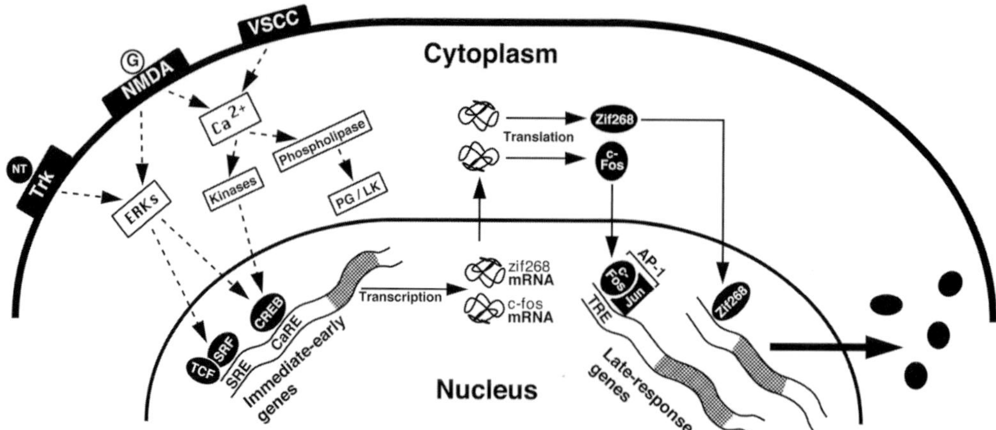

Fig. 1. Schematic representation of molecular pathways involved in induction of two immediate-early gene products, c-Fos and Zif268. Adapted from (Chaudhuri, 1997).

The ITFs c-Fos and Zif268 are both induced in neurons after extracellular stimulation by neurotransmitters and trophic substances (Fig. 1). The sequence of events that leads to ITF induction is largely coordinated by Ca^{2+} influx into the cell (Ghosh et al., 1994). This can occur either through the NMDA receptor–Ca^{2+} ionophore complex after glutamate binding or through voltage-sensitive calcium channels (VSCCs) following membrane depolarization. Thereafter, several different enzyme systems are marshaled by Ca^{2+}. These include various protein kinases that activate the transcription factor CREB (Sheng and Greenberg, 1990) and phospholipases that initiate the production of prostaglandins (PG) and leukotrienes (LK) (Lerea et al., 1995). NMDA receptor activation can also relay its effects through a second pathway involving extracellular signal-regulated kinases (ERKs) (Xia et al., 1996; Kaminska et al., 1999).

After transcription is completed, the ITF mRNAs are translated into a protein product (c-Fos and Zif268) in the cytosol. These products rapidly migrate into the nucleus where they themselves influence the expression of another set of genes, the late-response genes. In the case of c-Fos, it must dimerize with a member of the Jun phosphoprotein family (c-Jun, JunB, or JunD) to produce a functional transcription factor that is called activating-protein 1 (AP-1). By regulating the expression of a host of late-response genes, both AP-1 and Zif268 are able to have a commanding influence on short- and long-term cellular homeostasis. Regardless of what the specific molecular roles of ITFs may be, their accumulation in the neuron generally signifies a prior state of activity and thus forms the logic for obtaining functional maps based on ITF staining.

1.2. PROCEDURAL ASPECTS AND TECHNICAL CONSIDERATIONS

There are several technical approaches for studying IEG expression in the brain. In mapping studies, either in situ hybridization histochemistry (ISHH) or immunocytochemistry (ICC) may be employed to label the mRNA or protein products, respectively. The transient nature of IEG expression and the temporal differences between mRNA and protein accumulation are critical factors that affect labeling intensity. The two approaches for detecting IEG expression are not entirely equivalent in the information they represent. There may be instances where gene expression is induced leading to mRNA accumulation. However, the cellular processes necessary for translation into a protein product may not be engaged, leading to a situation where mRNA induction is not accompanied by a corresponding elevation of its encoded protein product. Even the cellular presence of the protein alone does not assure its functionality since this may rely on post-translational modification, such as phosphorylation and interaction with other partner proteins (Hunter and Karin, 1992; Karin and Hunter, 1995).

To obtain veridical labeling of a particular gene product, it is necessary to study expression at the mRNA level. Although ISHH provides considerable confidence for such detection, the dominant procedure for visualizing the signal relies on film autoradiography. This procedure suffers from poor spatial resolution and therefore cannot provide a direct display at the cellular level. Continuing advances in non-radioactive ISHH techniques, such as the recent development of tyramide signal amplification (TSA), should resolve this problem (Yang et al., 1999a; Holm, 2000). Northern blotting gives the most reliable identification of mRNA as well as quantitative assessment of its levels. However, this is not a mapping tool since its use relies on homogenized tissue as the starting material.

For mapping purposes, the procedural simplicity and cellular-level resolution that is offered by ICC has led to its widespread use. However, ICC applications have sometimes been complicated by the close homologies of proteins belonging to the same class. This has been especially problematic for the Fos family of proteins and was previously an obstacle to evaluating the expression of specific members since the specificity of the antigenic reaction was unclear. In the case of c-Fos, a generalized term — Fos-LI or FLI (Fos-like immunoreactivity) — has been used. The recent availability of antibodies that recognize epitopes specific to individual Fos members has largely resolved this problem. Throughout this chapter, we refer to c-Fos as the marker being detected, though in some older studies it is likely that some cross-reactivity would have stained for other Fos-family members, such as *fosB* and the fos-related antigens (Fra).

1.3. STRENGTHS AND CAVEATS OF MOLECULAR MAPPING

Molecular mapping strategies have distinct advantages over other functional imaging approaches and also suffer from some unique problems. The following are brief accounts of the key advantages offered by IEG activity mapping.

(1) *Rapid induction.* One of the striking advantages of IEG use in activity mapping is the short time course of induction. It appears that the intracellular mechanisms outlined in Fig. 1 operate in a very rapid manner such that mRNA levels are detectable in neurons within 20 min after stimulation. The time course for IEG protein expression is somewhat greater, most likely due to the added requirement for translation to be completed. However, depending on the particular brain structure examined, protein induction can be detected as early as 30 min after onset of stimulation and appears to peak at 60 to 90 min. The rapid induction of IEGs means that short stimulation times are sufficient, allowing its use in many behavioral situations.

(2) *Cellular resolution mapping.* Another key advantage is the superb spatial resolution of IEG activity maps. Histological labeling of IEG proteins, such as c-Fos and Zif268, produces staining that is confined to the nucleus. This provides a punctate visual display of immunostained neurons that actually extends the resolution even beyond the cellular level. This feature allows precise laminar and spatial localization, especially in complex structures or small nuclei within the brain.

(3) *Application with other histological techniques.* A particular advantage of the nuclear locus of the immunostained product is that double-labeling procedures can be employed

by counterstaining the same tissue section for various products that are confined to the cytoplasm. This allows one to correlate the expression of a particular IEG product with endogenous markers to reveal physiological or morphological features of activated neurons. Furthermore, molecular mapping can be used in tandem with tract-tracing procedures to identify input–output relationships.

(4) *Large-scale analysis.* Molecular mapping techniques require analysis of post-mortem tissue. Therefore, it is possible to undertake analysis on a large-scale basis to precisely identify the various cortical and subcortical brain structures that may be activated after sensory, behavioral, or pharmacological manipulation.

(5) *Behavioral assessment is straightforward.* As will be evident throughout this chapter, molecular mapping strategies have been used to explore neural activation after a variety of behavioral manipulations. The stimulation component of any experiment is quite straightforward because there is no need for a priori surgical, pharmacological, or anesthetic treatment.

The following are some of the disadvantages of molecular mapping approaches.

(1) *Temporal resolution.* The IEG induction process is not very dynamic and therefore much less sensitive to temporal changes than some other methods, such as electrophysiological recording and optical imaging. As a result, subtle changes in stimulation or behavioral treatment during an experiment can go undetected and become blurred in the resulting activity map.

(2) *Quantitative analysis.* While ICC staining provides good qualitative results, quantitative analysis is difficult because of the non-linear nature of substrate amplification and technical problems of accurately determining the signal strength. Some quantitation, however, is still possible by way of cell counts of immunostained or mRNA-stained neurons, allowing for a comparison of the effects of different stimulation or behavioral treatments.

(3) *Neuronal expression specificity.* It has been found that not all neurons express IEG products. For example, both c-*fos* and *zif268* are sparsely expressed in numerous subcortical structures and even in the neocortex appear to be largely restricted to excitatory neurons. Thus, activity maps do not reflect accumulation of induced IEG products across a broad range of neurons but rather only in a restricted set. The absence of labeling in a particular neural structure therefore cannot merely be taken as a lack of activation because certain IEGs may not be inducible in that structure.

(4) *False positives.* There have been several demonstrations of a dissociation between IEG expression and behavioral or physiological assay. In some cases, a positive IEG label has not been supported by other mapping techniques (e.g., 2-deoxyglucose accumulation) or behavioral condition of the animal (e.g., nociceptive response). False positive results have generally been attributed to other extraneous events, leading to the notion that neural activity does not have an exclusive association with IEG expression. Other factors, such as endocrine responses can sometimes be implicated in IEG induction that may either be in addition to or in lieu of the primary stimulus effects being mapped.

(5) *Multiple stimulation conditions.* It is extremely difficult to obtain activity maps in response to more than one stimulation sequence. This is because IEG products are unitary activity markers. If two compound or successive stimuli are applied, then neurons equally sensitive to either will express the IEG products in equal measure, making it difficult to differentiate the separate response to each stimulation sequence. Recent developments in double-labeling that exploit the differential time course of mRNA versus protein induction provide one solution to this problem. This will be discussed in Section 6.

2. THE SOMATOSENSORY SYSTEM

The use of immediate-early gene markers for studying somatosensory function can be traced to the very beginning of molecular mapping efforts in the brain. Much of this work has revolved around use of c-Fos, in particular detection of the protein through immunocytochemical procedures (Filipkowski, 2000). These studies have helped to clarify the roles of particular brain structures in somatosensory function by providing visualizable maps of neural activation. In general, molecular mapping efforts in the somatosensory system can be broadly categorized along three lines — direct tactile activation of somatosensory brain areas, activity maps after nociceptive stimulation, and neural substrates of complex somatosensory function such as sexual stimulation, maternal behavior, and stress responsivity.

2.1. MOLECULAR MAPS OF TACTILE STIMULATION

The rodent barrel cortex has been one of the preferred sites for ITF expression studies because of its large representation within the somatosensory cortex and the fact that physiological tactile stimulation can be accomplished easily and with precision through the whiskers. Early work with immediate-early genes had shown that induction of three transcription factors, NGFI-A, NGFI-B, and c-Fos, was generally related to electrical activity in the brain because of the large induction that was seen after seizures (Morgan et al., 1987; Saffen et al., 1988; Watson and Milbrandt, 1989; Mack et al., 1990). On the basis of this evidence, Mack and Mack (1992) examined the expression of all three ITFs in barrel cortex after 15 min of whisker stimulation. They found that this brief period of tactile stimulation was sufficient to induce all three ITFs in the contralateral barrel cortex after a period of 2 h. However, they noted a difference in the intensity of staining and laminar pattern of expression of the three transcription factors. Although immunostaining was most prominent in the thalamo-recipient layer IV, NGFI-A showed high expression in all other layers as well, whereas NGFI-B and c-Fos showed only sparse expression. The high basal expression of NGFI-A had already been described in an earlier in situ hybridization study of mRNA expression in somatosensory cortex (Schlingensiepen et al., 1991). Mack and Mack (1992) also noted that ITF induction after sensory stimulation was noticeably less intense than the expression pattern observed after seizure activity. In a subsequent study, Mack et al. (1995) showed that another ITF, NGFI-C, was induced in somatosensory cortex after both seizure activity and tactile stimulation, though the expression level was again significantly less in the latter condition.

The induction of ITFs after sensory stimulation revealed that expression of these factors can be rapidly up-regulated, possibly by way of neural activity. However, the converse process by which removal of activity down-regulates basal expression of an ITF has also been demonstrated in the barrel cortex. Steiner and Gerfen (1994) showed that clipping of whiskers on one side of the rat's snout produced an asymmetrical expression pattern in the barrel cortex of the two hemispheres. As shown in Fig. 2, the contralateral barrel cortex showed significantly reduced *zif268* mRNA expression in comparison to the ipsilateral cortex. This difference became especially marked after apomorphine, a dopamine receptor agonist, was injected systemically. Apomorphine produces increased orofacial motor activity with a corresponding emphasis on sniffing, whisking and licking (Szechtman et al., 1982; Beck et al., 1986).

The notion that changes in ITF expression that occurred after sensory stimulation and deprivation was linked to neural activity in barrel cortex was strengthened by studies that showed similar patterns in 2-deoxyglucose (2-DG) activity maps. A number of early reports

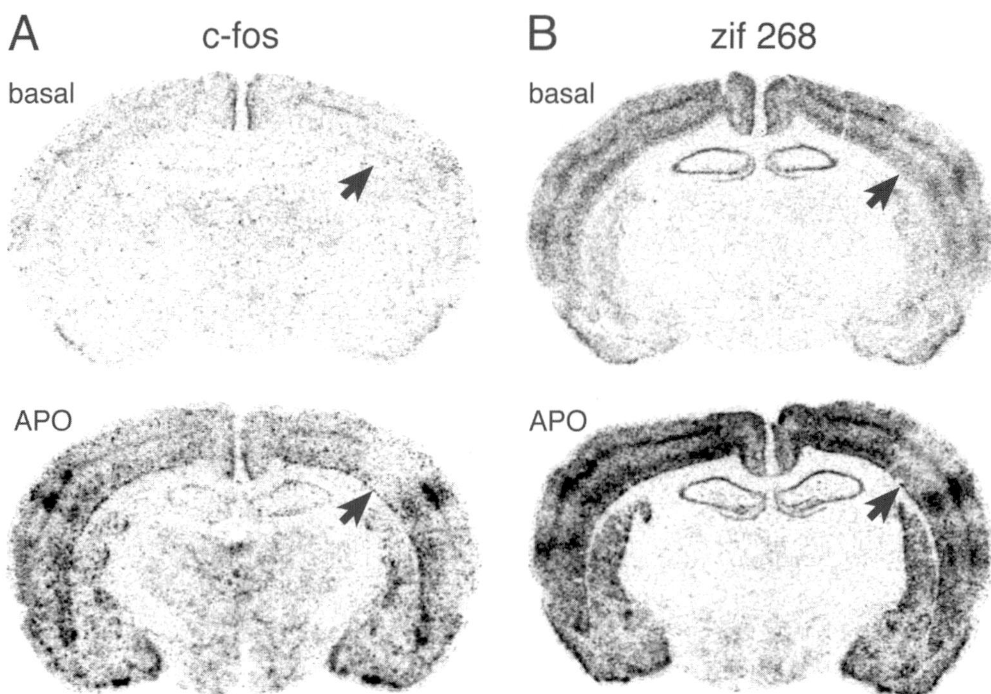

Fig. 2. Film autoradiograms depicting expression of c-*fos* (*A*) and *zif268* (*B*) mRNA in coronal sections at the level of the rat barrel cortex. Animals had their mystacial whiskers clipped on the left side of the snout. Four hours after whisker removal, they were either killed (basal, top), or received an injection of the D1/D2 dopamine receptor agonist apomorphine (5 mg/kg, 1 h survival; APO, bottom) (Steiner and Gerfen, 1994). Clipping the mystacial whiskers resulted in decreased basal levels of *zif268* mRNA in the corresponding part of the contralateral barrel cortex (arrow). Dopamine agonists such as apomorphine produce increased expression of c-*fos* and *zif268* throughout most of the cortex. In the sensory-deprived part of the barrel cortex, this increase was blocked, demonstrating that such gene regulation is dependent on sensory input. Courtesy of H. Steiner, Finch University of Health Sciences/Chicago Medical School.

had documented increased 2-DG accumulation after whisker stimulation and decrease after whisker removal (Durham and Wollsey, 1978; Melzer et al., 1985; Chmielowska et al., 1986). The fact that both up- and down-regulation can be visualized in 2-DG studies could be used as a benchmark for ITF mapping in somatosensory cortex. The study of Steiner and Gerfen (1994) clearly showed that ongoing neural activity was responsible for maintaining basal *zif268* expression in barrel cortex. However, without a detectable basal level, down-regulation effects would clearly not be observable. This appears to be the case with c-Fos. Steiner and Gerfen (1994) found that although deprivation effects were quite clear for *zif268*, they were less so for c-*fos* whose basal expression level is considerably lower (see Fig. 2). It was only through the stimulatory effects of apomorphine, which result in elevated basal c-*fos* mRNA levels, that the consequences of whisker removal could be clearly visualized.

The details of stimulus-induced c-*fos* and *zif268* expression have been further clarified in recent studies. Melzer and Steiner (1997) performed magnetic stimulation of whiskers equipped with metal filaments and then examined both c-*fos* and *zif268* mRNA levels. They found that brief whisker stimulation (5–15 min) produced increased expression of both products in radial columns that spanned barrels representing the stimulated whiskers. Although gene expression was generally increased throughout all layers, they found that

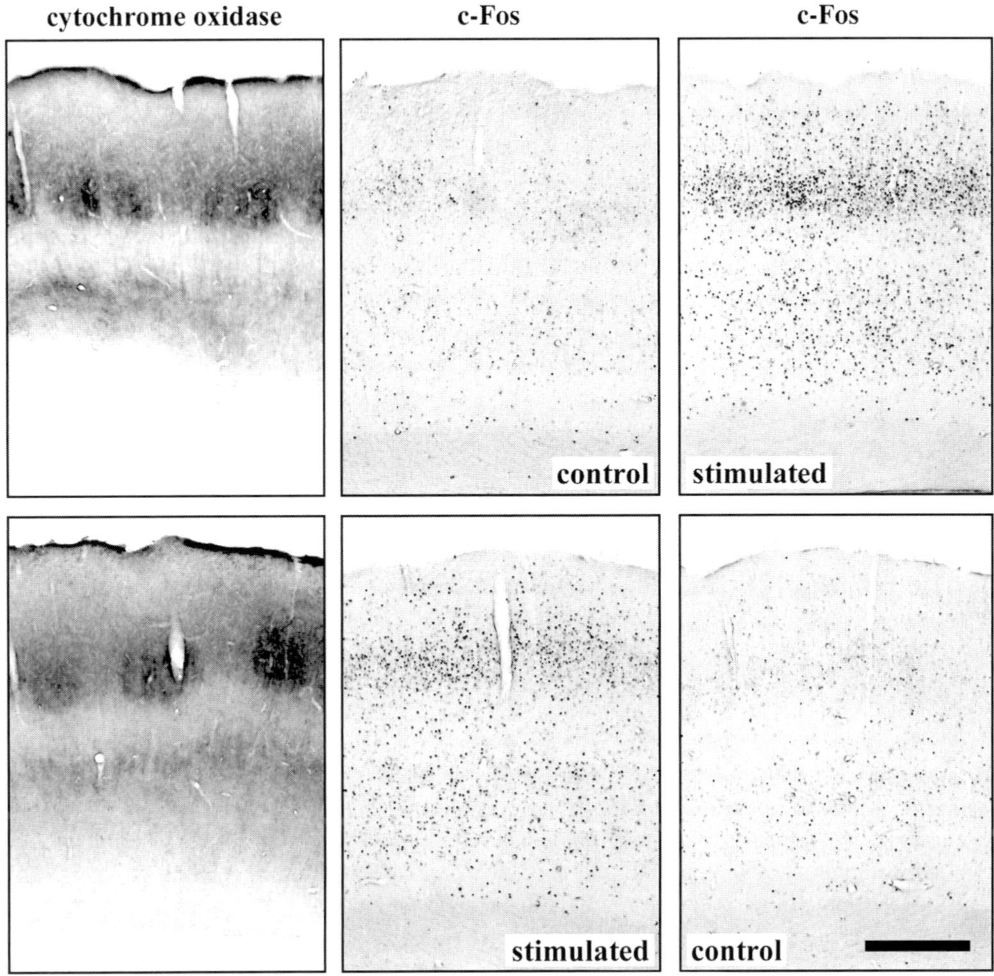

Fig. 3. Immunocytochemical staining for c-Fos protein shows an increase in expression in rat barrel cortex corresponding to stimulated whiskers. The panels show both left- and right-hemisphere sections of two rats which had their whiskers brushed on the left side (top) and the right side (bottom) of the snout. Cytochrome oxidase staining of neighboring sections are included to show the barrel profiles. Calibration bar 0.5 mm. Courtesy of R. Filipkowski and L. Kaczmarek.

maximal induction occurred in layer IV. Furthermore, it appeared that staining was largely confined to a subtype of stellate cells in layer IV and pyramidal cells in other layers. This result was corroborated by Filipkowski et al. (2000) who undertook an immunocytochemical study for c-Fos detection in awake, behaving rats whose whiskers were stimulated either by manual brushing or self-movement in a caged environment (Fig. 3). Filipkowski et al. (2000) also found that a subpopulation of neurons express c-Fos and that they were not likely to be inhibitory neurons based on the lack of co-occurrence with parvalbumin expression, a marker for a subset of GABAergic neurons. It would thus appear that ITF expression in somatosensory neocortex shows a similar preference for excitatory neurons that had been shown earlier for visual neocortex (Chaudhuri et al., 1995).

Another aspect of somatosensory function that has been explored by molecular mapping techniques is the interaction between the basal ganglia and neocortex. It has been recently shown that stimulation of D1 dopamine receptors in the striatum exerts a widespread fascilitatory effect on cortical function (Steiner and Kitai, 2000). The systemic administration of apomorphine resulted in increased c-*fos* and *zif268* expression in both striatum and throughout most of neocortex, including whisker stimulation-evoked expression in the barrel cortex. Infusion of a selective D1 receptor antagonist in the striatum blocked this response. The corresponding influence of cortical signals upon the striatum has been explored by Parthasarathy and Graybiel (1997). An unanswered question in this field is how cortical inputs are coordinated to bring about activation in striatal neurons. Parthasarathy and Graybiel (1997) employed focal microstimulation of sensorimotor cortex to show that clusters of neurons in the putamen contained increased levels of Fos- and Jun B-like proteins. Their molecular mapping study showed that modular activation of striatal neurons appeared to be consistent with anatomical input fiber clusters.

2.2. MOLECULAR MAPS OF NOCICEPTIVE STIMULATION

The use of c-Fos expression in nociception studies has provided much information on the neural substrates of pain processing (recently reviewed in Herdegen and Zimmermann, 1995; Harris, 1998). Much of the early work in this area concerned nociceptive networks in the spinal cord (Hunt et al., 1987; Bullitt, 1991; Herdegen et al., 1991). It was shown that positively stained neurons were primarily localized to laminae I, II, V and VI of the dorsal horn, thereby corresponding to the terminal fields of nociceptive afferent fibers as well as the loci of nociresponsive neurons identified by electrophysiological recordings (Hunt et al., 1987; Bullitt, 1991). The distribution pattern of tactile neurons activated by non-noxious stimulation, on the other hand, was localized to laminae II, III, and IV (Hunt et al., 1987; Jasmin et al., 1994a). Further work in this area with molecular mapping techniques has sought to clarify the distribution patterns of nociceptive neurons activated by different forms of noxious stimulation (Williams et al., 1990b; Lima et al., 1993; Meng and Bereiter, 1996).

Spinal mapping of nociceptive function has also been explored with regard to putative pain pathways in neonatal animals and the influence of analgesic agents. It has been shown that c-Fos can be induced in neonates in response to noxious stimulation (Williams et al., 1990a; Yi and Barr, 1995). Recent work in this area has been directed at mapping neural clusters that respond to stimulation through specific sets of afferent fibers, residual effects of nerve crush on spinal networks, and interactions between different pain-inducing agents (Jennings and Fitzgerald, 1996; Peterson et al., 1997; Sugimoto et al., 2000). The demonstration that pain-evoked neural activation can be visualized by c-Fos staining has led to numerous studies on the effects of analgesic agents on nociceptive neuron function at the spinal cord level (Tolle et al., 1990; Liu et al., 1997; Bhandari et al., 1999; Lin et al., 2000). One important question in this area has been the influence of selective analgesics on particular populations of spinal nociceptive neurons and how different neural populations may mediate chronic pain and hyperalgesia. For example, both morphine and NMDA antagonists exert powerful analgesic effects but appear to do so through different subpopulations of neurons. Morphine suppresses nociceptive responses primarily in the deep layers, whereas NMDA antagonists are more effective in the superficial layers (Presley et al., 1990; Jasmin et al., 1994b).

The association between c-Fos expression in the spinal cord and pain perception, as gauged by behavioral assessment, does not always appear to be absolutely firm. There have been reports that c-Fos expression levels under some circumstances do not correspond to the behavioral pattern, being either lower or higher than expected on the basis of pain behavior

(Presley et al., 1990; Harris et al., 1995). c-Fos expression patterns have also been shown to be inconsistent after applications of various anesthetics and analgesics which are well known to reduce pain signal transmission and yet continue to yield elevated expression (Yashpal et al., 1998; Gilron et al., 1999a,b). The accumulating evidence that c-Fos expression is not always predictive of behavioral analgesia and nociception has led to the suggestion that c-Fos induction may reflect processes that do not contribute directly to pain transmission (Harris, 1998).

Despite these reservations in the spinal cord, both c-Fos and Zif268 expression have been used to map nociceptive function in a number of subcortical brain structures. These include the thalamus, hypothalamus, and various brainstem nuclei (Bullitt, 1989; Lanteri-Minet et al., 1994; Pan et al., 1994; Carstens et al., 1995; Hermanson and Blomqvist, 1997; Baulmann et al., 2000). Molecular mapping has allowed a precise delineation of the neuronal populations that are activated by nociceptive stimulation in brain structures that are ordinarily too small to explore in a reliable way by electrophysiological means. Another advantage of using molecular mapping strategies for these structures is that nociceptive responses have been identified in nuclei that are not normally associated with pain processing, such as the locus coeruleus, anterior pituitary and superior colliculus (Bullitt, 1990; Pan et al., 1994; Smith and Day, 1994; Telford et al., 1996; Wang et al., 2000). The findings in the superior colliculus have confirmed and extended prior electrophysiological findings. Furthermore, these studies have provided more precise laminar localization of nociceptive responses and details of expression differences in awake versus anesthetized conditions (Telford et al., 1996; Wang et al., 2000).

There has been much less use of molecular mapping techniques for visualizing nociceptive function in somatosensory neocortex in comparison to the spinal cord or subcortical structures. The fact that nociceptive signals are represented in primary somatosensory cortex (area S1) is well established through electrophysiological studies (Jinghong and Guoxi, 1999; Kenshalo et al., 2000). The only comprehensive molecular mapping study to date in somatosensory cortex concerned the effects of repetitive nociceptive and tactile stimulation. Anand et al. (1999) showed that c-Fos expression was significantly reduced in somatosensory cortex of neonatal rats that had long-term exposure to nociceptive stimulation, suggesting that neuroplastic changes altered threshold levels of activation in these animals. However, there have been studies that showed elevated c-Fos expression in other neocortical areas, such as temporal and prefrontal cortex in response to nociceptive stimulation (Hirakawa and Kawata, 1993; Baulmann et al., 2000).

2.3. NEURAL SUBSTRATES OF COMPLEX SOMATOSENSORY FUNCTION

Inducible transcription factors have also been used to explore complex aspects of somatosensory behavior. For example, it is known that neuronal activity is triggered in selective brain sites in response to sexual stimulation. The molecular consequences of sexual behavior in a number of brain areas result from both synaptic and neuroendocrine stimulation, which in turn leads to induction of immediate-early genes (reviewed by Pfaus and Heeb, 1997). The use of c-Fos as a tool for mapping the neural substrates of sexual activity emerged quite early from the demonstration that it is rapidly induced during copulatory behavior (Robertson et al., 1991; Bialy et al., 1992). It was known that forebrain areas such as the nucleus accumbens (NAc) and medial preoptic area (MPOA) were susceptible to dopamine release that occurred with sexual behavior. Molecular mapping studies have shown that c-Fos expression increases markedly in both NAc and MPOA, confirming the notion that these areas are important substrates of sexual behavior (Bialy et al., 1992; Pfaus and Heeb, 1997; Veening and Coolen, 1998; Meddle et al., 1999; Yang et al., 1999b). Molecular mapping in conjunction with olfactory and somatosensory desensitization has been used to clarify the role of the MPOA in sexual activity. Baum and Everitt (1992) showed that removal of both olfactory and genital somatosensory input by simple sensory denervation failed to reduce c-Fos induction in MPOA in sexually interacting males, but removal of input from brain areas that process such information did affect the c-Fos levels. These results suggested that the MPOA is primarily a site for the integration of inputs from central processors and not through direct afferent input from sensory structures.

The stimulatory influences that arise from sexual activity reach the brain not only through sensory nerves that innervate the penis or vagina, but also by way of olfactory and endocrine influences. Among the uses of molecular mapping in this field have been efforts directed at sorting out the differential effects of the varied sensory influences (Coolen et al., 1997; Ohkura et al., 1997; Wersinger and Baum, 1997). The role of gender differences in activation patterns has also been examined with the view that it may reflect possible anatomical or functional dimorphism (Heeb and Yahr, 1996; Wersinger and Baum, 1997; Gill et al., 1998; Meddle et al., 1999). For example, the so-called sexually dimorphic nucleus (SDN) found in the hypothalamic preoptic area, which is believed to contribute to the development and expression of male-specific sex behaviors, has been studied with regard to anogenital grooming and whether this might contribute to its sexually dimorphic development (McCarthy et al., 1997). It was shown by way of c-Fos mapping that somatosensory stimulation did not yield significant activation of this nucleus, which suggests that neuroendocrine determinants instead may be critical to the fate of this nucleus during the critical period of development. This study is one of several where the role of endocrine influences in general has been critically assessed, especially within the context of sexual stimulation of hormonally sensitive brain regions and specific receptor systems (Greco et al., 1998; Foidart et al., 1999; Lumley and Hull, 1999; Wersinger and Rissman, 2000).

Another area in which molecular mapping has provided insight into complex somatosensory function is that of maternal behavior. Pronurturant behaviors such as licking, mouthing, and suckling produce elevated expression of c-Fos in a number of forebrain structures previously associated with maternal behavior (Fleming et al., 1994; Lonstein and Stern, 1997b, 1999; Katz et al., 1999). It has been demonstrated that activation of these areas can be specifically attributed to physical or somatosensory mother–litter interactions (Calamandrei and Keverne, 1994; Fleming and Walsh, 1994; Lonstein and Stern, 1997a). Molecular mapping studies of these areas in conjunction with various sensory desitizations have clarified the roles of the different structures in the so-called 'maternal circuit' (Walsh et al., 1996; Komisaruk et al., 2000). For example, although the MPOA shows elevated c-Fos expression in postpartum rats exposed to pups, this expression does not seem to be affected by removal of direct olfactory or somatosensory input (Walsh et al., 1996). This suggests that the MPOA may instead be largely involved as an effector system in the expression of maternal behavior rather than as a mere processor of sensory information, a role that appears to be consistent with that seen in conjunction with sexual behavior in the male.

A final area where the application of c-Fos mapping has provided insight is in delineating the neural substrates of stress. Much of this effort has been directed at the hypothalamic pituitary axis where it has shown that somatosensory and nociceptive aspects of mild and severe acute stress induce c-Fos expression in various hypothalamic nuclei (Pan et al., 1996, 1997; Li and Sawchenko, 1998; Sawchenko et al., 2000). It has been suggested that activation of this axis is responsible for driving adrenocortical expression of endocrine factors that mediate the stress response. It also appears on the basis of c-Fos expression maps that severe stress activates the locus coeruleus and amygdala much more so than mild stress, a result that may reveal a greater involvement of these areas in the perception of the severity of stress and the accompanying emotional responses (Chowdhury et al., 2000). One of the more intriguing uses of c-Fos mapping has been to determine the differential effects of processive (exteroceptive) versus systemic (interoceptive) stress and how they influence different brain areas (Emmert and Herman, 1999; Sawchenko et al., 2000). It appears that certain brain structures, such as the paraventricular hypothalamic nucleus, show similar c-Fos expression under both types of stressors, whereas other brain areas, such as the amygdala and frontoparietal cortex, show differential activation (Emmert and Herman, 1999).

In addition to mapping the neural substrates of stress, there has been some effort to ascertain whether stress can exert an unwanted influence and thereby interfere with the interpretation of molecular maps in brain areas devoted to the processing of sensory or mnemonic function. An early study by Anokhin et al. (1991) showed that c-Fos induction in chick forebrain areas could not be attributed to stress or arousal but rather only to the acquisition of new experience. The notion that a sensory novelty is an important factor for driving c-Fos expression in a number of different brain areas has been validated through several studies using different paradigms (Papa et al., 1993; Zhu et al., 1995; Montag-Sallaz et al., 1999).

3. THE CHEMOSENSORY SYSTEMS

Molecular mapping studies on both chemosensory systems — gustation (taste) and olfaction (smell) — have added significantly to our knowledge of the neural substrates of these sensory functions. In the gustatory system, much of the research has been aimed at delineating gustatory processing within two key brainstem areas, the nucleus of the solitary tract (NST) and the pontine parabrachial nucleus (PBN). Much of this effort has been aimed at mapping the neural bases of conditioned taste aversion, whereby an animal develops long-lasting taste preferences on the basis of just a single exposure to a toxin or emetic substance. In the olfactory system, molecular mapping has been used as tool to study a more diverse set of questions. These include visualizing sensory maps in neural structures such as the olfactory bulb and examining the neural substrates of circadian behavior, maternal communication, vomeronasal function and olfactory memory.

3.1. SENSORY MAPS OF GUSTATORY FUNCTION

Molecular mapping efforts in the gustatory system were initiated somewhat later than in the somatosensory system. The earliest studies concerned representation of taste quality largely within two brainstem areas, the NST and the PBN, although a number of other small nuclei have also been examined (Yamamoto et al., 1994a; Harrer and Travers, 1996; Streefland et al., 1996). A central question at the time, and one that continues to be discussed in the field, was the notion that the four different taste primaries are represented in an orderly fashion as a reflection of some kind of chemotopic organization (Fig. 4). Harrer and Travers (1996) provided evidence that the sweet and bitter tastes produced by sucrose and quinine respectively resulted in somewhat different patterns of c-Fos immunostaining in the NST. Although there was some regional difference in the staining pattern, they also noted that there was considerable intermingling of neurons activated by the two tastants. Similarly, DiNardo and Travers (1997) found that c-Fos staining in the medullary reticular formation

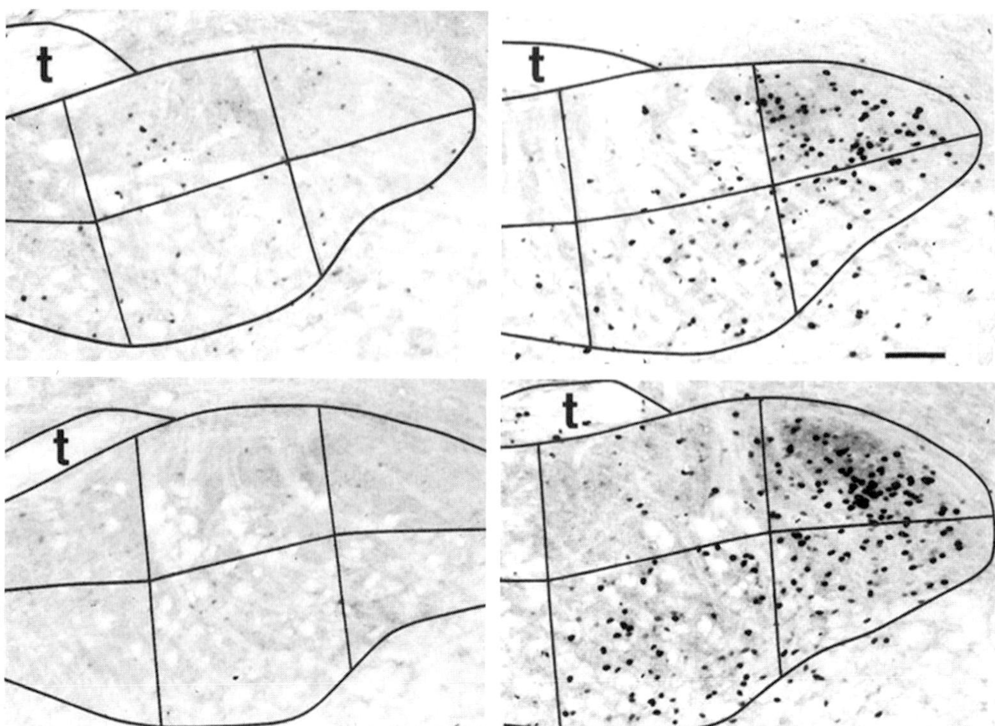

Fig. 4. c-Fos expression maps in the gustatory zone of the rat nucleus of solitary tract (gNST). The two left panels show that distilled water produced little or no c-Fos activation, unlike the effect seen after administration of 0.003 M quinine (two right panels). Much of the elevated c-Fos expression is distributed along a dorsomedial and midventral zone. t = solitary tract. Scale bar 100 μm. Adapted from King et al., 1999; image courtesy of A. Spector.

(RF) obtained after sucrose and quinine ingestion showed differential patterns of activation. They found increased c-Fos expression in three specific subregions of the RF in response to quinine but not sucrose, a result they attributed to the separate activation by quinine of a specific oral rejection circuit.

The notion of chemotopic organization has been more recently addressed by molecular mapping in conjunction with gustatory nerve transection. The rough chemotopy shown by Harrer and Travers (1996) was not entirely consistent with the terminal distribution of afferent taste fibers, a result that may have been compromised by non-gustatory factors and post-ingestive consequences. To clarify this further, King et al. (1999) have recently shown that transection of the glossopharyngeal nerve, one of two main sensory nerves carrying taste signals from the tongue, produces a dramatic decrease in quinine-evoked c-Fos expression in the NST. In comparison, transection of the other major afferent, the chorda tympani branch of the facial nerve, had no such effect. Thus, not only is the spatial distribution of quinine-evoked neural activation in NST suggestive of a chemotopic arrangement, these results further suggested that the bitter taste may be predominantly conveyed through the glossopharyngeal nerve. Further evidence infavor of this conclusion was shown by the reappearance of quinine-evoked c-Fos staining in the NST after the glossopharyngeal nerve had regenerated (King et al., 2000).

The question of whether NST activation is solely due to orosensory stimulation or other post-ingestive factors has continued to be of some interest (Emond and Weingarten, 1995; Yamamoto and Sawa, 2000a,b). These studies have shown that gastric infusion of certain chemical solutions can result in increased c-Fos expression in a number of brainstem areas, including the NST and PBN. It thus appears that these areas are not only reactive to orosensory stimulation but also to post-ingestive factors, including visceral signals. There also appears to be functional segregation in the PBN that suggests a differential pattern of activation based on the qualitative and hedonic aspects of various sweeteners, such as sucrose and saccharin (Yamamoto and Sawa, 2000b).

A number of other issues in which molecular mapping of gustatory function has been applied concern mapping of specific brain regions where opioid peptide secretion contributes to a positive hedonic response (Park and Carr, 1998), the influence of forebrain structures on brainstem gustatory processing (Travers et al., 1999), and the anatomical projections of c-Fos expressing neurons in the NST (Travers and Hu, 2000). The latter study provides an elegant example of how c-Fos mapping can be combined with retrograde tract tracing to reveal dual-labeled maps of activation/projection neurons. Travers and Hu (2000) showed by combined c-Fos immunostaining and Fluoro-Gold labeling that only a small subset of c-Fos-positive NST neurons project to the PBN or RF, leading to the suggestion that the majority of these neurons are involved in local, intranuclear projections.

3.2. NEURAL SUBSTRATES OF CONDITIONED TASTE AVERSION

An area of behavioral research where molecular mapping strategies have been effectively employed is in delineating the neural substrates of conditioned taste aversion (CTA). The early studies in this area convincingly showed that expression of several immediate-early genes, in particular c-*fos*, occurred in brainstem and forebrain areas during CTA learning (Reynolds et al., 1991; Yamamoto et al., 1992; Houpt et al., 1994; Swank and Bernstein, 1994; Lamprecht and Dudai, 1995). Much of the focus for these and subsequent experiments has been the NTS, PBN, and medial thalamus, which are all relay/processing stations in the gustatory transmission pathway. It had been previously shown that lesions produced in these areas impaired CTA learning (Ossenkopp and Giugno, 1985; Spector et al., 1992; Reilly, 1999). Consistent with this idea has been the demonstration that c-Fos expression maps show differential induction after a single CTA trial in comparison to controls (Yamamoto et al., 1994b; Swank et al., 1995; Thiele et al., 1996; Sakai and Yamamoto, 1997). c-Fos expression is generally compared after ingestion of a specific tastant versus ingestion of the same tastant but paired with a compound that had previously produced a transient illness.

The cellular-level resolution that molecular mapping offers has been especially useful in delineating the different subareas of activation within the small brainstem nuclei. For example, Yamamoto (1993) has shown that saccharin ingestion induces c-Fos expression in the central lateral subnucleus of the PBN, but that rats with CTA to saccharin showed elevated c-Fos staining in the ventral lateral subnucleus. Thus, the zone of activation in the PBN in response to saccharin changes from one subnucleus to another after acquisition of a specific CTA, suggesting that only one of the subareas is responsible for taste aversive learning. Similarly, it has been shown that CTA effects are largely confined to the intermediate nucleus of the NTS where oral infusion of a tastant induces c-*fos* expression only if given in conjunction with a conditioned aversive stimulus (Houpt et al., 1996a,b).

The more recent molecular mapping studies on the neural bases of CTA have explored the cortical and forebrain influences upon its induction in the NTS. In a series of studies, Schafe and colleagues (Schafe et al., 1995; Schafe and Bernstein, 1996, 1998) have shown that forebrain inputs, particularly from the amygdala, are necessary for c-Fos expression in the NTS after exposure to a conditioned aversive stimulus. Similarly, inputs from the primary gustatory cortex (insula) have an effect on both NTS c-Fos induction and the behavioral expression of CTA. The likelihood that c-Fos itself may be involved in mediating the CTA response was confirmed by direct injection of antisense oligodeoxynucleotides against c-*fos* mRNA (Lamprecht and Dudai, 1996; Swank et al., 1996). Although the link between c-*fos* expression and CTA learning is widely accepted, it has become recently clear that methodological factors can affect CTA driven c-Fos expression as well. For example, it has been shown that conditioning methods and context-dependent activation, such as the presence of novel stimulus cues, can influence CTA-driven c-Fos expression in the NTS (Spray et al., 2000; Swank, 2000). It has been suggested that a more cautious interpretation be applied during analysis of c-Fos activation maps to take into account any behavioral, environmental, or other confounding factors.

3.3. SENSORY MAPS OF OLFACTORY FUNCTION

One of the major research questions in the chemosensory system has concerned the way by which the mammalian olfactory bulb integrates odor information from the neuroepithelium to create a functional representation. Early work in this area employed metabolic markers such as 2-DG in behaving animals exposed to different odorants (Jourdan et al., 1980; Coopersmith et al., 1986; Slotnick et al., 1989). With the advent of molecular mapping techniques, an alternate approach was available for functional mapping in the olfactory bulb. The initial studies in this area were aimed at comparing c-Fos and 2-DG activation maps to ensure that similar spatial profiles were evident with both techniques (Onoda, 1992; Guthrie et al., 1993; Sallaz and Jourdan, 1993). These studies established that c-Fos was indeed triggered by odor stimulation and that the resulting pattern of glomerular activation was spatially coincident with focal regions of high 2-DG uptake.

With the validation that c-Fos maps were truly representative of odor-induced functional activity, a number of studies were undertaken to clarify the morphological specificity of the stained neurons and identify extrabulbar influences upon neuronal activity in the olfactory bulb. The morphological phenotype of c-Fos expressing cells was characterized by Guthrie and Gall (1995a,b). They found that c-Fos protein expression did not coincide with glial fibrillary acidic protein (GFAP), a marker of glial cells, but instead was co-localized with tyrosine hydroxylase immunostaining in a subpopulation of dopaminergic neurons. Sallaz and Jourdan (1996) showed that c-Fos expression in the olfactory bulb was not only triggered by odor stimulation but also by afferents of central origin. They showed through application of selective neurochemical agonists and antagonists that both noradrenergic and cholinergic centrifugal systems may be involved in the modulation of c-*fos* expression in the olfactory bulb.

It is well known that the olfactory bulb is a site where constant synaptic regeneration occurs due to the rapid turnover of olfactory receptor cells in the neuroepithelium. The reinnervation and reformation of synapses here has also been implicated in terms of the modifications that are necessary for learning. Thus, molecular mapping strategies have also been used to study neural events in the olfactory bulb that may be related to this process. The nature of this response has been characterized in both spatial and cellular terms. It has been shown that early olfactory preference training produces an increased number of juxtaglomerular cells expressing c-Fos protein but is accompanied by a reduction in c-Fos expression in the granule cell layer (Woo et al., 1996; McCollum et al., 1997). Furthermore, there appear to be certain regions within the glomerular layer that are susceptible to modification following odor learning. Johnson et al. (1995) have shown by way of c-Fos maps that midlateral regions of the olfactory bulb show greater responsivity to odor learning, suggesting that there may be regional differences in odor-induced plasticity. Odor discrimination learning also appears to engage different regions of the hippocampus, as shown by the significantly greater c-*fos* mRNA expression profiles in area CA3 than CA1 (Hess et al., 1995a,b).

3.4. SENSORY MAPS OF VOMERONASAL FUNCTION

The vomeronasal system represents one of the most intriguing, and controversial, aspects of chemosensory function in mammals. Although the presence of this system is well established in many species, its functionality in higher primates, including humans, remains disputed. The vomeronasal organ (VNO), which is a separate chemosensory organ from the main olfactory neuroepithelium, is largely responsible for detection of pheromones, leading to activation of the accessory olfactory bulb (AOB), amygdala, and certain hypothalamic nuclei (reviewed in Doving and Trotier, 1998; Johnston, 1998; Monti-Bloch et al., 1998; Keverne, 1999). Numerous studies have employed molecular mapping strategies in these areas in order to understand their role in pheromonal processing and how various chemosensory factors can influence their function (reviewed in Dudley et al., 1996; Pfaus and Heeb, 1997; Meredith, 1998).

The chemosensory cues that influence the AOB have been examined through mating behavior and exposure to fluids such as urine and vaginal secretions, both thought to contain pheromonal compounds. It has been shown that female rodents exposed to male pheromones during copulation display enhanced c-Fos expression in the AOB, amygdaloid nucleus, and hypothalamic nuclei such as the MPOA (Brennan et al., 1992; Rajendren and Moss, 1994; Halem et al., 1999). A similar activation of the AOB and other central areas has been shown in males that were exposed to female pheromones during mating (Fernandez-Fewell and Meredith, 1994, 1998; Meredith, 1998). Much of the c-Fos expression appears in the granule and mitral cell layers of the AOB (Brennan et al., 1992; Rajendren and Moss, 1994; Inamura et al., 1999). Similarly, *egr-1* expression has been shown to increase in both layers after mating behavior (Brennan et al., 1992). The notion that c-*fos* expression in the AOB in these situations is driven by VNO activity was shown through experiments in which the VNO was selectively removed. In both males (Fernandez-Fewell and Meredith, 1994) and females (Rajendren and Moss, 1994), removal of the VNO significantly reduced pheromone-driven c-*fos* expression in the AOB. The AOB has been further implicated in its specificity to the pheromonal response by studies that have documented elevated c-Fos staining in the AOB without a corresponding increase in the main olfactory bulb (Schellinck et al., 1993; Guo et al., 1997).

Activation of accessory olfactory structures through pheromonal stimulation has largely been studied by exposure to secreted fluids. Exposure of males to vaginal secretion induces c-Fos expression in multiple areas that include the AOB, MPOA, and amygdaloid nuclei (Fiber et al., 1993; Romeo et al., 1998). However, most of the mapping studies of the accessory olfactory system have been conducted by exposure to urine, which in many species is known to contain compounds with pheromonal action (Guo et al., 1997; Yamaguchi et al., 2000). A number of important properties of the accessory olfactory system have become apparent through studies of exposure to urine or soiled bedding odors. As with mating behavior, a set of early studies verified that exposure to urine induces c-*fos* expression in the AOB and in several central areas, a result that was interpreted in the context of pheromonal action on the vomeronasal projection circuit (Schellinck et al., 1993; Bressler and Baum, 1996; Guo et al., 1997). Furthermore, the action of urinary pheromonal compounds with regard to sex- or species-specific efficacy has been demonstrated in the AOB (Bressler and Baum, 1996; Tubbiola and Wysocki, 1997; Inamura et al., 1999; Matsuoka et al., 1999).

As with gustatory function, the cellular-level resolution that molecular mapping offers has been useful in delineating the different AOB regions of pheromonal activation. It has been reported that urinary chemosensory cues from female mice selectively activate a subpopulation of neurons in the anterior part of the AOB in the male (Dudley and Moss, 1999; Matsuoka et al., 1999). This difference in functional activation is consistent with a general neurochemical division of the VNO and its known anatomical relationship to the AOB. The laminar distribution of VNO receptor neurons appears to be distinct with regard to their expression of different classes of G proteins and pheromonal receptors (Dulac and Axel, 1995; Halpern et al., 1995; Jia and Halpern, 1996; Herrada and Dulac, 1997). This spatial segregation of the VNO is accompanied by different patterns of axonal projection to the AOB, a fact that may reflect detection and transmission of different pheromonal cues (Imamura et al., 1985; Taniguchi et al., 1993). Further insight into this has recently appeared from a study of Brennan et al. (1999), who showed by way of *egr-1* mapping that certain specific urinary proteins activate the posterior regions of the AOB, whereas whole urine and a preparation of major urinary proteins stripped of their ligands additionally activated the anterior AOB. Yamaguchi et al. (2000) have found that exposure to urinary substances separated into low and high molecular weight fractions failed to induce c-Fos expression in the AOB, whereas a mixture of these preparations produced strong induction.

3.5. NEURAL SUBSTRATES OF COMPLEX OLFACTORY FUNCTION

Although molecular mapping has been applied in many studies of conditioned taste aversion, there has been much less effort in examining its counterpart in the olfactory system. One of the few studies in this area is that of Funk and Amir (2000b), who showed that pairing an olfactory stimulus with footshock in adult rats produced subsequent activation in several olfactory structures when animals were exposed to the odorant alone. Because the c-Fos activation maps were more intense after forward conditioning than with the odorant alone, Funk and Amir (2000b) concluded that structures in the primary and accessory olfactory pathways can be modified through aversive conditioning.

Molecular mapping has recently been used to study the interaction between olfactory function and circadian rhythm. There is sufficient evidence to link the olfactory and circadian systems by virtue of olfactory modulation of circadian clock function (Goel and Lee, 1995; Friedman et al., 1997). Amir et al. (1999b) have shown that olfactory stimulation in conjunction with light exposure can dramatically enhance the circadian phase shifts and c-Fos expression in the suprachiasmatic nucleus (SCN). Their results further suggested that clock resetting by light can be facilitated by olfactory cues through structures within the olfactory pathways that showed enhanced activation. It has also been found that odor-induced c-Fos expression in several olfactory structures is modulated by the phase of the circadian clock (Fig. 5). Amir et al. (1999a) found that neural activity in response to a neutral odorant was greatly enhanced in the subjective night. c-Fos mapping has shown a similar circadian modulation in rat primary olfactory areas in response to a fear-induction stimulus (fox urine) (Funk and Amir, 2000a).

We have already discussed the use of molecular mapping in explorations of maternal-offspring behavior within the context of somatosensory interactions (Section 2.3). To complete

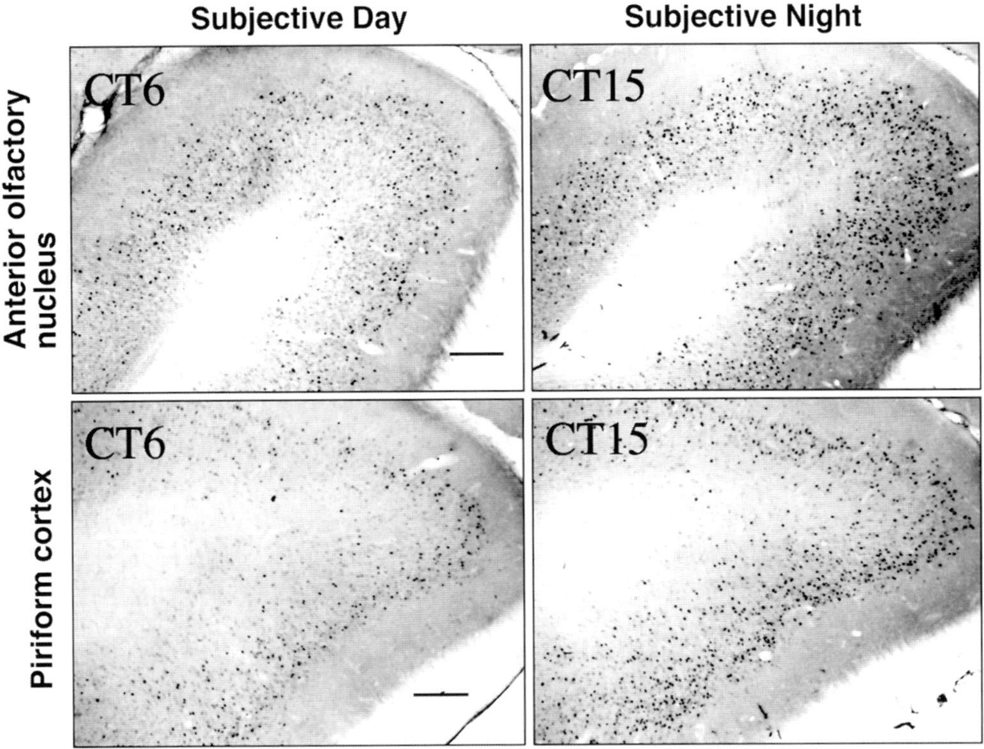

Fig. 5. c-Fos induction in the anterior olfactory nucleus and piriform cortex after exposure to a neutral odorant (cedar wood oil) shows different expression intensities during subjective day and night. Reprinted from Amir et al., 1999a, with permission.

our consideration of this issue, we briefly discuss here the mapping studies that have examined the olfactory components of maternal behavior. The well-known association between olfactory cues and maternal behavior suggested that certain neural sites, such as the MPOA, should show an associative relationship. However, early studies clearly established that removal of olfactory cues failed to reduce c-Fos expression in the MPOA, suggesting that this structure served primarily as an effector site (Calamandrei and Keverne, 1994; Fleming et al., 1994; Fleming and Walsh, 1994; Numan and Numan, 1995; Walsh et al., 1996). However, olfactory stimulation during mother–litter interactions in several different species has been shown by way of c-Fos mapping to activate multiple sites within the olfactory pathways as well as forebrain areas, such as the amygdaloid nucleus (Calamandrei and Keverne, 1994; Fleming et al., 1994; DaCosta et al., 1997; Katz et al., 1999). The effects of male paternal behavior have rarely been studied by comparison. Kirkpatrick et al. (1994) have shown elevated c-Fos expression in several areas, such as the AOB and several hypothalamic nuclei, in pup-exposed males in comparison to non-social olfactory stimulation.

4. THE AUDITORY SYSTEM

The use of IEG markers for activity labeling in the auditory system has become quite popular. The bulk of these studies have revolved around use of c-Fos as a marker of activity, with a smaller number of studies exploiting the mapping properties of other IEGs, such as Zif268. Similar to other sensory systems, these studies have provided important insight into the functional properties of the auditory system. In a broad sense, molecular mapping studies of auditory function can be grouped in four distinct categories — neural maps in response to electrical stimulation, molecular maps of brainstem activity, cortical maps of auditory function, and molecular maps of the song-control system.

4.1. NEURAL MAPS IN RESPONSE TO ELECTRICAL STIMULATION

Studies of IEG mapping of the auditory system in response to electrical stimulation are quite recent. The earliest study was by Vischer et al. (1994), who mapped Fos-like immunoreactivity (FLI) in the auditory pathway of Sprague–Dawley rats in response to unilateral electrical stimulation of the cochlea. Their results demonstrated functional changes that were present in and selective to various structures along the auditory pathway.

Later, Vischer et al. (1995) assessed the activity elicited by electrical stimulation of the cochlea in the auditory pathway in an animal model of cochlear implants by using c-Fos induction combined with single-unit recordings. They demonstrated that immunocytochemical maps were correlated to electrophysiological events in various nuclei of the auditory pathway after electrical stimulation of the cochlea. In a similar study, Zhang et al. (1996) attempted to establish the relationship between the duration and intensity of unilateral cochlear electrical stimulation, as well as the time-course of IEG induction after the offset of stimulation. The results of this study showed that the distribution and density of FLI in response to the two experimental conditions did not co-vary in an identical manner in the various auditory nuclei. In a follow-up study, Zhang et al. (1998) analyzed and quantified FLI in the auditory pathway using two different antibodies raised against Fos. Their results showed that the choice of Fos antibody can have significant impact on the resulting staining pattern observed in response to the same stimulus.

There have been several attempts to study deafness-induced plastic changes in the mature central auditory pathway by assessing the amount of evoked Fos-immunoreactivity (Bledsoe et al., 1995; Nagase et al., 2000). Bledsoe et al. (1995) found a net increase in Fos-immunoreactive (IR) neurons in the central nucleus of the inferior colliculus (CIC) in response to contralateral cochlear electrical stimulation after 21 days of bilateral deafness. In addition, they found a link between deafness and the population of CIC neurons that respond by suppression of activity to electrical stimulation. These results, combined with the observation of a marked decrease in gamma-aminobutyric acid (GABA) release from the CIC cells of the deafened animals, enabled them to evaluate the notion that local changes in inhibitory transmitters may be at the root of deafness-induced plastic changes.

Nagase et al. (2000) used Fos immunocytochemical mapping to study the changes in excitatory neuronal activity in the rat inferior colliculus (IC) in response to basal and apical monopolar cochlear electrical stimulation between normal hearing and 21-day deafened rats. They reported that, whereas acute cochlear stimulation produced Fos IR neuronal maps that revealed a linear relationship between the stimulus intensity and number of IR positive cells, chronic cochlear stimulation revealed an additional change in the pattern of Fos IR neurons compared to that of normal hearing animals. The authors concluded that chronic cochlear electrical stimulation induces complex plastic changes in the IC that are over and above the plastic changes effected by deafness alone.

Two recent studies have used molecular mapping strategies to examine in vivo functional properties of cochlear prostheses (Saito et al., 1999, 2000). Saito et al. (1999) used FLI to compare auditory brain stem excitation elicited by bipolar electrical stimulation of the cochlea at various current levels relative to the electrically evoked auditory brain stem response threshold. They found that discrete activation of cell populations within the central auditory pathways can occur with electrical stimulation up to the highest levels of stimulation typically useful in humans. The data also indicate a close, but not identical, quantitative relationship between Fos-like immunoreactivity and electrophysiological response amplitude. They concluded that Fos immunostaining can provide a powerful and quantitative tool for studying the dynamic response characteristics of cells of the central auditory system to electrical stimulation at suprathreshold levels.

In a follow-up study, Saito et al. (2000) evaluated FLI staining in the spiral ganglion and cochlear nuclei evoked by basal, second, or apical turn bipolar intracochlear electrical stimulation. Their results revealed a restricted topographic distribution of activity evoked by high levels of electrical stimulation which was interpreted to have been initiated at first-order primary neurons of this system. Their results indicate that place (channel) information can be maintained in the spiral ganglion and central nervous system even at very high levels of electrical stimulation, a result that may have practical implications for deaf individuals with cochlear implants.

4.2. MOLECULAR MAPS OF BRAINSTEM ACTIVITY

Molecular mapping has also been used to delineate the functional organization of the brainstem components in the auditory pathway. The early studies by Ehret and Fischer (1991) and Sato et al. (1993) allowed for the visualization of neural activity and tonotopy of the rat brainstem in response to auditory stimulation using Fos ICC. Ehret and Fischer (1991) concluded that such mapping could demonstrate stimulus-related local neuronal activation on a single-cell level and that it may be useful to complement other mapping techniques, such as electrophysiological recording or 2-DG autoradiography. Likewise, Sato et al. (1993) demonstrated Fos activity maps in the auditory brainstem in response to various pure tones and found different response patterns in the acoustically responsive nuclei of the brainstem.

Following this study, Brown and Liu (1995) and Friauf (1995) used Fos ICC to map the acoustic pathway of the rat and mouse brains, respectively. They both found that patterns of neuronal activation could be visualized in their maps by way of tone-evoked excitation and inhibition. Furthermore, Adams (1995) demonstrated an additional dimension of the usefulness of IEG mapping by showing that sound stimulation induces Fos-related antigens in cells having common morphological properties throughout the auditory brainstem.

In addition to these studies, several other investigators have also employed IEG activity mapping and have reported Fos protein upregulation in response to various forms of acoustic stimulation (Rouiller et al., 1992; Reimer, 1993; Pierson and Snyder-Keller, 1994; Keilmann and Herdegen, 1995, 1997; Chen et al., 2000). However, there seems to be some variability found in the brain regions and cell types expressing Fos protein in response to acoustic stimulation, with each study showing activity maps of only a subset of cell types in the auditory brainstem. On the contrary, Saint Marie et al. (1999) have shown that almost all the principal neuronal types found in the auditory brainstem are capable of expressing acoustically induced c-*fos* mRNA in both normal and unilaterally deafened rats. They concluded that activity mapping using c-*fos* mRNA rather than protein detection methods are more useful in delineating faithful and robust activity maps.

Another focus of research in the auditory system has been to understand the role of brainstem and other auditory responsive nuclei in the generation, modulation and propagation of audiogenic seizuresindexas, see audiogenic seizure@AS, see audiogenic seizure (ASs) (Simler et al., 1994, 1999; Kato et al., 1996). The use of Fos ICC mapping enabled Simler et al. (1994) to demonstrate that, whereas single AS induced the expression of c-*fos* only in the subcortical auditory nuclei, kindled AS induced a strong c-*fos* expression in the amygdala, piriform cortex, hippocampus, and the neocortex as well. The authors reached two conclusions from their activity maps. First, that audiogenic seizures are brainstem seizures related to dysfunction of auditory pathways, and second, that kindling of audiogenic seizures recruits forebrain and limbic structures into the seizure network. In an attempt to delineate the spatial and temporal relationships between c-Fos expression and kindling of ASs, Simler et al. (1999) have recently shown that the kindled ASs preferentially propagate from the brainstem, through the amygdala and the perirhinal cortex, to the motor cortex, with the piriform cortex and hippocampus being secondary targets. However, a recent study by Silveira et al. (1999) has questioned these results. They found that acoustic brainstem nuclei express Fos after flurothyl-induced general seizures. Because the flurothyl-induced activation of acoustic brainstem nuclei resembles that observed after AS, these authors reasoned that Fos expression in such nuclei is not specific to AS but to seizures in general.

Another issue of some interest in auditory brainstem research is the effect of acoustic deprivation on plastic events within the various nuclei. A recent study by Luo et al. (1999) has examined the effects of cochlear ablation on acoustically induced c-*fos* mRNA expression in the adult rat auditory brainstem. They found little evidence to suggest that neural plastic events were initatiated in the auditory brainstem after cochlear ablation in these animals.

4.3. CORTICAL MAPS OF AUDITORY FUNCTION

IEG activity maps have been useful in the study of auditory cortical function for delineating the cytoarchitecture and tonotopic organization of the primary auditory cortex (Qian and Jen, 1994; Zuschratter et al., 1995; Qian et al., 1996; Jen et al., 1997) and understanding the involvement of auditory cortex in higher cognitive functions such as recognition, learning and memory (Scheich and Zuschratter, 1995; Scheich et al., 1997; Carretta et al., 1999; Fichtel and Ehret, 1999).

Research on mapping cytoarchitectural and tonotopic features have been based on delineating the functional organization of the big brown bat *Eptesicus fuscus* (Qian and Jen, 1994; Jen et al., 1997). Qian and Jen (1994) used c-Fos ICC for the identification of acoustically responsive neurons in the cerebral cortex, as well as the cerebellum and subcortical nuclei. They found that a greater number of c-Fos-positive neurons were present along the auditory pathway when the animals were stimulated monaurally as opposed to binaurally. The authors attribute this finding to the removal of inhibition in the binaural EI neurons (i.e., neurons that receive excitation from one ear and inhibition from the other ear) that occurs under monaural stimulation conditions. This absence of inhibitory influence is suggested to lead to an overall activation of the EI neurons and increase the number of the c-Fos-positive population. Another focus of this line of research has been to compare the suitability of IEG mapping in conjunction with 2-DG (Zuschratter et al., 1995) or HRP tract-tracing approaches (Qian et al., 1996).

The second category of research using IEG mapping deals with the involvement of the auditory cortex (AC) in cognitive processes, such as recognition and learning. The AC has been implicated in the formation of sensory memory (Scheich and Zuschratter, 1995; Scheich et al., 1997), recognition (Fichtel and Ehret, 1999), and learning (Carretta et al., 1999). For example, Scheich et al. (1997) used c-Fos ICC maps combined with single-unit recording and fluoro-2-deoxyglucose (FDG) techniques to study the involvement of the primary auditory field (AI) of gerbils in learning and memory formation. They found that aversive tone conditioning paradigms remodeled frequency receptive fields of AI neurons and also altered the spatial representation of tones in FDG experiments. In addition, c-Fos ICC revealed an unusual spatial pattern of neural activation that resulted from repeated brief exposures of the animals to a tone in a new environment. As a result, Scheich et al. (1997) concluded that the AI of gerbils may contain representations of spectral features of sounds as well as aspects of the learned behavioural meaning of various sounds.

In a recent study Carretta et al. (1999) used c-Fos maps in rat auditory pathways to identify the areas involved in the encoding of the significance of acoustic signals during a sensory–motor acoustic learning. These investigators trained rats in a three-phase learning paradigm. The first phase contained an association between an auditory stimulus and a food reward; the second phase entailed a discrimination of two sounds with different frequency components; and the third phase involved a more complex discrimination of both spectral and spatial sound dimensions. The experiment also contained two control groups. One group received the same auditory stimuli as the 'learning' group but in a random order ('control' group) and another group remained naïve to the entire experimental protocol ('unstimulated' group).

After histological processing, they found FLI in the superior olivary complex, IC and medial geniculate body of the 'learning' group that was comparable to the same structures in the 'control' animals, but significantly greater than the 'unstimulated' animals. In the AC, however, and especially in the secondary area Te2, the number of FLI cells were found to be greater in the 'learning' animals than the 'control' group (see Fig. 6). This finding suggested that auditory cortical areas may be involved in encoding the behavioural significance of acoustic stimuli. Furthermore, they observed a relationship between FLI and performance (e.g., 'good learners' vs. 'bad learners' — see Fig. 6), suggesting that c-Fos mapping may be a suitable technique to reflect neuronal activity associated with the distinct steps in learning.

4.4. MOLECULAR MAPS OF THE SONG-CONTROL SYSTEM

There is now considerable interest in using molecular mapping techniques to study various aspects of the song-control system in songbirds. A series of elegant studies have revealed significant insight into the anatomical and functional circuits that are involved in the production and perception of birdsong in canaries and zebra finches. These studies have relied on detection of ZENK (Mello and Clayton, 1994, 1995; Chew et al., 1995; Jarvis et al., 1995, 1997, 1998; Mello et al., 1995; Jin and Clayton, 1997; Mello and Ribeiro, 1998; Ribeiro et al., 1998; Jarvis and Mello, 2000), c-*fos* (Kimpo and Doupe, 1997), or c-*jun* (Nastiuk et al., 1994) expression in the various brain structures involved in birdsong. A full account of these studies can be found in a separate chapter of this volume.

5. THE VISUAL SYSTEM

Molecular mapping efforts to study visual function can be broadly classified into four areas: activity-mapping of sensory function in the mammalian central nervous system, the influence of photoperiod on immediate-early gene induction, maps of developmental change and their postulated relationship to plasticity, and exploring the neural substrates of complex factors on visual processing. The basal distribution of several immediate-early genes is known for several visual structures in different species (Schlingensiepen et al., 1991; Herdegen et al.,

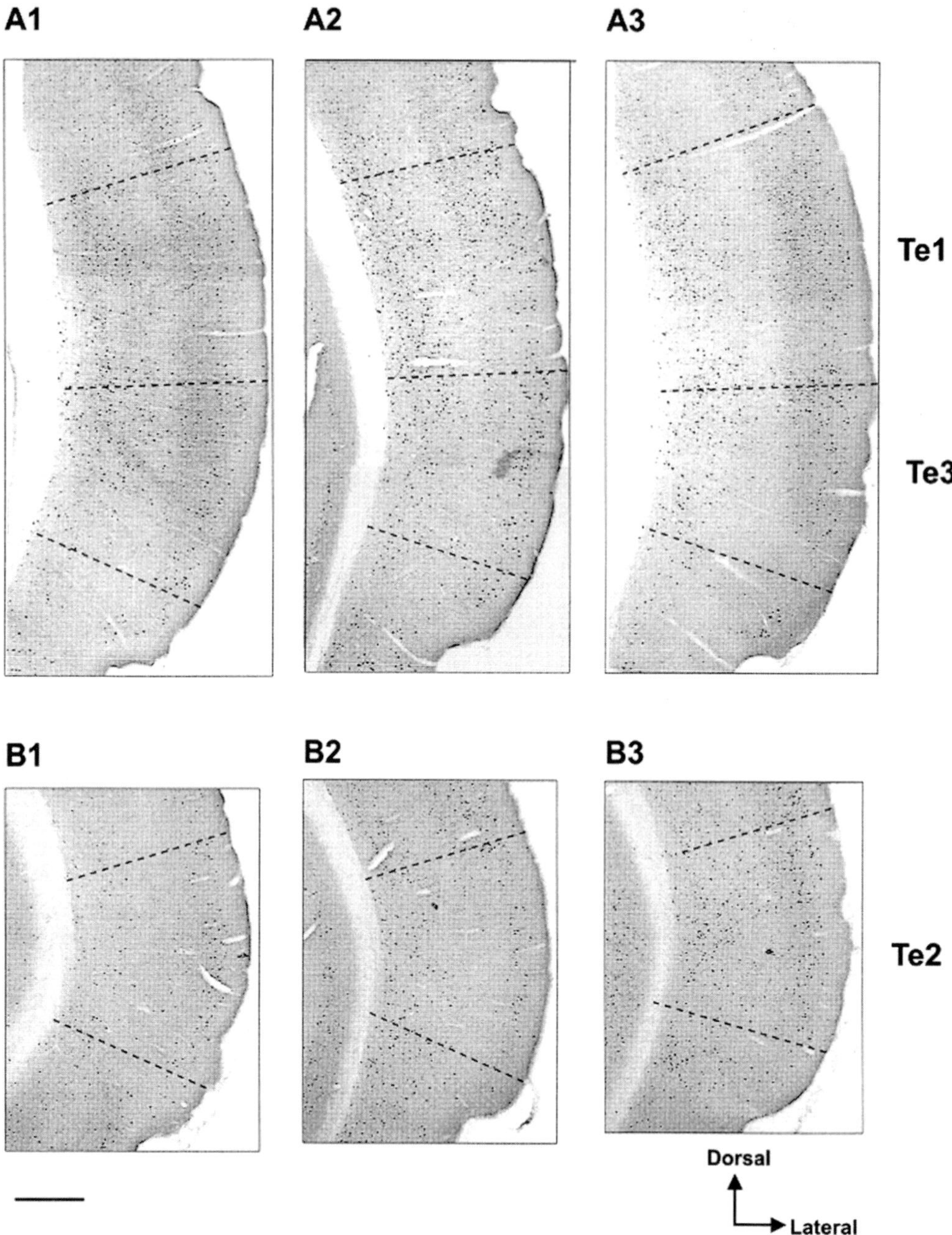

Fig. 6. Expression profiles for c-Fos in auditory cortex of control (*A1*, *B1*), 'bad learner' (*A2*, *B2*), and a 'good learner' animal (*A3*, *B3*). The images are taken from auditory cortical areas Te1, Te2, and Te3. The number of immunopositive neurons was lower in Te1 and Te2 of the 'bad learner' in comparison to the same locations of the 'good learner'. Reprinted from Carretta et al., 1999, with permission.

1993; Zhang et al., 1994; Kaminska et al., 1996). A number of studies have conclusively shown that the expression of several immediate-early genes is rapidly modulated by either application or removal of visual input. In this section, we discuss the various studies that have used these gene products to delineate visual function within the four general categories outlined above.

5.1. MOLECULAR MAPS OF VISUAL FUNCTION

The study of visual function with immediate-early gene products has been approached by way of either visual stimulation or deprivation (reviewed in: Chaudhuri, 1997; Kaczmarek and Chaudhuri, 1997; Herdegen and Leah, 1998; Chaudhuri et al., 2000). Deprivation studies have been undertaken by way of simple eyelid suture, administration of a sodium channel blocker such as tetrodotoxin (TTX) into the eye, or complete unilateral or bilateral enucleation. The latter approach raises the possibility that sensory effects in visual neurons may be compromised by response to nerve injury (Horton et al., 2000). In general, enucleation is not appropriate for short-term deprivation studies because of the associated trauma and possible upstream degenerative complications that may follow severing of the optic nerve.

Among the immediate-early genes, *zif268* has been a favored candidate for deprivation studies because of its normally high level of expression in neocortical areas (Kaczmarek and Chaudhuri, 1997; Caleo et al., 1999; Chaudhuri et al., 2000). Worley and colleagues (Worley et al., 1990, 1991) have shown that visual disruption in one eye leads to rapid down-regulation of *zif268* basal levels in the contralateral rodent visual neocortex, being most pronounced in the thalamorecipient layer IV. Similar results have been found in cat neocortical areas where visual deafferentation produces a time-dependent decrease of *zif268* mRNA (Zhang et al., 1995) and prolonged dark-rearing leads to a marked reduction in Zif268 protein levels (Kaplan et al., 1996). To the contrary, products of the c-*fos* gene are rather poor markers of deprivation effects because of their generally low level of basal expression (Dragunow and Robertson, 1988; Beaver et al., 1993; Kaczmarek and Chaudhuri, 1997). However, some decline in c-*fos* mRNA levels has been reported after visual deafferentation in the cat (Zhang et al., 1995).

One species where visual deprivation has been used to map cortical activity is the monkey. It has been shown that brief monocular deprivation, as little as two hours, is sufficient to reveal the details of ocular dominance architecture at cellular levels (Chaudhuri and Cynader, 1993; Chaudhuri et al., 1995). The interdigitated pattern of ocular dominance columns within each hemisphere provides a valuable internal control because it permits a more effective comparison between activity-induced expression and neural quiescence upon immediate-early gene expression (Fig. 7). Molecular maps of *zif268* expression within primate ocular dominance columns have been used to determine the temporal details of induction, assess the cellular morphology of immunopositive neurons, and explore developmental patterns and their possible relationship to ongoing neuroplastic events (Silveira et al., 1996; Chaudhuri et al., 1997; Markstahler et al., 1998).

A second, and more common, way to obtain IEG maps is by inducing their expression through visual stimulation. In general, these experiments require a period of dark-adaptation prior to stimulation in order to reduce basal expression and background staining. Furthermore, the transition from a quiescent period to intense synaptic stimulation affects the expression of ITFs, revealing the extent to which these genes are up-regulated in response. Visual stimulation experiments in rats have provided laminar details of c-*fos* expression in response to patterned visual stimulation (Montero and Jian, 1995), neural activity profiles in response to ultraviolet light (Amir and Robinson, 1996), effects of altering the temporal frequency of light presentation (Correa-Lacarcel et al., 2000), and the role of central noradrenergic input in guiding plastic change through modulation of c-*fos* expression (Yamada et al., 1999). One interesting application of c-Fos mapping has been to establish that functional connectivities

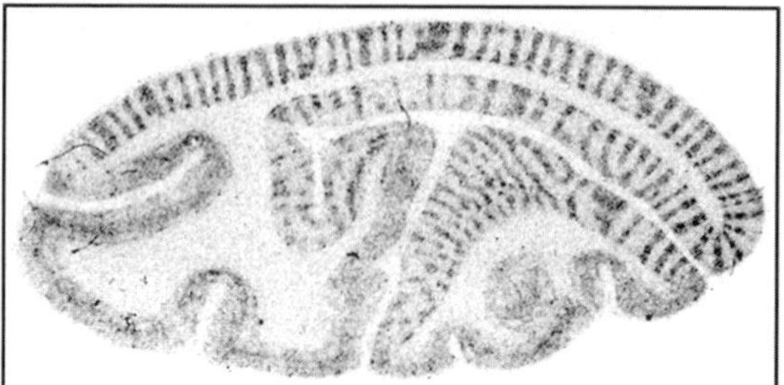

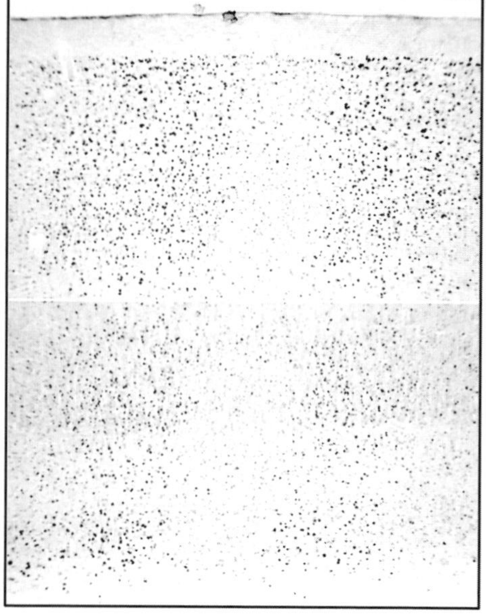

Fig. 7. An in situ hybridization map of c-*fos* expression shows a distinct ocular dominance pattern throughout an entire coronal section of monkey occipital cortex. The columns are visible at cellular resolution after immunostaining for Zif268 protein. Adapted from Kogan et al., 2000.

are made by transplanted fetal retinae onto the dorsal midbrain of neonatal rats. Craner et al. (1992) showed that transplanted retinae exposed to light stimulation produce patterns of c-Fos expression in visual neural structures that were similar to those seen upon stimulation of the intact eyes.

Induction experiments in higher animals have largely revolved around the issue of developmental sequences of cortical maturation and their corresponding effects upon IEG expression. Mower and colleagues (Mower, 1994; Kaplan et al., 1996) have shown that different laminar patterns of c-Fos expression are evident in kittens versus adult cats after brief visual stimulation. Furthermore, the induction effects in general were found to be greater in young animals, a result that was consistent with that found in rodents (Worley et al., 1990). A similar result has been reported for the *zif268* gene as well. Both mRNA and protein products are elevated after visual stimulation, the increase being more notable in young animals (Rosen et al., 1992; Nedivi et al., 1996). *zif268* induction after dark adaptation has also been used to show ocular dominance profiles in the striate cortex of developing and adult primates (Chaudhuri et al., 1995; Markstahler et al., 1998; Kaczmarek et al., 1999). The use of IEG markers to study developmental plasticity is taken up in greater detail in Section 5.3.

5.2. CIRCADIAN MODULATION OF NEURAL ACTIVITY

Circadian modulation of neural activity in certain brain structures is a well known phenomenon (reviewed in: Rusak and Zucker, 1979; Morin, 1994; Kennaway, 1998; Caputto and Guido, 2000). Molecular mapping techniques over the last decade have been used to uncover spatial details of this process and how it can be influenced by different factors (Caputto and Guido, 2000). The SCN has been the object of much attention in mapping circadian effects. The SCN is a hypothalamic nucleus that generates a circadian rhythm that is entrained to the environmental photoperiod via retinohypothalamic afferents (Rusak, 1977; Inouye and Kawamura, 1979; Ralph et al., 1990). The early research in this field conclusively established that SCN expression of IEGs such as c-*fos* are modulated in a circadian manner and governed by retinal illumination (Rea, 1989; Aronin et al., 1990; Earnest et al., 1990; Kononen et al., 1990; Kornhauser et al., 1990).

Given that photic entrainment must rely on retinal output, there has been some effort to show that environmental light/dark cycles can influence IEG expression in the mammalian retina as well. Here, light-induced c-*fos* expression has been demonstrated for a number of different cell types, such as photoreceptor, amacrine, and ganglion cells (Earnest et al., 1990; Sagar and Sharp, 1990; Chambille et al., 1993; Gudehithlu et al., 1993; Yoshida et al., 1993). The induction pattern in the output cells of the retina (ganglion cells) follows the light-dark cycle, being maximal shortly after onset of the light period. In other cell types, such as photoreceptors, c-*fos* expression remains elevated during the night cycle and therefore appears to follow an endogenously generated rhythm (Bussolino et al., 1998; Caputto and Guido, 2000).

The photic regulation of IEG expression and its relationship to pacemaker function in the SCN and other structures has been documented in a number of species in addition to rats (Wallman et al., 1994; Moffatt et al., 1995; King and Follett, 1997; Marchant and Morin, 1999; Bertolucci et al., 2000). The number of studies that have used molecular mapping tools to study circadian function is truly impressive and cannot be reviewed here in its entirety. One of the major insights that has emerged from these studies concerns differences in spatial activation across the SCN. Light-induced c-Fos expression is generally confined to the ventrolateral segment of the SCN (Rea, 1989; Earnest et al., 1990; Schwartz et al., 1994, 2000). Furthermore, the sensory induction of c-*fos* and other genes is limited to the subjective night portion of the circadian cycle and is not seen during the subjective day (Kornhauser et al., 1990; Rusak et al., 1992; Schwartz et al., 2000). In addition to light-induced c-*fos* expression, it has been shown that the SCN has an endogenous cycle of expression that is independent of sensory stimulation. This spontaneous rhythm of c-*fos* mRNA and protein expression occurs during the subjective day, beginning at dawn, and is largely localized to the dorsomedial segment of the SCN (Guido et al., 1999; Sumova et al., 2000).

A number of other immediate-early genes, such as *junB*, *fosB*, and *NGFI-A*, have been shown to exhibit both spontaneous and light-induced expression in the SCN and therefore may have some utility in mapping circadian activity. Expression of *junB* appears to be linked to light-stimulation as well as to a spontaneous circadian rhythm (Kornhauser et al., 1992; Guido et al., 1996a,c). The expression profile of *junB* therefore is quite similar to that of c-*fos*, not only in its sensory regulation feature, but also in the dorsomedial versus ventromedial pattern of expression. Constitutive expression of the *fosB* gene has been shown in the SCN and that it can be induced by a light-pulse during the subjective night (Guido et al., 1996b; Schwartz et al., 2000). Similarly, NGFI-A expression in response to light stimulation has been well documented (Lin et al., 1997; Tanaka et al., 1997, 1999). However, it appears that both of these genes are unlike c-*fos* in that they are not induced by a spontaneous circadian rhythm in the dorsomedial SCN. In fact, the results of one recent study with NGFI-A deficient mice suggest that induction of NGFI-A is not required for photic entrainment or phase shifting of the mouse circadian system (Kilduff et al., 1998).

5.3. DEVELOPMENTAL MAPS AND IMPLICATIONS FOR NEUROPLASTICITY

The expression profiles of immediate-early genes in visual cortex have been documented in a number of studies with the aim of providing further insight into developmental maturation and cortical plasticity. The role of certain IEGs, especially c-*fos*, has been speculated in terms of orchestrating activity-dependent change of neuronal events that mediate plasticity (Chiasson et al., 1997; Kaczmarek and Chaudhuri, 1997; Hughes et al., 1999). The fact that neuronal activation provokes expression of a number of IEGs has focused attention on their role in regulating long-term changes in brain function.

One way to obtain insight into the possible role of IEGs in maturation and plasticity is to map the developmental profiles of their basal expression in visual neocortex. The developmental expression of c-Fos and the fos-related antigens (Fras) has been examined in the rat (Alcantara and Greenough, 1993). Immunostaining for c-Fos protein showed negligible expression during the entire developmental period, though restricted expression of Fras was observed up to postnatal day nine (PD9) in the deep layers of visual cortex. A second peak of Fras expression is evident throughout all layers at PD12 and PD15. This has been associated with eye opening, which occurs around this time. Similarly, *zif268* mRNA and protein levels rise sharply in rat visual cortex between PD10 and PD16, reaching peak values by PD21 (Worley et al., 1990; Herms et al., 1994). As with expression of Fras, the *zif268* developmental profile has been attributed to eye opening and subsequent light-driven gene induction.

In the cat visual cortex, a significant increase in c-*fos* and *zif268* mRNA levels is observed between the first and fifth postnatal weeks (McCormack et al., 1992). The elevated levels of both gene products continue throughout development and begin to decrease by the 20th postnatal week to reach adult levels. In terms of mapping the developmental profiles, particular attention has been paid to the pattern in different cortical layers because of their known roles in receiving thalamic afferents as well as intracortical processing and projections to other cortical areas. Immunodetection for c-Fos protein shows that stained cells are present in the primary visual cortex of 30-days-old kittens but not adults (Beaver et al., 1993). The immunopositive cells are most abundant in layers II/III, VI and the upper half of layer IV. In general, c-fos expression has been noted in layer IV for developing cats but appears to be absent in adults (Kaplan et al., 1996).

Developmentally regulated expression of Zif268 protein also shows striking differences in laminar expression (Kaplan et al., 1995; Kaplan et al., 1996). At 0.5 weeks of age, Zif268 expression is localized almost exclusively in layer VI/subplate. Thereafter, immunopositive cells spread out from a dense band in the deep layers to more superficial cortical layers, being distributed principally in two distinct bands, one in layer VI/subplate, and another in the lower part of the cortical plate. At later time points, Zif268 immunopositive cells are more widely distributed across cortical layers IV–VI, with no expression in layers I–III. By 5 weeks of age, Zif268 is observed to be highly expressed in neurons of all cortical layers, with reduced immunolabeling in layers I and V. In cats of 20 weeks, the staining in layer IV becomes decreased and, as in adults, Zif268 immunopositive cells are located almost exclusively above and below layer IV.

While the implications for basal c-Fos or Zif268 in coordinating developmental plasticity remains an open question, it is generally agreed that any such role would have to be tied to ongoing visual stimulation. Thus, there has been significant effort toward delineating activity-driven induction of various IEGs in the developing visual cortex. It has been shown that induction of both c-*fos* and *junB* is quite robust in young rats that were dark-reared from birth and then exposed to light stimulation (Worley et al., 1990; Sato et al., 2000). However, such increases are not as apparent in adult rats. A significant increase in *zif268* mRNA and protein levels is also evident in rats that are exposed to light after a period of dark-rearing (Worley et al., 1990, 1991; Nedivi et al., 1996). As with c-*fos* and *junB*, there is a more robust response of *zif268* mRNA in young rats compared to adult animals.

A high degree of c-*fos* and *junB* mRNA induction is also observed in the developing cat visual cortex after brief light stimulation (Rosen et al., 1992). However, light exposure for 6 h per day for two successive days after dark-rearing resulted in returning the mRNA levels of both genes to the values observed in dark-reared controls. Light exposure in dark-reared kittens shows robust c-Fos protein induction throughout visual cortex (Beaver et al., 1993; Mitchell et al., 1995). Layers II, III, upper half of layer IV, and VI were the most densely labeled. In addition, a strip of immunopositive cells was also seen at the very bottom of layer IV. As with the rat results, prolonged light-exposure times resulted in decreased c-Fos immunoreactivity. Similarly, Mower and colleagues have found that c-Fos induction in dark-reared kittens is much more robust compared to the adult (Mower, 1994; Kaplan et al., 1996; Mower and Kaplan, 1999). An accumulation of c-Fos was observed in all layers of kitten primary visual cortex, though at somewhat lower levels in layer V. In adults, however, immunoreactive cells were concentrated in layers II, III and VI, with only faint labeling in layers IV and V. The main difference between adult and kitten c-Fos expression is evident in layer IV, where adults show negligible induction and kittens robust induction.

zif268 mRNA expression is also elevated by light stimulation following dark-rearing after birth (Rosen et al., 1992). Exposure to light for 1 h produces marked accumulation of *zif268* mRNA, whereas exposure for 6 h each day for two successive days resulted in a return to basal levels that are ordinarily seen in dark-reared kittens. It has been found using the same paradigm as described above for c-Fos that a sustained increase in Zif268 immunostaining occurs in dark-reared kitten visual cortex following brief light exposure (Kaplan et al., 1996). Immunostained cells are evident in all layers in kitten primary visual cortex, though at a reduced intensity in layers I and V. In adults, in addition to the poor labeling in layers I and V, layer IV also remained weakly stained except for a narrow band of positive cells at the very bottom of the layer. This pattern of expression persists even after 4 h of stimulation and was essentially similar to that observed with light-adapted control animals.

While it is known that Old World monkeys are born with segregated ocular dominance columns (ODC), it was unclear for some time whether neurons within these columns are capable of activity-dependent IEG expression during development. Studies on New World monkeys had questioned the contribution of Zif268 to ODC formation (Silveira et al., 1996; Fonta et al., 1997). These studies failed to reveal columns of Zif268 expression after monocular deprivation during the critical period, although they were clearly visible in adults (Silveira et al., 1996; Fonta et al., 1997; Markstahler et al., 1998). Thus, the conjecture that both c-*fos*, and *zif268* may play important roles in postnatal cortical development has not been supported by evidence that these IEGs are linked to one of the most intriguing developmental events in primates, namely ODC formation and stabilization. To clarify the role of IEGs in this process, a recent study examined c-*fos* and *zif268* expression in developing primary visual cortex of Old World monkeys (Kaczmarek et al., 1999). It was found that well delineated patterns of ODCs could indeed be visualized by sensory induction of both IEGs throughout the critical period, starting as early as the first postnatal day. The expression levels are similar in layers II, III, IVC, and VI throughout development, with no selective decline in the thalamorecepient layer (layer IVC) of adult monkeys. A narrow strip of non-columnar c-Fos expression was observed at the border of layers IVC and V. These results showed that neurons in monkey visual cortex are equipped at birth with the molecular machinery for coupling sensory inputs to active genomic responses and that this responsivity extends throughout the critical period.

5.4. NEURAL SUBSTRATES OF COMPLEX VISUAL FUNCTION

The use of molecular mapping strategies to observe activation in response to complex visual stimuli represents a new and exciting direction for this field (Miyashita et al., 1998; Broad et al., 2000). Miyashita et al. (1998) have used *zif268* mapping in monkey inferotemporal cortex to show that visual paired associate learning produces increased immunostaining in superficial and deep layers. The study by Broad et al. (2000) examined c-*fos* mRNA expression in several brain areas, including inferotemporal cortex, amygdala, and hippocampus, of sheep after they were exposed to normal and inverted facial stimuli. They found that face recognition in sheep preferentially engages the right temporal cortex. There also appeared to be activation of the amygdala and hippocampus that reflected downstream activation after active choice between upright faces. Interestingly, inverted faces, which are known to be perceived in a less veridical manner in humans, did not show enhanced activation of either the inferotemporal cortex, amygdala, or hippocampus. The lateralization effects observed in this animal model are consistent with human data. A wide range of studies involving assessment of pathological changes, neuroimaging, and psychophysical experimentation have shown hemispheric specialization for face recognition and a preferential engagement of the right hemisphere (Broman, 1978; Sergent and Bindra, 1981; De Renzi et al., 1994).

Behavioral experiments suggest that IEG induction in visual cortex may be linked to stimulus novelty and can often be critical for c-*fos* activation. Quantitative analysis of *zif268* expression in visual cortex has shown that it is expressed at significantly higher levels in rats exposed to a complex housing environment in comparison to those housed individually and without any handling (Wallace et al., 1995). It may be that the accumulation of *zif268* mRNA in animals reared in an enriched environment reflects a higher level of neuronal activity that correlates with plastic changes of sensory cortex. Several research groups have investigated the possible effects of behavioral training with complex stimuli on IEG expression. A single training session of 2-way active avoidance behavior with darkness being used as a conditioned stimulus (CS) resulted in a large accumulation of c-*fos* and *zif268* mRNA levels in various brain regions, including the visual cortex (Nikolaev et al., 1992). Long-term training with darkness as CS for 9 days up to an asymptotic level of performance produced negligible c-*fos* activation. Active exploration has been shown to produce marked increase in c-*fos* mRNA in visual cortex of rats, a result that has been attributed to arousal, stress, and stimulus novelty (Hess et al., 1995a).

A significant increase in c-Fos immunostaining has been observed in occipital and temporal neocortex and hippocampus after rats were exposed to novel objects (Zhu et al., 1995, 1996, 1997). Different groups of animals were exposed to either novel objects, familiar objects, or to the same pattern of illumination without any objects being shown. c-Fos expression was elevated when animals were exposed only to novel objects. Similarly, an enhancing effect of exposure to a novel environment on c-Fos and c-Jun immunoreactivity has also been observed by Papa et al. (1993). The enhanced activation in all of these cases has been attributed to processing and storage of information about the novelty of visual stimuli as a prelude to recognition memory.

The role of stimulus novelty in selective cortical activation has been explored in further detail by Montero (1997, 1999, 2000). He has suggested that attentional factors play a dominant role in the enhancement of c-*fos* induction that is seen in rat visual and somatosensory cortices after exploring a novel environment. Furthermore, the known interaction between cortical and subcortical sites was postulated to produce an attention-dependent gating of thalamocortical transmission. Attentional activation of the thalamic reticular nucleus (TRN) was significantly diminished in animals with monocular postnatal deprivation (Montero, 1999). Furthermore, c-Fos mapping showed that activation of the TRN was reduced after unilateral visual cortex lesions but sparing activity in the LGN. This suggests that top-down control from sensory neocortex is responsible for activation of TRN by attentive exploration of a novel complex environment (Montero, 2000).

6. MULTISENSORY PROCESSING

One of the brain regions where multisensory processing has been well characterized is the superior colliculus (SC). A number of unresolved issues in SC multisensory processing concern the ability to which they integrate information across diverse sensory cues. A further elucidation of response characteristics among multisensory neurons here would benefit from imaging studies that provide information on sensory integration and spatial layout within the structure. However, this would require that separate activity maps be obtained for each stimulation condition so that they can reveal both modality preference as well as incidences of cross-modal responsivity.

A double-labeling technique that was recently developed can be used to study this issue (Chaudhuri, 1997; Chaudhuri et al., 1997, 2000). The technique is based on exploiting the different time course of IEG mRNA versus protein induction. By timing the stimulation sequences appropriately, it is possible to obtain a map of mRNA staining in response to the first stimulus and protein staining in response to the second stimulus. The staining of both products within the same tissue section can also reveal instances of co-activation because such neurons will be labeled for both products.

A recent study on the visual and auditory neocortex and SC used the double-label mapping technique with *zif268* to obtain activity-induced expression under two separate conditions of sensory (visual and auditory) stimulation (Zangenehpour and Chaudhuri, 1999; Zangenehpour et al., 2000). It was found that activity maps of the SC superficial layers contain an exclusively visual unimodal map, whereas activity maps in deep SC contain multimodal neurons with varying degrees of sensory convergence. The activity maps of the deep SC also showed that auditory processing was largely carried out by a small, bimodal group of neurons, whereas visual processing was coordinated by both a large unimodal and a small bimodal pool of multisensory neurons (Fig. 8).

A similar approach has been taken by Mikula and Hendry (2000). They examined the topographic distribution of unimodal visual and somatosensory activated neurons within the monkey prefrontal cortex. Rather than staining for mRNA and protein products of the same immediate-early gene, they have exploited the differential time course of c-Fos versus Fra-2 induction. They found that certain areas of the prefrontal cortex were largely unimodal for either visual or somatosensory preference, with activated neurons appearing in a columnar fashion spanning the thickness of the cortex. However, there were also regions of prefrontal cortex where multisensory responses where noted. These regions showed either overlapping or interdigitated profiles of visual and somatosensory responsivity.

7. ABBREVIATIONS

2-DG	2-deoxyglucose
AOB	accessory olfactory bulb

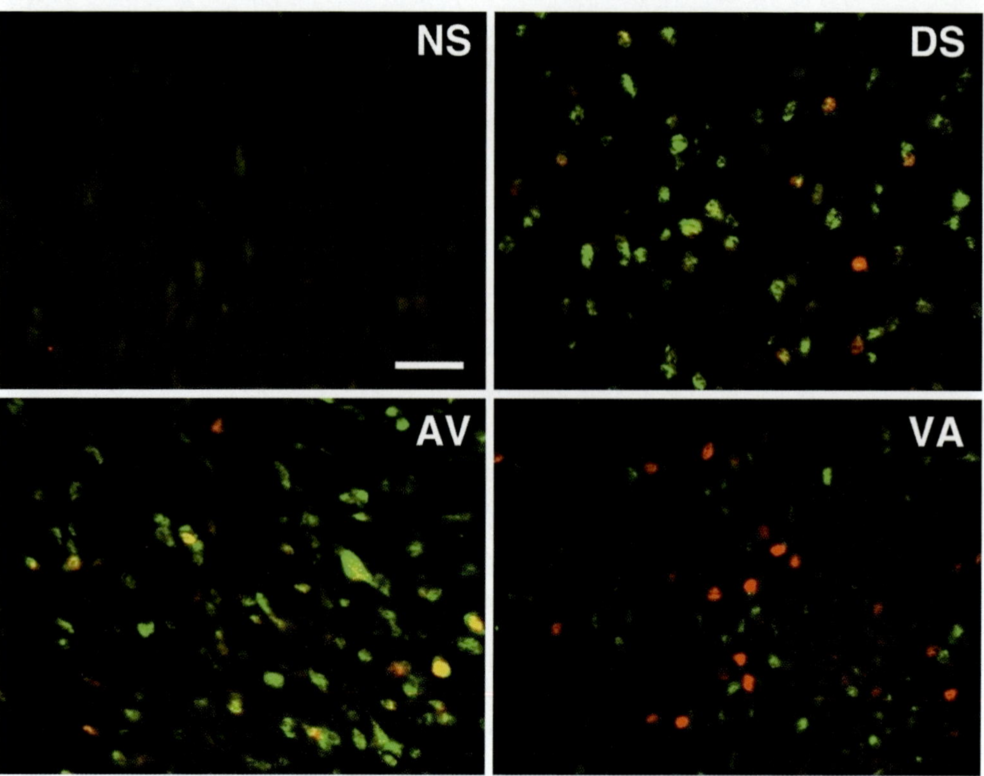

Fig. 8. Fluorescent labeling of *zif268* mRNA and protein products by simultaneous non-radioactive ISHH and ICC processing on the same tissue section of the superior colliculus. The mRNA product was tagged with a green chromophore, whereas the protein product was tagged with a red chromophore. The control condition (no stimulus, NS) showed minimal/baseline level of *zif268* expression present both as mRNA and protein forms. When exposed to a compound stimulus of both visual and auditory signals (dual stimulus, DS), the entire responsive population is both mRNA- and protein-positive. When exposed to 2 h of auditory following by 30 min of visual stimulation (AV) or vice versa (VA), the mRNA-positive neurons represent visually or acoustically driven neurons, respectively. The protein-positive neurons in both cases represent acoustically or visually driven neurons, respectively. A sample of double-labeled neurons in the structure shows a sub-population of bimodal (multisensory) neurons. Adapted from Zangenehpour and Chaudhuri, 1999.

AP-1	activating-protein 1
AS	audiogenic seizure
CIC	central nucleus of the inferior colliculus
CTA	conditioned taste aversion
ERK	extracellular signal-regulated kinase
FDG	fluoro-2-deoxyglucose
FLI	Fos-like immunoreactivity
GABA	gamma-aminobutyric acid
GFAP	glial fibrillary acidic protein
IC	inferior colliculus
ICC	immunocytochemistry
IEG	immediate-early gene
IR	immunoreactive
ISHH	in situ hybridization histochemistry

ITF	inducible transcription factor
LK	leukotrienes
MPOA	medial preoptic area
NAc	nucleus accumbens
NST	nucleus of the solitary tract
ODC	ocular dominance columns
PBN	pontine parabrachial nucleus
PG	prostaglandins
RF	reticular formation
SC	superior colliculus
SCN	suprachiasmatic nucleus
SDN	sexually dimorphic nucleus
TRN	thalamic reticular nucleus
VNO	vomeronasal organ
VSCC	voltage-sensitive calcium channel

8. ACKNOWLEDGEMENTS

We are grateful to the following people for providing images and/or critical comments: Shimon Amir, Robert Filipkowski, Cary Kogan, Alan Spector, and Heinz Steiner. We thank Michelle To for assistance. Supported by MRC-CIHR (MOP-42514, MOP-14776, MT-14526) and NSERC (155482-98, 228230-99) operating grants to AC and an NSERC PGS-B Fellowship to SZ.

9. REFERENCES

Adams JC (1995): Sound stimulation induces Fos-related antigens in cells with common morphological properties throughout the auditory brainstem. *J Comp Neurol 361*:645–668.

Alcantara AA, Greenough WT (1993): Developmental regulation of Fos and Fos-related antigens in cerebral cortex, striatum, hippocampus, and cerebellum of the rat. *J Comp Neurol 334*:75–85.

Amir S, Robinson B (1996): Fos expression in rat visual cortex induced by ocular input of ultraviolet light. *Brain Res 716*:213–218.

Amir S, Cain S, Sullivan J, Robinson B, Stewart J (1999a): In rats, odor-induced Fos in the olfactory pathways depends on the phase of the circadian clock. *Neurosci Lett 272*:175–178.

Amir S, Cain S, Sullivan J, Robinson B, Stewart J (1999b): Olfactory stimulation enhances light-induced phase shifts in free-running activity rhythms and Fos expression in the suprachiasmatic nucleus. *Neuroscience 92*:1165–1170.

Anand KJ, Coskun V, Thrivikraman KV, Nemeroff CB, Plotsky PM (1999): Long-term behavioral effects of repetitive pain in neonatal rat pups. *Physiol Behav 66*:627–637.

Anokhin KV, Mileusnic R, Shamakina IY, Rose SP (1991): Effects of early experience on *c-fos* gene expression in the chick forebrain. *Brain Res 544*:101–107.

Aronin N, Sagar SM, Sharp FR, Schwartz WJ (1990): Light regulates expression of a Fos-related protein in rat suprachiasmatic nuclei. *Proc Natl Acad Sci USA 87*:5959–5962.

Baulmann J, Spitznagel H, Herdegen T, Unger T, Culman J (2000): Tachykinin receptor inhibition and c-Fos expression in the rat brain following formalin-induced pain. *Neuroscience 95*:813–820.

Baum MJ, Everitt BJ (1992): Increased expression of *c-fos* in the medial preoptic area after mating in male rats: role of afferent inputs from the medial amygdala and midbrain central tegmental field. *Neuroscience 50*:627–646.

Beaver CJ, Mitchell DE, Robertson HA (1993): Immunohistochemical study of the pattern of rapid expression of c-Fos protein in the visual cortex of dark-reared kittens following initial exposure to light. *J Comp Neurol 333*:469–484.

Beck CHM, Chow HL, Cooper SJ (1986): Dose-related response of male rats to apomorphine: snout contact int he open-field. *Physiol Behav 37*:819–825.

Bertolucci C, Sovrano VA, Magnone MC, Foa A (2000): Role of suprachiasmatic nuclei in circadian and light-entrained behavioral rhythms of lizards. *Am J Physiol: Regulatory, Integrative and Comparative Physiology 279*:2121–2131.

Bhandari RN, Ogilvie J, Clarke RW (1999): Differences in opioidergic inhibition of spinal reflexes and Fos expression evoked by mechanical and chemical noxious stimuli in the decerebrated rabbit. *Neuroscience 90*:177–189.

Bialy M, Nikolaev E, Beck J, Kaczmarek L (1992): Delayed *c-fos* expression in sensory cortex following sexual learning in male rats. *Brain Res Mol Brain Res 14*:352–356.

Bledsoe Jr. SC, Nagase S, Miller JM, Altschuler RA (1995): Deafness-induced plasticity in the mature central auditory system. *Neuroreport 7*:225–229.

Brennan PA, Hancock D, Keverne EB (1992): The expression of the immediate-early genes *c-fos*, *egr-1* and *c-jun* in the accessory olfactory bulb during the formation of an olfactory memory in mice. *Neuroscience 49*:277–284.

Brennan PA, Schellinck HM, Keverne EB (1999): Patterns of expression of the immediate-early gene *egr-1* in the accessory olfactory bulb of female mice exposed to pheromonal constituents of male urine. *Neuroscience 90*:1463–1470.

Bressler SC, Baum MJ (1996): Sex comparison of neuronal Fos immunoreactivity in the rat vomeronasal projection circuit after chemosensory stimulation. *Neuroscience 71*:1063–1072.

Broad KD, Mimmack ML, Kendrick KM (2000): Is right hemisphere specialization for face discrimination specific to humans?. *Eur J Neurosci 12*:731–741.

Broman M (1978): Reaction time differences between left and right hemispheres for face and letter discrimination in children and adults. *Cortex 14*:578–591.

Brown MC, Liu TS (1995): Fos-like immunoreactivity in central auditory neurons of the mouse. *J Comp Neurol 357*:85–97.

Bullitt E (1989): Induction of *c-fos*-like protein within the lumbar spinal cord and thalamus of the rat following peripheral stimulation. *Brain Res 493*:391–397.

Bullitt E (1990): Expression of *c-fos*-like protein as a marker for neuronal activity following noxious stimulation in the rat. *J Comp Neurol 296*:517–530.

Bullitt E (1991): Somatotopy of spinal nociceptive processing. *J Comp Neurol 312*:279–290.

Bussolino DF, de Arriba Zerpa GA, Grabois VR, Conde CB, Guido ME, Caputto BL (1998): Light affects *c-fos* expression and phospholipid synthesis in both retinal ganglion cells and photoreceptor cells in an opposite way for each cell type. *Brain Res Mol Brain Res 58*:10–15.

Calamandrei G, Keverne EB (1994): Differential expression of Fos protein in the brain of female mice dependent on pup sensory cues and maternal experience. *Behav Neurosci 108*:113–120.

Caleo M, Lodovichi C, Pizzorusso T, Maffei L (1999): Expression of the transcription factor Zif268 in the visual cortex of monocularly deprived rats: effects of nerve growth factor. *Neuroscience 91*:1017–1026.

Caputto BL, Guido ME (2000): Immediate early gene expression within the visual system: light and circadian regulation in the retina and the suprachiasmatic nucleus. *Neurochem Res 25*:153–162.

Carretta D, Herve-Minvielle A, Bajo VM, Villa AE, Rouiller EM (1999): c-Fos expression in the auditory pathways related to the significance of acoustic signals in rats performing a sensory–motor task. *Brain Res 841*:170–183.

Carstens E, Saxe I, Ralph R (1995): Brainstem neurons expressing c-Fos immunoreactivity following irritant chemical stimulation of the rat's tongue. *Neuroscience 69*:939–953.

Chambille I, Doyle S, Serviere J (1993): Photic induction and circadian expression of Fos-like protein. Immunohistochemical study in the retina and suprachiasmatic nuclei of hamster. *Brain Res 612*:138–150.

Chaudhuri A (1997): Neural activity mapping with inducible transcription factors. *Neuroreport 8*:v–ix.

Chaudhuri A, Cynader MS (1993): Activity-dependent expression of the transcription factor Zif268 reveals ocular dominance columns in monkey visual cortex. *Brain Res 605*:349–353.

Chaudhuri A, Matsubara JA, Cynader MS (1995): Neuronal activity in primate visual cortex assessed by immunodetection for the transcription factor Zif268. *Vis Neurosci 12*:35–50.

Chaudhuri A, Nissanov J, Larocque S, Rioux L (1997): Dual activity maps in primate visual cortex produced by different temporal patterns of *zif268* mRNA and protein expression. *Proc Natl Acad Sci USA 94*:2671–2675.

Chaudhuri A, Zangenehpour S, Rahbar-Dehgan F, Ye F (2000): Molecular maps of neural activity and quiescence. *Acta Neurobiol Exp 60*:403–410.

Chen T, Chen S, Hsieh Y (2000): Evaluating the protective role of the olivocochlear bundle against acoustic overexposure in rats by using fos immunohistochemistry. *J Neurol Sci 177*:104–113.

Chew SJ, Mello C, Nottebohm F, Jarvis E, Vicario DS (1995): Decrements in auditory responses to a repeated conspecific song are long-lasting and require two periods of protein synthesis in the songbird forebrain. *Proc Natl Acad Sci USA 92*:3406–3410.

Chiasson BJ, Hong MG, Robertson HA (1997): Putative roles for the inducible transcription factor *c-fos* in the central nervous system: studies with antisense oligonucleotides. *Neurochem Int 31*:459–475.

Chmielowska J, Kossut J, Chmielowski M (1986): Single vibrissal cortical column in the mouse labeled with 2-deoxyglucose. *Exp Brain Res 63*:607–619.

Chowdhury GMI, Fujioka T, Nakamura S (2000): Induction and adaptation of fos expression in the rat brain by two types of acute restraint stress. *Brain Res Bull 52*:171–182.

Coolen LM, Peters HJ, Veening JG (1997): Distribution of Fos immunoreactivity following mating versus anogenital investigation in the male rat brain. *Neuroscience 77*:1151–1161.

Coopersmith R, Lee S, Leon M (1986): Olfactory bulb responses after odor aversion learning by young rats. *Brain Res 389*:271–277.

Correa-Lacarcel J, Pujante MJ, Terol FF, Almenar-Garcia V, Puchades-Orts A, Ballesta JJ, Lloret J, Robles JA, Sanchez-del-Campo F (2000): Stimulus frequency affects *c-fos* expression in the rat visual system. *J Chem Neuroanat 18*:135–146.

Craner SL, Hoffman GE, Lund JS, Humphrey AL, Lund RD (1992): c-Fos labeling in rat superior colliculus: activation by normal retinal pathways and pathways from intracranial retinal transplants. *Exp Neurol 117*:219–229.

DaCosta APC, Broad KD, Kendrick KM (1997): Olfactory memory and maternal behaviour-induced changes in *c-fos* and zif/268 mRNA expression in the sheep brain. *Mol Brain Res 46*:63–76.

De Renzi E, Perani D, Carlesimo GA, Silveri MC, Facio F (1994): Prosopagnosia can be associated with damage confined to the right hemisphere: an MRI and PET study and a review of the literature. *Neuropsychologia 179*:893–902.

DiNardo LA, Travers JB (1997): Distribution of fos-like immunoreactivity in the medullary reticular formation of the rat after gustatory elicited ingestion and rejection behaviors. *J Neurosci 17*:3826–3839.

Doving KB, Trotier D (1998): Structure and function of the vomeronasal organ. *J Exp Biol 201*:2913–2925.

Dragunow M, Robertson HA (1988): Localization and induction of *c-fos* protein-like immunoreactive material in the nuclei of adult mammalian neurons. *Brain Res 440*:252–260.

Dudley CA, Moss RL (1999): Activation of an anatomically distinct subpopulation of accessory olfactory bulb neurons by chemosensory stimulation. *Neuroscience 91*:1549–1556.

Dudley CA, Rajendren G, Moss RL (1996): Signal processing in the vomeronasal system: modulation of sexual behavior in the female rat. *Crit Rev Neurobiol 10*:265–290.

Dulac C, Axel R (1995): A novel family of genes encoding putative pheromone receptors in mammals. *Cell 83*:195–206.

Durham D, Wollsey TA (1978): Acute whisker removal reduces neuronal activity in barrels of mouse SmI cortex. *J Comp Neurol 178*:629–644.

Earnest DJ, Iadarola M, Yeh HH, Olschowka JA (1990): Photic regulation of *c-fos* expression in neural components governing the entrainment of circadian rhythms. *Exp Neurol 109*:353–361.

Ehret G, Fischer R (1991): Neuronal activity and tonotopy in the auditory system visualized by *c-fos* gene expression. *Brain Res 567*:350–354.

Emmert MH, Herman JP (1999): Differential forebrain *c-fos* mRNA induction by ether inhalation and novelty: evidence for distinctive stress pathways. *Brain Res 845*:60–67.

Emond MH, Weingarten HP (1995): Fos-like immunoreactivity in vagal and hypoglossal nuclei in different feeding states: a quantitative study. *Physiol Behav 58*:459–465.

Fernandez-Fewell GD, Meredith M (1994): *c-fos* expression in vomeronasal pathways of mated or pheromone-stimulated male golden hamsters: contributions from vomeronasal sensory input and expression related to mating performance. *J Neurosci 14*:3643–3654.

Fernandez-Fewell GD, Meredith M (1998): Olfactory contribution to Fos expression during mating in inexperienced male hamsters. *Chem Senses 23*:257–267.

Fiber JM, Adames P, Swann JM (1993): Pheromones induce *c-fos* in limbic areas regulating male hamster mating behavior. *Neuroreport 4*:871–874.

Fichtel I, Ehret G (1999): Perception and recognition discriminated in the mouse auditory cortex by c-Fos labeling. *Neuroreport 10*:2341–2345.

Filipkowski RK (2000): Inducting gene expression in barrel cortex — focus on immediate early genes. *Acta Neurobiol Exp 60*:411–418.

Filipkowski RK, Rydz M, Berdel B, Morys J, Kaczmarek L (2000): Tactile experience induces *c-fos* expression in rat barrel cortex. *Learning Mem 7*:116–122.

Fleming AS, Walsh C (1994): Neuropsychology of maternal behavior in the rat: *c-fos* expression during mother–litter interactions. *Psychoneuroendocrinology 19*:429–443.

Fleming AS, Suh EJ, Korsmit M, Rusak B (1994): Activation of Fos-like immunoreactivity in the medial preoptic area and limbic structures by maternal and social interactions in rats. *Behav Neurosci 108*:724–734.

Foidart A, Meddle SL, Balthazart J (1999): Mating-induced Fos and aromatase are not co-localized in the preoptic area. *Neuroreport 10*:907–912.

Fonta C, Chappert C, Imbert M (1997): *N*-methyl-D-aspartate subunit R1 involvement in the postnatal organization of the primary visual cortex of *Callithrix jacchus*. *J Comp Neurol 386*:260–276.

Friauf E (1995): *c-fos* immunocytochemical evidence for acoustic pathway mapping in rats. *Behav Brain Res 66*:217–224.

Friedman D, Haim A, Zisapel N (1997): Temporal segregation in coexisting spiny mice (genus Acomys): role of photoperiod and heterospecific odor. *Physiol Behav 62*:407–411.

Funk D, Amir S (2000a): Circadian modulation of Fos responses to odor of the red fox, a rodent predator, in the rat olfactory system. *Brain Res 866*:262–267.

Funk D, Amir S (2000b): Enhanced fos expression within the primary olfactory and limbic pathways induced by an aversive conditioned odor stimulus. *Neuroscience 98*:403–406.

Ghosh A, Ginty DD, Bading H, Greenberg ME (1994): Calcium regulation of gene expression in neuronal cells. *J Neurobiol 25*:294–303.

Gill CJ, Wersinger SR, Veney SL, Rissman EF (1998): Induction of fos-like immunoreactivity in musk shrews after mating. *Brain Res 811*:21–28.

Gilron I, Quirion R, Coderre TJ (1999a): Pre- versus postformalin effects of ketamine or large-dose alfentanil in the rat: discordance between pain behavior and spinal Fos-like immunoreactivity. *Anesth Analg 89*:128–135.

Gilron I, Quirion R, Coderre TJ (1999b): Pre- versus postinjury effects of intravenous GABAergic anesthetics on formalin-induced Fos immunoreactivity in the rat spinal cord. *Anesth Analg 88*:414–420.

Goel N, Lee TM (1995): Sex differences and effects of social cues on daily rhythms following phase advances in *Octodon degus*. *Physiol Behav 58*:205–213.

Greco B, Edwards DA, Michael RP, Clancy AN (1998): Androgen receptors and estrogen receptors are colocalized in male rat hypothalamic and limbic neurons that express Fos immunoreactivity induced by mating. *Neuroendocrinology 67*:18–28.

Gudehithlu KP, Neff NH, Hadjiconstantinou M (1993): *c-fos* and NGFI-A mRNA of rat retina: evidence for light-induced augmentation and a role for cholinergic and glutamate receptors. *Brain Res 631*:77–82.

Guido ME, Goguen D, Robertson HA, Rusak B (1996a): Spontaneous and light-evoked expression of JunB-like protein in the hamster suprachiasmatic nucleus near subjective dawn. *Neurosci Lett 217*:9–12.

Guido ME, Rusak B, Robertson HA (1996b): Expression of fosB mRNA in the hamster suprachiasmatic nucleus is induced at only selected circadian phases. *Brain Res 739*:132–138.

Guido ME, Rusak B, Robertson HA (1996c): Spontaneous circadian and light-induced expression of *junB* mRNA in the hamster suprachiasmatic nucleus. *Brain Res 732*:215–222.

Guido ME, Goguen D, De Guido L, Robertson HA, Rusak B (1999): Circadian and photic regulation of immediate-early gene expression in the hamster suprachiasmatic nucleus. *Neuroscience 90*:555–571.

Guo J, Zhou AW, Moss RL (1997): Urine and urine-derived compounds induce *c-fos* mRNA expression in accessory olfactory bulb. *Neuroreport 8*:1679–1683.

Guthrie KM, Gall CM (1995a): Functional Mapping of Odor-Activated Neurons in the Olfactory- Bulb. *Chem Senses 20*:271–282.

Guthrie KM, Gall CM (1995b): Odors increase Fos in olfactory bulb neurons including dopaminergic cells. *Neuroreport, 6*:2145–2149; discussion 2103.

Guthrie KM, Anderson AJ, Leon M, Gall C (1993): Odor-induced increases in *c-fos* mRNA expression reveal an anatomical 'unit' for odor processing in olfactory bulb. *Proc Natl Acad Sci USA 90*:3329–3333.

Halem HA, Cherry JA, Baum MJ (1999): Vomeronasal neuroepithelium and forebrain Fos responses to male pheromones in male and female mice. *J Neurobiol 39*:249–263.

Halpern M, Shapiro LS, Jia C (1995): Differential localization of G proteins in the opossum vomeronasal system. *Brain Res 677*:157–161.

Harrer MI, Travers SP (1996): Topographic organization of Fos-like immunoreactivity in the rostral nucleus of the solitary tract evoked by gustatory stimulation with sucrose and quinine. *Brain Res 711*:125–137.

Harris JA (1998): Using *c-fos* as a neural marker of pain. *Brain Res Bull 45*:1–8.

Harris JA, Westbrook RF, Duffield TQ, Bentivoglio M (1995): Fos expression in the spinal cord is suppressed in rats displaying conditioned hypoalgesia. *Behav Neurosci 109*:320–328.

Heeb MM, Yahr P (1996): c-Fos immunoreactivity in the sexually dimorphic area of the hypothalamus and related brain regions of male gerbils after exposure to sex-related stimuli or performance of specific sexual behaviors. *Neuroscience 72*:1049–1071.

Herdegen T, Leah JD (1998): Inducible and constitutive transcription factors in the mammalian nervous system: control of gene expression by Jun, Fos and Krox, and CREB/ATF proteins. *Brain Res Brain Res Rev 28*:370–490.

Herdegen T, Zimmermann M (1995): Immediate early genes (IEGs) encoding for inducible transcription factors (ITFs) and neuropeptides in the nervous system: functional network for long-term plasticity and pain. *Prog Brain Res 104*:299–321.

Herdegen T, Kovary K, Leah J, Bravo R (1991): Specific temporal and spatial distribution of JUN, FOS, and KROX-24 proteins in spinal neurons following noxious transsynaptic stimulation. *J Comp Neurol 313*:178–191.

Herdegen T, Sandkuhler J, Gass P, Kiessling M, Bravo R, Zimmermann M (1993): JUN, FOS, KROX, and CREB transcription factor proteins in the rat cortex: basal expression and induction by spreading depression and epileptic seizures. *J Comp Neurol 333*:271–288.

Hermanson O, Blomqvist A (1997): Subnuclear localization of FOS-like immunoreactivity in the parabrachial nucleus after orofacial nociceptive stimulation of the awake rat. *J Comp Neurol 387*:114–123.

Herms J, Zurmohle U, Schlingensiepen R, Brysch W, Schlingensiepen KH (1994): Developmental expression of the transcription factor *zif268* in rat brain. *Neurosci Lett 165*:171–174.

Herrada G, Dulac C (1997): A novel family of putative pheromone receptors in mammals with a topographically organized and sexually dimorphic distribution. *Cell 90*:763–773.

Herrera DG, Robertson HA (1996): Activation of *c-fos* in the brain. *Prog Neurobiol 50*:83–107.

Hess US, Lynch G, Gall CM (1995a): Regional patterns of *c-fos* mRNA expression in rat hippocampus following exploration of a novel environment versus performance of a well-learned discrimination. *J Neurosci 15*:7796–7809.

Hess US, Lynch G, Gall CM (1995b): Changes in *c-fos* mRNA expression in rat brain during odor discrimination learning: differential involvement of hippocampal subfields CA1 and CA3. *J Neurosci 15*:4786–4795.

Hirakawa M, Kawata M (1993): Distribution pattern of c-Fos expression induced by sciatic nerve sectioning in the rat central nervous system. *J Hirnforsch 34*:431–444.

Holm R (2000): A highly sensitive nonisotopic detection method for in situ hybridization. *Appl Immunohistochem Mol Morphol 8*:162–165.

Horton JC, Hocking DR, Adams DL (2000): Rapid identification of ocular dominance columns in macaques using cytochrome oxidase, Zif268, and dark-field microscopy. *Vis Neurosci 17*:495–508.

Houpt TA, Philopena JM, Wessel TC, Joh TH, Smith GP (1994): Increased c-Fos expression in nucleus of the solitary tract correlated with conditioned taste-aversion to sucrose in rats. *Neurosci Lett 172*:1–5.

Houpt TA, Philopena JM, Joh TH, Smith GP (1996a): c-Fos induction in the rat nucleus of the solitary tract correlates with the retention and forgetting of a conditioned taste aversion. *Learn Mem 3*:25–30.

Houpt TA, Philopena JM, Joh TH, Smith GP (1996b): c-Fos induction in the rat nucleus of the solitary tract by intraoral quinine infusion depends on prior contingent pairing of quinine and lithium chloride. *Physiol Behav 60*:1535–1541.

Hughes PE, Alexi T, Walton M, Williams CE, Dragunow M, Clark RG, Gluckman PD (1999): Activity and injury-dependent expression of inducible transcription factors, growth factors and apoptosis-related genes within the central nervous system. *Prog Neurobiol 57*:421–450.

Hunt SP, Pini A, Evan G (1987): Induction of *c-fos*-like protein in spinal cord neurons following sensory stimulation. *Nature 328*:632–634.

Hunter T, Karin M (1992): The regulation of transcription by phosphorylation. *Cell 70*:375–387.

Imamura K, Mori K, Fujita SC, Obata K (1985): Immunochemical identification of subgroups of vomeronasal nerve fibers and their segregated terminations in the accessory olfactory bulb. *Brain Res 328*:362–366.

Inamura K, Kashiwayanagi M, Kurihara K (1999): Regionalization of Fos immunostaining in rat accessory olfactory bulb when the vomeronasal organ was exposed to urine. *Eur J Neurosci 11*:2254–2260.

Inouye ST, Kawamura H (1979): Persistence of circadian rhythmicity in a mammalian hypothalamic 'island' containing the suprachiasmatic nucleus. *Proc Natl Acad Sci USA 76*:5962–5966.

Jarvis ED, Mello CV (2000): Molecular mapping of brain areas involved in parrot vocal communication. *J Comp Neurol 419*:1–31.

Jarvis ED, Mello CV, Nottebohm F (1995): Associative learning and stimulus novelty influence the song-induced expression of an immediate early gene in the canary forebrain. *Learn Mem 2*:62–80.

Jarvis ED, Schwabl H, Ribeiro S, Mello CV (1997): Brain gene regulation by territorial singing behavior in freely ranging songbirds. *Neuroreport 8*:2073–2077.

Jarvis ED, Scharff C, Grossman MR, Ramos JA, Nottebohm F (1998): For whom the bird sings: context-dependent gene expression [see comments]. *Neuron 21*:775–788.

Jasmin L, Gogas KR, Ahlgren SC, Levine JD, Basbaum AI (1994a): Walking evokes a distinctive pattern of Fos-like immunoreactivity in the caudal brainstem and spinal cord of the rat. *Neuroscience 58*:275–286.

Jasmin L, Wang H, Tarczy-Hornoch K, Levine JD, Basbaum AI (1994b): Differential effects of morphine on noxious stimulus-evoked fos-like immunoreactivity in subpopulations of spinoparabrachial neurons. *J Neurosci 14*:7252–7260.

Jen PH, Sun X, Shen JX, Chen QC, Qian Y (1997): Cytoarchitecture and sound activated responses in the auditory cortex of the big brown bat, *Eptesicus fuscus. Acta Otolaryngol Suppl 532*:61–67.

Jennings E, Fitzgerald M (1996): *c-fos* can be induced in the neonatal rat spinal cord by both noxious and innocuous peripheral stimulation. *Pain 68*:301–306.

Jia C, Halpern M (1996): Subclasses of vomeronasal receptor neurons: differential expression of G proteins (Gi alpha 2 and G(o alpha)) and segregated projections to the accessory olfactory bulb. *Brain Res 719*:117–128.

Jin H, Clayton DF (1997): Localized changes in immediate-early gene regulation during sensory and motor learning in zebra finches. *Neuron 19*:1049–1059.

Jinghong C, Guoxi T (1999): Morphological characteristics and electrophysiological responses of visceral nociceptive neurons in somatosensory cerebral cortex of cat. *Brain Res 846*:243–252.

Johnson BA, Woo CC, Duong HC, Nguyen V, Leon M (1995): A learned odor evokes an enhanced Fos-like glomerular response in the olfactory-bulb of young rats. *Brain Res 699*:192–200.

Johnston RE (1998): Pheromones, the vomeronasal system, and communication. From hormonal responses to individual recognition. *Ann NY Acad Sci 855*:333–348.

Jourdan F, Duveau A, Astic L, Holley A (1980): Spatial distribution of [^{14}C]2-deoxyglucose uptake in the olfactory bulbs of rats stimulated with two different odours. *Brain Res 188*:139–154.

Kaczmarek L, Chaudhuri A (1997): Sensory regulation of immediate-early gene expression in mammalian visual cortex: implications for functional mapping and neural plasticity. *Brain Res Brain Res Rev 23*:237–256.

Kaczmarek L, Zangenehpour S, Chaudhuri A (1999): Sensory regulation of immediate-early genes *c-fos* and *zif268* in monkey visual cortex at birth and throughout the critical period. *Cereb Cortex 9*:179–187.

Kaminska B, Kaczmarek L, Chaudhuri A (1996): Visual stimulation regulates the expression of transcription factors and modulates the composition of AP-1 in visual cortex. *J Neurosci 16*:3968–3978.

Kaminska B, Kaczmarek L, Zangenehpour S, Chaudhuri A (1999): Rapid phosphorylation of Elk-1 transcription factor and activation of MAP kinase signal transduction pathways in response to visual stimulation. *Mol Cell Neurosci 13*:405–414.

Kaplan IV, Guo Y, Mower GD (1995): Developmental expression of the immediate early gene EGR-1 mirrors the critical period in cat visual cortex. *Brain Res Dev Brain Res 90*:174–179.

Kaplan IV, Guo Y, Mower GD (1996): Immediate early gene expression in cat visual cortex during and after the critical period: differences between EGR-1 and Fos proteins. *Brain Res Mol Brain Res 36*:12–22.

Karin M, Hunter T (1995): Transcriptional control by protein phosphorylation: signal transmission from the cell surface to the nucleus. *Curr Biol 5*:747–757.

Kato N, Ishida N, Kanai H, Watanabe Y, Kuroda Y, McEwen BS (1996): Expression of *c-fos* mRNA after audiogenic seizure in adult rats with neonatal hypothyroidism. *Brain Res Mol Brain Res 38*:85–90.

Katz LF, Ball GF, Nelson RJ (1999): Elevated Fos-like immunoreactivity in the brains of postpartum female prairie voles, *Microtus ochrogaster. Cell Tissue Res 298*:425–435.

Keilmann A, Herdegen T (1995): Expression of the *c-fos* transcription factor in the rat auditory pathway following postnatal auditory deprivation. *Eur Arch Otorhinolaryngol 252*:287–291.

Keilmann A, Herdegen T (1997): The c-Fos transcription factor in the auditory pathway of the juvenile rat: effects of acoustic deprivation and repetitive stimulation. *Brain Res 753*:291–298.

Kennaway DJ (1998): Generation and entrainment of circadian rhythms. *Clin Exp Pharmacol Physiol 25*:862–865.

Kenshalo DR, Iwata K, Sholas M, Thomas DA (2000): Response properties and organization of nociceptive neurons in area 1 of monkey primary somatosensory cortex. *J Neurophysiol 84*:719–729.

Keverne EB (1999): The vomeronasal organ. *Science 286*:716–720.

Kilduff TS, Vugrinic C, Lee SL, Milbrandt JD, Mikkelsen JD, O'Hara BF, Heller HC (1998): Characterization of the circadian system of NGFI-A and NGFI-A/NGFI-B deficient mice. *J Biol Rhythms 13*:347–357.

Kimpo RR, Doupe AJ (1997): FOS is induced by singing in distinct neuronal populations in a motor network. *Neuron 18*:315–325.

King VM, Follett BK (1997): *c-fos* expression in the putative avian suprachiasmatic nucleus. *J Comp Physiol [A] 180*:541–551.

King CT, Travers SP, Rowland NE, Garcea M, Spector AC (1999): Glossopharyngeal nerve transection eliminates quinine-stimulated fos-like immunoreactivity in the nucleus of the solitary tract: implications for a functional topography of gustatory nerve input in rats. *J Neurosci 19*:3107–3121.

King CT, Garcea M, Spector AC (2000): Glossopharyngeal nerve regeneration is essential for the complete recovery of quinine-stimulated oromotor rejection behaviors and central patterns of neuronal activity in the nucleus of the solitary tract in the rat. *J Neurosci 20*:8426–8434.

Kirkpatrick B, Kim JW, Insel TR (1994): Limbic system fos expression associated with paternal behavior. *Brain Res 658*:112–118.

Kogan C, Zangenehpour S, Dehgan F, Kaczmarek L, Chaudhuri A (2000): Enucleation reveals persistent sensory-driven *c-fos* expression in monkey but not rat visual cortex. *Soc Neurosci Abstr 26*:2193.

Komisaruk BR, Rosenblatt JS, Barona ML, Chinapen S, Nissanov J, O'Bannon RT, Johnson BM, Del Cerro MC (2000): Combined *c-fos* and 14C-2-deoxyglucose method to differentiate site-specific excitation from disinhibition: analysis of maternal behavior in the rat. *Brain Res 859*:262–272.

Kononen J, Koistinaho J, Alho H (1990): Circadian rhythm in *c-fos*-like immunoreactivity in the rat brain. *Neurosci Lett 120*:105–108.

Kornhauser JM, Nelson DE, Mayo KE, Takahashi JS (1990): Photic and circadian regulation of *c-fos* gene expression in the hamster suprachiasmatic nucleus. *Neuron 5*:127–134.

Kornhauser JM, Nelson DE, Mayo KE, Takahashi JS (1992): Regulation of jun-B messenger RNA and AP-1 activity by light and a circadian clock. *Science 255*:1581–1584.

Lamprecht R, Dudai Y (1995): Differential modulation of brain immediate early genes by intraperitoneal LiCl. *Neuroreport 7*:289–293.

Lamprecht R, Dudai Y (1996): Transient expression of c-Fos in rat amygdala during training is required for encoding conditioned taste aversion memory. *Learn Mem 3*:31–41.

Lanteri-Minet M, Weil-Fugazza J, de Pommery J, Menetrey D (1994): Hindbrain structures involved in pain processing as revealed by the expression of c-Fos and other immediate early gene proteins. *Neuroscience 58*:287–298.

Lerea LS, Carlson NG, McNamara JO (1995): *N*-methyl-D-aspartate receptors activate transcription of *c-fos* and NGFI-A by distinct phospholipase A2-requiring intracellular signaling pathways. *Mol Pharmacol 47*:1119–1125.

Li HY, Sawchenko PE (1998): Hypothalamic effector neurons and extended circuitries activated in 'neurogenic' stress: a comparison of footshock effects exerted acutely, chronically, and in animals with controlled glucocorticoid levels. *J Comp Neurol 393*:244–266.

Lima D, Avelino A, Coimbra A (1993): Differential activation of *c-fos* in spinal neurones by distinct classes of noxious stimuli. *Neuroreport 4*:747–750.

Lin JT, Kornhauser JM, Singh NP, Mayo KE, Takahashi JS (1997): Visual sensitivities of nur77 (NGFI-B) and *zif268* (NGFI-A) induction in the suprachiasmatic nucleus are dissociated from *c-fos* induction and behavioral phase-shifting responses. *Brain Res Mol Brain Res 46*:303–310.

Lin Y, Mather LE, Power I, Cousins MJ (2000): The effect of diclofenac on the expression of spinal cord *c-fos*-like immunoreactivity after ischemia-reperfusion-induced acute hyperalgesia in the rat tail. *Anesth Analg 90*:1141–1145.

Liu RJ, Wang R, Nie H, Zhang RX, Qiao JT, Dafny N (1997): Effects of intrathecal monoamine antagonists on the nociceptive c-Fos expression in a lesioned rat spinal cord. *Int J Neurosci 91*:169–180.

Lonstein JS, Stern JM (1997a): Somatosensory contributions to *c-fos* activation within the caudal periaqueductal gray of lactating rats: effects of perioral, rooting, and suckling stimuli from pups. *Horm Behav 32*:155–166.

Lonstein JS, Stern JM (1997b): Role of the midbrain periaqueductal gray in maternal nurturance and aggression: *c-fos* and electrolytic lesion studies in lactating rats. *J Neurosci 17*:3364–3378.

Lonstein JS, Stern JM (1999): Effects of unilateral suckling on nursing behavior and *c-fos* activity in the caudal periaqueductal gray in rats. *Dev Psychobiol 35*:264–275.

Lumley LA, Hull EM (1999): Effects of a D1 antagonist and of sexual experience on copulation-induced Fos-like immunoreactivity in the medial preoptic nucleus. *Brain Res 829*:55–68.

Luo L, Ryan AF, Saint Marie RL (1999): Cochlear ablation alters acoustically induced *c-fos* mRNA expression in the adult rat auditory brainstem. *J Comp Neurol 404*:271–283.

Mack KJ, Mack PA (1992): Induction of transcription factors in somatosensory cortex after tactile stimulation. *Brain Res Mol Brain Res 12*:141–147.

Mack KJ, Day M, Millbrandt J, Gottlieb DI (1990): Localization of the NGFI-A protein in rat brain. *Mol Brain Res 8*:177–180.

Mack KJ, Yi SD, Chang S, Millan N, Mack P (1995): NGFI-C expression is affected by physiological stimulation and seizures in the somatosensory cortex. *Brain Res Mol Brain Res 29*:140–146.

Marchant EG, Morin LP (1999): The hamster circadian rhythm system includes nuclei of the subcortical visual shell. *J Neurosci 19*:10482–10493.

Markstahler U, Bach M, Spatz WB (1998): Transient molecular visualization of ocular dominance columns (ODCs) in normal adult marmosets despite the desegregated termination of the retino-geniculo-cortical pathways. *J Comp Neurol 393*:118–134.

Matsuoka M, Yokosuka M, Mori Y, Ichikawa M (1999): Specific expression pattern of Fos in the accessory olfactory bulb of male mice after exposure to soiled bedding of females. *Neurosci Res 35*:189–195.

McCarthy MM, Besmer HR, Jacobs SC, Keidan GM, Gibbs RB (1997): Influence of maternal grooming, sex and age on Fos immunoreactivity in the preoptic area of neonatal rats: implications for sexual differentiation. *Dev Neurosci 19*:488–496.

McCollum JF, Woo CC, Leon M (1997): Granule and mitral cell densities are unchanged following early olfactory preference training. *Brain Res Dev Brain Res 99*:118–120.

McCormack MA, Rosen KM, Villa-Komaroff L, Mower GD (1992): Changes in immediate early gene expression during postnatal development of cat cortex and cerebellum. *Brain Res Mol Brain Res 12*:215–223.

Meddle SL, Foidart A, Wingfield JC, Ramenofskyand M, Balthazart J (1999): Effects of sexual interactions with a male on fos-like immunoreactivity in the female quail brain. *J Neuroendocrinol 11*:771–784.

Mello CV, Clayton DF (1994): Song-induced ZENK gene expression in auditory pathways of songbird brain and its relation to the song control system. *J Neurosci 14*:6652–6666.

Mello CV, Clayton DF (1995): Differential induction of the ZENK gene in the avian forebrain and song control circuit after metrazole-induced depolarization. *J Neurobiol 26*:145–161.

Mello CV, Ribeiro S (1998): ZENK protein regulation by song in the brain of songbirds. *J Comp Neurol 393*:426–438.

Mello C, Nottebohm F, Clayton D (1995): Repeated exposure to one song leads to a rapid and persistent decline in an immediate early gene's response to that song in zebra finch telencephalon. *J Neurosci 15*:6919–6925.

Melzer P, Steiner H (1997): Stimulus-dependent expression of immediate-early genes in rat somatosensory cortex. *J Comp Neurol 380*:145–153.

Melzer P, Van der Loos H, Dorfl J, Welker E, Robert P, Emery D, Berrini J-C (1985): A magnetic device to stimulate selected whiskers of freely moving or restrained small rodents: its application in a deoxyglucose study. *Brain Res 348*:229–240.

Meng ID, Bereiter DA (1996): Differential distribution of Fos-like immunoreactivity in the spinal trigeminal nucleus after noxious and innocuous thermal and chemical stimulation of rat cornea. *Neuroscience 72*:243–254.

Meredith M (1998): Vomeronasal, olfactory, hormonal convergence in the brain. Cooperation or coincidence?. *Ann NY Acad Sci 855*:349 361.

Mikula SA, Hendry SHC (2000): Visual and somatosensory representation in the macaque prefrontal cortex: an immediate-early gene double-label functional mapping study. *Soc Neurosci Abstr 26*:1221.

Mitchell DE, Beaver CJ, Ritchie PJ (1995): A method to study changes in eye-related columns in the visual cortex of kittens during and following early periods of monocular deprivation. *Can J Physiol Pharmacol 73*:1352–1363.

Miyashita Y, Kameyama M, Hasegawa I, Fukushima T (1998): Consolidation of visual associative long-term memory in the temporal cortex of primates. *Neurobiol Learn Mem 70*:197–211.

Moffatt CA, Ball GF, Nelson RJ (1995): The effects of photoperiod on olfactory c-Fos expression in prairie voles, Microtus-Ochrogaster. *Brain Res 677*:82–88.

Montag-Sallaz M, Welzl H, Kuhl D, Montag D, Schachner M (1999): Novelty-induced increased expression of immediate-early genes *c-fos* and arg 3.1 in the mouse brain. *J Neurobiol 38*:234–246.

Montero VM (1997): *c-fos* induction in sensory pathways of rats exploring a novel complex environment: shifts of active thalamic reticular sectors by predominant sensory cues. *Neuroscience 76*:1069–1081.

Montero VM (1999): Amblyopia decreases activation of the corticogeniculate pathway and visual thalamic reticularis in attentive rats: a 'focal attention' hypothesis. *Neuroscience 91*:805–817.

Montero VM (2000): Attentional activation of the visual thalamic reticular nucleus depends on 'top-down' inputs from the primary visual cortex via corticogeniculate pathways. *Brain Res 864*:95–104.

Montero VM, Jian S (1995): Induction of *c-fos* protein by patterned visual stimulation in central visual pathways of the rat. *Brain Res 690*:189–199.

Monti-Bloch L, Jennings-White C, Berliner DL (1998): The human vomeronasal system. A review. *Ann NY Acad Sci 855*:373–389.

Morgan JI, Cohen DR, Hempstead JL, Curran T (1987): Mapping patterns of *c-fos* expression in the central nervous system after seizure. *Science 237*:192–197.

Morin LP (1994): The circadian visual system. *Brain Res Brain Res Rev 19*:102–127.

Mower GD (1994): Differences in the induction of Fos protein in cat visual cortex during and after the critical period. *Brain Res Mol Brain Res 21*:47–54.

Mower GD, Kaplan IV (1999): Fos expression during the critical period in visual cortex: differences between normal and dark reared cats. *Brain Res Mol Brain Res 64*:264–269.

Nagase S, Miller JM, Dupont J, Lim HH, Sato K, Altschuler RA (2000): Changes in cochlear electrical stimulation induced Fos expression in the rat inferior colliculus following deafness. *Hear Res 147*:242–250.

Nastiuk KL, Mello CV, George JM, Clayton DF (1994): Immediate-early gene responses in the avian song control system: cloning and expression analysis of the canary *c-jun* cDNA. *Brain Res Mol Brain Res 27*:299–309.

Nedivi E, Fieldust S, Theill LE, Hevroni D (1996): A set of genes expressed in response to light in the adult cerebral cortex and regulated during development. *Proc Natl Acad Sci USA 93*:2048–2053.

Nikolaev E, Kaminska B, Tischmeyer W, Matthies H, Kaczmarek L (1992): Induction of expression of genes encoding transcription factors in the rat brain elicited by behavioral training. *Brain Res Bull 28*:479–484.

Numan M, Numan MJ (1995): Importance of pup-related sensory inputs and maternal performance for the expression of Fos-like immunoreactivity in the preoptic area and ventral bed nucleus of the stria terminalis of postpartum rats. *Behav Neurosci 109*:135–149.

Ohkura S, Fabre-Nys C, Broad KD, Kendrick KM (1997): Sex hormones enhance the impact of male sensory cues on both primary and association cortical components of visual and olfactory processing pathways as well as in limbic and hypothalamic regions in female sheep. *Neuroscience 80*:285–297.

Onoda N (1992): Odor-induced fos-like immunoreactivity in the rat olfactory bulb. *Neurosci Lett 137*:157–160.

Ossenkopp KP, Giugno L (1985): Taste aversions conditioned with multiple exposures to gamma radiation: abolition by area postrema lesions in rats. *Brain Res 346*:1–7.

Pan B, Castro-Lopes JM, Coimbra A (1994): *c-fos* expression in the hypothalamo-pituitary system induced by electroacupuncture or noxious stimulation. *Neuroreport 5*:1649–1652.

Pan B, Castro-Lopes JM, Coimbra A (1996): Activation of anterior lobe corticotrophs by electroacupuncture or noxious stimulation in the anaesthetized rat, as shown by colocalization of Fos protein with ACTH and beta-endorphin and increased hormone release. *Brain Res Bull 40*:175–182.

Pan B, Castro-Lopes JM, Coimbra A (1997): Chemical sensory deafferentation abolishes hypothalamic pituitary activation induced by noxious stimulation or electroacupuncture but only decreases that caused by immobilization stress. A c-fos study. *Neuroscience 78*:1059–1068.

Papa M, Pellicano MP, Welzl H, Sadile AG (1993): Distributed changes in c-Fos and c-Jun immunoreactivity in the rat brain associated with arousal and habituation to novelty. *Brain Res Bull 32*:509–515.

Park TH, Carr KD (1998): Neuroanatomical patterns of fos-like immunoreactivity induced by a palatable meal and meal-paired environment in saline- and naltrexone-treated rats. *Brain Res 805*:169–180.

Parthasarathy HB, Graybiel AM (1997): Cortically driven immediate-early gene expression reflects modular influence of sensorimotor cortex on identified striatal neurons in the squirrel monkey. *J Neurosci 17*:2477–2491.

Peterson MA, Basbaum AI, Abbadie C, Rohde DS, McKay WR, Taylor BK (1997): The differential contribution of capsaicin-sensitive afferents to behavioral and cardiovascular measures of brief and persistent nociception and to Fos expression in the formalin test. *Brain Res 755*:9–16.

Pfaus JG, Heeb MM (1997): Implications of immediate-early gene induction in the brain following sexual stimulation of female and male rodents. *Brain Res Bull 44*:397–407.

Pierson M, Snyder-Keller A (1994): Development of frequency-selective domains in inferior colliculus of normal and neonatally noise-exposed rats. *Brain Res 636*:55–67.

Presley RW, Menetrey D, Levine JD, Basbaum AI (1990): Systemic morphine suppresses noxious stimulus-evoked Fos protein-like immunoreactivity in the rat spinal cord. *J Neurosci 10*:323–335.

Qian Y, Jen PH (1994): Fos-like immunoreactivity elicited by sound stimulation in the auditory neurons of the big brown bat *Eptesicus fuscus*. *Brain Res 664*:241–246.

Qian Y, Wu M, Jen PH (1996): Tracing the auditory pathways to electrophysiologically characterized neurons with HRP and Fos double-labeling technique. *Brain Res 731*:241–245.

Rajendren GV, Moss RL (1994): Vomeronasal organ-mediated induction of fos in the central accessory olfactory pathways in repetitively mated female rats. *Brain Res Bull 34*:53–59.

Ralph MR, Foster RG, Davis FC, Menaker M (1990): Transplanted suprachiasmatic nucleus determines circadian period. *Science 247*:975–978.

Rea MA (1989): Light increases Fos-related protein immunoreactivity in the rat suprachiasmatic nuclei. *Brain Res Bull 23*:577–581.

Reilly S (1999): The parabrachial nucleus and conditioned taste aversion. *Brain Res Bull 48*:239–254.

Reimer K (1993): Simultaneous demonstration of Fos-like immunoreactivity and 2-deoxy-glucose uptake in the inferior colliculus of the mouse. *Brain Res 616*:339–343.

Reynolds DJ, Barber NA, Grahame-Smith DG, Leslie RA (1991): Cisplatin-evoked induction of *c-fos* protein in the brainstem of the ferret: the effect of cervical vagotomy and the anti-emetic 5-HT3 receptor antagonist granisetron (BRL 43694). *Brain Res 565*:231–236.

Ribeiro S, Cecchi GA, Magnasco MO, Mello CV (1998): Toward a song code: evidence for a syllabic representation in the canary brain. *Neuron 21*:359–371.

Robertson GS, Pfaus JG, Atkinson LJ, Matsumura H, Phillips AG, Fibiger HC (1991): Sexual behavior increases *c-fos* expression in the forebrain of the male rat. *Brain Res 564*:352–357.

Romeo RD, Parfitt DB, Richardson HN, Sisk CL (1998): Pheromones elicit equivalent levels of Fos-immunoreactivity in prepubertal and adult male Syrian hamsters. *Horm Behav 34*:48–55.

Rosen KM, McCormack MA, Villa-Komaroff L, Mower GD (1992): Brief visual experience induces immediate early gene expression in the cat visual cortex. *Proc Natl Acad Sci USA 89*:5437–5441.

Rouiller EM, Wan XS, Moret V, Liang F (1992): Mapping of *c-fos* expression elicited by pure tones stimulation in the auditory pathways of the rat, with emphasis on the cochlear nucleus. *Neurosci Lett 144*:19–24.

Rusak B (1977): The role of the suprachiasmatic nuclei in the generation of circadian rhythms in the golden hamster, *Mesocricetus auratus*. *J Comp Physiol 118*:145–164.

Rusak B, Zucker I (1979): Neural regulation of circadian rhythms. *Physiol Rev 59*:449–526.

Rusak B, McNaughton L, Robertson HA, Hunt SP (1992): Circadian variation in photic regulation of immediate-early gene mRNAs in rat suprachiasmatic nucleus cells. *Brain Res Mol Brain Res 14*:124–130.

Saffen DW, Cole AJ, Worley PF, Christy BA, Ryder K, Baraban JM (1988): Convulsant-induced increase in transcription factor message RNAs in rat brain. *Proc Natl Acad Sci USA 85*:7795–7799.

Sagar SM, Sharp FR (1990): Light induces a Fos-like nuclear antigen in retinal neurons. *Brain Res Mol Brain Res 7*:17–21.

Saint Marie RL, Luo L, Ryan AF (1999): Effects of stimulus frequency and intensity on *c-fos* mRNA expression in the adult rat auditory brainstem. *J Comp Neurol 404*:258–270.

Saito H, Miller JM, Pfingst BE, Altschuler RA (1999): Fos-like immunoreactivity in the auditory brainstem evoked by bipolar intracochlear electrical stimulation: effects of current level and pulse duration. *Neuroscience 91*:139–161.

Saito H, Miller JM, Altschuler RA (2000): Cochleotopic fos immunoreactivity in cochlea and cochlear nuclei evoked by bipolar cochlear electrical stimulation. *Hear Res 145*:37–51.

Sakai N, Yamamoto T (1997): Conditioned taste aversion and *c-fos* expression in the rat brainstem after administration of various USs. *Neuroreport 8*:2215–2220.

Sallaz M, Jourdan F (1993): *c-fos* expression and 2-deoxyglucose uptake in the olfactory bulb of odour-stimulated awake rats. *Neuroreport 4*:55–58.

Sallaz M, Jourdan F (1996): Odour-induced *c-fos* expression in the rat olfactory bulb: involvement of centrifugal afferents. *Brain Res 721*:66–75.

Sato K, Houtani T, Ueyama T, Ikeda M, Yamashita T, Kumazawa T, Sugimoto T (1993): Identification of rat brainstem sites with neuronal Fos protein induced by acoustic stimulation with pure tones. *Acta Otolaryngol Suppl 500*:18–22.

Sato MT, Tokunaga A, Kawai Y, Shimomura Y, Tano Y, Senba E (2000): The effects of binocular suture and dark rearing on the induction of *c-fos* protein in the rat visual cortex during and after the critical period. *Neurosci Res 36*:227–233.

Sawchenko PE, Li HY, Ericsson A (2000): Circuits and mechanisms governing hypothalamic responses to stress: a tale of two paradigms. *Prog Brain Res 122*:61–78.

Schafe GE, Bernstein IL (1996): Forebrain contribution to the induction of a brainstem correlate of conditioned taste aversion, I. The amygdala. *Brain Res 741*:109–116.

Schafe GE, Bernstein IL (1998): Forebrain contribution to the induction of a brainstem correlate of conditioned taste aversion, II. Insular (gustatory) cortex. *Brain Res 800*:40–47.

Schafe GE, Seeley RJ, Bernstein IL (1995): Forebrain contribution to the induction of a cellular correlate of conditioned taste aversion in the nucleus of the solitary tract. *J Neurosci 15*:6789–6796.

Scheich H, Zuschratter W (1995): Mapping of stimulus features and meaning in gerbil auditory cortex with 2-deoxyglucose and *c-Fos* antibodies. *Behav Brain Res 66*:195–205.

Scheich H, Stark H, Zuschratter W, Ohl FW, Simonis CE (1997): Some functions of primary auditory cortex in learning and memory formation. *Adv Neurol 73*:179–193.

Schellinck HM, Smyth C, Brown R, Wilkinson M (1993): Odor-induced sexual maturation and expression of *c-fos* in the olfactory system of juvenile female mice. *Brain Res Dev Brain Res 74*:138–141.

Schlingensiepen KH, Luno K, Brysch W (1991): High basal expression of the zif/268 immediate early gene in cortical layers IV and VI, in CA1 and in the corpus striatum — an in situ hybridization study. *Neurosci Lett* *122*:67–70.

Schwartz WJ, Takeuchi J, Shannon W, Davis EM, Aronin N (1994): Temporal regulation of light-induced Fos and Fos-like protein expression in the ventrolateral subdivision of the rat suprachiasmatic nucleus. *Neuroscience* *58*:573–583.

Schwartz WJ, Carpino Jr. A, de la Iglesia HO, Baler R, Klein DC, Nakabeppu Y, Aronin N (2000): Differential regulation of fos family genes in the ventrolateral and dorsomedial subdivisions of the rat suprachiasmatic nucleus. *Neuroscience 98*:535–547.

Sergent J, Bindra D (1981): Differential hemispheric processing of faces: methodological considerations and reinterpretation. *Psychol Bull 89*:541–554.

Sheng M, Greenberg ME (1990): The regulation and function of *c-fos* and other immediate early genes in the nervous system. *Neuron 4*:477–485.

Silveira LC, de Matos FM, Pontes-Arruda A, Picanco-Diniz CW, Muniz JA (1996): Late development of Zif268 ocular dominance columns in primary visual cortex of primates [erratum appears in *Brain Res*, 1997 May 16; *757*:164–165]. *Brain Res 732*:237–241.

Silveira DC, Schachter SC, Schomer DL (1999): Acoustic brainstem nuclei express Fos after flurothyl-induced generalized seizures in rats. *Epilepsy Res 34*:49–55.

Simler S, Hirsch E, Danober L, Motte J, Vergnes M, Marescaux C (1994): *c-fos* expression after single and kindled audiogenic seizures in Wistar rats. *Neurosci Lett 175*:58–62.

Simler S, Vergnes M, Marescaux C (1999): Spatial and temporal relationships between c-Fos expression and kindling of audiogenic seizures in Wistar rats. *Exp Neurol 157*:106–119.

Slotnick BM, Panhuber H, Bell GA, Laing DG (1989): Odor-induced metabolic activity in the olfactory bulb of rats trained to detect propionic acid vapor. *Brain Res 500*:161–168.

Smith DW, Day TA (1994): *c-fos* expression in hypothalamic neurosecretory and brainstem catecholamine cells following noxious somatic stimuli. *Neuroscience 58*:765–775.

Spector AC, Norgren R, Grill HJ (1992): Parabrachial gustatory lesions impair taste aversion learning in rats. *Behav Neurosci 106*:147–161.

Spray KJ, Halsell CB, Bernstein IL (2000): c-Fos induction in response to saccharin after taste aversion learning depends on conditioning method. *Brain Res 852*:225–227.

Steiner H, Gerfen CR (1994): Tactile sensory input regulates basal and apomorphine-induced immediate-early gene expression in rat barrel cortex. *J Comp Neurol 344*:297–304.

Steiner H, Kitai ST (2000): Regulation of rat cortex function by D1 dopamine receptors in the striatum. *J Neurosci 20*:5449–5460.

Streefland C, Farkas E, Maes FW, Bohus B (1996): *c-fos* expression in the brainstem after voluntary ingestion of sucrose in the rat. *Neurobiology (Bp) 4*:85–102.

Sugimoto T, Li YL, Kishimoto H, Fujita M, Ichikawa H (2000): Compensatory projection of primary nociceptors and *c-fos* induction in the spinal dorsal horn following neonatal sciatic nerve lesion. *Exp Neurol 164*:407–414.

Sumova A, Travnickova Z, Illnerova H (2000): Spontaneous c-Fos rhythm in the rat suprachiasmatic nucleus: location and effect of photoperiod. *Am J Physiol: Regulatory, Integrative and Comparative Physiology 279*:2262–2269.

Swank MW (2000): Conditioned c-Fos in mouse NTS during expression of a learned taste aversion depends on contextual cues. *Brain Res 862*:138–144.

Swank MW, Bernstein IL (1994): c-Fos induction in response to a conditioned stimulus after single trial taste aversion learning [erratum appears in *Brain Res*, 1994 May 9; *645*:359]. *Brain Res 636*:202–208.

Swank MW, Schafe GE, Bernstein IL (1995): c-Fos induction in response to taste stimuli previously paired with amphetamine or LiCl during taste aversion learning. *Brain Res 673*:251–261.

Swank MW, Ellis AE, Cochran BN (1996): c-Fos antisense blocks acquisition and extinction of conditioned taste aversion in mice. *Neuroreport 7*:1866–1870.

Szechtman H, Ornstein K, Teitelbaum P, Golani I (1982): Snout contact fixation, climbing and gnawing during apomorphine stereotype in rats from two substrains. *Eur J Pharmacol 80*:385–392.

Tanaka M, Amaya F, Tamada Y, Okamura H, Hisa Y, Ibata Y (1997): Induction of NGFI-A gene expression in the rat suprachiasmatic nucleus by photic stimulation. *Brain Res 756*:305–310.

Tanaka M, Iijima N, Amaya F, Tamada Y, Ibata Y (1999): NGFI-A gene expression induced in the rat suprachiasmatic nucleus by photic stimulation: spread into hypothalamic periventricular somatostatin neurons and GABA receptor involvement. *Eur J Neurosci 11*:3178–3184.

Taniguchi K, Nii Y, Ogawa K (1993): Subdivisions of the accessory olfactory bulb, as demonstrated by lectin-histochemistry in the golden hamster. *Neurosci Lett 158*:185–188.

Telford S, Wang S, Redgrave P (1996): Analysis of nociceptive neurones in the rat superior colliculus using *c-fos* immunohistochemistry. *J Comp Neurol 375*:601–617.

Thiele TE, Roitman MF, Bernstein IL (1996): c-Fos induction in rat brainstem in response to ethanol- and lithium chloride-induced conditioned taste aversions. *Alcohol Clin Exp Res 20*:1023–1028.

Tolle TR, Castro-Lopes JM, Coimbra A, Zieglgansberger W (1990): Opiates modify induction of *c-fos* proto-oncogene in the spinal cord of the rat following noxious stimulation [erratum appears in *Neurosci Lett*, 1990 Jul 3; *114*:239]. *Neurosci Lett 111*:46–51.

Travers SP, Hu H (2000): Extranuclear projections of rNST neurons expressing gustatory-elicited Fos. *J Comp Neurol 427*:124–138.

Travers JB, Urbanek K, Grill HJ (1999): Fos-like immunoreactivity in the brain stem following oral quinine stimulation in decerebrate rats. *Am J Physiol 277*:R384–394.

Tubbiola ML, Wysocki CJ (1997): FOS immunoreactivity after exposure to conspecific or heterospecific urine: where are chemosensory cues sorted?. *Physiol Behav 62*:867–870.

Veening JG, Coolen LM (1998): Neural activation following sexual behavior in the male and female rat brain. *Behav Brain Res 92*:181–193.

Vischer MW, Hausler R, Rouiller EM (1994): Distribution of Fos-like immunoreactivity in the auditory pathway of the Sprague–Dawley rat elicited by cochlear electrical stimulation. *Neurosci Res 19*:175–185.

Vischer MW, Bajo-Lorenzana V, Zhang J, Hausler R, Rouiller EM (1995): Activity elicited in the auditory pathway of the rat by electrical stimulation of the cochlea. *ORL J Otorhinolaryngol Relat Spec 57*:305–309.

Wallace CS, Withers GS, Weiler IJ, George JM, Clayton DF, Greenough WT (1995): Correspondence between sites of NGFI-A induction and sites of morphological plasticity following exposure to environmental complexity. *Brain Res Mol Brain Res 32*:211–220.

Wallman J, Saldanha CJ, Silver R (1994): A putative suprachiasmatic nucleus of birds responds to visual motion. *J Comp Physiol [A] 174*:297–304.

Walsh CJ, Fleming AS, Lee A, Magnusson JE (1996): The effects of olfactory and somatosensory desensitization on Fos-like immunoreactivity in the brains of pup-exposed postpartum rats. *Behav Neurosci 110*:134–153.

Wang S, Wang H, Niemi-Junkola U, Westby GW, McHaffie JG, Stein BE, Redgrave P (2000): Parallel analyses of nociceptive neurones in rat superior colliculus by using *c-fos* immunohistochemistry and electrophysiology under different conditions of anaesthesia. *J Comp Neurol 425*:599–615.

Watson MA, Milbrandt J (1989): The NGFI-B gene, a transcriptionally inducible member of the steroid receptor gene superfamily: genomic structure and expression in rat brain after seizure induction. *Mol Cell Biol 9*:4213–4219.

Wersinger SR, Baum MJ (1997): Sexually dimorphic processing of somatosensory and chemosensory inputs to forebrain luteinizing hormone-releasing hormone neurons in mated ferrets. *Endocrinology 138*:1121–1129.

Wersinger SR, Rissman EF (2000): Oestrogen receptor alpha is essential for female-directed chemo-investigatory behaviour but is not required for the pheromone-induced luteinizing hormone surge in male mice. *J Neuroendocrinol 12*:103–110.

Williams S, Evan G, Hunt SP (1990a): Spinal *c-fos* induction by sensory stimulation in neonatal rats. *Neurosci Lett 109*:309–314.

Williams S, Evan GI, Hunt SP (1990b): Changing patterns of *c-fos* induction in spinal neurons following thermal cutaneous stimulation in the rat. *Neuroscience 36*:73–81.

Woo CC, Oshita MH, Leon M (1996): A learned odor decreases the number of Fos-immunopositive granule cells in the olfactory bulb of young rats. *Brain Res 716*:149–156.

Worley PF, Cole AJ, Murphy TH, Christy BA, Nakabeppu Y, Baraban JM (1990): Synaptic regulation of immediate-early genes in brain. *Cold Spring Harb Symp Quant Biol 55*:213–223.

Worley PF, Christy BA, Nakabeppu Y, Bhat RV, Cole AJ, Baraban JM (1991): Constitutive expression of *zif268* in neocortex is regulated by synaptic activity. *Proc Natl Acad Sci USA 88*:5106–5110.

Xia Z, Dudek H, Miranti CK, Greenberg ME (1996): Calcium influx via the NMDA receptor induces immediate early gene transcription by a MAP kinase/ERK-dependent mechanism. *J Neurosci 16*:5425–5436.

Yamada Y, Hada Y, Imamura K, Mataga N, Watanabe Y, Yamamoto M (1999): Differential expression of immediate-early genes, *c-fos* and *zif268*, in the visual cortex of young rats: effects of a noradrenergic neurotoxin on their expression. *Neuroscience 92*:473–484.

Yamaguchi T, Inamura K, Kashiwayanagi M (2000): Increases in Fos-immunoreactivity after exposure to a combination of two male urinary components in the accessory olfactory bulb of the female rat. *Brain Res 876*:211–214.

Yamamoto T (1993): Neural mechanisms of taste aversion learning. *Neurosci Res 16*:181–185.

Yamamoto T, Sawa K (2000a): c-Fos-like immunoreactivity in the brainstem following gastric loads of various chemical solutions in rats. *Brain Res 866*:135–143.

Yamamoto T, Sawa K (2000b): Comparison of c-Fos-like immunoreactivity in the brainstem following intraoral and intragastric infusions of chemical solutions in rats. *Brain Res 866*:144–151.

Yamamoto T, Shimura T, Sako N, Azuma S, Bai WZ, Wakisaka S (1992): *c-fos* expression in the rat brain after intraperitoneal injection of lithium chloride. *Neuroreport 3*:1049–1052.

Yamamoto T, Shimura T, Sakai N, Ozaki N (1994a): Representation of hedonics and quality of taste stimuli in the parabrachial nucleus of the rat. *Physiol Behav 56*:1197–1202.

Yamamoto T, Shimura T, Sako N, Yasoshima Y, Sakai N (1994b): Neural substrates for conditioned taste aversion in the rat. *Behav Brain Res 65*:123–137.

Yang H, Wanner IB, Roper SD, Chaudhari N (1999a): An optimized method for in situ hybridization with signal amplification that allows the detection of rare mRNAs. *J Histochem Cytochem 47*:431–446.

Yang S, Lee Y, Voogt JL (1999b): Fos expression in the female rat brain during the proestrous prolactin surge and following mating. *Neuroendocrinology 69*:281–289.

Yashpal K, Mason P, McKenna JE, Sharma SK, Henry JL, Coderre TJ (1998): Comparison of the effects of treatment with intrathecal lidocaine given before and after formalin on both nociception and Fos expression in the spinal cord dorsal horn. *Anesthesiology 88*:157–164.

Yi DK, Barr GA (1995): The induction of Fos-like immunoreactivity by noxious thermal, mechanical and chemical stimuli in the lumbar spinal cord of infant rats. *Pain 60*:257–265.

Yoshida K, Kawamura K, Imaki J (1993): Differential expression of *c-fos* mRNA in rat retinal cells: regulation by light/dark cycle. *Neuron 10*:1049–1054.

Zangenehpour S, Chaudhuri A (1999): Multisensory neurons of the superior colliculus revealed by dual fluorescent ICC/ISH labeling using *c-fos* and *zif268* expression profiles. *Soc Neurosci Abstr 25*:1414.

Zangenehpour S, Dehgan F, Darghouth S, Chaudhuri A (2000): Dual activity maps with inducible transcription factors: combined fluorescent immunocytochemistry and in situ hybridization. *Soc Neurosci Abstr 26*:876.

Zhang F, Halleux P, Arckens L, Vanduffel W, Van Bree L, Mailleux P, Vandesande F, Orban GA, Vanderhaeghen JJ (1994): Distribution of immediate early gene zif-268, *c-fos*, *c-jun* and jun-D mRNAs in the adult cat with special references to brain region related to vision. *Neurosci Lett 176*:137–141.

Zhang F, Vanduffel W, Schiffmann SN, Mailleux P, Arckens L, Vandesande F, Orban GA, Vanderhaeghen JJ (1995): Decrease of zif-268 and *c-fos* and increase of *c-jun* mRNA in the cat areas 17, 18 and 19 following complete visual deafferentation. *Eur J Neurosci 7*:1292–1296.

Zhang JS, Haenggeli CA, Tempini A, Vischer MW, Moret V, Rouiller EM (1996): Electrically induced fos-like immunoreactivity in the auditory pathway of the rat: effects of survival time, duration, and intensity of stimulation. *Brain Res Bull 39*:75–82.

Zhang JS, Vischer MW, Moret V, Roulin C, Rouiller EM (1998): Antibody-dependent Fos-like immunoreactivity (FLI) in the auditory pathway of the rat in response to electric stimulation of the cochlea. *J Hirnforsch 39*:21–35.

Zhu XO, Brown MW, McCabe BJ, Aggleton JP (1995): Effects of the novelty or familiarity of visual stimuli on the expression of the immediate early gene *c-fos* in rat brain. *Neuroscience 69*:821–829.

Zhu XO, McCabe BJ, Aggleton JP, Brown MW (1996): Mapping visual recognition memory through expression of the immediate early gene *c-fos*. *Neuroreport 7*:1871–1875.

Zhu XO, McCabe BJ, Aggleton JP, Brown MW (1997): Differential activation of the rat hippocampus and perirhinal cortex by novel visual stimuli and a novel environment. *Neurosci Lett 229*:141–143.

Zuschratter W, Gass P, Herdegen T, Scheich H (1995): Comparison of frequency-specific c-Fos expression and fluoro-2-deoxyglucose uptake in auditory cortex of gerbils (*Meriones unguiculatus*). *Eur J Neurosci 7*:1614–1626.

CHAPTER VI

Immediate-early gene expression in the analysis of circadian rhythms and sleep

BENJAMIN RUSAK, MARIO E. GUIDO AND KAZUE SEMBA

1. INTRODUCTION

1.1. TEMPORAL STRUCTURE OF PHYSIOLOGY AND BEHAVIOR

The temporal structure of mammalian physiology and behavior is dominated by two major kinds of rhythms: 'circadian' (about a day) rhythms, related to the cycle of day and night; and cycles of alternating states of sleep and waking. Since in humans these rhythms coincide generally (but not universally), they are sometimes considered aspects of a single process. In many other species, however, daily rhythms in a host of other functions provide a temporal framework for higher frequency sleep–wake oscillations, indicating that each of these cycles is controlled by at least partially independent mechanisms. Sleep itself is also comprised of cyclic alternations among different states, marked by different electrophysiological, subjective and behavioral characteristics. The most prominent of these is the alternation between rapid eye movement (REM) sleep and non-REM sleep. Non-REM sleep comprises several distinguishable stages, including both light states of sleep and deeper sleep marked by prominent slow waves in the cortical electroencephalogram (EEG).

Studies of immediate-early genes (IEGs) and other transcription factors in the brain have been conducted in order to investigate the mechanisms underlying sleep, waking and circadian cyclicity. Such studies have correlated changes in gene expression with changes in behavioral states, as well as with responses to environmental or physiological signals that affect these states. These studies have used gene expression to identify neural regions that may be relevant to regulation of sleep, waking or circadian timing, as well as to provide insights into their underlying physiological mechanisms. To understand the use of these methods, it is necessary first to be familiar with a few important features of the structure and regulation of sleep–wake cyclicity and of circadian rhythmicity.

1.2. BASIC FEATURES OF CIRCADIAN AND SLEEP–WAKE CYCLES

Daily rhythms are evident in virtually every feature of mammalian physiology and behavior, including enzyme activity, immune system function, cell division, hormone secretion, feeding, cognitive performance, sensory processing, and motor function, among others (Hastings et al., 1991). These rhythms are the product of an endogenous central clock ('pacemaker') that functions in the absence of any rhythmicity in the external environment and runs with a

Handbook of Chemical Neuroanatomy Vol. 19: Immediate Early Genes and Inducible Transcription Factors in Mapping of the Central Nervous System Function and Dysfunction
L. Kaczmarek and H.A. Robertson, editors

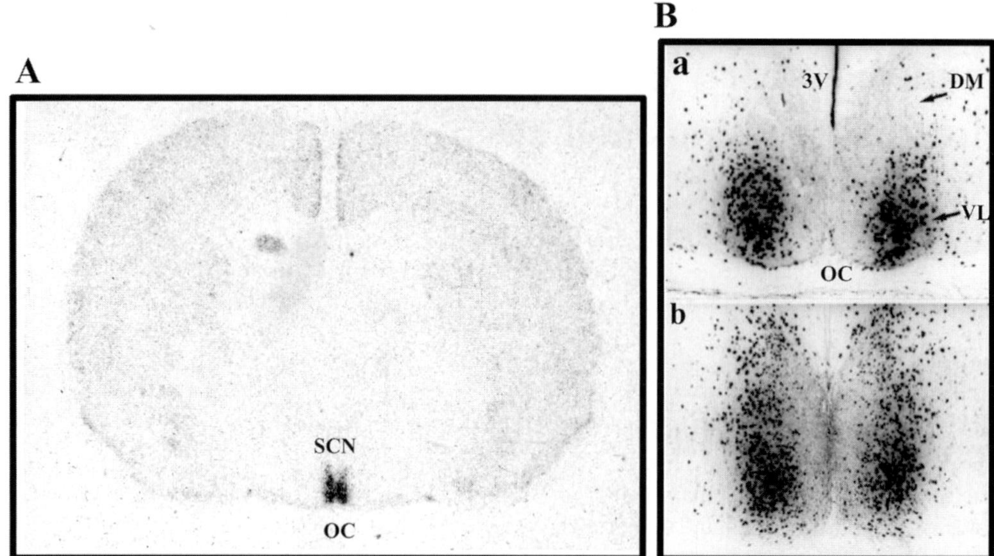

Fig. 1. Coronal sections through the mid-caudal SCN region from hamsters maintained in constant dark and exposed to a 30-min light pulse during the subjective night. (*A*) Autoradiogram of a brain section showing the SCN labeled by radioactive in situ hybridization of *junB* mRNA. (*B*) Coronal sections including the SCN stained immunohistochemically for c-Fos. Hamsters were exposed to a 30-min light pulse and then returned to the dark for another 30 min (*a,b*) and killed in the early night (Zeitgeber time, ZT 15) (*a*) and late night (ZT 22) (*b*). No c-Fos immunoreactivity was observed at these phases in controls kept in the dark (not shown). SCN, suprachiasmatic nucleus; OC, optic chiasm; 3V, third ventricle; DM, dorsomedial SCN; VL, ventrolateral SCN. Modified from Neuroscience 90, Guido ME, Goguen D, de Guido LB, Robertson HA, Rusak B. Circadian and photic regulation of immediate-early gene expression in the hamster suprachiasmatic nucleus. pp. 555–571, Copyright (1999), with permission from Elsevier Science.

stable period near 24 h in length, which differs among individuals and species. This clock is regulated by daily signals from the environment that synchronize ('entrain') the endogenously generated rhythms to a precise 24-h periodicity. The most effective and important of these external cues is the pattern of exposure to light and dark during the day–night cycle.

One critical feature of all circadian clocks is that their sensitivity to the phase-resetting, and thus entraining, effects of light is different at different times of the circadian cycle (Pittendrigh and Daan, 1976). Light exposure during the subjective day (i.e., the time of its endogenous cycle when the organism behaves in artificially constant conditions as it normally does during the natural day) has little or no resetting effect on the central pacemaker. Light exposure early in the subjective night tends to delay the circadian clock, while light late in the subjective night advances it. These delay and advance phase shifts are the basis for entrainment by lighting cycles. Studies of IEG expression have been used to gain insight into the physiological mechanisms responsible for both entrainment by light and other cues, and for the endogenous generation of circadian rhythms.

The major circadian pacemaker in mammals has been identified as the suprachiasmatic nuclei (SCN) of the anterior hypothalamus (Fig. 1) (Rusak and Zucker, 1979; Meijer and Rietveld, 1989; Moore, 1995). These tiny midline nuclei lie immediately dorsal to the optic chiasm, and receive a direct retinal projection from ganglion cell axons rising out of the optic chiasm (Morin, 1994; Moore, 1995). A broad array of evidence demonstrates convincingly that the SCN are the pacemaker for most mammalian circadian rhythms and are the major target for entraining photic information (Hastings et al., 1991; Moore, 1995; Silver et al., 1996).

Both sleep and waking are regulated by neural and other mechanisms that initiate and sustain these states. Although these states are mutually exclusive, physiologically one is not simply the absence of the other state; rather, each is positively regulated by active mechanisms, as well as negatively by the mutual inhibition of these mechanisms (Steriade and McCarley, 1990; Lydic and Baghdoyan, 1999). Sleep–wake alternation is regulated by a circadian clock, as are other physiological functions, but, it is also affected by a homeostatic regulatory mechanism. Thus, both the phase of the circadian cycle and the duration of prior waking contribute interactively to the drive to sleep. Similarly, circadian phase and length of prior sleep interact to determine the probability of waking up from sleep (Borbély, 1982; Dijk and Edgar, 1999).

The physiological mechanisms regulating sleep and waking are diverse and widely distributed within the nervous system. Major regulatory centers for sleep and waking are found in the preoptic area (rostral to the SCN), the posterior hypothalamus, the basal forebrain, the mesencephalic reticular formation and in the pontine tegmentum (Steriade and McCarley, 1990; Lydic and Baghdoyan, 1999). These systems interact in a complex and as yet poorly understood fashion to regulate the onset and duration of sleep, its termination, and the cycling of its internal stage structure. Since sleep and waking are also constrained within a circadian framework because of regulation by the SCN circadian pacemaker, some or all of these systems must also receive input from the SCN. Waking at unusual times can also alter the phase of the SCN pacemaker, indicating that there are reciprocal functional connections between these systems (Antle and Mistlberger, 2000).

Molecular studies must account for a number of features of the temporal organization of sleep and waking, including: the fact that the probability of transition to either state depends strongly on the length of time the organism has spent in the other state; that circadian phase strongly modulates the probability of sleep and waking; sleep and waking alternate regularly in different ways in different species and their total durations during 24 h differ among species; and, within sleep, REM and non-REM states alternate with different frequencies in different species.

2. IEG EXPRESSION IN THE SCN

2.1. LIGHT-INDUCED IEG EXPRESSION

The first reports of IEG expression associated with circadian rhythm regulation appeared simultaneously as published abstracts in 1989. These were followed by a series of papers demonstrating that c-Fos immunoreactivity in the SCN increases from baseline to very high levels within an hour of nocturnal light exposure (Rea, 1989; Rusak et al., 1990, 1992; Kornhauser et al., 1990, 1992; Aronin et al., 1990). These effects were especially dramatic since c-Fos protein in the SCN is virtually undetectable during the night in baseline dark conditions. This demonstration that an environmental stimulus acting through the usual sensory pathways evoked c-Fos expression in target neurons was reinforced by evidence that the patterns of expression within the SCN of different species matched closely the anatomical differences among species in the patterns of termination of retinal afferents to the SCN (Rusak et al., 1992).

These results were interpreted to reflect a possible role for IEG expression in circadian clock resetting (Rusak et al., 1990), because these effects appeared to be specific to the retinorecipient SCN and to the circadian interval during which light exposure can reset the phase of the circadian clock. Other retinorecipient structures, such as the dorsal lateral geniculate nucleus, did not show similar responses to nocturnal light exposure, and the SCN did not respond to light during the mid-subjective day, a time at which light does not reset the clock (Rusak et al., 1990). (For these tests, animals are first transferred to constant darkness for varying lengths of time and exposed to light pulses at different circadian phases.) In addition, the threshold sensitivity for phase shifting and c-Fos expression were similar and intensity of expression tracked increasing light intensity and size of phase shifts over at least part of the range tested (Kornhauser et al., 1990).

The early evidence that light could induce c-Fos expression during the subjective night, but not during the subjective day, was extended to demonstrate that, as the lengths of the subjective day and night phases change (as happens annually in non-equatorial regions), sensitivity to light tracks the duration of the subjective night (Trávníčková et al., 1996). These changes in the duration of the light-responsive phase are preserved immediately after transfer from a lighting cycle to constant darkness. c-Fos expression changes in the SCN may reflect processes involved in the measurement of day length and in the myriad of seasonal changes in physiology triggered by photoperiod changes (Sumová et al., 1995).

Since c-Fos expression is triggered by light transduced through a pathway originating in the eye, it is possible that the rhythmicity of response originates not in the SCN, but in the retina. This possibility is not unreasonable, given that the retina is itself known to contain a circadian clock mechanism (Terman et al., 1993; Yamazaki et al., 2000). Studies of the SCN in an in vitro slice preparation, however, showed that the ability to elicit c-Fos protein expression was still rhythmic and restricted to the subjective night (Bennett et al., 1996). Thus, while gating of photic input to the SCN by the retina is not ruled out, the SCN itself is intrinsically rhythmic in its response to stimuli that induce c-Fos expression.

Another important finding was that c-Fos expression is not increased in the SCN in response to environmental cues other than light that shift the circadian clock (Janik and Mrosovsky, 1992; Mead et al., 1992). The same non-photic stimuli do, however, induce c-Fos expression in the intergeniculate leaflet of the thalamus, a structure that has been linked to mediating non-photic effects on circadian phase. These observations support the idea that IEG expression in the SCN is closely related to the responses of clock cells in the SCN to photic input.

In subsequent studies, the range of genes shown to be regulated rapidly in SCN cells by light exposure was expanded to include other members of the c-Fos family (c-*fos*, *fosB*, *ΔfosB*, *fra-1* and *fra-2*), the Jun family (*junB*, c-*jun* and *junD*), activator protein-1 (AP-1) complexes, and other inducible transcription factors, such as *NGFI-A* (*zif268*, *egr1*, *krox-24*) and *NGFI-B* (*nur77*). Evidence with respect to regulation of FosB protein levels in the SCN was initially unclear. FosB-like immunoreactivity was reported to be present constitutively in the SCN at all circadian phases, and to be either increased or unaffected by light at night (Peters et al., 1994; Ebling et al., 1996). A subsequent report demonstrated that FosB mRNA is not detectable in any part of the SCN in constant darkness, but that its levels are upregulated rapidly by nocturnal light pulses, as for other IEGs (Guido et al., 1996b). This discrepancy between apparent levels of mRNA and protein was attributed to the possible presence in the SCN of two forms of FosB, the full-length protein and an alternative splice version (ΔFosB) in which the carboxyl terminus is lost (Guido et al., 1996b). This hypothesis was subsequently confirmed by demonstrating that ΔFosB protein, not the full-length FosB, is present constitutively at all times in the SCN (Schwartz et al., 2000).

In addition, some so-called 'clock genes' which were originally identified as components of intracellular clocks in non-mammalian species, were subsequently shown to oscillate in SCN cells and to respond to retinal illumination (see King and Takahashi, 2000 for review). Light-induced changes in levels and distribution of clock gene protein products appear to be critical to phase-shifting effects of light (King and Takahashi, 2000, but see below). The time course of light-evoked changes in clock gene expression appears generally to be slower than that of IEG expression, and the functional relations between their gene products remain unclear.

2.2. TEMPORAL ASPECTS OF LIGHT-INDUCED IEG EXPRESSION

The most dramatic temporal effect related to SCN gene expression is the gating by the circadian clock of its own sensitivity to light, so that it responds primarily during the subjective night with IEG expression. But responsiveness during the subjective night is not uniform. Early in the subjective night, when light causes delay shifts, IEG expression is restricted to the most ventral region of the SCN (where the strongest retinal innervation terminates), and relatively long periods of light exposure are needed to induce substantial gene expression (Figs. 1 and 2) (Rea, 1992; Guido et al., 1996a,c, 1999a,b,c). By contrast, in the second half of the night, when light elicits phase advances, IEG expression is induced in cells throughout the SCN, and even very brief exposure to bright (100 lux) or dim light (5 lux) is highly effective (Guido et al., 1999a,b). Thus, it appears that different cells and different neurochemical systems (Ralph and Menaker, 1985; Rusak and Bina, 1990; Melo et al., 1997) are involved in mediating the effects of light at different circadian phases and that the threshold of light sensitivity for these cells varies during the night.

The patterns of expression of different IEGs vary during the night. Levels of mRNAs for *NGFI-A* and *fosB* show a monophasic peak of photically induced expression near the middle of the subjective night; i.e., 4–6 h after activity onset (Guido et al., 1996b, 1999a). Both *junB* and c-*fos* mRNAs, however, show bimodal peaks of sensitivity at 6 and 10/11 h after activity onset, with a distinct trough between these phases (Guido et al., 1996a, 1999a). The threshold for light induction of *NGFI-A* and *NGFI-B* mRNAs is also 1–2 log units lower than for c-*fos* and *junB*, and therefore similarly lower than the threshold for behavioral phase shifts (Lin et al., 1997).

2.3. SCN CELL TYPES EXPRESSING IEGS

The SCN is an extremely heterogeneous nucleus, despite its small size. In most species studied, but most clearly in rats, cells in the retinorecipient ventrolateral SCN (Fig. 1B) are immunoreactive for vasoactive intestinal polypeptide (VIP) and peptide histidine isoleucine (PHI), while dorsomedial SCN cells are immunoreactive for vasopressin and somatostatin (see van den Pol, 1991; Inouye and Shibata, 1994 for review). Cells containing gastrin-releasing peptide (GRP) are found primarily ventrally in the SCN (Mikkelsen et al., 1991). SCN cells can also be differentiated based on markers such as glial fibrillary acidic protein (GFAP, characteristic of some astroglia) and calmodulin.

An early report indicated that VIP cells in the SCN respond to retinal input with increased c-Fos expression (Daikoku et al., 1992). Later studies showed that early in the subjective night 33% of rat SCN cells induced by light exposure to express c-Fos were also immunoreactive for PHI, GRP or VIP, while late in the night, the percentage of colocalization had doubled to approximately 65% (Romijn et al., 1996). The most dramatic increases in co-localization from

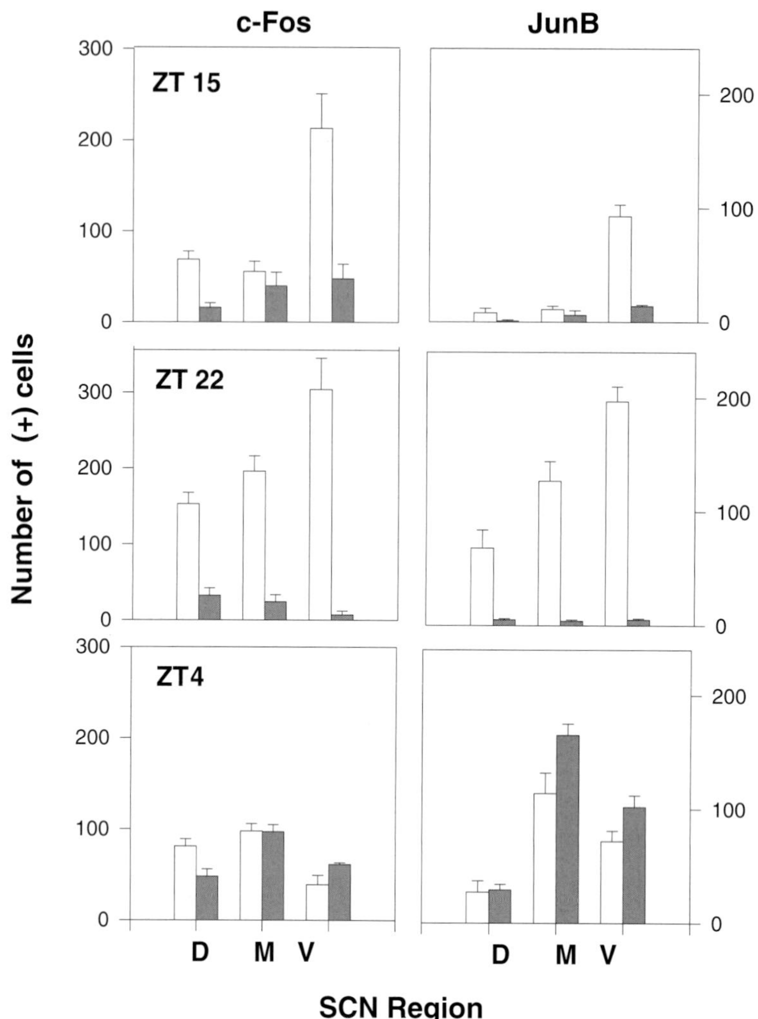

Fig. 2. Distribution of light-induced and spontaneous circadian expression of c-Fos and JunB-like proteins in the hamster SCN at a mid-caudal level in the dorsal (D), middle (M) and ventral (V) thirds of the nucleus. Animals maintained in constant darkness were exposed to a 30 min light pulse (white columns) or kept in darkness (hatched columns) at the same ZTs. Data represent mean (± SEM) number of cells immunostained for both IEG proteins. Light-evoked expression of c-Fos and JunB proteins at ZT 15 is concentrated in the ventral SCN; however, at ZT 22, positive cells are present throughout the SCN with the highest number in the ventral third after light exposure. Spontaneous expression of both proteins in constant conditions at ZT 4 is present throughout the nucleus with the highest number of positive cells in the middle third of the region sampled, corresponding to the dorsomedial SCN. The most dorsal third of the region has a relatively sparse cell content and cell counts are therefore not dramatically elevated. Reprinted from Neuroscience 90, Guido ME, Goguen D, de Guido LB, Robertson HA, Rusak B. Circadian and photic regulation of immediate-early gene expression in the hamster suprachiasmatic nucleus. pp.555–571, Copyright (1999), with permission from Elsevier Science.

early to late subjective night were for GRP, and to a lesser extent, VIP. These results further emphasize the distinctness of the patterns and extent of c-Fos expression at different times in the subjective night, and may especially link GRP-immunoreactive cells to phase advance shifts, while PHI cells may be more involved in mediating phase delay shifts.

In a study of mice, more than 56% of SCN c-Fos-expressing cells (induced by light at night) had an unknown phenotype (although some were positive for nitric oxide synthase), 24% were immunoreactive for VIP, 13% for VP-neurophysin and 7% for GFAP (Castel et al., 1997). The last result is consistent with an earlier finding in rats that some glial cells increased c-Fos expression in response to retinal illumination (Bennett and Schwartz, 1994). In Syrian hamsters, a nocturnal light pulse induced c-Fos protein mostly in cells immunoreactive for GRP, as well as in some VIP-positive cells (Aioun et al., 1998).

Whether there are significant species differences in the types of cells expressing IEGs and whether all IEGs are co-expressed in the same cells cannot be determined based on the data presently available. It is clear that the pattern of expression and of co-localization changes with circadian phase during the night as more cells are recruited with late-night than with early-night light exposure.

3. SPONTANEOUS RHYTHMS OF IEG EXPRESSION

While the first studies of c-Fos expression in the SCN did not detect any baseline level of protein in the absence of light exposure at the times studied, subsequent studies did indicate that at certain circadian phases c-Fos immunoreactivity might increase even in the absence of light (Koibuchi et al., 1992; Sutin and Kilduff, 1992). Differences reported among studies may have been related to differences in the sensitivity and specificity of antibodies used or to other technical issues. More recent studies have converged on the view that there are consistent spontaneous rhythms of IEG expression in the SCN that are restricted to certain circadian phases and specific sub-regions of the SCN.

In our studies, the most striking rhythms of expression are for the IEG *junB* and its protein in both the hamster and the rat SCN, with weaker expression of c-*fos* at the same circadian phase (Fig. 3) (Guido et al., 1996a, 1999a,b,c). Other laboratories have also reported spontaneous expression of c-*fos* mRNA or protein (Chambille et al., 1993; Sumová et al., 1995; Schwartz et al., 2000). In contrast to *junB* and c-*fos*, other IEGs, such as *NGFI-A*, *NGFI-B* and *fosB*, whose expression is regulated by light, do not show rhythms of expression in the rat or hamster SCN, and appear to be strictly under photic control (Rusak et al., 1992; Guido et al., 1996b, 1999a,b,c).

Spontaneous expression of *junB* and c-*fos* mRNAs is strikingly distinct from light-evoked expression in two respects: It begins quite abruptly in total darkness around subjective dawn, when sensitivity to photic stimuli is still present, but waning rapidly (Guido et al., 1999a,c). Second, it is concentrated in the dorsal half of the SCN, while photically evoked expression is stronger in the ventral half (Fig. 3). In Syrian hamsters, these same features are found for both c-*fos* and *junB*, but the distribution of spontaneous c-*fos* expression is less strictly confined to the dorsal half of the SCN (Guido et al., 1996a,c). The restriction of spontaneous expression of *junB* mRNA to the dorsal SCN in hamsters is even more clearly seen at the level of protein, because immunohistochemistry provides improved anatomical resolution. At this level, JunB protein is expressed in a very distinctive pocket of dorsomedial SCN cells; this region shows virtually no response to nocturnal light exposure that evokes robust expression in other parts of the SCN (Fig. 3) (Guido et al., 1996c, 1999a).

The dorsomedial SCN is enriched primarily in arginine-vasopressin (AVP) and somatostatin (SS)-containing neurons (van den Pol, 1980), and is sparsely innervated by retinal fibers. Levels of both AVP and SS undergo circadian variations in the rat SCN, with peaks in the early subjective day (van den Pol, 1980; Majzoub et al., 1991). Since the genes coding for

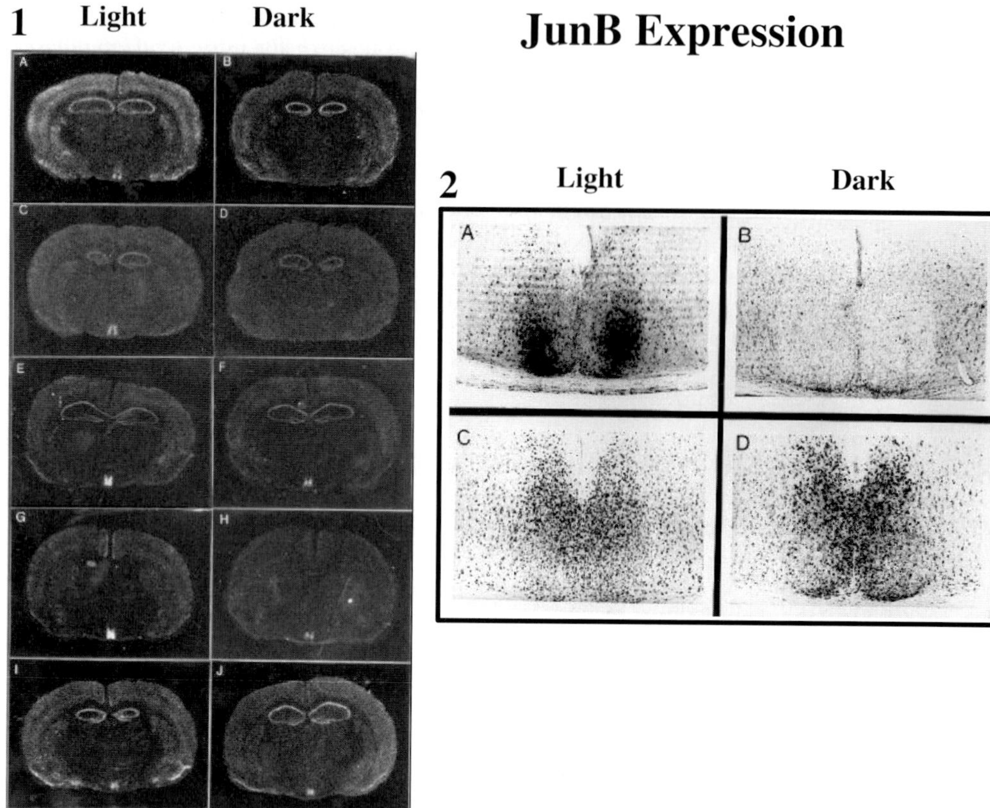

Fig. 3. Light-evoked and spontaneous circadian expression of *junB* mRNA (*1*) and protein (*2*) in the hamster SCN at a mid-caudal level. (*1*) Autoradiograms of sections of hamster brain through the SCN from in situ hybridization studies of *junB* mRNA. Animals were transferred from a LD 14 : 10 h cycle to constant darkness for 48 h, and then exposed to a 30-min light pulse (left column) or kept in darkness (right column). Light-exposed animals were killed at the end of the light pulse and dark controls were killed at the same projected ZT. The times of sacrifice were ZT 14.5 (*A, B*), ZT 18 (*C, D*), ZT 23.5 (*E, F*) during the projected night and ZT 0.5 (*G, H*) and ZT 4 (*I, J*) during the projected day. (*2*) Coronal sections through the mid-caudal SCN region from hamsters maintained for at least 10 days in constant darkness, stained immunohistochemically for JunB-lir. Hamsters were exposed to a 30-min light pulse and then returned to the dark for another 30 min (*A, C*). Controls were kept in the dark (*B, D*) and killed at the same times as the light-exposed animals at CT 16 (*A, B*) and CT 4 (*C, D*). Modified from Neuroscience 90, Guido ME, Goguen D, de Guido LB, Robertson HA, Rusak B. Circadian and photic regulation of immediate-early gene expression in the hamster suprachiasmatic nucleus. pp. 555–571, Copyright (1999), with permission from Elsevier Science.

AVP and SS mRNAs include a regulatory motif (TGAC/GTCA) that can bind IEG proteins (Sassone-Corsi et al., 1988), it is possible that the dawn peak of IEG expression contributes to the initiation of transcription leading to the production of these peptides. A recent study using double-label immunohistochemistry demonstrated that some dorsomedial cells showing a spontaneous circadian rhythm of c-Fos expression also synthesize VP (Schwartz et al., 2000), thus lending support to this possible mechanism. Future studies will need to establish whether these spontaneous rhythms of IEG expression play a role in any aspect of overt circadian rhythmicity.

These descriptions of spontaneous rhythmicity of IEG expression are based on studies of Syrian hamsters and rats. While most of these features also appear to characterize diurnal rodents that have been studied, differences have been noted in these species in relation to IEG expression in the SCN or in the area immediately dorsal to the SCN (Abe et al., 1995; Krajnak et al., 1997; Smale et al., 2001).

A recent finding in mice (C57Bl/6J) is also not consistent with some of the conclusions based on data in rats and hamsters. We found that NGFI-A immunoreactivity in the mouse was elevated at night (without light exposure) in cells located in a distinctive zone at the lateral border of the SCN, while similar immunoreactivity was not observed in hamsters. In addition, light exposure during the subjective day, when no spontaneous expression was observed, increased NGFI-A immunoreactivity in the same region in mice, but not in hamsters (Dong et al., in press). These observations suggest that the observed patterns of IEG expression in rats and hamsters may not be general across other mammals, and individual IEGs may be regulated differently in different species. The functional implications of these differences remain unknown.

4. REGULATION OF IEG EXPRESSION

4.1. NEUROTRANSMITTERS

Excitatory amino acids (EAAs) have been identified as the principal small-molecule neurotransmitters mediating effects of retinal input on the SCN via the retinohypothalamic tract (RHT; Kim and Dudek, 1991; Ebling, 1996). EAAs play a key role in photic entrainment and control of gene expression in SCN cells, but recent evidence suggests that retinal peptides conveyed to the SCN via the RHT may also modulate or play an independent role in affecting the SCN (Harrington et al., 1999; Hannibal et al., 2000, 2001; Kim et al., 2001). MK-801 (dizocilpine), a non-competitive NMDA antagonist that blocks Ca^{2+} channels, has been shown to inhibit light-induced phase advances and delays (Colwell et al., 1990; Rea et al., 1993), as well as the photic induction of c-Fos-like immunoreactivity (lir) in the ventral SCN, but not in the dorsolateral SCN (Abe et al., 1991, 1992; Rea et al., 1993). This dorsolateral region is not identical to the dorsomedial region in which spontaneous expression occurs, but it does overlap it in part.

The transmitter/receptor combination involved in mediating photic induction of gene expression in the dorsolateral SCN remains unidentified. Binding assays indicate that radiolabeled MK-801 is fairly uniformly distributed across the SCN region, with binding levels unaffected by circadian time or lighting condition, so the receptor distribution for MK-801 is not likely to be an important factor (Hartgraves and Fuchs, 1994). The possibility that an AMPA/kainate receptor might be differentially involved in different SCN regions was ruled out by the use of the AMPA/kainate antagonist CNQX, which also failed to block dorsolateral expression while blocking ventral expression (see Abe and Rusak, 1994 for review).

We asked whether spontaneous expression of IEGs in the dorsal SCN is also under the control of glutamatergic input. We examined the role of NMDA-type glutamate receptors in the expression of several IEG mRNAs either induced by light pulses or expressed spontaneously (Guido et al., 1999b). Hamsters were injected with dizocilpine (MK-801, 3 mg/kg, i.p.), prior to light exposure (two different intensities) either during the night (Zeitgeber time [ZT] 18, where ZT 0 is defined as subjective dawn), or at subjective dawn (ZT 0) to examine light-evoked expression, or before the onset of spontaneous expression near ZT 0 without any subsequent light exposure to examine spontaneous expression.

Both dim and bright light induced robust expression of *junB* and *NGFI-A* mRNAs throughout the SCN, and the degree of blockade by MK-801 in the ventral portion was similar with both light intensities. Expression of *fosB* mRNA was weaker in response to dim light, and this less extensive expression was only slightly reduced by MK-801 pretreatment. Cells in the dorsolateral SCN continued to express all three mRNAs in response to light at ZT 18, despite pretreatment with MK-801, as was reported previously for c-*fos* mRNA and protein (Abe et al., 1991, 1992; Rea et al., 1993; Abe and Rusak, 1994).

Levels of *junB* mRNA at dawn (ZT 0) after a light pulse reflect a combination of spontaneous and light-evoked expression. MK-801 injections before dawn were able to reduce mRNA levels in both the dorsal and ventral SCN. In the absence of light, mRNA expression at dawn is restricted largely to the dorsomedial SCN, and MK-801 treatment reduced these levels of expression, with the most dramatic effect at the lateral edge of this subdivision. Since there was little spontaneous expression in the ventral SCN, it is not surprising that MK-801 had little effect there, just as it had little effect on background levels of mRNAs in the absence of light induction at ZT 18. Thus, a glutamatergic mechanism acting via NMDA receptors is somehow involved in regulation of spontaneous gene expression in the dorsal SCN around subjective dawn. Involvement of other EAA receptor subtypes has not been investigated. These observations may reflect a role for glutamatergic projections to the SCN, of which there appear to be several (Ebling, 1996), in triggering the dawn expression peak. Alternatively, activation of NMDA receptors, in the SCN or elsewhere, may serve a permissive function for mechanisms intrinsic to the cells that generate the mRNA expression peak.

The threshold intensity for *fosB* mRNA activation by light appeared to be higher than for other IEGs at ZT 18, so that activation mechanisms were not saturated by dim light (5 lux), as they were for *junB* and *NGFI-A*. Lin et al. (1997) also reported that there are differences in threshold for light to evoke *NGFI-A* and c-*fos* expression. Thus, threshold light intensities may differ for different IEGs at a single phase as well as for a single IEG at different times during the night (Guido et al., 1999a,b).

Studies using glutamate receptor antagonists have identified at least three functionally distinct SCN regions: one is in the ventrolateral SCN, which shows sensitivity to light at all times during the subjective night, and in which NMDA receptor blockade blocks photic induction. Another is in the dorsomedial SCN, where spontaneous expression occurs beginning near subjective dawn, and which is also sensitive to NMDA antagonist actions. The third is in the dorsolateral SCN, in which light activation occurs primarily during the latter half of the subjective night, and where NMDA antagonists are ineffective in blocking light-evoked expression during the subjective night, but have a small effect around dawn.

4.2. SIGNALING MECHANISMS REGULATING IEG EXPRESSION

Release of glutamate or other neurotransmitters in the SCN may act via several mechanisms to induce gene expression and mediate phase shifts. The diffusible gas nitric oxide (NO) is released by activation of glutamate receptors and contributes to activation of a cyclic-GMP-dependent protein kinase (PKG) involved in mediating phase advances to light pulses (Ding et al., 1994, 1997). In addition, NO stimulates release of intracellular calcium from the endoplasmic reticulum via the ryanodine receptor calcium channel to mediate phase delays (Ding et al., 1998).

The role of the cyclic-AMP response element binding protein (CREB) in the signaling pathway mediating photic induction of c-*fos* in the SCN has been examined (Ginty et al., 1993). CREB is a transcription factor constitutively expressed in SCN cells that is activated by phosphorylation of a consensus serine (Ser-133) via the cyclic-AMP dependent protein kinase (PKA). The phosphorylation of CREB occurs in the SCN very rapidly after light stimulation during the subjective night, preceding the rise in IEG mRNA expression, and occurring in the same SCN regions. Light pulses during the subjective day do not activate phosphorylation of CREB in the SCN, suggesting that CREB phosphorylation may be a critical step regulated by the circadian clock that gives rise to rhythmicity in photic induction. It is not known whether CREB phosphorylation is involved in the spontaneous oscillation of IEG expression in the SCN.

The CREB protein is a point of convergence for various signal transduction pathways. For example, activation of the pathway involving mitogen-activated protein kinase (MAPK)/extracellular signal-regulated kinase (ERK) results in phosphorylation of CREB on Ser-133. In the rodent SCN, MAPKs are activated by nocturnal light pulses and phosphorylation of ERK and CREB converge partially in terms of timing and anatomical distribution (Obrietan et al., 1998). Moreover, ERK phosphorylation varies rhythmically under constant conditions (Obrietan et al., 1998). A recent report indicates that light-induced phase delays and altered *per* gene expression in the SCN are mediated specifically by calcium/calmodulin dependent kinase II (CaMKII) rather than MAPK activation (Yokota et al., 2001).

Different classes of tyrosine kinases that activate transcription factors by phosphorylation of tyrosine residues have been shown to be involved in the regulation of *junB* expression. These tyrosine kinases are sensitive to different and specific types of pharmacological treatments (Rezzonico et al., 1995). A recent study examined the effect of two kinds of tyrosine kinase inhibitors, genistein and tyrphostin, on spontaneous and light-evoked expression of *junB* in the SCN (Dong et al., 2000). Genistein was ineffective in all regions of the SCN and regardless of whether the expression was spontaneous or light-evoked. Tyrphostin pretreatment reduced both spontaneous and light-induced *junB* mRNA expression primarily in the dorsal, and not the ventral portion of the hamster SCN. Because spontaneous expression around subjective dawn is virtually restricted to the dorsal SCN, tyrphostin had a potent effect at that time.

The results suggest that there are two distinct transduction pathways in the dorsal and ventral SCN involving different protein tyrosine kinases, which control the expression of *junB* mRNA. These results reinforce other evidence that the dorsal and ventral halves of the SCN are anatomically, physiologically and biochemically different. They may also suggest approaches to the analysis of the functional significance of these differences.

A recent study examined the role of chromatin remodeling in regulation of IEG expression in the SCN. Regulation of gene expression has been proposed to be related to changes in chromatin architecture, such as post-translational modifications of histones (phosphorylation, acetylation, methylation, ubiquitination, and ADP-ribosylation). Light was reported to induce chromatin modification in SCN cells, specifically light pulses at night that induce c-*fos* expression also cause histone H3 phosphorylation in SCN cells (Crosio et al., 2000). These findings introduce a new level of complexity in the regulation by light of gene expression in the SCN.

5. FUNCTIONAL ROLES OF IEG PROTEINS

Members of the Jun and c-Fos families of transcription factors are components of the AP-1 complexes that bind in a sequence-specific manner to a DNA recognition site. *NGFI-A* and *NGFI-B* are characterized by their 'zinc-finger' motifs that also bind to DNA, but at a different recognition site. As reviewed above, different IEGs may be transcribed in response to photic input in different SCN cells, depending on circadian phase and light intensity or spontaneously at different circadian times, thus generating distinct patterns of transcription factor availability. These IEG proteins may form different combinatorial complexes with members of AP-1 or CREB/ATF families with distinct DNA binding activities (Hai and Curran, 1991) so as to dynamically regulate transcription of various target genes involved in the function of the circadian clock (Takeuchi et al., 1993).

Among the IEGs expressed in the mammalian SCN, c-*fos* has been studied most extensively. One study reported that when antisense oligonucleotides to both c-*fos* and *junB* were injected into the third ventricle, light could no longer evoke phase shifts of circadian activity rhythms in rats (Wollnik et al., 1995). Despite this demonstration that the products of these IEGs were critical to mediating light-evoked phase shifts, other evidence does not strongly support the conclusion that expression of specific IEGs is critical to, or rate-limiting for, circadian rhythm shifts.

There are a number of situations in which there is a dissociation between induction of c-Fos expression and circadian rhythm phase shifts. For example, one study used a novel method of studying c-Fos expression during entrainment and did not observe the expected difference between c-Fos expression in response to light stimuli that were and were not involved in entraining circadian activity rhythms (Schwartz et al., 1994). The threshold light intensity for c-Fos induction is very similar to that for phase shifting (Kornhauser et al., 1990), but the thresholds for induction of *NGFI-A* and *NGFI-B* may be 10–100-fold lower (Lin et al., 1997), indicating that they are not rate-limiting for phase-shifting. In addition, while they show some changes in photic entrainment, mice lacking the c-*fos* gene are still capable of responding to entraining light pulses (Honrado et al., 1996).

This demonstration that c-Fos is not required for entrainment in these knock-out mice does not, however, rule out a role for this protein in normal animals. A number of gene products may be able to form AP-1 complexes, and it is possible that compensatory responses involving other genes occur that obviate the need for c-Fos in these animals. Similarly, increases in nitric oxide synthase (NOS) activity appear to be critical to mediating photically induced phase shifts and gene expression in the SCN (Ding et al., 1994, 1997); however, knockout mice lacking either neuronal or epithelial isoforms of NOS still show normal photic responses (Kriegsfeld et al., 1999, 2001). It will be important in future studies to consider the possible existence of redundant molecular routes for photic regulation of the SCN, as well as the role of developmental compensatory mechanisms in knockout animals.

The target genes on which c-Fos and other IEG products may act to regulate later transcriptional events remain largely unknown. Thus, there is little basis on which to speculate on the functions served by the elaborate, anatomically and temporally differentiated patterns of IEG expression that occur in SCN cells spontaneously or in response to retinal illumination.

Functional studies of gene expression changes have focused on the clock genes that appear to be central to circadian function in SCN cells (King and Takahashi, 2000), but the relations between these genes and IEGs have not been characterized. A recent study has, however, raised questions about the role of two genes that were considered to be central to mediating photically induced phase shifts, namely the *period 1* and *2* genes (*per1* and *per2*). In mice lacking Type 1 receptors for the peptide PACAP, light induced phase delays that were larger than those of controls, but also induced much lower levels of expression of *per1*, *per2,* and c-*fos* mRNAs. Light-induced phase advances were blunted or reversed in direction, while expression of these genes was normal or enhanced (Hannibal et al., 2001). Thus, a molecular account of circadian phase shifts in the SCN based solely on changes in levels of these clock gene products does not appear to be complete, nor does one based on the coordinate expression of the IEG products c-Fos and JunB. A good deal more information will be required to untangle these complex gene expression webs and to establish their functional relations to each other and to the regulation of behavioral and physiological functions.

6. INTERACTION OF CIRCADIAN AND SLEEP–WAKE MECHANISMS

6.1. INTRODUCTION

Sleep patterns differ strongly among species, with humans and other primates showing a consolidated sleep phase with a unique or at least principal sleep period during one circadian phase, and other species, such as laboratory rodents, showing rapid cycles of alternating sleep and waking within each circadian period. Nevertheless, a circadian framework still constrains the expression of sleep, so that despite the polycyclic structure of sleep, rats will sleep more during the day phase while humans concentrate sleep in the night. The role of the circadian clock in regulating the expression of sleep and waking is revealed after ablation of the SCN. Rats without an SCN continue to show rapidly alternating periods of sleep and waking, but the circadian constraint on this pattern is lost and the distribution of sleep, waking and activity becomes random with respect to time of day (Coindet et al., 1975; Ibuka and Kawamura, 1975; Mistlberger et al., 1983). A similar loss of daily rhythms of sleep has been documented in SCN-ablated squirrel monkeys (Edgar et al., 1993) and in humans with hypothalamic damage that likely destroyed the SCN (Schwartz et al., 1986; Cohen and Albers, 1991).

The nature of circadian influences on the homeostatic regulation of sleep varies among different species. In rodents, the homeostatic regulation of sleep appears to be intact after SCN ablation and loss of circadian rhythmicity, so that amounts of sleep and waking are normal but distributed differently throughout the day and night (Ibuka and Kawamura, 1975; Mistlberger et al., 1983). In squirrel monkeys, however, waking is reduced and sleep durations increased by SCN ablation (Edgar et al., 1993), implying a role for the SCN in promoting waking in this species. This role might be important in species with lengthy consolidated sleep and waking phases in order to counteract the accumulating homeostatic drive for sleep near the end of the waking phase.

While the details of the rodent and primate systems differ, they share the common feature that a homeostatic mechanism regulating sleep and wake durations interacts with a separate circadian mechanism responsible for their distribution throughout the day–night cycle. This conclusion implies that circadian phase and prior sleep/wake duration should interact to regulate the activity of systems responsible for initiating and maintaining sleep and waking. A number of studies have examined gene expression patterns in the nervous system associated with different states of sleep and waking, but few have examined the interaction of these mechanisms with circadian phase. We begin with a brief summary of current knowledge of gene expression in the brain in relation to sleep and waking, and then review studies that used gene expression to investigate the interactions between circadian and sleep regulatory mechanisms.

6.2. GENE EXPRESSION DURING SLEEP AND WAKEFULNESS

A considerable body of knowledge has been accumulated about the neural systems that control behavioral states, using neuroanatomical, neurophysiological, neurochemical, and behavioral approaches (Steriade and McCarley, 1990; Lydic and Baghdoyan, 1999). These systems are distributed throughout the brain, and use a number of transmitters including noradrenaline, serotonin, acetylcholine, histamine and glutamate; hypocretins/orexins have been added to this list recently (Kilduff and Peyron, 2000). Studies of gene expression during sleep and wakefulness began more recently, but have expanded rapidly. Currently available data provide a number of interesting insights into the cellular and molecular mechanisms of behavioral state control, and into the functions of sleep, as summarized below.

Historically, two approaches have been used to study patterns of gene expression following periods of sleep and wakefulness. The first approach used in situ hybridization and immunohistochemical studies aimed at specific candidate genes. Subtractive hybridization techniques were also used, but this approach is not particularly efficient, since eventually all genes in a species' genome would have to be screened to identify those that are selectively activated in a given behavioral state. The second, more efficient, approach screens large numbers of genes simultaneously using mRNA differential display and microarray technologies.

Pioneering work in this field has been reviewed in a number of recent papers (Pompeiano et al., 1995; Shiromani and Schwartz, 1995; Tononi et al., 1995; Cirelli and Tononi, 1998, 1999a,b, 2000a,b; Bentivoglio and Grassi-Zucconi, 1999). We will summarize the main conclusions that have emerged from this recent work. Many of these studies analyzed gene expression patterns in rats that had spent most of the last 3–12 h in spontaneous sleep, spontaneous wakefulness, or forced wakefulness. As there are no reported differences in gene expression between spontaneous and forced wakefulness, these are discussed together. In these studies, the circadian time was typically kept constant or similar to allow direct comparison between behavioral states without the potential confound of variation in circadian phase. Most of the studies using candidate gene approaches analyzed gene expression in the whole brain, whereas studies using more systematic approaches, i.e., mRNA differential display or cDNA microarray technologies, were primarily focused on the cerebral cortex. Specific references and details are provided in the reviews cited above.

One main conclusion of these studies is that a small number of genes show different levels of expression during sleep and wakefulness, while the vast majority of genes do not change their expression levels. A major group of responsive genes includes the IEG transcription factors that have been discussed with respect to circadian function. Thus, c-*fos*, *junB*, *NGFI-A*, *NGFI-B*, *IER5* and *Arc* strongly increase their levels of expression in some brain regions after a number of hours of wakefulness, as compared to during sleep.

The activation of these IEGs following periods of wakefulness has been observed in rats, mice, ground squirrels, Syrian hamsters, and cats, and has been demonstrated for both mRNA and protein. The brain areas where c-*fos* and *NFGI-A* are activated selectively following wakefulness include the cerebral cortex, caudate-putamen, hippocampal formation, medial and lateral preoptic areas, and various thalamic and brainstem nuclei, some of which are known to be involved in regulation of behavioral and EEG arousal. Genes encoding other transcription factors that are activated following a few hours of wakefulness include the rat homologue of the human Zn-15 related zinc finger gene *rlf*. Activation of IEGs in the cortex appears to be dependent on intact noradrenergic innervation.

Other genes that are activated following a number of hours of wakefulness are part of the mitochondrial genome, including genes encoding subunit I of cytochrome *c* oxidase, subunit 2 of NADH (nicotinamide adenine dinucleotide, reduced form) dehydrogenase, and 12S ribosomal (r)RNA. Cerebral glucose utilization is 20–30% higher during wakefulness compared to during non-REM sleep. Activation of mitochondrial genes following sustained wakefulness suggests that brain cells respond to increased metabolic demand during wakefulness by activating mitochondrial genes related to cellular metabolism and energy homeostasis. These results imply that one function served by sleep may involve a response to these changing metabolic demands.

Because a major group of genes that are activated after a few hours of wakefulness are transcription factors, it is likely that their expression is followed by changes in the expression of 'late' genes that play a role in the cellular and molecular functions of sleep or wakefulness. Thus, longer (e.g., 8 or 12 h) periods of spontaneous or forced wakefulness are accompanied by the activation of a number of genes that are not significantly activated after shorter periods of wakefulness. These include genes related to energy metabolism, growth factors/adhesion molecules, molecular chaperones, vesicle and synapse-related genes, genes for neurotransmitter receptors and transporters, and genes coding for enzymes involved in degradation of monoamines or in signal transduction pathways (Cirelli and Tononi, 2000b). In addition, IEGs including *NGFI-A*, *NGFI-B*, *Arc*, and *IER5* are expressed at higher levels after 8 h of wakefulness, whereas c-*fos* expression, which is elevated after 3 h of wakefulness, returns to baseline levels, probably as a result of inhibition of transcription of its own mRNA by c-Fos protein. These different patterns likely reflect differences among responsive genes in the time course or kinetics of their expression following a stimulus.

These results with 8-h wakefulness suggest that the brain responds to prolonged neuronal activity by replenishing energy supply and molecules involved in neurotransmission. Furthermore, several genes that are activated during waking, including c-*fos*, *NGFI-A*, *Arc* as well as Ca^{2+}/calmodulin-dependent protein kinase II, are known to be involved in neuronal plasticity. The expression of these genes, therefore, suggests that mechanisms for neuronal plasticity including learning and memory might be activated during wakefulness.

Less is known about gene expression following periods of sleep. IEG expression is generally low during sleep, and the only brain region that has been identified as expressing IEGs after periods of sleep is the ventrolateral preoptic area (VLPO), which contains a cluster of neurons that express c-*fos* after a short period (1 h) of sleep (Sherin et al., 1996). Sleep deprivation does not induce c-*fos* expression in this nucleus, suggesting that these neurons are involved in the maintenance of sleep, and not the initiation of sleep nor the assessment of sleep need. Studies using differential display and microarrays have identified several other genes that are activated following periods of sleep, but these genes remain to be characterized (Cirelli and Tononi, 1999b).

Although most studies on gene expression after sleep did not distinguish REM sleep from non-REM sleep, REM sleep in the species studied tends to be brief (up to approximately 2 min in rats), thus not allowing sufficient time for gene expression selectively related to this state to take place. A number of studies have, however, examined gene expression after periods of enhanced REM sleep following selective REM sleep deprivation, or pharmacologically induced REM sleep, such as after microinjection of the cholinergic agonist carbachol into the pontine reticular formation (reviewed in Cirelli and Tononi, 2000a). Most of these studies used c-Fos as a marker for neuronal activation, and identified a group of neurons that express this IEG following periods of increased REM sleep. These neurons included cholinergic, GABAergic and other neurons that are thought to be involved in expression of REM sleep features, such as muscle atonia (Maloney et al., 1999, 2000).

In summary, recent findings on gene expression during sleep or wakefulness are exciting and suggest that cellular and molecular events that are associated with prolonged wakefulness include processes related to homeostasis at the cellular level and to synaptic plasticity. Only a small number of genes have so far been studied, however, and relatively little is known about genes that are selectively activated during sleep. Furthermore, more recent studies using systematic gene screening approaches have been focused primarily on the cerebral cortex, and there is a need to extend this knowledge further to non-cortical regions of the brain. Finally, investigation into interactions among diverse types of genes would lead to a better understanding of the molecular and cellular mechanisms of behavioral state control.

6.3. CIRCADIAN MODULATION OF GENE EXPRESSION ASSOCIATED WITH WAKEFULNESS

Although the circadian timing of sleep and waking is phenomenologically obvious, the mechanisms that link the circadian and sleep–wake regulatory systems are poorly understood. One feature of the output organization of the SCN is that direct projections do not reach those brain structures known to be involved in sleep/wake control (Watts et al., 1987; Leak and Moore, 2001). One exception is the sparse projection from the SCN to the VLPO, which is known to contain sleep-active neurons (Novak and Nunez, 2000; Sun et al., 2000, 2001). These considerations suggest that most of the output from the SCN to the sleep–wake regulatory system must be conveyed indirectly via intermediary nuclei. This possibility is being investigated using anatomical, neurophysiological and behavioral approaches. Recent studies suggest the presence of disynaptic pathways from the SCN to sleep/wake-regulatory neurons via intermediary nuclei in the preoptic area and hypothalamus, including one from the SCN to the locus coeruleus via several hypothalamic nuclei, especially the dorsomedial nucleus (Aston-Jones et al., 2001). It is also important to note that some of the neurons known to be involved in wakefulness, REM sleep or both also project to the SCN, and probably provide feedback from the sleep–wake regulatory system to the circadian system (Bina et al., 1993), which may be functionally important in regulation of circadian rhythms (Antle and Mistlberger, 2000).

Given the functional and anatomical evidence for a neural network that links the SCN with the sleep–wake regulatory system, it is likely that these systems also interact at the level of gene expression; more specifically, sleep/wake-related gene expression is likely to be modulated by circadian input as well as by homeostatic factors. Analysis of this issue should also provide insight into how the circadian signal interacts with homeostatic factors to time the occurrence of sleep.

To address these issues, we examined c-Fos and JunB immunoreactivity in several brain regions of rats after 3 or 6 h of sleep deprivation that was initiated at four different time points during the daily light phase (Semba et al., 2001). We used forced slow locomotion in a motorized wheel as a means of sleep deprivation. It is important to separate the effects of sleep loss from any non-specific stress associated with the method used to deprive animals of sleep. We therefore habituated animals to the situation by housing them in the wheels for 2–3 weeks before deprivation and compared them to animals that were similarly locked in their wheels, but without rotation. We also assessed levels of c-Fos immunoreactivity in the paraventricular hypothalamic nucleus (PVH), a key structure involved in response to stress. No significant increase in c-Fos immunoreactivity was observed in the PVH, indicating that there was little stress associated with the deprivation process over 3–6 h. Sleep deprivation was initiated at ZT 0 (Zeitgeber time 0 = onset of light of the daily 12 h light phase), ZT 3, ZT 6, and ZT 9. The structures analyzed for IEG expression included the medial preoptic area (MPA), anterior (PVA) and posterior paraventricular thalamic nuclei (PVP), cortex, amygdala, caudate-putamen, and laterodorsal tegmental nucleus. These regions have been shown to display a significant increase in c-Fos expression following spontaneous (Pompeiano et al., 1994) or forced wakefulness (Cirelli et al., 1995b).

Sleep deprivation for either 3 or 6 h starting during the light phase induced strong c-Fos expression in all brain regions investigated, as anticipated. There were, however, differences in levels of induction depending on both the duration of deprivation (the homeostatic factor) and on the timing of the onset of the deprivation period during the light phase (the circadian factor) (Fig. 4). One important observation was that only the PVP and the amygdala showed significantly greater c-Fos induction after 6 h than after 3 h of sleep deprivation beginning at a single clock time. This observation suggests that these regions might serve a crucial function in responding to the duration of prior wakefulness (homeostatic sleep need). The strength of c-Fos expression in PVP and amygdala may reflect the intensity of sleep need, and deprivation-induced increases in neural output from these regions might influence other structures and contribute to initiating sleep.

Another result of this study was the demonstration of temporal variations in the extent of c-Fos expression after 3 h sleep deprivation initiated at different times in the light phase. The most robust c-Fos expression in most regions occurred after 3 h sleep deprivation beginning at ZT 3. In the MPA and PVA, similarly robust expression occurred when sleep deprivation started at ZT 0. In MPA and cortex, similarly timed peaks were observed after 6 h of sleep deprivation. This increased sensitivity to sleep deprivation near the onset of the daily light phase parallels the changes in EEG delta power in the light phase in rat (Frank and Heller, 1997). Delta power represents the extent of low frequencies in the cortical EEG and is considered a measure of sleep pressure (Benington and Heller, 1995). The temporal association between maximal EEG delta power and maximal sensitivity to sleep deprivation in terms of induced gene expression suggests that the MPA and PVA are sensitive to the homeostatic features of both the amount and intensity of sleep.

If these regions were involved in mediating homeostatic aspects of sleep, it is surprising that we found the amount of c-Fos expression was not significantly greater after 6 h than after 3 h of deprivation. One potential explanation is that both homeostatic and circadian aspects of sleep regulation act on these structures. Since, unlike 3 h of deprivation, 6 h of deprivation would result in studying rats during the second half of the light phase, a circadian influence might be important. Based on the reduced tendency to show slow-wave sleep during the latter half of the light phase, and the physiological preparations for arousal around dusk, it is probable that there is a reduced circadian tendency to promote sleep in the second half of the light phase. This change is likely to be mediated by an altered signal from the SCN circadian pacemaker to sleep-regulatory mechanisms. This inhibition of homeostatically induced activation might be mediated by direct, probably inhibitory, projections from the SCN to the paraventricular thalamic nucleus and MPA (Watts et al., 1987; Sun et al., 2000; Leak and Moore, 2001). A failure to observe further c-Fos expression in response to lengthened deprivation might be the result of a balance between increased homeostatic drive and decreased circadian drive for sleep. In addition, the self-inhibition of c-*fos* gene expression by c-Fos protein might be a factor.

JunB expression after 3 or 6 h sleep deprivation was generally lower than that of c-Fos in all regions studied. There was, however, a generally similar pattern of expression of JunB with statistically significant increases observed in the MPA, cortex, amygdala and caudate-putamen. Similar patterns of expression of JunB and c-Fos proteins indicates that both would be available simultaneously to form heterodimers that are capable of binding to AP-1 binding sites on DNA, via which the expression of other genes can be regulated.

These results indicate that both duration of prior wakefulness and time of day influence the extent of IEG expression differentially in brain regions responsive to sleep deprivation. The MPA is particularly interesting potentially as a site of interaction between homeostatic

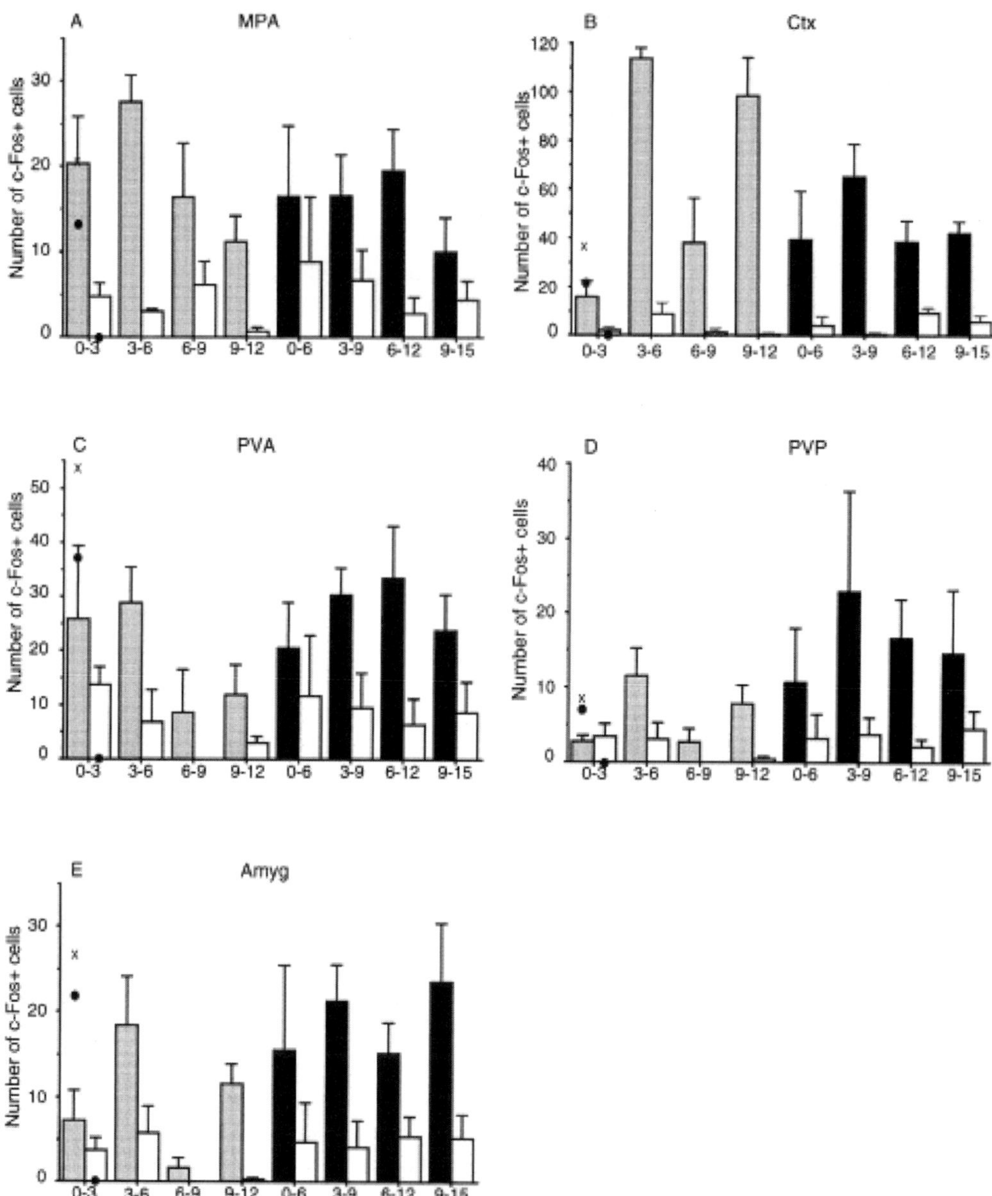

Fig. 4. The number of c-Fos-expressing cells (mean ± SEM) in selected regions of the brain after 3 h (gray) or 6 h (black) sleep deprivation starting at ZT 0, ZT 3, ZT 6, and ZT 9. Values obtained from respective non-deprived controls are shown as white bars. Note the peak of expression for sleep deprivation starting at ZT 3 in all regions shown, and generally high levels of c-Fos expression after 6 h, compared to 3 h, deprivation in the PVP and amygdala. 3 h deprivation by gentle handling (• and ×: means and SEM) induced levels of c-Fos expression similar to those induced by forced locomotion (see text for details about the deprivation procedure). Amyg, amygdala; Ctx, cortex; MPA, medial preoptic area; PVA, anterior paraventricular thalamic nucleus; PVP, posterior paraventricular thalamic nucleus. Reproduced from Semba et al. (2001). Copyright with Elsevier Science B.V.

and circadian signals. Interestingly, unlike the VLPO (Sherin et al., 1996), the MPA does not express c-Fos after sleep (Pompeiano et al., 1994). Infusions of c-*fos* anti-sense oligonucleotides into the MPA reduce amounts of sleep in rats (Cirelli et al., 1995a). Together, these findings suggest that the MPA is involved in the initiation, but not the maintenance, of sleep. It is possible that medial preoptic neurons monitor both homeostatic need for sleep and the circadian signal from the SCN, and thus, can gate the occurrence of sleep differentially according to circadian phase.

7. CONCLUSIONS

Studies of IEG expression in the circadian regulatory system and in structures involved in sleep–wake control have been useful in identifying anatomical sites that play a role in these mechanisms. They have also helped to identify candidate molecular mechanisms which mediate environmental and endogenous regulation of sleep–wake and circadian rhythmicity at the cellular level. Studies of IEGs have taken advantage of their rapid, brief expression changes in response to signals to better understand processes involved in responding to photic input and to sleep deprivation. Studies of patterns of activation using IEG expression have also provided interesting hints about how different brain regions respond to differently timed periods of sleep deprivation. These results may provide an indication of the differential roles of these structures in evaluating and responding to sleep loss, and in integrating the influences of homeostatic and circadian regulators of sleep onset and maintenance.

It will be important in future studies to take account of the likelihood that mechanisms identified through molecular approaches as playing a role in circadian and sleep–wake regulation may be part of complex regulatory webs. It is likely that regulation of these systems involves multiple, partially redundant mechanisms, and mechanisms that are developmentally plastic in highly modified organisms, such as knock-out mice. These features will continue to complicate analyses of these systems at the molecular level, but novel and increasingly efficient and systematic approaches are becoming available to identify genes that are activated in response to particular stimuli or behavioral conditions. These approaches may provide a more comprehensive picture of these complex regulatory systems. Future studies should be able to address at the molecular level how the temporal structure of behavior, including sleep and wakefulness, is organized in relation to environmental cycles and internal regulatory needs.

8. ABBREVIATIONS

AMPA	α-amino-3-hydroxy-5-methylisoxazole-4-propionic acid
AP	activator protein
AVP	arginine-vasopressin
CaMKII	calcium/calmodulin dependent kinase II
CREB	cyclic-AMP response element binding protein
EAA	excitatory amino acid
EEG	electroencephalogram
ERK	extracellular signal-regulated kinase
GFAP	glial fibrillary acidic protein
GRP	gastrin-releasing peptide

IEG	immediate-early gene
lir	like immunoreactivity
MAPK	mitogen-activated protein kinase
MPA	medial preoptic area
NMDA	*N*-methyl-D-aspartate
NO	nitric oxide
NOS	nitric oxide synthase
PACAP	pituitary adenylate cyclase-activating polypeptide
PHI	peptide histidine isoleucine
PKA	cyclic-AMP-dependent protein kinase (protein kinase A)
PVA	anterior paraventricular thalamic nucleus
PVH	paraventricular hypothalamic nucleus
PVP	posterior paraventricular thalamic nucleus
REM	rapid eye movement
RHT	retinohypothalamic tract
SCN	suprachiasmatic nuclei
SS	somatostatin
VIP	vasoactive intestinal polypeptide
VLPO	ventrolateral preoptic nucleus
VP	vasopressin
ZT	Zeitgeber time

9. REFERENCES

Abe H, Rusak B (1994): Physiological mechanisms regulating photic induction of Fos-like protein in hamster suprachiasmatic nucleus. *Neurosci Biobehav Rev 18*:531–536.

Abe H, Rusak B, Robertson HA (1991): Photic induction of Fos protein in the suprachiasmatic nucleus is inhibited by the NMDA receptor antagonist MK-801. *Neurosci Lett 127*:9–12.

Abe H, Rusak B, Robertson HA (1992): NMDA and non-NMDA receptor antagonists inhibit photic induction of Fos protein in the hamster suprachiasmatic nucleus. *Brain Res Bull 28*:831–835.

Abe H, Honma S, Shinohara K, Honma KI (1995): Circadian modulation in photic induction of Fos-like immunoreactivity in the suprachiasmatic nucleus cells of diurnal chipmunk, *Eutamias asiaticus*. *J Comp Physiol A 176*:159–167.

Aioun J, Chambille I, Peytevin J, Martinet L (1998): Neurons containing gastrin-releasing peptide and vasoactive intestinal polypeptide are involved in the reception of the photic signal in the suprachiasmatic nucleus of the Syrian hamster: an immunocytochemical ultrastructural study. *Cell Tissue Res 291*:239–253.

Antle MC, Mistlberger RE (2000): Circadian clock resetting by sleep deprivation without exercise in the Syrian hamster. *J Neurosci 20*:9326–9332.

Aronin N, Sagar S, Sharp F, Schwartz W (1990): Light regulates expression of a Fos-related protein in rat suprachiasmatic nucleus. *Proc Natl Acad Sci USA 87*:5959–5962.

Aston-Jones G, Chen S, Zhu Y, Oshinsky ML (2001): A neural circuit for circadian regulation of arousal. *Nat Neurosci 4*:732–738.

Benington JH, Heller HC (1995): Restoration of brain energy metabolism as the function of sleep. *Prog Neurobiol 45*:347–360.

Bennett MR, Schwartz W (1994): Are glia among the cells that express immunoreactive c-Fos in the suprachiasmatic nucleus?. *NeuroReport 5*:1737–1740.

Bennett MR, Aronin N, Schwartz W (1996): In vitro stimulation of c-Fos protein expression in the suprachiasmatic nucleus of hypothalamic slices. *Brain Res Mol Brain Res 42*:140–144.

Bentivoglio M, Grassi-Zucconi G (1999): Immediate early gene expression in sleep and wakefulness. In: Lydic R, Baghdoyan HA (Eds), *Handbook of Behavioral State Control. Cellular and Molecular Mechanisms*. Boca Raton, FL: CRC Press, pp. 235–253.

Bina K, Rusak B, Semba K (1993): Localization of cholinergic neurons in the forebrain and brainstem that project to the suprachiasmatic nucleus of the hypothalamus in rat. *J Comp Neurol 335*:295–307.

Borbély A (1982): A two hypothesis model of sleep regulation. *Hum Neurobiol 1*:195–204.

Castel M, Belenky M, Cohen S, Wagner S, Schwartz WJ (1997): Light-induced c-Fos expression in the mouse suprachiasmatic nucleus: immunoelectron microscopy reveals co-localization in multiple cell types. *Eur J Neurosci 9*:1950–1960.

Chambille I, Doyle S, Servière J (1993): Photic induction and circadian expression of Fos-like protein. Immunohistochemical study in the retina and suprachiasmatic nuclei of hamster. *Brain Res 612*:138–150.

Cirelli C, Tononi G (1998): Differences in gene expression between sleep and waking as revealed by mRNA differential display. *Brain Res Mol Brain Res 56*:293–305.

Cirelli C, Tononi G (1999a): Differences in brain gene expression between sleep and waking as revealed by mRNA differential display and cDNA microarray technology. *J Sleep Res 8 (Suppl 1)*:44–52.

Cirelli C, Tononi G (1999b): Differences in gene expression during sleep and wakefulness. *Ann Med 31*:117–124.

Cirelli C, Tononi G (2000a): On the functional significance of c-*fos* induction during the sleep–waking cycle. *Sleep 23*:453–469.

Cirelli C, Tononi G (2000b): Gene expression in the brain across the sleep–wake cycle. *Brain Res 885*:303–321.

Cirelli C, Pompeiano M, Tononi G (1995a): Sleep deprivation and c-*fos* expression in the rat brain. *J Sleep Res 4*:92–106.

Cirelli C, Pompeiano M, Arrighi P, Tononi G (1995b): Sleep–waking changes after c-*fos* antisense injections in the medial preoptic area. *NeuroReport 6*:801–805.

Cohen RA, Albers HE (1991): Disruption of human circadian and cognitive regulation following a discrete hypothalamic lesion: a case study. *Neurology 41*:726–729.

Coindet J, Chouvet G, Mouret J (1975): Effects of lesions of the suprachiasmatic nuclei on paradoxical sleep and slow wave sleep circadian rhythms in the rat. *Neurosci Lett 1*:243–247.

Colwell C, Ralph M, Menaker M (1990): Do NMDA receptors mediate the effects of light on circadian behavior?. *Brain Res 523*:117–120.

Crosio C, Cermakian N, Allis CD, Sassone-Corsi P (2000): Light induces chromatin modification in cells of the mammalian circadian clock. *Nat Neurosci 3*:1241–1247.

Daikoku S, Yokote R, Aizawa T, Kawano H (1992): Light stimulation of the hypothalamic neuroendocrine system. *Arch Histol Cytol 55*:67–76.

Dijk D-J, Edgar DM (1999): Circadian and homeostatic control of wakefulness and sleep. In: Turek FW, Zee PC (Eds), *Regulation of Sleep and Circadian Rhythms*. New York: Marcel Dekker, pp. 111–147.

Ding JM, Chen D, Weber ET, Faiman LE, Rea MA, Gillette MU (1994): Resetting the biological clock: mediation of nocturnal circadian shifts by glutamate and NO. *Science 266*:1713–1717.

Ding JM, Faiman LE, Hurst WJ, Kuriashkina LR, Gillette MU (1997): Resetting the biological clock: mediation of nocturnal CREB phosphorylation via light, glutamate, and nitric oxide. *J Neurosci 17*:667–675.

Ding JM, Buchanan GF, Tischkau SA, Chen D, Kuriashkina L, Faiman LE, Alster JM, McPherson PS, Campbell KP, Gillette MU (1998): A neuronal ryanodine receptor mediates light-induced phase delays of the circadian clock. *Nature 394*:381–384.

Dong Y, Guido ME, Robertson HA, Rusak B (2000): Selective regional blockade of *junB* gene expression in the hamster suprachiasmatic nucleus by a tyrosine kinase inhibitor. *Brain Res Mol Brain Res 77*:29–36.

Dong Y, Goguen D, Robertson HA, Rusak B (in press): Anatomical and temporal differences in the regulation of ZIF268 (NGFI-A) protein in the hamster and mouse suprachiasmatic nucleus. *Neuroscience*, in press.

Ebling FJP (1996): The role of glutamate in the photic regulation of the suprachiasmatic nucleus. *Prog Neurobiol 50*:109–132.

Ebling FJP, Maywood ES, Mehta M, Hancock DC, McNulty S, De Bono J, Bray SJ, Hastings MH (1996): FosB in the suprachiasmatic nucleus of the Syrian and Siberian hamster. *Brain Res Bull 41 (5)*:257–268.

Edgar DM, Dement WC, Fuller CA (1993): Effects of SCN lesions on sleep in squirrel monkeys: Evidence for opponent processes in sleep–wake regulation. *J Neurosci 13*:1065–1079.

Frank MG, Heller HC (1997): Development of diurnal organization of EEG slow-wave activity and slow-wave sleep in the rat. *Am J Physiol 273*:R472–R478.

Ginty DD, Kornhauser JM, Thompson MA, Bading H, Mayo KE, Takahashi JS, Greenberg ME (1993): Regulation of CREB phosphorylation in the suprachiasmatic nucleus by light and a circadian clock. *Science 260*:238–241.

Guido ME, Rusak B, Robertson HA (1996a): Spontaneous circadian and light-induced expression of *junB* mRNA in the hamster suprachiasmatic nucleus. *Brain Res 732*:215–222.

Guido ME, Rusak B, Robertson HA (1996b): Expression of *fosB* mRNA in the hamster suprachiasmatic nucleus is induced at only selected circadian phases. *Brain Res 739*:132–138.

Guido ME, Goguen D, Robertson HA, Rusak B (1996c): Spontaneous and light-evoked expression of JunB-like protein in the hamster suprachiasmatic nucleus near subjective dawn. *Neurosci Lett 217*:9–12.

Guido ME, Goguen DM, de Guido L, Robertson HA, Rusak B (1999a): Circadian and photic regulation of immediate-early gene expression in the hamster suprachiasmatic nucleus. *Neuroscience 90 (2)*:555–571.

Guido ME, de Guido L, Goguen DM, Robertson HA, Rusak B (1999b): Differential effects of glutamatergic blockade on circadian and photic regulation of gene expression in the hamster suprachiasmatic nucleus. *Brain Res Mol Brain Res 67*:247–257.

Guido ME, de Guido L, Goguen DM, Robertson HA, Rusak B (1999c): Daily rhythms of spontaneous immediate-early gene expression in the rat suprachiasmatic nucleus. *J Biol Rhythms 14 (4)*:275–280.

Hannibal J, Moller M, Ottersen OP, Fahrenkrug J (2000): PACAP and glutamate are co-stored in the retinohypothalamic tract. *J Comp Neurol 418 (2)*:147–155.

Hannibal J, Jamen F, Nielsen HS, Journot L, Brabet P, Fahrenkrug J (2001): Dissociation between light-induced phase shift of the circadian rhythm and clock gene expression in mice lacking the pituitary adenylate cyclase activating polypeptide type 1 receptor. *J Neurosci 21*:4883–4890.

Hai T, Curran T (1991): Cross-family dimerization of transcription factors Fos/Jun and ATF/CREB alters DNA binding specificity. *Proc Natl Acad Sci USA 88*:3720–3724.

Hartgraves MD, Fuchs JL (1994): NMDA receptor binding in rodent suprachiasmatic nucleus. *Brain Res 640*:113–118.

Harrington ME, Hoque S, Hall A, Golombek D, Biello S (1999): Pituitary adenylate cyclase activating peptide phase shifts circadian rhythms in a manner similar to light. *J Neurosci 19*:6637–6642.

Hastings JW, Rusak B, Boulos Z (1991): Circadian rhythms: The physiology of biological timing. In: Prosser CL (Ed), *Neural and Integrative Animal Physiology*. New York: Wiley-Liss, pp. 435–546.

Honrado G, Jonhson R, Golombek D, Spiegelman B, Papaionnou V, Ralph M (1996): The circadian system of c-*fos* deficient mice. *J Comp Physiol A 178 (4)*:563–570.

Ibuka N, Kawamura H (1975): Loss of circadian rhythm in sleep–wakefulness cycle in the rat by suprachiasmatic nucleus lesions. *Brain Res 96*:76–81.

Inouye SIT, Shibata S (1994): Neurochemical organization of circadian rhythm in the suprachiasmatic nucleus. *Neurosci Res 20*:109–130.

Janik D, Mrosovsky N (1992): Gene expression in the geniculate induced by a nonphotic circadian phase shifting stimulus. *NeuroReport 3*:575–578.

Kilduff TS, Peyron C (2000): The hypocretin/orexin ligand-receptor system: implications for sleep and sleep disorders. *Trends Neurosci 23*:359–365.

Kim YI, Dudek FE (1991): Intracellular electrophysiological study of suprachiasmatic nucleus neurons in rodents: excitatory synaptic mechanisms. *J Physiol 444*:268–287.

Kim DY, Kang HC, Shin HC, Lee KJ, Yoon YW, Han HC, Na HS, Hong SK, Kim YI (2001): Substance P plays a critical role in photic resetting of the circadian pacemaker in the rat hypothalamus. *J Neurosci 21 (11)*: 4026–4031.

King DP, Takahashi JS (2000): Molecular genetics of circadian rhythms in mammals. *Annu Rev Neurosci 23*:713–742.

Koibuchi N, Sakai M, Watanabe K, Yamkaoka S (1992): Change in Fos-like immunoreactivity in the suprachiasmatic nucleus in the adult male rat. *NeuroReport 3 (6)*:501–504.

Kornhauser JM, Nelson DE, Mayo KE, Takahashi JS (1990): Photic and circadian regulation of c-*fos* gene expression in the hamster suprachiasmatic nucleus. *Neuron 5*:127–134.

Kornhauser JM, Nelson DE, Mayo KE, Takahashi JS (1992): Regulation of *junB* messenger RNA and AP-1 activity by light and a circadian clock. *Science 255*:1581–1584.

Krajnak K, Dickenson L, Lee TM (1997): The induction of Fos-like proteins in the suprachiasmatic nuclei and intergeniculate leaflet by light pulses in degus (Octodon degus) and rats. *J Biol Rhythms 12*:401–412.

Kriegsfeld LJ, Demas GE, Lee Jr. SE, Dawson TM, Dawson VL, Nelson RJ (1999): Circadian locomotor analysis of male mice lacking the gene for neuronal nitric oxide synthase (nNOS−/−). *J Biol Rhythms 14*:20–27.

Kriegsfeld LJ, Drazen DL, Nelson RJ (2001): Circadian organization in male mice lacking the gene for endothelial nitric oxide synthase (eNOS−/−). *J Biol Rhythms 16*:142–148.

Leak RK, Moore RY (2001): Topographic organization of suprachiasmatic nucleus projection neurons. *J Comp Neurol 433*:312–334.

Lin JT, Kornhauser JM, Singh NP, Mayo KE, Takahashi JS (1997): Visual sensitivities of nur77 (NGFI-B) and *zif268* (NGFI-A) induction in the suprachiasmatic nucleus are dissociated from c-*fos* induction and behavioral phase-shifting responses. *Brain Res Mol Brain Res 46*:303–310.

Lydic R, Baghdoyan HA (1999): *Handbook of Behavioral State Control. Cellular and Molecular Mechanisms.* Boca Raton, FL: CRC Press.

Maloney KJ, Mainville L, Jones BE (1999): Differential c-Fos expression in cholinergic, monoaminergic, and GABAergic cell groups of the pontomesencephalic tegmentum after paradoxical sleep deprivation and recovery. *J Neurosci 19*:3057–3072.

Maloney KJ, Mainville L, Jones BE (2000): c-Fos expression in GABAergic, serotonergic, and other neurons of the pontomedullary reticular formation and raphe after paradoxical sleep deprivation and recovery. *J Neurosci 20*:4669–4679.

Majzoub JA, Robinson BC, Emanuel RL (1991): Suprachiasmatic nuclear rhythms of vasopressin mRNA in vivo. In: Klein DC, Moore RY, Reppert SM (Eds), *Suprachiasmatic Nucleus. The Mind's Clock.* New York: Oxford University Press, pp. 177–190.

Mead S, Ebling FJP, Maywood ES, Humby T, Herbert J, Hastings MH (1992): A nonphotic stimulus causes instantaneous phase advances of the light-entrainable circadian oscillator of the Syrian hamster but does not induce the expression of c-*fos* in the suprachiasmatic nuclei. *J Neurosci 12*:2516–2522.

Meijer JH, Rietveld WJ (1989): Neurophysiology of the suprachiasmatic circadian pacemaker in rodents. *Physiol Rev 69*:671–707.

Melo L, Golombek DA, Ralph MR (1997): Regulation of circadian photic responses by nitric oxide. *J Biol Rhythms 12*:319–326.

Mikkelsen JD, Larsen PJ, O'Hare MM, Wiegand SJ (1991): Gastrin releasing peptide in the rat suprachiasmatic: an immunohistochemical, chromatographic and radioimmunological study. *Neuroscience 40*: 55–66.

Mistlberger RE, Bergmann BM, Waldenar W, Rechtshaffen A (1983): Recovery sleep following sleep deprivation in intact and suprachiasmatic nuclei-lesioned rats. *Sleep 6*:217–233.

Morin LP (1994): The circadian visual system. *Brain Res Brain Res Rev 67*:102–127.

Moore RY (1995): Organization of the mammalian circadian system. In: Chadwick DJ, Ackrill K (Ed), *Ciba Foundation Symposium on Circadian Clocks and Their Adjustment*, Vol 183. Chichester: Wiley, pp. 88–106.

Novak CM, Nunez AA (2000): A sparse projection from the suprachiasmatic nucleus to the sleep active ventrolateral preoptic area in the rat. *NeuroReport 11*:93–96.

Obrietan K, Impey S, Storm DR (1998): Light and circadian rhythmicity regulate MAP kinase activation in the suprachiasmatic nuclei. *Nat Neurosci 1 (8)*:693–700.

Peters RV, Aronin N, Schwartz WJ (1994): Circadian regulation of FosB is different from c-Fos in the rat suprachiasmatic nucleus. *Brain Res Mol Brain Res 27*:243–248.

Pittendrigh CS, Daan S (1976): A functional analysis of circadian pacemakers in nocturnal rodents. IV. Entrainment: Pacemaker as clock. *J Comp Physiol 106*:291–331.

Pompeiano M, Cirelli C, Tononi G (1994): Immediate-early genes in spontaneous wakefulness and sleep: expression of c-*fos* and NGFI-A mRNA and protein. *J Sleep Res 3*:80–96.

Pompeiano M, Cirelli C, Arrighi P, Tononi G (1995): c-Fos expression during wakefulness and sleep. *Neurophysiol Clin 25*:329–341.

Ralph MR, Menaker M (1985): Bicuculline blocks circadian phase delays but not advances. *Brain Res 325*:362–365.

Rea M (1989): Light increases Fos-related protein immunoreactivity in the rat suprachiasmatic nucleus. *Brain Res Bull 23*:577–581.

Rea M (1992): Different populations of cells in the suprachiasmatic nuclei express c-*fos* in association with light-induced phase delays and advances of the free-running activity rhythm in hamsters. *Brain Res 579*:107–112.

Rea MA, Buckley B, Lutton LM (1993): Local administration of EAA antagonists blocks light-induced phase shifts and c-*fos* expression in hamster SCN. *Am J Physiol 265*:R1191–R1198.

Rezzonico R, Ponzio G, Loubat A, Lallemand D, Proudfoot A, Rossi B (1995): Two distinct signalling pathways are involved in the control of the biphasic *junB* transcription induced by interleukin-6 in the B cell hybridoma 7TD1. *J Biol Chem 270*:1261–1268.

Romijn HJ, Sluiter AA, Pool CW, Wortel J, Buijis RM (1996): Differences in colocalization between Fos and PHI, GRP, VIP and VP in neurons of the rat suprachiasmatic nucleus after a light stimulus during the phase delay versus the phase advance period of the night. *J Comp Neurol 372*:1–8.

Rusak B, Zucker I (1979): Neural regulation of circadian rhythms. *Physiol Rev 59*:449–527.

Rusak B, Bina KG (1990): Neurotransmitters in the mammalian circadian system. *Annu Rev Neurosci 13*:387–401.

Rusak B, Robertson HA, Wisden W, Hunt SP (1990): Light pulses that shift rhythms induce gene expression in the suprachiasmatic nucleus. *Science 248*:1237–1240.

Rusak B, McNaughton L, Robertson HA, Hunt SP (1992): Circadian variation in photic regulation of immediate-early gene mRNAs in rat suprachiasmatic nucleus cells. *Brain Res Mol Brain Res 14*:124–130.

Sassone-Corsi P, Lamph WW, Kamps M, Verma IM (1988): Fos associated p39 is related to nuclear transcription factor AP-1. *Cell 54*:553–560.

Schwartz WJ, Busis NA, Hedley-Whyte ET (1986): A discrete lesion of ventral hypothalamus and optic chiasm that disturbed the daily temperature rhythm. *J Neurol 233*:1–4.

Schwartz WJ, Takeuchi J, Shannon W, Davis EM, Aronin N (1994): Temporal regulation of light-induced Fos and Fos-like protein expression in the ventrolateral subdivision of the rat suprachiasmatic nucleus. *Neuroscience 58*:573–583.

Schwartz WJ, Carpino Jr. A, De la Iglesia H, Baler R, Klein DC, Nakabeppu Y, Aronin N (2000): Differential regulation pf fos family genes in the ventrolateral and dorsomedial subdivisions of the rat suprachiasmatic nucleus. *Neuroscience 98*:535–547.

Semba K, Pastorius J, Wilkinson M, Rusak B (2001): Sleep deprivation-Induced c-*fos* and *junB* expression in the rat brain: effects of duration and timing. *Behav Brain Res 120*:75–86.

Sherin JE, Shiromani PJ, McCarley RW, Saper CB (1996): Activation of ventrolateral preoptic neurons during sleep. *Science 271*:216–219.

Shiromani P, Schwartz WJ (1995): Towards a molecular biology of the circadian clock and sleep of mammals. *Adv Neuroimmunol 5*:217–230.

Silver R, LeSauter J, Tresco P, Lehman N (1996): A diffusible coupling signal from the transplanted suprachiasmatic nucleus controlling circadian locomotor rhythms. *Nature 382*:810–813.

Smale L, Castleberry C, Nunez AA (2001): Fos rhythms in the hypothalamus of Rattus and Arvicanthis that exhibit nocturnal and diurnal patterns of rhythmicity. *Brain Res 899*:101–105.

Steriade M, McCarley RW (1990): *Brainstem Control of Wakefulness and Sleep*. New York: Plenum Press.

Sun X, Rusak B, Semba K (2000): Electrophysiology and pharmacology of projections from suprachiasmatic nucleus to ventromedial preoptic area in rat. *Neuroscience 98*:715–728.

Sun X, Whitefield S, Rusak B, Semba K (2001): Electrophysiological analysis of suprachiasmatic nucleus projections to the ventrolateral preoptic area in rat: Implications for circadian control of behavioral state. *Eur J Neurosci 14*:1257–1274.

Sutin EL, Kilduff TS (1992): Circadian and light-induced expression of immediate-early gene mRNAs in the rat suprachiasmatic nucleus. *Brain Res Mol Brain Res 15*:281–290.

Sumová E, Trávnícková Z, Peters R, Schwartz WJ, Illnerová H (1995): The rat suprachiasmatic nucleus is a clock for all seasons. *Proc Natl Acad Sci USA 92*:7754–7758.

Takeuchi J, Shannon W, Aronin N, Schwartz WJ (1993): Compositional changes of AP-1 DNA-binding proteins are regulated by light in a mammalian circadian clock. *Neuron 11*:825–836.

Terman JS, Reme CE, Terman M (1993): Rod outer segment disk shedding in rats with lesions of the suprachiasmatic nucleus. *Brain Res 605*:256–264.

Tononi G, Cirelli C, Pompeiano M (1995): Changes in gene expression during the sleep–waking cycle: a new view of activating systems. *Arch Ital Biol 134*:21–37.

Trávnícková Z, Sumová E, Peters R, Schwartz WJ, Illnerová H (1996): Photoperiod-dependent correlation between light-induced SCN c-*fos* expression and resetting of circadian phase. *Am J Physiol 271(Regulatory x Integrative Comp. Physiol. 40)*:R825–R836.

van den Pol A (1980): The hypothalamic suprachiasmatic nucleus of the rat: intrinsic anatomy. *J Comp Neurol 191*:661–702.

van den Pol A (1991): The suprachiasmatic nucleus: morphological and cytochemical substrates for cellular interaction. In: Klein DC, Moore RY, Reppert SM (Eds), *Suprachiasmatic Nucleus. The Mind's Clock*. New York: Oxford University Press, pp. 17–50.

Watts AG, Swanson LW, Sanchez-Watts G (1987): Efferent projections of the suprachiasmatic nucleus: I. Studies using anterograde transport of Phaseolus vulgaris leucoagglutinin in the rat. *J Comp Neurol 258*:204–229.

Wollnik F, Brysch W, Uhlmann E, Gillardon F, Bravo R, Zimmermann M, Schlingensiepen KH, Herdegen T (1995): Block of c-Fos and JunB expression by antisense oligonucleotides inhibits light-induced phase shifts of the mammalian circadian clock. *Eur J Neurosci 7*:388–393.

Yamazaki S, Numano R, Abe M, Hida A, Takkahashi R, Ueda M, Block GD, Sakaki Y, Menaker M, Tei H (2000): Resetting central and peripheral circadian oscillators in transgenic rats. *Science 288*:682–685.

Yokota S, Yamamoto M, Moriya T, Akiyama M, Fukunaga K, Miyamoto E, Shibata S (2001): Involvement of calcium-calmodulin protein kinase but not mitogen-activated protein kinase in light-induced phase delays and Per gene expression in the suprachiasmatic nucleus of the hamster. *J Neurochem 77*:618–627.

CHAPTER VII

The expression of c-Fos in the spinal cord: mapping of nociceptive pathways

HERVÉ BESTER AND STEPHEN P. HUNT

1. INTRODUCTION

The use of c-Fos protein expression as a mapping tool in the central nervous system derived from an immunohistochemical study that demonstrated that noxious stimulation of peripheral tissues always resulted in the expression of c-Fos protein in postsynaptic neurones within the spinal cord (Hunt et al., 1987). Since that time, many hundreds of studies have confirmed and extended this observation and the technique is widely used in experimental pain research. Following the initial observation, a number of studies attempted to characterise the necessary stimulus and time course of the response as well as the appearance of c-Fos in the brainstem and forebrain. These studies have revealed unexpected complexity in the response and there is now evidence that the pattern of expression of c-Fos in the central nervous system changes with time following the noxious stimulus. This chapter will describe the c-Fos response in the spinal cord and how it has been used to map nociceptive pathways and contribute to our understanding of nociceptive processing.

2. PRIMARY AFFERENT TERMINATION WITHIN THE SPINAL CORD

A great deal is known about the biochemical characterisation of primary afferents, their mode of termination within the dorsal horn and periphery and the type of nociceptive information they relay to the neurones of the spinal cord. The expression of c-Fos in dorsal horn neurones is initially a direct consequence of activation of these afferents and the pattern of c-Fos-positive neurones reflects both the subset of primary afferents stimulated and the time that has elapsed since stimulation.

The spinal cord has been divided into 10 laminae or layers in the cat (Rexed, 1952), and this nomenclature was adapted to the rat by Molander (Molander et al., 1984, 1989). Grossly the grey matter is divided into a ventral and a dorsal horn. The dorsal horn, made of laminae I to VI and X (Fig. 1A), has a receptive and integrative role. The superficial laminae I–II receive primary afferent input from peripheral thinly myelinated Aδ- and non-myelinated C-fibres (Fig. 1A) that convey nociceptive and thermal information. Laminae III–IV receive Aβ- and Aδ-fibres from hair follicles and low threshold mechanical afferents that convey innocuous stimuli of the mechanical modality. Laminae V–VI receive input from Aβ-, Aδ- and C-fibres, which convey both innocuous and noxious information.

Handbook of Chemical Neuroanatomy Vol. 19: Immediate Early Genes and Inducible Transcription Factors in Mapping of the Central Nervous System Function and Dysfunction
L. Kaczmarek and H.A. Robertson, editors

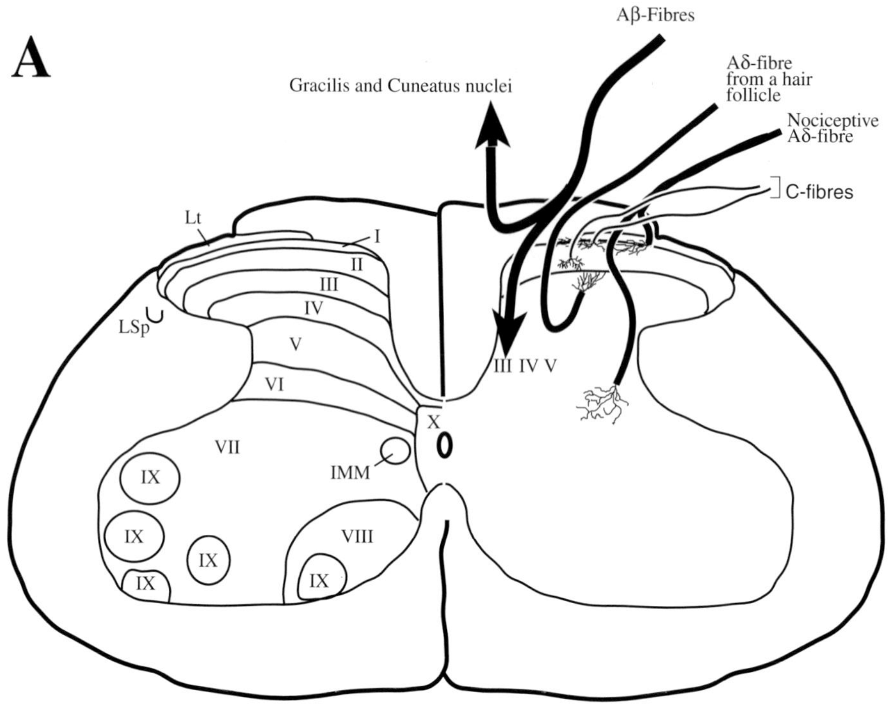
A
Aβ-Fibres
Aδ-fibre from a hair follicle
Gracilis and Cuneatus nuclei
Nociceptive Aδ-fibre
C-fibres
Lt
I
II
III
IV
V
VI
VII
VIII
IX
X
LSp
IMM
III IV V

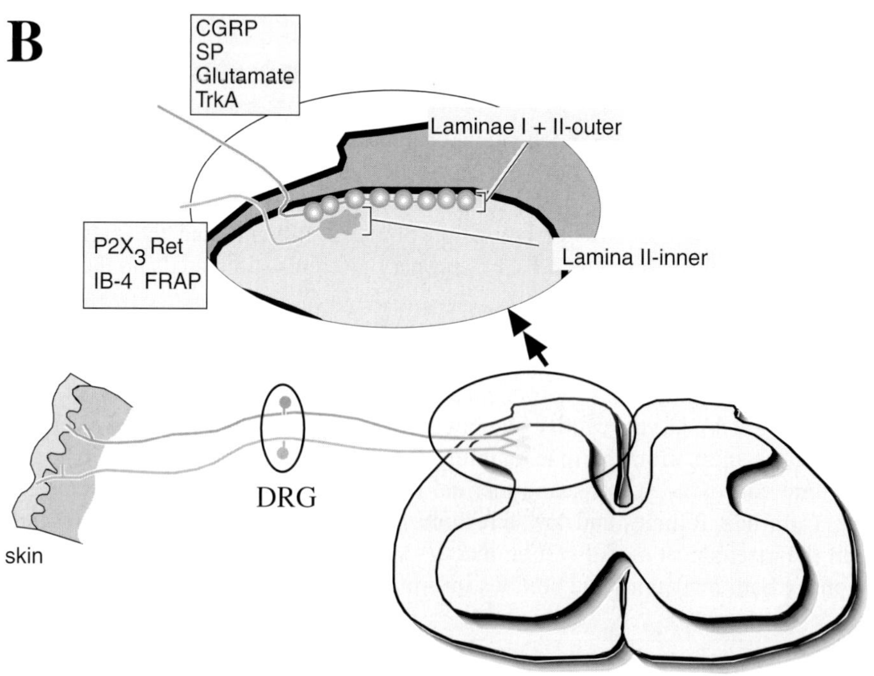
B
CGRP
SP
Glutamate
TrkA
Laminae I + II-outer
P2X3 Ret
IB-4 FRAP
Lamina II-inner
DRG
skin

Nociceptive information is relayed through subsets of C- and Aδ- primary afferent fibres. C-fibres arise from cutaneous sites, but also from deep tissues, such as viscera, muscles and joints (Alarcon and Cervero, 1990; Povlsen and Hildebrand, 1993; Sugiura et al., 1993; Povlsen et al., 1994; Birder and de Groat, 1998). Small unmyelinated fibres innervate essentially laminae I and II and V of the dorsal horn, but their pattern of termination differs accordingly to their peripheral targets. Lamina II has been divided into two sublaminae, II inner and II outer and C-fibres have been divided into two roughly equal groups. One subset of C-fibres contains peptides, express the tyrosine kinase (trk) A receptor and terminate in laminae I and II outer (Fig. 1B) and primarily on neurones that project to the brainstem as components of the spinoparabrachial and spinothalamic pathways. The other subset of C-fibres tends not to contain known peptide neurotransmitters, expresses the binding site for the IB4 lectin (isolectin B4 from *Griffonia simplicifolia*) and is responsive to the glial cell line-derived neurotrophic factor (GDNF) through the GDNF receptor tyrosine kinase (Ret) and associated receptor subunits (Fig. 1B). These fibres tend to innervate cutaneous tissue and terminate almost exclusively within lamina II inner in relation to a protein kinase C (PKC) γ-expressing group of spinal interneurones.

In the lumbar spinal cord, of the total unmyelinated fibres innervating the dorsal horn, roughly 80% correspond to polymodal C-fibres responding to all forms of noxious stimulation (Sugiura, 1996). But the pattern of termination of small unmyelinated fibres differs accordingly to their peripheral targets. Somatic (cutaneous) afferents extend caudorostrally over several hundreds of micrometres and over 200 μm mediolaterally usually in a highly topographic fashion especially in lamina II. However, visceral afferents are more diffusely organised and can extend over 2 or 3 segments, largely within lamina I (Sugiura, 1996). In Sugiura's study, it was reported that 60% of single visceral C-fibres terminate within lamina I. Other studies have confirmed that visceral afferents terminate superficially in the lamina I (Morgan et al., 1981; Cervero and Connell, 1984a,b; Cervero et al., 1984; Kuo and de Groat, 1985) and that muscle and joint afferent fibres have a similar termination pattern (Janig, 1987; McMahon and Wall, 1985; Heppelmann et al., 1988; Mense and Craig, 1988; Gillette et al., 1993; Furukawa et al., 2000).

This pattern of termination of primary afferents is reflected in the physiological properties of postsynaptic neurones. Neurones in the different laminae of the dorsal horn respond differently to innocuous and noxious peripheral stimuli and the responsiveness of dorsal horn neurones can also be modulated by descending controls originating in the brainstem (Urban and Gebhart, 1999). Those neurones involved in nociception are mainly located in the superficial laminae I–II and in the deep laminae V–VI. Neurones in the superficial third of the grey matter were first recognised as nociceptive by Kolmodin and Skoglund (1960), and 10 years later the 'nociceptive region' was further restricted to the superficial layers I–II (Christensen and Perl, 1970). Recent work has clearly detailed the electrophysiological properties of lamina I neurones that project to the parabrachial (PB) area. These lamina I

⟵
Fig. 1. Dorsal horn termination pattern of primary afferents. (*A*) Schematic representation of a transverse section of the L4 segment of the spinal cord showing the 10 laminae (I–X), the lateral spinal nucleus (LSp) and the Lissauer tract (Lt). On the right hand side, distribution pattern of primary afferent terminals. (*B*) Different aspects of C-fibre arborisation within the superficial laminae of the dorsal horn. In dark grey, peptidergic C-fibres arborise in the laminae I and II outer. These fibres express CGRP, substance P (SP), glutamate, and the TrkA receptor. Non-peptidergic fibres bind the IB4 lectin, express FRAP, and the $P2X_3$ and Ret receptors (modified from Hunt and Mantyh, 2001).

spino-PB neurones are bi- or poly-modal (mechanical, and thermal) and, in the rat, respond only to stimuli in the noxious range (Bester et al., 2000b). Because C-fibres originating in muscles and viscera also reach lamina I, it is not surprising that neurones responding to noxious somatic and visceral input (somatovisceral neurones) were also recorded in lamina I (Cervero and Tattersall, 1987; Alarcon and Cervero, 1990).

3. c-FOS EXPRESSION IN THE DORSAL HORN

Immunohistochemical detection of c-Fos in dorsal horn neurones in response to noxious stimuli (heat, mechanical or chemical) was first reported in 1987 (Hunt et al., 1987). It has become clear that at least within the first 2 h following stimulation, the pattern of c-Fos expression faithfully reflected, at least in part, the pattern of primary afferents that have been stimulated, but that this correlation was lost at longer survival times. At 2 h following noxious stimulation, Fos appears largely within superficial laminae of the spinal cord (Fig. 2A1,B). When the stimulation has a prolonged, inflammatory or neuropathic aspect, labelling within deeper laminae V–VI occurs. Deeper labelling may be bilateral and as primary afferent input is exclusively ipsilateral this would imply that while most c-Fos-positive neurones are monosynaptically activated by sensory neurones, many contralateral ones are activated through polysynaptic pathways either through commissural pathways (for which there is very little evidence) or through ascending–descending loops that include the brainstem. By 8 h after noxious heat stimulation, the pattern of c-Fos activation is purely in deep laminae and bilateral (Fig. 2B) (Williams et al., 1990). Many of these deep ipsilateral as well as contralaterally expressing neurones are at a distance from sites of primary afferent termination implying that c-Fos activation within the spinal cord is changing over considerable periods of time following a brief noxious stimulus and recruits neurones that are at a distance from the primary site of activation.

A variety of different types of stimulus applied to various parts of the body have been used to study Fos expression. Broadly speaking, noxious stimuli, including stimulation of deep tissue such as joints, muscles and viscera, results in the appearance of c-Fos in lamina I and lamina V neurones. Cutaneous stimulation resulted in c-Fos activation in neurones within laminae I–II and V. Fos staining of the neurones within lamina II inner has never been reported and may represent an exception. One study aiming at analysing NK1 receptor expressing neurones in lamina I showed that these neurones were activated by all sorts of peripheral noxious injuries: joint pain, muscle pain, subcutaneous formalin, sciatic nerve crush, noxious heat and topical mustard oil (Doyle and Hunt, 1999b) and that the number of Fos-positive NK1 neurones was directly proportional to the intensity (as estimated from

⟶

Fig. 2. Examples of the distribution of c-Fos expression upon noxious stimuli. (*A*) Photomicrographs of the hemi-dorsal horn of wild-type and NK1$^{-/-}$ mice (A1 and A2, respectively) showing c-Fos immunoreactive neurones (black dots) in the superficial laminae, in response to noxious heat stimulation applied to the ipsilateral hindpaw. (*B*) Camera lucida drawings of rat L4 segments showing the time course of c-Fos expression following noxious heat stimulation of the left hindpaw (modified from Williams et al., 1990). Note the progressive shift from the superficial towards the deep laminae of the grey matter and the bilateralisation of the c-Fos expression. (*C*) Photomicrograph illustrating c-Fos immunoreactive neurones (black dots) at the L6 segment, following cyclophosphamide-induced cystitis in the rat (modified from Lanteri-Minet et al., 1995). Note the expression in the sacral parasympathetic nucleus (arrow heads), and in the dorsal grey commissure (arrow).

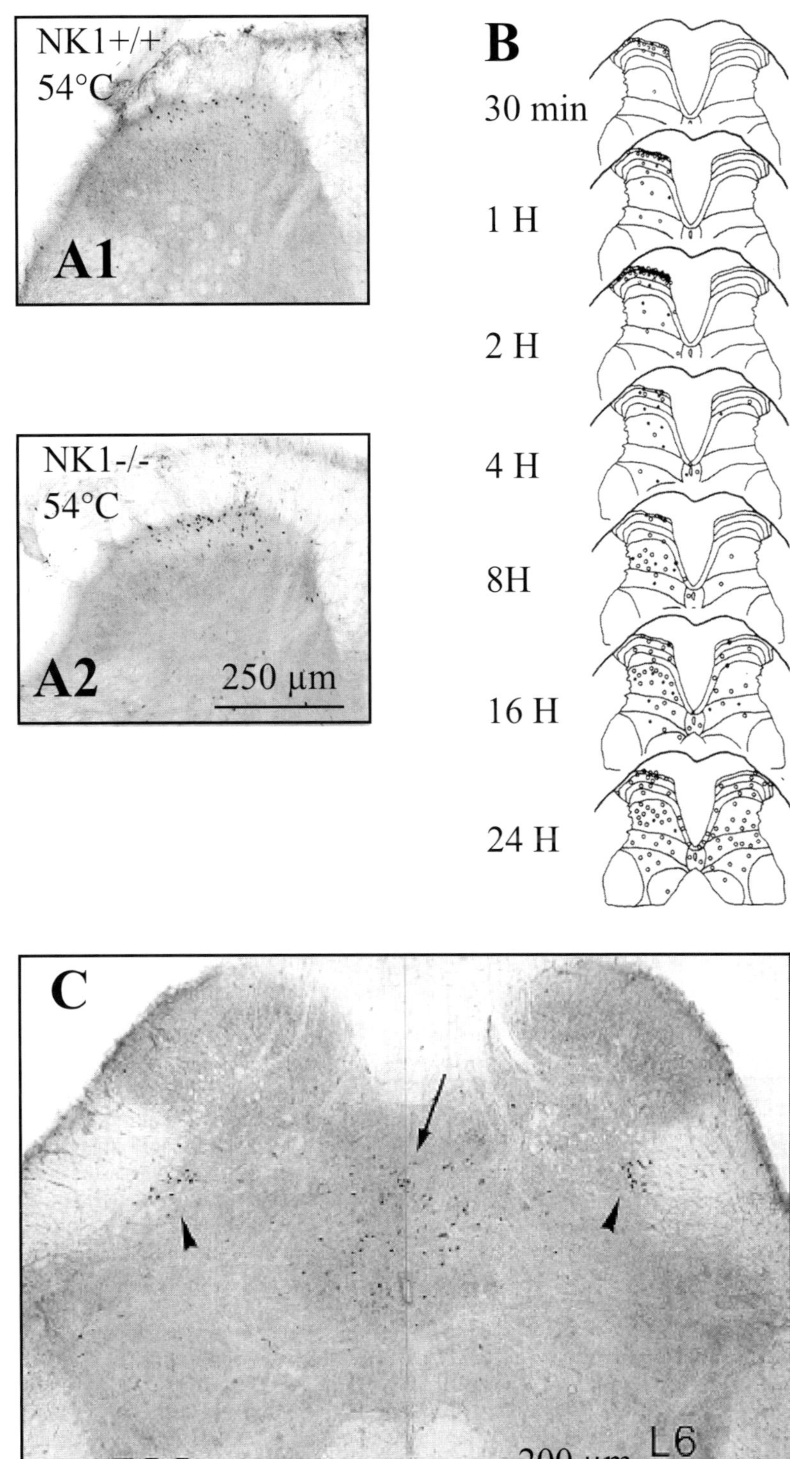
NK1+/+
54°C
A1
NK1-/-
54°C
A2
250 μm
B
30 min
1 H
2 H
4 H
8H
16 H
24 H
C
c-FOS
200 μm
L6

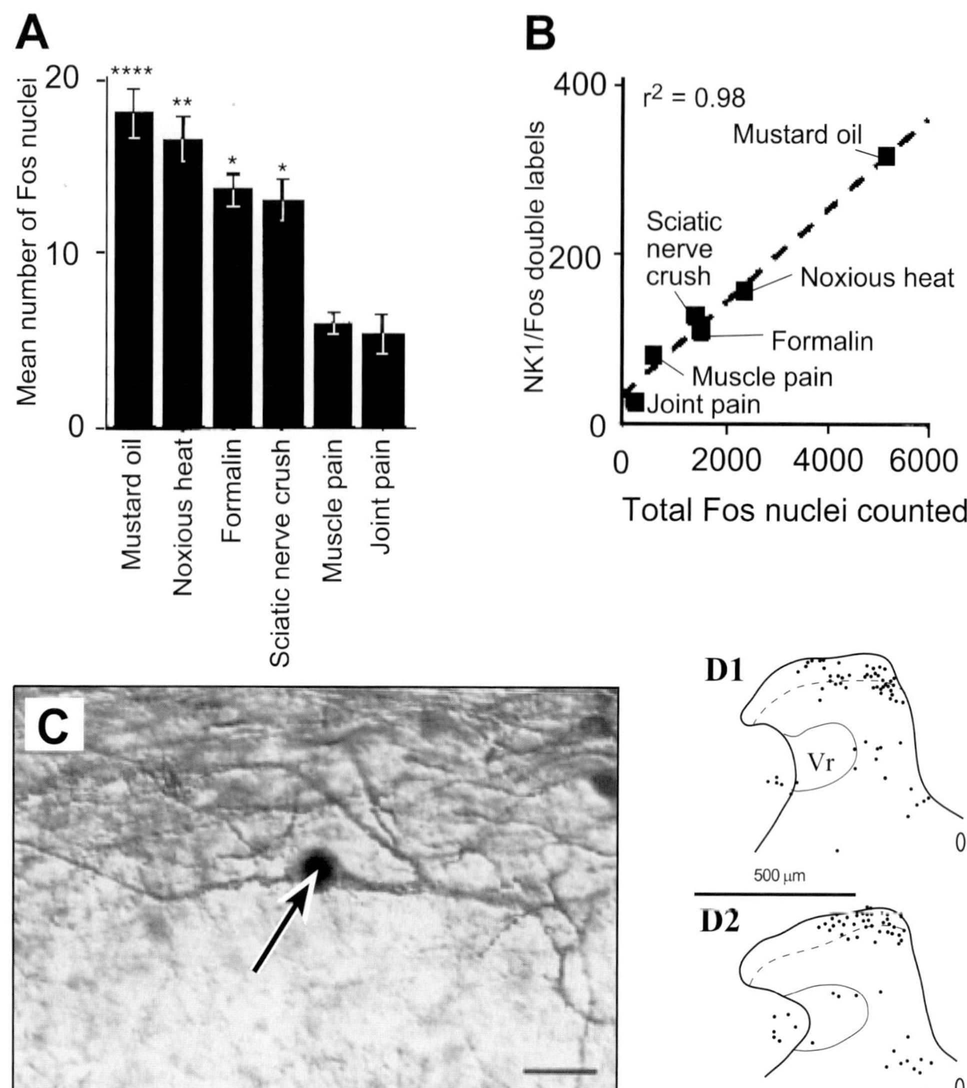

Fig. 3. c-Fos expression and NK1 receptor immunoreactivity. (*A*) Average number of Fos immunoreactive neurones in the superficial laminae of the dorsal horn in response to various types of noxious stimuli (modified from Doyle and Hunt, 1999b). (*B*) Correlation plot of the number of laminae I–II neurones double stained for c-Fos and NK1, and total number of c-Fos expression in response to different noxious stimuli (modified from Doyle and Hunt, 1999b). (*C*) Photomicrograph of a sagittal section of the lumbar segment of the spinal cord. This illustrates a neurone double stained for NK1 and c-Fos (arrow), following hindpaw stimulation with mustard oil. Scale bar: 40 µm (modified from Doyle and Hunt, 1999b). (*D*) Camera lucida drawings of c-Fos immunoreactive neurones at the lumbar level of a naive wild-type mouse and of a wild-type mouse treated with the NK1 antagonist RP67580 (D1 and D2, respectively) following noxious heat stimulation of the ipsilateral hindpaw. Note that blockade of the NK1 receptor does not alter c-Fos expression following acute noxious heat stimulation.

the total c-Fos count) of the noxious stimulation rather than the type of tissue stimulated (Fig. 3A,B). In contrast, discrete cutaneous stimulation revealed a precise topographical mapping within lamina II. For example, stimulation of individual digits by injection of 5 µl of formalin revealed discrete and in part overlapping populations of Fos-expressing neurones in lamina II. Similarly, the somatotopic and laminar organisation of neurones expressing c-Fos in the medullary and upper cervical dorsal horn induced by noxious facial stimulation have been described (Strassman and Vos, 1993).

The effects of visceral stimuli on c-Fos expression in the cord has also been reported. Spinal cord neurones were shown to express c-Fos in response to colorectal distension, i.p. acetic acid injections (Lanteri-Minet et al., 1993), or bladder stimulation (Fig. 2C) (Bon et al., 1996; Birder et al., 1997; Vizzard, 2000). It has been possible to correlate the number of Fos-stained nuclei with the intensity of the stimulus defined by either the intensity of the stimulus (colorectal distension) or with the severity of the bladder inflammation determined histologically (Bon et al., 1996). c-Fos expression in the dorsal horn has also been induced following deep tissue, such as muscle or joint, stimulation (Hunt et al., 1987; Menetrey et al., 1989; Yang et al., 1997; Doyle and Hunt, 1999b; Bereiter and Bereiter, 2000). Generally, c-Fos expression was more restricted to the lamina I than when induction was elicited by visceral, muscular or joint stimulation (Hunt et al., 1987; Menetrey et al., 1989; Hammond et al., 1992; Bon et al., 1996; Taylor et al., 1998).

Although it is difficult to quantify the absolute levels of stimulation, the most effective stimulus to elicit c-Fos in the dorsal horn is chemical, particularly the injection of formalin or capsaicin into the hind paw or topical application of mustard oil. This is followed by mechanical stimulation and heat stimulation, which appears to be the least effective. Nevertheless, many studies have used noxious heat (Williams et al., 1990; Abbadie et al., 1994c; Bester et al., 1997), while fewer have used noxious cold (Abbadie et al., 1994b; Doyle and Hunt, 1999a). The heat threshold to evoke significant c-Fos expression in laminae I–II is around 44°C (Abbadie et al., 1994c; Bester et al., 1997). This threshold seems to be somewhat higher (between 44 and 48°C) in the mouse (unpublished data). The threshold for cold stimulation is less clear, varying between 10°C (Doyle and Hunt, 1999a) and −15°C (Abbadie et al., 1994b). It is possible that this discrepancy lies in the differences in primary antibodies raised against the c-Fos protein, in the differences in the anaesthesia regimen or in the way in which thermal stimuli were applied. Non-noxious stimulation over 10–30 min, such as continuous brushing, does evoke c-Fos expression in the non-nociceptive laminae III–IV of the spinal cord.

In summary, the pattern of c-Fos expression 2 h following noxious stimulation follows that predicted by the termination of subsets of primary afferents. At longer survival times, c-Fos-expressing neurones become less tightly connected to the primary afferent input and the pattern shifts to deeper laminae on both sides of the spinal cord.

4. PRIMARY AFFERENT NEUROTRANSMITTERS AND FOS ACTIVATION

Primary afferent nociceptors release a number of neurotransmitters, including glutamate and substance P that mediate the postsynaptic activation of dorsal horn neurones. Selective antagonism of AMPA or NMDA glutamate receptors or substance P (NK1) receptors reduces the expression of c-Fos following acute or prolonged noxious peripheral stimulation (Chapman et al., 1996; Kakizaki et al., 1996; Munglani et al., 1999; Bereiter and Bereiter, 2000; King et al., 2000). Substance P is released in the superficial dorsal horn by unmyelinated primary afferents in response to noxious stimulation. Remarkably, there is a very good correlation between the number of NK1-positive neurones expressing c-Fos in the superficial dorsal horn and the modality of the noxious stimulus applied (Fig. 3) (Doyle and Hunt, 1999a,b). Electrophysiological data have shown that substance P antagonists may attenuate neuronal responses and studies using c-Fos pointed in the same direction, at least following inflammatory pain (Cutrer et al., 1995; Chapman et al., 1996; Bereiter et al., 1998). However, when the noxious stimulation is of short duration (50°C for 10 s), the substance P antagonist RP56780 at an intravenous dose (0.5 mg/kg) that proved to be efficient in an inflammatory model (Chapman et al., 1996) did not modify the level of c-Fos expression in laminae I–II of the dorsal horn (Fig. 3D1,D2) (Bester et al., 2001). This is confirmed in the $NK1^{-/-}$ mice (Fig. 2A1,A2). NMDA receptor antagonism also reduced noxiously induced c-Fos expression in the dorsal horn and trigeminal nucleus (Chapman et al., 1995; Buritova et al., 1996a; Chapman et al., 1996; Le Guen et al., 1999; Bereiter and Bereiter, 2000).

4.1. EXPERIMENTAL MODELS

c-Fos mapping in the dorsal horn has been recognised as a useful tool to study pathological pain. There are numerous experimental 'pain models' including inflammation, nerve section, multiple loose ligatures of a peripheral nerve and tight ligature of spinal nerves as they leave the dorsal root ganglia. Apart from nerve section, all treatments tend to result in a change in pain thresholds described as allodynia or hyperalgesia. However, even though there are well-documented changes in pain sensitivity following these peripheral nerve treatments, constitutive expression of Fos is similar to that seen in control animals. At early time points (2 h) following any of these peripheral manipulation c-Fos expression is seen particularly in superficial and deeper laminae. At 6–24 h, Fos is seen in deeper neurones and often bilaterally (Fig. 2B) and this pattern is maintained over the next week or so before disappearing completely. A noxious stimulation at any point after the initial induction of c-Fos has occurred results in the reappearance of Fos in superficial laminae (Catheline et al., 1999; Munglani et al., 1999). Non-noxious stimuli may also induce c-Fos expression in the superficial laminae of the dorsal horn in pain models (Fig. 4) (Vos and Strassman, 1995; Ma and Woolf, 1996; Bester et al., 2000a). In the case of inflammation, the mechanisms underlying c-Fos expression in nociceptive spinal regions upon innocuous stimulation are not clearly understood. On the

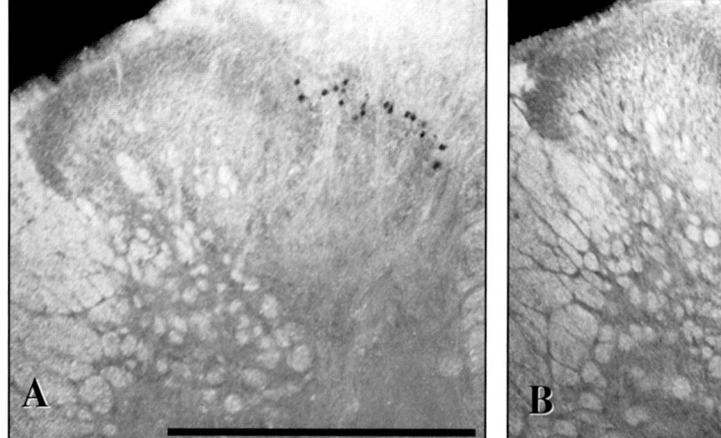

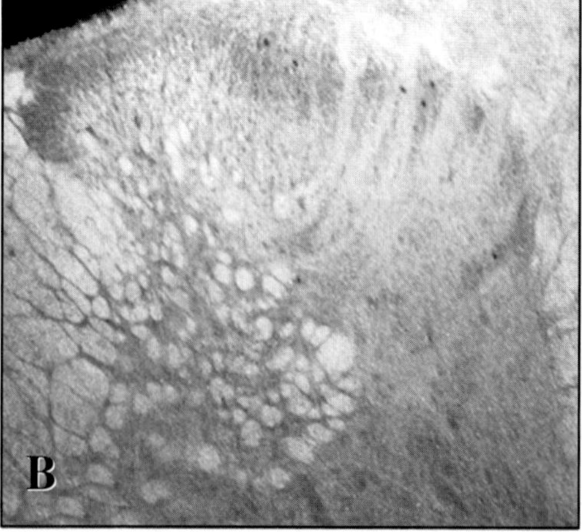

Fig. 4. Mechanical allodynia and c-Fos expression. Photomicrographs of a spinal cord transverse section of rats. (*A*) 15 weeks following sciatic nerve crush, prolonged (8 min) light touch stimulation of the distal sciatic nerve territory in awake rats results in c-Fos expression in laminae I–II of the ipsilateral dorsal horn. (*B*) In sham operated rats, this increased c-Fos expression cannot be observed (from Bester et al., 2000a).

other hand, in the case of neuropathy, among several hypothesis, it be suggested that A (β and/or δ)-fibres, that sprout from laminae III–IV into laminae I–II (Woolf et al., 1992, 1995), now convey messages of a non-noxious nature (light touch) onto nociceptive spinal neurones. This misdirection of the messages may lead to an aberrant interpretation (noxious instead of innocuous) of the stimulation modality by the sensory systems and be responsible for mechanical allodynia (Bester et al., 2000a). Thus while c-Fos is always expressed in neurones for short periods following noxious stimulation, it does not appear to tonically regulate pain sensitivity over long periods of time. It should, however, be mentioned that application of antisense c-Fos mRNA intrathecally following the establishment of a pain state has been reported to both antagonise and increase the sensitivity to noxious stimulation, implying a regulatory role for c-Fos in setting pain sensitivity (Hunter et al., 1995; Hou et al., 1997).

4.2. EFFECTS OF DRUGS

Many inflammatory pain models have been used to assess the potency of various drugs in reducing c-Fos and thus by implication, the pain associated with inflammation. These include subcutaneous injections, generally in the plantar part of the hindpaw, of CFA (complete Freunds adjuvant) (Ma and Woolf, 1996; Martin et al., 1999), formalin (Abbadie et al., 1992; Lima et al., 1993; Herdegen et al., 1994; Jasmin et al., 1994; Chapman et al., 1996), carrageenan (see for example (Honore et al., 1996b), or topical application of mustard oil (Doyle and Hunt, 1999b). Others have chosen to inject inflammatory drugs into the joint (Menetrey et al., 1989; Doyle and Hunt, 1999b). Finally, a general inflammation, such as a polyarthritis, has also been used (Abbadie and Besson, 1992; Abbadie et al., 1994a). In all these models, inflammatory insult of the periphery led to massive c-Fos expression in the superficial laminae of the dorsal horn, but also in the deep dorsal horn (laminae V–VI). The number of c-Fos labelled neurones was thought to be a good means of assessing the effects of different drugs in reducing pain. For example, opiates administered by injection of agonists subcutaneously (Hammond et al., 1992; Presley et al., 1990; Jasmin et al., 1994), intracerebroventricularly (Gogas et al., 1991), intravenously (Abbadie et al., 1994c; Honore et al., 1995c), or locally at the site of the injury (Honore et al., 1996a) can significantly reduce c-Fos expression in the dorsal horn. Besson and colleagues have used the carrageenan-induced inflammation model extensively for these studies. In this model, they have tested the efficiency of a large number of drugs in reducing the carrageenan-induced c-Fos expression in the dorsal horn of the rat. For example, inflammation-induced pain can be reduced by treatment with anti-inflammatory drugs: piroxicam, diclofenac, dexamethasone, cyclooxygenase 2 inhibitors, lornoxicam, flurbiprofen, aspirin or indomethacin (Buritova and Besson, 1998a,b; Buritova et al., 1995a,b, 1996b,c; Honore et al., 1995a,b). Overall, these anti-inflammatory drugs affect c-Fos expression in laminae I–II and laminae V–VI of the dorsal horn. The effect of these drugs could either be peripheral (accompanied by a reduction of the hindpaw oedema) and/or central. A central effect could involve spinal modulation and/or brain structures acting at the spinal level via descending controls. Thus, these studies cannot help in deciding how strongly brain structures involved in descending controls of spinal cord excitability are recruited to reduce the pain sensitivity.

4.3. ENDOGENOUS PAIN CONTROL SYSTEMS

The level of responsiveness to noxious stimuli is set by descending excitatory and inhibitory influences originating in the brainstem. Stress-induced hypoalgesia, stress induced hyperalgesia, counter irritation, and diffuse noxious inhibitory controls (DNIC) are thought to be the physiological and behavioural sequel of the actions of these descending controls. DNIC is implied when a conditioning noxious stimulation can be shown to reduce the consequences of a second remote noxious stimulation (a pain masking another pain). DNIC has been analysed with both physiological methods and c-Fos immunohistochemistry (Le Bars et al., 1979; Morgan et al., 1994; Fields et al., 1995; Danziger et al., 1999; Bester et al., 2000b, 2001). Histochemical studies have shown that DNIC acts both on laminae V–VI and laminae I–II neuronal populations (Fig. 5A,B) (Morgan et al., 1994; Bester et al., 2001). This is supported by electrophysiological studies performed at the level of laminae V–VI (Le Bars et al., 1979), and lamina I (Bester et al., 2000b). Brainstem structures involved in the mechanisms of descending controls, such as the raphe nuclei, have been established in the rat (Jones and Light, 1990; Wei et al., 1999) and in the mouse (Fig. 5C,D) (Bester et al., 2001). The full development of DNIC requires the NK1 receptor (Bester et al., 2001).

Descending controls do not only have an inhibitory effect on dorsal horn activity. Indeed, it has been shown that c-Fos expression induced in the dorsal horn ipsilateral to a noxious stimulation applied to the hindpaw could be enhanced by a first noxious stimulus applied to the contralateral hindpaw 1–1.5 h prior to this stimulus (Leah et al., 1992). This increase in responsiveness (sensitisation) was shown to last several hours after the first stimulus, and is likely to involve descending controls.

The major role of supraspinal structures in the control of dorsal horn excitability is now well established as mentioned above and are thought to be one site of action of morphine (Duggan et al., 1980; Jensen and Yaksh, 1986a,b; Ossipov et al., 1995). Interestingly, it has recently been shown that the pattern of lumbar spinal cord c-Fos expression induced by withdrawal to chronic morphine is primarily a consequence of an increased activity in opioid receptor-containing circuits intrinsic to the dorsal horn (Rohde et al., 1997). This study indicated that the magnitude of c-Fos expression is normally regulated by supraspinal and primary afferent-derived inhibitory inputs.

⟶

Fig. 5. Descending inhibitory controls. In a situation where the hindpaw only (HP) or the forepaw plus the hindpaw (FP + HP) are stimulated with noxious heat, NK1 transmission appears to be critical for the development of pain controls. (*A*) Photomicrographs illustrating the noxiously evoked controls of dorsal horn neurones activation. In naive wild-type mice, the c-Fos expression induced by noxious heat applied to the hindpaw (A1) is significantly reduced when concurrent conditioning noxious heat is applied to the forepaw (A2). (*B*) Histograms of the average numbers of c-Fos-expressing neurones in the superficial laminae of the ipsilateral cord at the lumbar level. In $NK1^{-/-}$ (ko) mice and in wild-type mice treated with the NK1 antagonist RP67580, the concurrent forepaw stimulation had no effect on the c-Fos expression induced by a hindpaw noxious heat stimulation. In naive wild-type (wt or $NK1^{+/+}$) and wild mice treated with the inactive isomer of the NK1 antagonist (RP68651), the forepaw stimulation significantly reduced the number of neurones expressing c-Fos upon noxious heat stimulation of the hindpaw. (*C*) Histograms of average numbers of Fos-ir neurones observed in different raphe nuclei viewed on a schematic representation of the medulla from the Franklin and Watson atlas (1997). ** $P < 0.01$; *** $P < 0.001$. In the rostroventral medulla, nuclei known to be involved in pain control fewer neurones expressing c-Fos were counted in $NK1^{-/-}$ mice upon the double forepaw plus hindpaw noxious stimulation. In fact, mice that expressed less c-Fos in the raphe nuclei (RPa + RMg) had a higher levels of c-Fos expression in the lumbar cord, and these were $NK1^{-/-}$ mice (*D*). Conversely, mice that expressed more c-Fos in the raphe nuclei (RPa + RMg) had lower levels of c-Fos expression in the lumbar cord, and these were $NK1^{+/+}$ mice. 4V, 4th ventricle; Gi, gigantocellular reticular; GiA, gigantocellular reticular pars alpha; icp, inferior cerebellar peduncle; LPGi, lateral paragigantocellular reticular; P7, perifacial zone; py, pyramidal tract; RMg, raphe magnus; RPa, raphe pallidus; sp5, spinal trigeminal tract; Sp5, spinal trigeminal nucleus. Filled symbols correspond to $NK1^{-/-}$ mice; open symbols correspond to wild-type mice (modified from Bester et al., 2001).

4.4. FOS EXPRESSION DURING POSTNATAL DEVELOPMENT

A number of studies have analysed the pattern of c-Fos expression induced by noxious and/or non-noxious stimulation in rat pups and during postnatal development. Dorsal horn circuits are immature up to postnatal day 17, i.e. the adult pattern is only seen after day 17 (Golden et al., 1997). The receptive field properties and evoked activity of newborn dorsal horn cells to single repetitive and persistent innocuous and noxious inputs are developmentally regulated and reflect the maturation of excitatory transmission within the spinal cord (Fitzgerald and Jennings, 1999). In newborn rat pups, repeated mechanical stimulation induced sensitization, which changed at 4 postnatal weeks to habituation (as seen in 30- and 37.5-week preterm

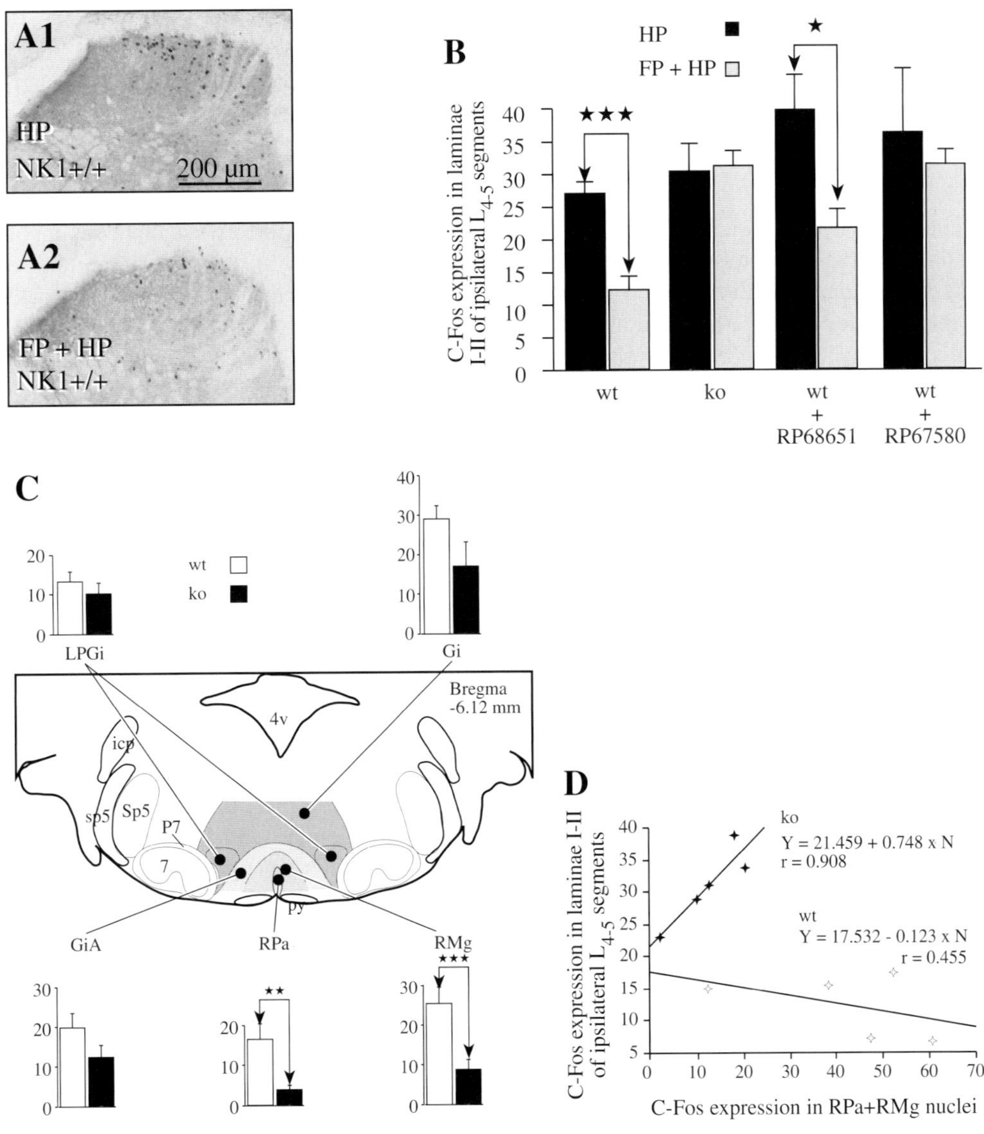

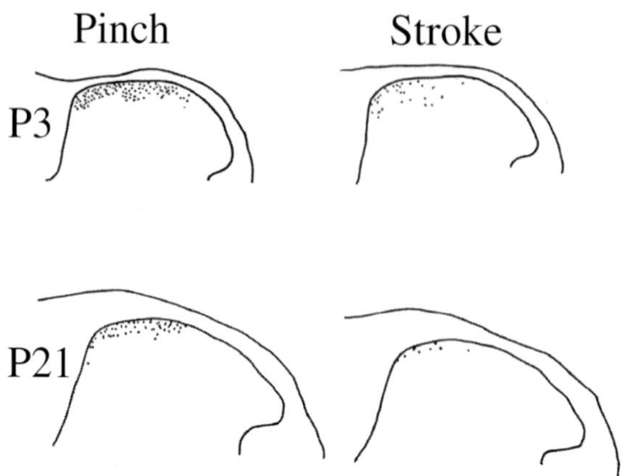

Fig. 6. Development. Camera lucida drawings of the lumbar dorsal horn ipsilateral to pinch or stroke stimulation of the hindpaw in a P3 or P21 rat (modified from Jennings and Fitzgerald, 1996). Note that light mechanical stimulation can evoke significant c-Fos expression in the superficial laminae of neonates (P3) which then disappears with age.

infants) (Fitzgerald et al., 1988). Thus, the sensitivity of spinal reflexes to cutaneous inputs changes in the neonate with development. This may result from lack of inhibitory or descending control in the immature spinal cord. In neonates, thermal nociceptive thresholds in the tail-flick test are considerably lowered, relative to adults, up to postnatal day 12 (Falcon et al., 1996). Indeed, thresholds vary between 37.5 and 47.2°C, in rats aged of 3–90 days. Following spinal transections at different ages, these authors suggested that changes in responsiveness to suprathreshold noxious stimuli involved maturation of both spinal and descending supraspinal structures. Descending inhibitory controls depending on PAG activation are also developmentally regulated. Indeed, morphine or glutamate PAG injections known to produce antinociception in the adult have a differential age-related potency of action (Tive and Barr, 1992). Other supraspinal systems also play a critical role in the postnatal tuning of spinal nociceptive systems (Levinsson et al., 1999).

Noxious pinch of the hindpaw evoked a clear c-Fos response in the newborn rat pup which was not significantly different from that seen at postnatal day (P) 21 (Jennings and Fitzgerald, 1996). This study also showed that low intensity touch stimulation also produced a significant c-Fos response in laminae I–II at P3 and P10 that was 60 and 27% of that of the pinch response, respectively. This touch-evoked c-Fos induction in the superficial laminae of the dorsal horn was gone by P21 (Fig. 6). Electrical stimulation showed that Aβ-fibre stimulation produced a c-Fos response at P3 that was not significantly different from that produced by C-fibre stimulation. This strong Aβ-induced c-Fos expression lasted, with some intensity variations, for up to P21 and was thought to mimic the pattern of maturation of primary afferents as they rearrange after birth. Initially, unlike the adult situation, Aβ-fibres terminate throughout laminae I–II and over the first 3 postnatal weeks gradually retract to terminate entirely within laminae III–IV. Thus, in the neonatal spinal cord, low threshold A-fibres are able to activate neurones in laminae I–II of the dorsal horn that in the adult receive predominantly nociceptive input. Paradoxically, at P21, light touch activating Aβ-fibres fail to induce significant c-Fos expression in laminae I–II (Torsney et al., 2000).

In a recent study (Boucher et al., 1998), DNIC were studied in postnatal rats aged 12, 21 and 42-days-old. It was shown that in the lumbar cord, the number of neurones expressing c-Fos following a standardised pinch of the hindpaw stimulus was significantly reduced when this stimulus was applied seconds after a conditioning noxious stimulus of the forepaw (formalin injection) was first applied. Significant reductions were seen at P42 (15% reduction) and P21 (17% reduction), but concurrent stimulation had no significant effect at P12. These results suggest that the system subserving DNIC is functionally maturing between P12 and P21. This delayed maturation of an inhibitory system may underlie the extreme sensitivity to somatosensory stimulation seen in neonatal pups and premature infants.

Overall, c-Fos studies performed in neonates, have confirmed that nociceptive systems are maturing postnatally and apparently processing nociceptive information in a qualitatively different way from the adult.

4.5. TYPES OF DORSAL HORN NEURONES EXPRESSING c-FOS

At the spinal level, c-Fos is expressed in different types of neurones in response to noxious stimulation. It is possible to characterise these neurones by referring: (1) to the modality of the activating stimulus; (2) to molecules or receptors expressed; (3) whether they are interneurones or projecting neurones; (4) to the supraspinal targets they project to; and (5) whether they are excitatory or inhibitory. All noxious peripheral stimuli, if sustained enough, elicit a dorsal horn c-Fos response (Harris, 1998).

Lamina I neurones of the dorsal horn are of particular interest because they have been shown to regulate spinal excitability through ascending–descending loops involving the brainstem. These lamina I neurones express the NK1 receptor and respond to somatic or visceral stimuli (Fig. 3C) (Lu et al., 1995; Doyle and Hunt, 1999a,b). These observations support earlier studies showing that substance P containing primary afferents are closely related to c-Fos-expressing neurones (Pretel and Piekut, 1991; Tao and Zhao, 1997; Hunt and Mantyh, 2001). Interestingly, NK1 expressing lamina I neurones are projection neurones, the majority of which (80%) target the PB area (Ding et al., 1995a,b; Todd et al., 2000).

Fos is also induced in inhibitory spinal neurones (local interneurones and/or propriospinal neurones) (Todd et al., 1994; Sandkuhler, 1996). It is very likely that these neurones are involved in a fine tuning of the spinal cord excitability, and that they are activated, either directly from peripheral stimuli, or via ascending–descending loops involving brainstem sites such as the raphe or noradrenergic nuclei (A5, A7), or lateral nuclei of the ventral medulla. In the spinal cord, 5–10% of c-Fos immunoreactive neurones also display glucocorticoid receptor immunoreactivity (Cintra et al., 1993), making them targets for the analgesic effects of corticotherapy.

5. CONCLUSION

c-Fos has proven to be a valuable tool to investigate the spinal processing of nociceptive information. The consequences of c-Fos activation remain unclear and there is also some evidence that certain neuronal subpopulations (such as those PKC-γ-positive neurones within lamina II inner and motor neurones) do not express c-Fos following stimulation. Nevertheless c-Fos has proved valuable in mapping the activity over time of various stimuli applied to the periphery and for estimating the analgesic potency of certain drugs. Future issues to be investigated will include the downstream events which follow the activation of the transcription factor and the presence of other immediate early genes either in the same or overlapping populations of neurones.

6. ABBREVIATIONS

4V	4th ventricle
CGRP	calcitonin gene related peptide
DNIC	diffuse noxious inhibitory controls
GDNF	glial cell line-derived neurotrophic factor
Gi	gigantocellular reticular
GiA	gigantocellular reticular pars alpha
IB4	isolectin B4 from *Griffonia simplicifolia*
icp	inferior cerebellar peduncle
LPGi	lateral paragigantocellular reticular
LSp	lateral spinal nucleus
Lt	Lissauer tract
NK1	neurokinin 1 receptor for substance P
P7	perifacial zone
PKC	protein kinase C
py	pyramidal tract
Ret	GDNF receptor tyrosine kinase
RMg	raphe magnus
RPa	raphe pallidus
SP	substance P
Sp5	spinal trigeminal nucleus
sp5	spinal trigeminal tract
TrkA	tyrosine kinase

7. REFERENCES

Abbadie C, Besson JM (1992): c-Fos expression in rat lumbar spinal-cord during the development of adjuvant-induced arthritis. *Neuroscience 48*:985–993.

Abbadie C, Lombard MC, Morain F, Besson JM (1992): Fos-like immunoreactivity in the rat superficial dorsal horn induced by formalin injection in the forepaw — effects of dorsal rhizotomies. *Brain Res 578*:17–25.

Abbadie C, Besson JM, Calvino B (1994a): c-Fos expression in the spinal cord and pain-related symptoms induced by chronic arthritis in the rat are prevented by pretreatment with Freund adjuvant. *J Neurosci 14*:5865–5871.

Abbadie C, Honore P, Besson JM (1994b): Intense cold noxious-stimulation of the rat hindpaw induces c-Fos expression in lumbar spinal-cord neurons. *Neuroscience 59*:457–468.

Abbadie C, Honore P, Fournie-Zaluski MC, Roques BP, Besson JM (1994c): Effects of opioids and non-opioids on c-Fos-like immunoreactivity induced in rat lumbar spinal cord neurons by noxious heat stimulation. *Eur J Pharmacol 258*:215–227.

Alarcon G, Cervero F (1990): The effects of electrical stimulation of A and C visceral afferent fibres on the excitability of viscerosomatic neurones in the thoracic spinal cord of the cat. *Brain Res 509*:24–30.

Bereiter DA, Bereiter DF (2000): Morphine and NMDA receptor antagonism reduce c-fos expression in spinal trigeminal nucleus produced by acute injury to the TMJ region. *Pain 85*:65–77.

Bereiter DA, Bereiter DF, Tonnessen BH, Maclean DB (1998): Selective blockade of substance P or neurokinin A receptors reduces the expression of c-fos in trigeminal subnucleus caudalis after corneal stimulation in the rat. *Neuroscience 83*:525–534.

Bester H, Matsumoto N, Besson JM, Bernard JF (1997): Further evidence for the involvement of the spinoparabrachial pathway in nociceptive processes: a c-Fos study in the rat. *J Comp Neurol 383*:439–458.

Bester H, Beggs S, Woolf CJ (2000a): Changes in tactile stimuli-induced behavior and c-Fos expression in the superficial dorsal horn and in parabrachial nuclei after sciatic nerve crush. *J Comp Neurol 428*:45–61.

Bester H, Chapman V, Besson JM, Bernard JF (2000b): Physiological properties of the lamina I spinoparabrachial neurons in the rat. *J Neurophysiol 83*:2239–2259.

Bester H, De Felipe C, Hunt SP (2001): The NK1 receptor is essential for the full expression of noxious inhibitory controls in the mouse. *J Neurosci 21*:1039–1046.

Birder LA, de Groat WC (1998): Contribution of C-fiber afferent nerves and autonomic pathways in the urinary bladder to spinal c-fos expression induced by bladder irritation. *Somatosens Mot Res 15*:5–12.

Birder LA, Kanai AJ, de Groat WC (1997): DMSO: effect on bladder afferent neurons and nitric oxide release. *J Urol 158*:1989–1995.

Bon K, Lanteri Minet M, de Pommery J, Michiels JF, Menetrey D (1996): Cyclophosphamide cystitis as a model of visceral pain in rats. A survey of hindbrain structures involved in visceroception and nociception using the expression of c-Fos and Krox-24 proteins. *Exp Brain Res 108*:404–416.

Boucher T, Jennings E, Fitzgerald M (1998): The onset of diffuse noxious inhibitory controls in postnatal rat pups: a c-Fos study. *Neurosci Lett 257*:9–12.

Buritova J, Besson JM (1998a): Dose-related anti-inflammatory/analgesic effects of lornoxicam: a spinal c-Fos protein study in the rat. *Inflamm Res 47*:18–25.

Buritova J, Besson JM (1998b): Peripheral and/or central effects of racemic-, S(+)- and R(−)-flurbiprofen on inflammatory nociceptive processes: a c-Fos protein study in the rat spinal cord. *Br J Pharmacol 125*:87–101.

Buritova J, Honore P, Chapman V, Besson JM (1995a): Carrageenan oedema and spinal Fos-LI neurones are reduced by piroxicam in the rat. *NeuroReport 6*:1385–1388.

Buritova J, Honore P, Chapman V, Besson JM (1995b): Concurrent reduction of inflammation and spinal Fos-LI neurons by systemic diclofenac in the rat. *Neurosci Lett 188*:175–178.

Buritova J, Chapman V, Honore P, Besson JM (1996a): Interactions between NMDA- and prostaglandin receptor-mediated events in a model of inflammatory nociception. *Eur J Pharmacol 303*:91–100.

Buritova J, Chapman V, Honore P, Besson JM (1996b): Selective cyclooxygenase-2 inhibition reduces carrageenan oedema and associated spinal c-Fos expression in the rat. *Brain Res 715*:217–220.

Buritova J, Honore P, Chapman V, Besson JM (1996c): Enhanced effects of co-administered dexamethasone and diclofenac on inflammatory pain processing and associated spinal c-Fos expression in the rat. *Pain 64*:559–568.

Catheline G, Le Guen S, Honore P, Besson JM (1999): Are there long-term changes in the basal or evoked Fos expression in the dorsal horn of the spinal cord of the mononeuropathic rat?. *Pain 80*:347–357.

Cervero F, Connell LA (1984a): Distribution of somatic and visceral primary afferent fibres within the thoracic spinal cord of the cat. *J Comp Neurol 230*:88–98.

Cervero F, Connell LA (1984b): Fine afferent fibers from viscera do not terminate in the substantia gelatinosa of the thoracic spinal cord. *Brain Res 294*:370–374.

Cervero F, Tattersall JE (1987): Somatic and visceral inputs to the thoracic spinal cord of the cat: marginal zone (lamina I) of the dorsal horn. *J Physiol 388*:383–395.

Cervero F, Connell LA, Lawson SN (1984): Somatic and visceral primary afferents in the lower thoracic dorsal root ganglia of the cat. *J Comp Neurol 228*:422–431.

Chapman V, Honore P, Buritova J, Besson JM (1995): The contribution of NMDA receptor activation to spinal c-Fos expression in a model of inflammatory pain. *Br J Pharmacol 116*:1628–1634.

Chapman V, Buritova J, Honore P, Besson JM (1996): Physiological contributions of neurokinin 1 receptor activation, and interactions with NMDA receptors, to inflammatory-evoked spinal c-Fos expression. *J Neurophysiol 76*:1817–1827.

Christensen BN, Perl ER (1970): Spinal neurons specifically excited by noxious or thermal stimuli: marginal zone of the dorsal horn. *J Neurophysiol 33*:293–307.

Cintra A, Molander C, Fuxe K (1993): Colocalization of Fos- and glucocorticoid receptor-immunoreactivities is present only in a very restricted population of dorsal horn neurons of the rat spinal cord after nociceptive stimulation. *Brain Res 632*:334–338.

Cutrer FM, Moussaoui S, Garret C, Moskowitz MA (1995): The non-peptide neurokinin-1 antagonist, RPR 100893, decreases c-fos expression in trigeminal nucleus caudalis following noxious chemical meningeal stimulation. *Neuroscience 64*:741–750.

Danziger N, Weil-Fugazza J, Le Bars D, Bouhassira D (1999): Alteration of descending modulation of nociception during the course of monoarthritis in the rat. *J Neurosci 19*:2394–2400.

Ding YQ, Takada M, Shigemoto R, Mizumo N (1995a): Spinoparabrachial tract neurons showing substance P receptor-like immunoreactivity in the lumbar spinal cord of the rat. *Brain Res 674*:336–340.

Ding YQ, Takada M, Shigemoto R, Mizuno N (1995b): Trigeminoparabrachial projection neurons showing substance P receptor-like immunoreactivity in the rat. *Neurosci Res 23*:415–418.

Doyle CA, Hunt SP (1999a): A role for spinal lamina I neurokinin-1-positive neurons in cold thermoreception in the rat. *Neuroscience 91*:723–732.

Doyle CA, Hunt SP (1999b): Substance P receptor (neurokinin-1)-expressing neurons in lamina I of the spinal cord encode for the intensity of noxious stimulation: a c-Fos study in rat. *Neuroscience 89*:17–28.

Duggan AW, Griersmith BT, North RA (1980): Morphine and supraspinal inhibition of spinal neurones: evidence that morphine decreases tonic descending inhibition in the anaesthetized cat. *Br J Pharmacol 69*:461–466.

Falcon M, Guendellman D, Stolberg A, Frenk H, Urca G (1996): Development of thermal nociception in rats. *Pain 67*:203–208.

Fields HL, Malick A, Burstein R (1995): Dorsal horn projection targets of ON and OFF cells in the rostral ventromedial medulla. *J Neurophysiol 74*:1742–1759.

Fitzgerald M, Jennings E (1999): The postnatal development of spinal sensory processing. *Proc Natl Acad Sci USA 96*:7719–7722.

Fitzgerald M, Shaw A, MacIntosh N (1988): Postnatal development of the cutaneous flexor reflex: comparative study of preterm infants and newborn rat pups. *Dev Med Child Neurol 30*:520–526.

Franklin K, Paxinos G (1997): The Mouse Brain Atlas in Stereotaxic Coordinates. New York.

Furukawa M, Wada N, Tokuriki M, Miyata H (2000): Spinal projections of cat primary afferent fibers innervating caudal facet joints. *J Vet Med Sci 62*:1005–1007.

Gillette RG, Kramis RC, Roberts WJ (1993): Spinal projections of cat primary afferent fibers innervating lumbar facet joints and multifidus muscle. *Neurosci Lett 157*:67–71.

Gogas KR, Presley RW, Levine JD, Basbaum AI (1991): The antinociceptive action of supraspinal opioids results from an increase in descending inhibitory control: correlation of nociceptive behavior and c-fos expression. *Neuroscience 42*:617–628.

Golden JP, Demaro JA, Jacquin MF (1997): Postnatal development of terminals and synapses in laminae I and II of the rat medullary dorsal horn. *J Comp Neurol 383*:326–338.

Hammond DL, Presley R, Gogas KR, Basbaum AI (1992): Morphine or U-50,488 suppresses Fos protein-like immunoreactivity in the spinal cord and nucleus tractus solitarii evoked by a noxious visceral stimulus in the rat. *J Comp Neurol 315*:244–253.

Harris JA (1998): Using c-fos as a neural marker of pain. *Brain Res Bull 45*:1–8.

Heppelmann B, Messlinger K, Schmidt RF (1988): Morphological characteristics of the innervation of the cat's knee joint. *Agents Actions 25*:225–227.

Herdegen T, Rudiger S, Mayer B, Bravo R, Zimmermann M (1994): Expression of nitric oxide synthase and colocalisation with Jun, Fos and Krox transcription factors in spinal cord neurons following noxious stimulation of the rat hindpaw. *Brain Res Mol Brain Res 22*:245–258.

Honore P, Buritova J, Besson JM (1995a): Aspirin and acetaminophen reduced both Fos expression in rat lumbar spinal cord and inflammatory signs produced by carrageenin inflammation. *Pain 63*:365–375.

Honore P, Buritova J, Besson JM (1995b): Carrageenin-evoked c-Fos expression in rat lumbar spinal cord: the effects of indomethacin. *Eur J Pharmacol 272*:249–259.

Honore P, Chapman V, Buritova J, Besson JM (1995c): When is the maximal effect of pre-administered systemic morphine on carrageenin evoked spinal c-Fos expression in the rat?. *Brain Res 705*:91–96.

Honore P, Buritova J, Besson JM (1996a): Intraplantar morphine depresses spinal c-Fos expression induced by carrageenin inflammation but not by noxious heat. *Br J Pharmacol 118*:671–680.

Honore P, Buritova J, Besson JM (1996b): The effects of morphine on carrageenin-induced spinal c-Fos expression are completely blocked by beta-funaltrexamine, a selective mu-opioid receptor antagonist. *Brain Res 732*:242–246.

Hou WY, Shyu BC, Chen TM, Lee JW, Shieh JY, Sun WZ (1997): Intrathecally administered c-fos antisense oligodeoxynucleotide decreases formalin-induced nociceptive behavior in adult rats. *Eur J Pharmacol 329*:17–26.

Hunt SP, Mantyh PW (2001): The molecular dynamics of pain control. *Nat Rev Neurosci 2*:83–91.

Hunt SP, Pini A, Evan G (1987): Induction of c-fos-like protein in spinal cord neurons following sensory stimulation. *Nature 328*:632–634.

Hunter JC, Woodburn VL, Durieux C, Pettersson EK, Poat JA, Hughes J (1995): c-fos antisense oligodeoxynucleotide increases formalin-induced nociception and regulates preprodynorphin expression. *Neuroscience 65*:485–492.

Janig W (1987): Neuronal mechanisms of pain with special emphasis on visceral and deep somatic pain. *Acta Neurochir Suppl (Wien) 38*:16–32.

Jasmin L, Wang H, Tarczy Hornoch K, Levine JD, Basbaum AI (1994): Differential effects of morphine on noxious stimulus-evoked fos-like immunoreactivity in subpopulations of spinoparabrachial neurons. *J Neurosci 14*:7252–7260.

Jennings E, Fitzgerald M (1996): C-fos can be induced in the neonatal rat spinal cord by both noxious and innocuous peripheral stimulation. *Pain 68*:301–306.

Jensen TS, Yaksh TL (1986a): Comparison of antinociceptive action of morphine in the periaqueductal gray, medial and paramedial medulla in rat. *Brain Res 363*:99–113.

Jensen TS, Yaksh TL (1986b): Comparison of the antinociceptive action of mu and delta opioid receptor ligands in the periaqueductal gray matter, medial and paramedial ventral medulla in the rat as studied by the microinjection technique. *Brain Res 372*:301–312.

Jones SL, Light AR (1990): Electrical stimulation in the medullary nucleus raphe magnus inhibits noxious heat-evoked fos protein-like immunoreactivity in the rat lumbar spinal cord. *Brain Res 530*:335–338.

Kakizaki H, Yoshiyama M, de Groat WC (1996): Role of NMDA and AMPA glutamatergic transmission in spinal c-fos expression after urinary tract irritation. *Am J Physiol 270*:R990–R996.

King TE, Cheng J, Wang S, Barr GA (2000): Maturation of NK1 receptor involvement in the nociceptive response to formalin. *Synapse 36*:254–266.

Kolmodin C, Skoglund C (1960): Analysis of spinal interneurones activated by tactile and nociceptive stimulation. *Acta Physiol Scand 50*:337–355.

Kuo DC, de Groat WC (1985): Primary afferent projections of the major splanchnic nerve to the spinal cord and gracile nucleus of the cat. *J Comp Neurol 231*:421–434.

Lanteri-Minet M, Isnardon P, de Pommery J, Menetrey D (1993): Spinal and hindbrain structures involved in visceroception and visceronociception as revealed by the expression of Fos, Jun and Krox-24 proteins. *Neuroscience 55*:737–753.

Lanteri-Minet M, Bon K, de Pommery J, Michiels JF, Menetrey D (1995): Cyclophosphamide cystitis as a model of visceral pain in rats: model elaboration and spinal structures involved as revealed by the expression of c-Fos and Krox-24 proteins. *Exp Brain Res 105*:220–232.

Le Bars D, Dickenson AH, Besson JM (1979): Diffuse noxious inhibitory controls (DNIC). I. Effects on dorsal horn convergent neurones in the rat. *Pain 6*:283–304.

Le Guen S, Catheline G, Besson JM (1999): Effects of NMDA receptor antagonists on morphine tolerance: a c-Fos study in the lumbar spinal cord of the rat. *Eur J Pharmacol 373*:1–11.

Leah JD, Sandkuhler J, Herdegen T, Murashov A, Zimmermann M (1992): Potentiated expression of FOS protein in the rat spinal cord following bilateral noxious cutaneous stimulation. *Neuroscience 48*:525–532.

Levinsson A, Luo XL, Holmberg H, Schouenborg J (1999): Developmental tuning in a spinal nociceptive system: effects of neonatal spinalization. *J Neurosci 19*:10397–10403.

Lima D, Avelino A, Coimbra A (1993): Differential activation of c-fos in spinal neurones by distinct classes of noxious stimuli. *NeuroReport 4*:747–750.

Lu Y, Jin SX, Xu TL, Qin BZ, Li JS, Ding YQ, Shigemoto R, Mizuno N (1995): Expression of c-fos protein in substance P receptor-like immunoreactive neurons in response to noxious stimuli on the urinary bladder: an observation in the lumbosacral cord segments of the rat. *Neurosci Lett 198*:139–142.

Ma QP, Woolf CJ (1996): Basal and touch-evoked fos-like immunoreactivity during experimental inflammation in the rat. *Pain 67*:307–316.

Martin WJ, Loo CM, Basbaum AI (1999): Spinal cannabinoids are anti-allodynic in rats with persistent inflammation. *Pain 82*:199–205.

McMahon SB, Wall PD (1985): The distribution and central termination of single cutaneous and muscle unmyelinated fibres in rat spinal cord. *Brain Res 359*:39–48.

Menetrey D, Gannon A, Levine JD, Basbaum AI (1989): Expression of c-fos protein in interneurons and projection neurons of the rat spinal cord in response to noxious somatic, articular, and visceral stimulation. *J Comp Neurol 285*:177–195.

Mense S, Craig AD (1988): Spinal and supraspinal terminations of primary afferent fibers from the gastrocnemius-soleus muscle in the cat. *Neuroscience 26*:1023–1035.

Molander C, Xu Q, Grant G (1984): The cytoarchitectonic organization of the spinal cord in the rat. I. The lower thoracic and lumbosacral cord. *J Comp Neurol 230*:133–141.

Molander C, Xu Q, Rivero Melian C, Grant G (1989): Cytoarchitectonic organization of the spinal cord in the rat: II. The cervical and upper thoracic cord. *J Comp Neurol 289*:375–385.

Morgan C, Nadelhaft I, de Groat WC (1981): The distribution of visceral primary afferents from the pelvic nerve to Lissauer's tract and the spinal gray matter and its relationship to the sacral parasympathetic nucleus. *J Comp Neurol 201*:415–440.

Morgan MM, Gogas KR, Basbaum AI (1994): Diffuse noxious inhibitory controls reduce the expression of noxious stimulus-evoked Fos-like immunoreactivity in the superficial and deep laminae of the rat spinal cord. *Pain 56*:347–352.

Munglani R, Hudspith MJ, Fleming B, Harrisson S, Smith G, Bountra C, Elliot PJ, Birch PJ, Hunt SP (1999): Effect of pre-emptive NMDA antagonist treatment on long-term Fos expression and hyperalgesia in a model of chronic neuropathic pain. *Brain Res 822*:210–229.

Ossipov MH, Kovelowski CJ, Nichols ML, Hruby VJ, Porreca F (1995): Characterization of supraspinal antinociceptive actions of opioid delta agonists in the rat. *Pain 62*:287–293.

Povlsen B, Hildebrand C (1993): Axonal regeneration in an articular branch following rat sciatic nerve lesions. *J Neurol Sci 120*:153–158.

Povlsen B, Stankovic N, Danielsson P, Hildebrand C (1994): Fiber composition of the lateral plantar and superficial peroneal nerves in the rat foot. *Anat Embryol Berl 189*:393–399.

Presley RW, Menetrey D, Levine JD, Basbaum AI (1990): Systemic morphine suppresses noxious stimulus-evoked Fos protein-like immunoreactivity in the rat spinal cord. *J Neurosci 10*:323–335.

Pretel S, Piekut DT (1991): Enkephalin, substance P, and serotonin axonal input to c-fos-like immunoreactive neurons of the rat spinal cord. *Peptides 12*:1243–1250.

Rexed B (1952): The cytoarchitectonic organisation of the spinal cord in the cat. *J Comp Neurol 96*:415–495.

Rohde DS, McKay WR, Chang DS, Abbadie C, Basbaum AI (1997): The contribution of supraspinal, peripheral and intrinsic spinal circuits to the pattern and magnitude of Fos-like immunoreactivity in the lumbar spinal cord of the rat withdrawing from morphine. *Neuroscience 80*:599–612.

Sandkuhler J (1996): The organization and function of endogenous antinociceptive systems. *Prog Neurobiol 50*:49–81.

Strassman AM, Vos BP (1993): Somatotopic and laminar organization of fos-like immunoreactivity in the medullary and upper cervical dorsal horn induced by noxious facial stimulation in the rat. *J Comp Neurol 331*:495–516.

Sugiura Y (1996): Spinal organization of C-fiber afferents related with nociception or non-nociception. *Prog Brain Res 113*:320–339.

Sugiura Y, Terui N, Hosoya Y, Tonosaki Y, Nishiyama K, Honda T (1993): Quantitative analysis of central terminal projections of visceral and somatic unmyelinated (C) primary afferent fibers in the guinea pig. *J Comp Neurol 332*:315–325.

Tao YX, Zhao ZQ (1997): Ultrastructure of Fos-labeled neurons relating to nociceptive primary afferent and substance P terminals in rat spinal superficial laminae. *Neuropeptides 31*:327–332.

Taylor J, Mellstrom B, Fernaud I, Naranjo JR (1998): Metamizol potentiates morphine effects on visceral pain and evoked c-Fos immunoreactivity in spinal cord. *Eur J Pharmacol 351*:39–47.

Tive LA, Barr GA (1992): Analgesia from the periaqueductal gray in the developing rat: focal injections of morphine or glutamate and effects of intrathecal injection of methysergide or phentolamine. *Brain Res 584*:92–109.

Todd AJ, Spike RC, Brodbelt AR, Price RF, Shehab SA (1994): Some inhibitory neurons in the spinal cord develop c-fos-immunoreactivity after noxious stimulation. *Neuroscience 63*:805–816.

Todd AJ, McGill MM, Shehab SA (2000): Neurokinin 1 receptor expression by neurons in laminae I, III and IV of the rat spinal dorsal horn that project to the brainstem. *Eur J Neurosci 12*:689–700.

Torsney C, Meredith-Middleton J, Fitzgerald M (2000): Neonatal capsaicin treatment prevents the normal postnatal withdrawal of A fibres from lamina II without affecting fos responses to innocuous peripheral stimulation. *Brain Res Dev Brain Res 121*:55–65.

Urban MO, Gebhart GF (1999): Supraspinal contributions to hyperalgesia. *Proc Natl Acad Sci USA 96*:7687–7692.

Vizzard MA (2000): Alterations in spinal cord Fos protein expression induced by bladder stimulation following cystitis. *Am J Physiol Regul Integr Comp Physiol 278*:R1027–R1039.

Vos BP, Strassman AM (1995): Fos expression in the medullary dorsal horn of the rat after chronic constriction injury to the infraorbital nerve. *J Comp Neurol 357*:362–375.

Wei F, Dubner R, Ren K (1999): Nucleus reticularis gigantocellularis and nucleus raphe magnus in the brain stem exert opposite effects on behavioral hyperalgesia and spinal Fos protein expression after peripheral inflammation [published erratum appears in Pain 1999 May;81(1-2):215-9]. *Pain 80*:127–141.

Williams S, Evan GI, Hunt SP (1990): Changing patterns of c-fos induction in spinal neurons following thermal cutaneous stimulation in the rat. *Neuroscience 36*:73–81.

Woolf CJ, Shortland P, Coggeshall RE (1992): Peripheral nerve injury triggers central sprouting of myelinated afferents. *Nature 355*:75–78.

Woolf CJ, Shortland P, Reynolds M, Ridings J, Doubell T, Coggeshall RE (1995): Reorganization of central terminals of myelinated primary afferents in the rat dorsal horn following peripheral axotomy. *J Comp Neurol 360*:121–134.

Yang LC, Marsala M, Orendacova J, Yaksh TL (1997): Knee joint inflammation attenuates spinal FOS expression after unilateral paw formalin injection in rat. *Neurosci Lett 225*:89–92.

CHAPTER VIII

c-Fos in learning: beyond the mapping of neuronal activity

LESZEK KACZMAREK

1. INTRODUCTION

As is extensively discussed in many other chapters of this volume, c-Fos became a prototypical immediate early gene product to be employed as an almost universal neuronal activity marker. However, in order to understand fully this experimental tool and to appreciate the meaning of the results achieved by this means, one would like to know the functional significance of c-Fos in the cells of the central nervous system. This book is focused on gene expression patterns. Learning experience provides especially complicated, challenging, and difficult to interpret conditions in this regard. Before, however, going into description of the learning-related c-*fos* expression, it is worthwhile to present the basic features of this gene and its products. It should also be noted that throughout this chapter, the standard convention of denoting genes and their mRNA products in italics (c-*fos*) and the proteins encoded by them with a single capital letter (c-Fos) will be used.

A useful starting point for all functional considerations is a reminder how c-*fos* became so important to study. Originally, the close homologue — v-*fos* was identified as a tumorigenic component of oncogenic viruses (so-called FBJ or FBR osteosarcoma viruses, Curran et al., 1983). Later on, an alteration at the carboxy terminus of the protein was identified to be responsible for generating the neoplastic phenotype (Van Beveren et al., 1983; Curran et al., 1984). Otherwise, the c-*fos* (cellular homologue of v-*fos*) is very similar to its deadly counterpart. Thus, one has always to bear in mind that relatively small change in gene structure of c-*fos* (or even its overexpression) (Miller et al., 1984) may result in dramatic effects on cell functioning. Next, it has been found out that c-Fos protein is a component of transcription factor (named AP-1; activatory protein-1; Angel and Karin, 1991). This appears to explain why c-*fos* is so important — it may regulate (orchestrate) expression of a number of other genes. The name 'master switch' was even employed to underline the significance of this gene and its product.

2. c-FOS: ITS ACTIVATION, ITS INVOLVEMENT IN AP-1 TRANSCRIPTION FACTOR, AND ITS TARGET GENES

c-*fos* mRNA and c-Fos protein levels are very low at basal conditions, i.e., in a variety of non-stimulated cells, including neurons. Increased neuronal expression results from stimulation

Handbook of Chemical Neuroanatomy Vol. 19: Immediate Early Genes and Inducible Transcription Factors in Mapping of the Central Nervous System Function and Dysfunction
L. Kaczmarek and H.A. Robertson, editors

of membrane receptors and subsequent rise in second messengers and kinase activation. The temporal pattern of this activation is quite uniform. An increase in mRNA is observed within a few minutes after the signal arrives at the cell membrane, next the protein is accumulated, this occurs roughly between 30 and 90 min. Both mRNA and protein increases are transient. c-*fos* transcription is soon shut off and the mechanisms of this phenomenon are not fully understood, although c-Fos itself, as well as, e.g., CREM/ICER (see Chapter XIII), have been implicated in negative regulation of c-*fos* transcription (Sassone-Corsi et al., 1988; Foulkes et al., 1991; Molina et al., 1993; Konopka et al., 1998). Furthermore, the c-*fos* products' half-lives (as measured in cultured cells) are very short: less than 20 min for c-*fos* mRNA and less than 2 h for the protein (Vosatka et al., 1989; Greenberg et al., 1990; Werlen et al., 1993). More protracted time-courses of c-*fos* expression were sometimes reported for the brain; however, they most probably result from sequential activation of various cells within the heterogeneous nervous tissue.

A number of neurotransmitter, neuropeptide, and neurotrophin receptors were implicated in driving c-*fos* expression up (see Hughes and Dragunow, 1995; Kaczmarek and Chaudhuri, 1997; Herdegen and Leah, 1998; Platenik et al., 2000 for reviews). In the field of neuronal plasticity, glutamate receptors, especially calcium-permeable NMDA receptors and AMPA receptors — producing depolarization and subsequent opening of the voltage gated calcium channel — were the most often implicated in controlling c-*fos* expression. In both cases Ca^{2+} is believed to be the major second messenger. Next to its rise, however, the situation get so complicated that it is difficult at present to decipher a variety of kinases involved. Then, however, the c-*fos* regulation is apparently simplified at the level of the gene regulatory regions. Two major promoter sequences were identified in control of c-*fos*: CRE (Ca^{2+}/cAMP responsive element) occupied by CREB (CRE binding proteins, see Chapter XIII) regulated by phosphorylation and SRE (serum responsive element) to which SRF) (SRE-binding factors) together with TCF (ternary complex factors) bind. Recently, Elk-1 transcription factor protein was identified as the major TCF in the brain, activated most probably by ERK kinase pathways (see Chapter XI).

As soon as c-Fos is produced, it may interact with Jun proteins to form AP-1 transcription factor that acts as a protein dimer. There are three Juns identified to date: c-Jun, JunB and JunD. Their ability to form dimers with c-Fos in the brain has been investigated only marginally. However, the studies on cultured (mostly non-neuronal) cells suggest that c-Fos may interact with each of the Jun proteins. It has also been observed that AP-1 of various composition may bring about differential effects on DNA binding and gene expression (see Chapter III; Kaminska et al., 2000; Szabowski et al., 2000). The functional significance of this phenomenon in the brain is virtually unknown.

Despite multiple efforts, very little is known about AP-1 target genes in the brain. They are so difficult to identify, because there are no very good methods available to affect c-Fos expression in the living animal. We have recently combined a variety of approaches to show that gene encoding TIMP-1 (tissue inhibitor of metalloproteinases-1, involved in reorganization of extracellular matrix) may be an AP-1 target in the rodent hippocampus in response to seizures (Jaworski et al., 1999). The evidence has been based on the fact that *timp-1* mRNA expression is: (1) promoter activity-driven (alternatively it could be via mRNA stabilization); (2) subsequent to c-*fos* mRNA accumulation; (3) prior protein synthesis dependent; (4) spatially correlated with c-Fos protein expression; furthermore (5) the AP-1 transcription factor, containing c-Fos, was accumulated in stimulated hippocampi and was capable of binding the *timp-1* AP-1-responsive DNA regulatory element in a sequence-dependent manner; and finally (6) glutamate was capable to activate expression of a reporter gene driven by *timp-1* promoter containing intact AP-1 site in cultured neurons of the dentate gyrus, and mutation of this promoter element abolished the expression.

The list of other genes possibly driven by c-Fos/AP-1 in the nerve cells includes those encoding: nerve growth factor, proenkephalin, prodynorphin, endorphin, neurotensin, tyrosine hydroxylase, neuropeptide Y, and others (see Herdegen and Leah, 1998; Kovacs, 1998). It is clear that this list contains many factors that may further carry on the c-Fos driven signal within the activated neurons, and even to their neighboring cells. Thus, we are by no means short of possibilities to explain the far-reaching significance of c-Fos in the brain. However, it has to be stressed that each of the aforementioned target gene examples still requires further studies.

An additional facet of c-Fos/AP-1 function is its ability to interact with other transcription factors. For instance glucocorticoid receptors, NFAT, ETS, ZIF268, etc. were all reported to interact with AP-1 (either directly or indirectly) and to be capable to modify AP-1 activity (Diamond et al., 1990; Jain et al., 1992; Kerppola et al., 1993; Logan et al., 1996; Papanikolaou and Sabban, 2000; Schule et al., 1990; Yang-Yen et al., 1990). This is not surprising, as regulation of every single gene studied to date was found to be dependent on a number of positively and/or negatively interacting transcription factors. However, it is then important to bear in mind that AP-1 activation alone cannot be taken as an indicator of subsequent specific genomic events. On the contrary, it should be expected that the final genomic response is a consequence of intricate play of a number of transcriptional activators and repressors performing their duties in a concerted manner (see Robertson et al., 1995).

3. PATTERNS OF EXPRESSION IN LEARNING

The great interest that has surrounded c-*fos* in the brain has originally been spurred by the hypothesis that its protein product ('master switch') may control a variety of phenomena of long-term change in cellular (and thus also neuronal) functioning, including learning and memory (Berridge, 1986; Goelet et al., 1986; Kaczmarek, 1986; Curran and Morgan, 1987; Kaczmarek and Kaminska, 1989). Very soon after these hypotheses were formulated, first experimental studies showed that indeed c-*fos* activation follows behavioral training and correlates with an acquisition of the behaviorally measured response, i.e., with learning (Maleeva et al., 1989; Kaczmarek, 1990; Kaczmarek and Nikolajew, 1990; Maleeva et al., 1990; Tischmeyer et al., 1990). These results were then reproduced and extended in dozens of research reports (see below, as well as recent comprehensive reviews: Hughes and Dragunow, 1995; Dragunow, 1996; Anokhin, 1997; Kaczmarek and Chaudhuri, 1997; Herdegen and Leah, 1998; Kovacs, 1998; Tischmeyer and Grimm, 1999).

Importantly, whereas the mRNA studies appear to be fully reliable as far as c-*fos* identity is concerned, it should be noted that some of the immunocytochemical studies employed antibodies that were not fully specific against c-Fos, but recognized its cognates (there are at least four additional members of this family: FosB, ΔFosB, Fra-1, and Fra-2) as well, and hence some of them may require to be reproduced, if c-Fos is to be specifically implicated. For the sake of simplicity, this issue has been neglected in this review.

Since the results on instrumental, aversively motivated conditioning-related c-*fos* expression (Kaczmarek, 1990; Kaczmarek and Nikolajew, 1990; Maleeva et al., 1990; Maleeva et al., 1989; Tischmeyer et al., 1990) laid the foundation for this field, they will be considered first in the more detailed presentation that follows.

3.1. INSTRUMENTAL CONDITIONING IN MAMMALS

3.1.1. Aversive reinforcement

3.1.1.1. Escape and avoidance

Two-way (active) avoidance behavior is being acquired in a shuttle box apparatus. The animal is originally placed in one of the two compartments of the training apparatus that are both equipped with a source of conditioned stimulus, CS (e.g., a lamp and/or a speaker) and gridded floor through which unconditioned stimulus, US (a footshock) can be delivered. The session is usually composed of a number of trials, each initiated with a CS, followed within a few seconds by the US. The animals are supposed to learn to avoid the US by reacting to CS by way of moving to the opposite compartment.

Our group reported (Kaczmarek and Nikolajew, 1990; Nikolaev et al., 1992a,b; Lukasiuk et al., 1999) that a single training session of two-way avoidance reaction resulted in a marked increase of c-*fos* mRNA and protein in rat sensory and limbic cortices. In the hippocampus, enhanced mRNA levels were not accompanied by comparable c-Fos protein induction. Furthermore, in the animals that were initially trained to visual stimulus acting as CS, c-*fos* mRNA was found to be highly elevated after the first such session and then was down to the basal levels after the ninth session, when the animals were performing the already well-learned tasks. However, exposure to the auditory + visual compound CS — that resulted in increased performance levels, apparently because of the more salient nature of the auditory component — again evoked c-*fos* expression in specific brain regions. Thus performance of the well-trained behavior did not evoke c-*fos* expression that, however, remained responsive to alteration in experimental conditions that involved additional learning.

A more detailed analysis of c-Fos expression was carried out for the amygdala by Savonenko et al. (1999) who applied factor analysis to separate and group a number of behavioral components of the two-way active avoidance acquisition. Next, these factors were investigated in the context of c-Fos accumulation in various subdivisions of the amygdala. It should be noted that the acquisition of the avoidance reaction is preceded first by fleeing and staying behavior, and then by escape reaction (movement to the opposite compartment not during the CS alone, but only after the US is turned on). In the experiments of Savonenko et al. (1999), rats were exposed to 50-trial session, with each trial separated by 14- or 20- or 26-s intervals presented in a semi-random order. During the inter-trial intervals (ITI) the animals could move freely between the compartments. A number of behavioral variables was scored when observing the animals. Statistical analysis allowed five major behavioral factors to be identified (avoidance, escape, aggression, grooming and inter-trial responses or ITRs, i.e., crossings between compartments during ITIs). Contrary to the most simplistic expectations, c-Fos expression in amygdala correlated with none of the first three factors. In particular, no correlation was found between c-Fos levels in any of the 13 amygdalar subdivisions that were investigated and either avoidance reaction or sum of the shock received. On the other hand, c-Fos expression within each of two separately analyzed components of cortical amygdala was found to correlate significantly with the grooming behavior, whereas c-Fos levels in basolateral and medial amygdala correlated with the ITRs. Since both grooming — reflecting most probably lack of fear, and ITRs — reflecting anticipatory anxiety (see Savonenko et al., 1999 for discussion) are learned reactions during this procedure and both are emotionally motivated behaviors, this study, using an unbiased approach, shows that emotional components of acquisition the two-way avoidance reaction are specifically linked to amygdalar c-Fos. It remains to be seen in which brain regions c-Fos expression could be linked to the instrumental components of this behavior (such as avoidance reaction) (see Lukasiuk et al., 1999).

Anokhin et al. (2000) followed the original observation by Maleeva et al. (1989, 1990) and studied by northern hybridization the levels of c-*fos* mRNA expression in mouse cerebral cortex and hippocampus at different stages of escape and avoidance learning. In the first series of experiments, mice were presented with electric footshocks in a chamber where they could escape from the floor by jumping on the safe platform attached to the wall. A large increase in c-*fos* mRNA level in the cerebral cortex and hippocampus was observed following the first training session. Mice that were trained for 9 consecutive days and acquired a footshock escape reaction showed no elevation of c-*fos* expression in the brain as compared to the quiet control group. In the second series of experiments, the levels of c-*fos* expression were compared in individual mice trained to avoid the footshock by jumping on the platform in response to an auditory conditioned stimulus. Mice which acquired avoidance behavior more rapidly had lower c-*fos* mRNA levels than slow learners. There was no such difference between the corresponding yoked control groups, which consisted of animals matching the rapid and slow learners by the number of footshocks received.

Duncan et al. (1996) employed immunohistochemical staining of the multiple brain regions after footshock avoidance training in a shuttle box and found prominent c-Fos expression in a number of selected brain regions, including the medial prefrontal, retrosplenial, cingulate, and ventrolateral orbital cortices (the frontal parietal cortex was conspicuously low labeled), taenia tecta, nucleus accumbens, bed nucleus of stria terminalis, paraventricular nucleus of the hypothalamus (PVN), amygdala (with the exception of the central amygdala), lateral septum, locus coeruleus, inferior colliculus, central gray, and cuneiform nucleus.

Tischmeyer et al. (1990), and Grimm and Tischmeyer (1997) employed northern analysis of c-*fos* mRNA levels following training of rats on a footshock-motivated brightness discrimination task in a Y-maze. In the initial study, hippocampal c-*fos* mRNA accumulation was observed. A similar increase was also obtained when rats were subjected to a pseudotraining with an equal number of runs, but with random pairing of the choice of bright and dark alleys with foot shock. In a follow-up experiment, c-*fos* mRNA upregulation was also noted in other brain regions, such as cerebral cortex and cerebellum.

Castro-Alamancos et al. (1992) studied c-Fos expression in two different areas of the motor cortex and in the hippocampus of the rat after performance of an already trained escape task in a Skinner box. Importantly, in these experiments, the performance testing session differed significantly in its duration from the training sessions, and thus involved an important novel component to be learned by the animals. The authors found an increase in the number of cells showing c-Fos immunoreactivity in layers V and VI of the forelimb sensorimotor cortex with respect to yoked animals which had received the same amount, frequency and duration of aversive stimulation and manipulation as the trained animals. In the hindlimb motor-sensory cortex, there were no differences between the trained and the yoked animals in any of the cortical layers. No differences were observed in the dentate gyrus of the hippocampus between trained and yoked animals.

3.1.1.2. Passive avoidance

Passive avoidance, also named dark avoidance or inhibitory avoidance is a procedure in which the animal is originally placed in a white-painted compartment of the training apparatus and is allowed to enter the dark compartment. In the latter, the animal is treated to aversive stimuli, such as inescapable footshocks. Then, the trained animal avoids entering the dark compartment, after being placed in the white one.

Zhang et al. (2000) investigated the dark-avoidance training-induced c-Fos expression in the nucleus basalis of Meynert (nbM) in rats. The study demonstrated that c-Fos expression in the nbM was significantly increased at 2 h after dark-avoidance training. Notably, the increase of c-Fos expression was also observed after pseudotraining, although the number of c-Fos-immunoreactive neurons in pseudotrained rats was significantly less than that in the trained rats. Using a similar training procedure, Cammarota et al. (2000) found an increase in the c-Fos levels in the rat hippocampus.

3.1.2. Appetitive reinforcement

Heurteaux et al. (1993) originally reported an elevated c-*fos* expression in the hippocampus after training of an appetitive bar-pressing task in a Skinner box. Then, Bertaina and Destrade (1995) studied spatiotemporal patterns of c-*fos* mRNA expression in the mouse brain at various stages of acquisition of the same task. Two experimental situations were used: during the initial acquisition of the task and during the retention test. In both conditions, the c-*fos* in situ hybridization signal was exclusively located in the hippocampus and the posterior cingulate cortex. In a follow-up study, Bertaina-Anglade et al. (2000) further analyzed the effects of the stage of learning of this task on the spatial and temporal patterns of c-Fos protein levels in the mouse brain as assessed by immunohistochemistry at various times after either the first, the second or the fifth daily training session and compared the results with those of non-trained animals. Following the first acquisition session, significant increases in c-Fos-positive neurons were observed in the dorsal hippocampus (CA3), anterior cingulate, occipital and parietal cortices. Following the second daily training session, c-Fos was highly expressed in some subcortical regions, the hippocampus, the subiculum, the entorhinal, and posterior cingulate areas. Moreover, a significant correlation was found between the progression of performance from day 1 to day 2 and c-Fos expression on the hippocampal CA1 subfield. Following complete acquisition, no further task-dependent increases in c-Fos-labeled nuclei was observed in any brain region sampled.

Hess et al. (1995a,b, 1997); see also Gall et al., 1998) measured levels of c-*fos* mRNA by in situ hybridization in amygdala, piriform cortex and hippocampus as well as in components of the olfactory and visual systems in rats that were trained a nose-poke response for water reward. The authors attempted to examine c-*fos* expression associated with various stages of acquisition of the reaction, odor discrimination and performance of a well-learned odor discrimination. However, from the description of the behavioral procedures provided in the paper, it appears that the performance of a well-learned task was tested in a session that was not exactly the same as the previous training sessions, thus suggesting that a novelty component has to be taken into account in considering the results presented below. The authors found that after the training c-*fos* mRNA levels were broadly increased — relative to values in home cage-control rats — in the hippocampus, amygdala (medial and basolateral, with no activation in the central), visual cortex, superior colliculus, and olfactory bulb. In rats first trained on the nose-poke behavior and then required to discriminate between two odors for water reward, the increase in c-*fos* mRNA was generally not as great and was more regionally differentiated. Thus, in olfactory bulb, hybridization signal was more elevated in lateral than medial fields in accordance with the topographic representation of the predominant odor sampled in the discrimination task. In the hippocampus of odor-discriminating rats, c-*fos* mRNA levels were far greater in CA3 region than the CA1 region, but in dentate gyrus were not elevated. Interestingly, c-*fos* mRNA levels in each hippocampal subdivision were highly correlated with levels in other regions (e.g., visual cortex) for home cage controls, but not for rats in the two behavioral groups. Hippocampal labeling in exploration and performance rats differed in that: (1) hybridization was greater in the CA1, CA3, and dentate gyrus in the former group; and (2) a tendency for labeled cells to occur in clusters was more evident in exploration animals. Levels of c-*fos* mRNA in olfactory and visual structures were not predictive of expression patterns within hippocampus, although labeling in the piriform cortex and dentate gyrus was correlated in rats performing a well-practiced discrimination. Moreover, the pattern of hybridization in the olfactory bulb was found to be behaviorally dependent. In the rats engaged in familiar nose-poke responses, the ratio of hippocampal to amygdala labeling was at control levels. Rats required to switch from ad libitum responding to cued responding to odors had high amygdalar basolateral to medial labeling ratios. This was in marked contrast to the medial amygdala dominance found in control and exploration rats. In situ hybridization signal was substantially more pronounced in basolateral amygdala than in hippocampal CA1; this imbalance was unique to the group required to form first associations between odors and rewards. Rats performing an overtrained odor discrimination had the least differentiation between amygdalar subdivisions of any behavioral group. The hippocampus-to-amygdala labeling ratio favored hippocampus and was nearly identical to the ratio in exploratory rats.

Carretta et al. (1999a,b) investigated c-Fos expression in rats that were trained to associate an acoustic stimulus with a reward and then to discriminate between two sounds. Rats were trained in three consecutive phases. The first phase was an association between an auditory stimulus and a food reward; the second phase, a simple discrimination between two sounds of different frequency components; and the third phase, a more complex discrimination involving both spectral and spatial sound dimensions. The rats undergoing this protocol were progressively sacrificed at the beginning, middle and end of each phase. For comparison, c-Fos levels were also investigated in control rats exposed to the same stimuli delivered pseudorandomly. Furthermore, there were also rats unexposed to controlled acoustic stimulation. In the superior olivary complex, inferior colliculus, medial geniculate body, and cochlear nucleus the number of c-Fos immunopositive cells was comparable in learning and pseudotrained animals, but higher than in the unstimulated rats. Notably, in the auditory cortex, most prominently in the secondary area Te2, the number of Fos-like positive cells differed between learning and pseudotrained rats.

Duncan et al. (1996) trained the rats on the bar pressing reaction for a food reward and did not notice any significant difference in c-Fos levels between the trained rats and those that were either just placed into the operant chamber or that were trained without the response contingency. Unfortunately, no information is provided by Duncan et al. (1996) as to the c-Fos difference with the handled animals, well habituated to the training environment. Thus, it is difficult to estimate whether the results describe the bona fide lack of c-Fos activation or rather they reflect a similar level of activation achieved by various forms of learning (association of the appropriate contingency vs. lack of such contingency, see below, vs. spatial learning of a novel environment).

3.2. CLASSICAL CONDITIONING

3.2.1. Appetitive reinforcement

Young rats exposed to peppermint odor and reinforcing tactile stimulation from postnatal days (PND) 1–18 increase their preference for that odor relative to controls. Johnson et al. (1995) reported that this early olfactory memory is accompanied by a marked increase in the density of glomerular layer cells displaying c-Fos immunoreactivity in response to the learned odor on PND 19. The difference was observed in midlateral portions of the olfactory bulb that aligned with foci of 2-deoxyglucose (2-DG) uptake in adjacent sections. Trained and control animals were not different in the c-Fos response of juxtaglomerular cells within ventrolateral 2-DG foci. Ratios of midlateral/ventrolateral response differed significantly between trained and control animals and included differences among cells of three staining intensities. These ratios were correlated with ratios of 2-DG uptake (midlateral/ventrolateral foci), which also differed significantly between trained and control rats. In the further study, Woo et al. (1996) characterized the response of the olfactory bulb granule cells to an odor for which young rats had acquired a preference. c-Fos-immunoreactive granule cells were quantified in a region previously shown to be responsive to peppermint odor. The authors found that odor-trained pups had about half the number of c-Fos-immunopositive superficial granule cells which responded to a learned odor than did control pups. Then the authors determined whether there was a correlation between the juxtaglomerular cell response and the response of the superficial granule cells deep to those glomerular layer cells. They found a positive correlation between the number of juxtaglomerular cells and the number of granule cells demonstrating c-Fos immunoreactivity in both control and trained pups, a relationship that changed with early olfactory training.

Allingham et al. (1999) examined expression of c-Fos in the olfactory bulbs of 3-day-old rabbits after they had been presented with the odor of garlic as a novel stimulus, as a learned odor, or during conditioning, and this expression was also compared with baseline levels in non-stimulated controls. Exposure to garlic odor resulted in substantial and widespread increases in c-Fos expression in the olfactory bulbs of all animals. However, although conditioned pups showed a specific behavioral response to the learned garlic odor, neither the amount nor pattern of c-Fos expression differed compared to pups exposed to garlic as a novel odor. The odor-induced expression of c-Fos was not well localized, although there was a significant increase in the number of granule cells expressing c-Fos in the ventrolateral region of the bulb.

Environmental light is the dominant temporal cue for the entrainment of circadian rhythms. In mammals, light entrains circadian rhythms by daily resetting a pacemaker located in the hypothalamic suprachiasmatic nucleus (SCN). Although it is widely held that phase resetting by light involves cellular elements within the SCN that are responsive only to photic cues, Amir and Stewart (1996, 1998a,b,c, 1999) reported, however, using Pavlovian conditioning procedures, that a non-photic stimulus that has been repeatedly paired with light can – in the absence of light – affect c-Fos expression in the suprachiasmatic nucleus.

3.2.2. Fear conditioning

Fear conditioning is a training paradigm in which the animals are placed for a short time in a chamber in which they receive an inescapable footshock (US). Then, after various delays they spend in the home cages, the animals are tested for the memory of the situation (the memory is expressed by a freezing reaction) by placing the subjects in the same apparatus but without the US. Sometimes, context conditioning is being distinguished from cue conditioning (in the latter case, original exposure to US is accompanied by clear sensory stimulus, i.e., a tone CS and then testing is carried out in a different environment than training apparatus, but in the presence of the tone). Importantly, exposing the animals to a footshock immediately after placement in the chamber precludes effective contextual fear conditioning and produces a freezing deficit.

It has repeatedly been reported that footshocks alone activate c-Fos expression in a number of brain structures, such as the amygdala (including the central nucleus), hippocampus, PVN, parietal, frontal, occipital, temporal and cingulate cortices, locus coeruleus, nucleus of the solitary tract (NTS, nucleus tractus solitarius), ventral lateral medulla, dorsal and ventral subdivisions of the periaqueductal gray (PAG), dorsal raphe nuclei (Campeau et al., 1991; Smith et al., 1992; Pezzone et al., 1992, 1993; Carrive et al., 1997a,b; Li et al., 1996; Milanovic et al., 1998; Radulovic et al., 1998; Rosen et al., 1998; Bubser and Deutch, 1999; Morrow et al., 1999).

Exposure to either a novel environment or a tone that may be next used as CS in the fear-conditioning paradigm, also results in elevated c-Fos expression in various brain regions, such as the medial amygdala, cingulate, parietal, and piriform cortices (see, e.g., Smith et al., 1992; Rosen et al., 1998). Although it should be noted that this response appears to be much more limited in terms of both range and magnitude, when compared to the footshock-evoked c-Fos activation.

The most extensive analysis of c-Fos expression following re-exposure to the environment in which the rats had previously received footshock was performed by Beck and Fibiger (1995). The authors found a significantly increased number of c-Fos-immunoreactive neurons in nearly 50 brain regions, both cortical and subcortical, among 60 areas analyzed. Among the structures showing the most dramatic increases in fear-induced c-Fos expression were the cingulate, piriform, infralimbic, and retrosplenial cortices, the anterior olfactory nucleus, claustrum, endopiriform nucleus, nucleus accumbens shell, lateral septal nucleus, various amygdalar nuclei, paraventricular thalamic nucleus, ventral lateral geniculate nucleus, the ventromedial, lateral, and dorsal hypothalamic nuclei, the ventral tegmental area, and the supramammillary area. These results are in rather good agreement with other studies on a more limited number of brain regions (Campeau et al., 1991; Carrive et al., 1997b; Milanovic et al., 1998; Radulovic et al., 1998; Rosen et al., 1998). Also employing the tone as CS in pairing with a footshock and re-exposing the animals to the tone, produced elevated c-Fos expression in various brain areas; however, the number of data reported is apparently much more limited (see, e.g., Pezzone et al., 1992, 1993; Smith et al., 1992; Hall et al., 2001).

There was also a number of studies on fear conditioning that employed quite sophisticated training/testing paradigms to relate the observed c-Fos activation to the memory formation. For instance, Sotty et al. (1996) investigated a latent inhibition phenomenon that refers to the fact that the formation of a conditioned association between a conditioned and an unconditioned stimulus is delayed by prior exposure to the conditioned stimulus. Latent inhibition is often investigated in the context of the conditioned emotional response, in which a tone serves as the conditioned stimulus and a footshock as the unconditioned stimulus. Such a paradigm was used for the experiments in which some rats had been pre-exposed to the tone. Two hours after a subsequent exposure to the tone, c-Fos immunocytochemistry was used to map brain areas involved in audition, fear, stress and memory. For the basic conditioning group, pre-exposure to the tone decreased the number of c-Fos-positive cells in the auditory system, areas involved in fear and stress and a number of limbic areas, namely the amygdala, the hippocampus, and the entorhinal cortex. In contrast, the c-Fos signal density increased in the dentate gyrus, subiculum and nucleus accumbens.

Milanovic et al. (1998) determined c-Fos production in mouse brain following acquisition of context-dependent fear and after re-exposure to the conditioning context. Fear-conditioning was induced by a single exposure of mice to a context followed by an electric shock. Control groups consisted of mice exposed to context only (context group) or to an immediate electric shock. After the training, higher c-Fos levels were observed in the hippocampus, medial amygdala and parietal somatosensory cortex of the fear-conditioned mice than mice exposed to an immediate shock. Significantly higher c-Fos expression was observed in the central nucleus of the amygdala of mice belonging to both shock groups than the context group. Twenty-four hours after conditioning and treatment to retention test, fear-conditioned mice generated more c-Fos in the hippocampus and central amygdaloid nucleus than the two control groups. However, all groups exhibited similarly low c-Fos production after animals were exposed to a different context from days 2 to 5 after conditioning and then tested for retention on day 6.

In a following study, Radulovic et al. (1998) investigated the relationship between c-Fos levels in the sensory cortex and limbic system and the ability of mice to acquire context- and tone-dependent freezing after fear conditioning, which was achieved by exposure of mice to context only or context and tone as CS paired with an electric footshock as US. High simultaneous c-Fos expression in the parietal cortex, hippocampus, and amygdala was found to parallel the ability of mice to acquire strong freezing responses to novel CS. After contextual pre-exposure (latent inhibition), c-Fos expression could be elicited in the central amygdala only by shock and in the basolateral amygdala only by tone. Under these conditions, the ability of mice to acquire contextual freezing was almost abolished, whereas tone-dependent freezing was reduced. Lack of c-Fos accumulation in the central amygdala after pre-exposure to context followed by shock (US pre-exposure effect) paralleled the inability of mice to acquire tone-dependent freezing, although the tone elicited c-Fos production in the basolateral amygdala.

Rosen et al. (1998) investigated the increase in expression of c-*fos* mRNA in the amygdala following contextual fear conditioning. Rats were either handled (handled group), placed in a test chamber without receiving footshock (context–no-footshock group), received footshock immediately upon being placed in the chamber (immediate-footshock group), or received footshock after a 1-min delay (delayed-footshock group). Only the delayed-footshock group displayed a fear response (freezing behavior). Rats were sacrificed either 15 min after the experience or after a retention test 24 h later. The c-*fos* mRNA levels were increased in the medial nucleus of the amygdala in all of the groups that were placed in the test chamber. However, rats that received footshock (immediate- and delayed-footshock groups) had greater levels of c-*fos* mRNA expression than rats of the context–no-footshock group. On the other hand, the c-*fos* mRNA expression in the immediate- and delayed-footshock groups did not differ. After the retention test, the expression of c-*fos* mRNA in the medial nucleus of the amygdala did not differ between groups.

Funk and Amir (2000) employed an interesting behavioral analysis to investigate c-Fos responses within the olfactory and limbic systems by comparing three groups of animals, namely, odor-exposed, backward conditioned and forward conditioned. The authors found that presentation of the conditioned odor stimulus alone resulted in significantly higher c-Fos levels in forward conditioned group in the main olfactory bulb, accessory olfactory bulb, and accessory olfactory nucleus, infralimbic cortex, orbital cortex, perirhinal–entorhinal cortex, basolateral amygdala but neither in the piriform cortex nor the central amygdala.

3.2.3. Classical conditioning of the rabbit's nictitating membrane reflex

Irwin et al. (1992) investigated distribution of neuronal c-Fos expression associated with the classical conditioning of the rabbit nictitating membrane reflex (NMR). Rabbits were divided into two groups: a conditioning group that received paired tone and airpuff stimuli in a traditional delay NMR conditioning paradigm and a pseudo-conditioning group in which the same number of tone and airpuff stimuli were applied, but without being paired. c-Fos labeling was present in similar brainstem nuclei in both groups of animals. The labeled sites included trigeminal and auditory nuclei in the classical pathway for the nictitating membrane reflex as well as other nuclei, such as the raphe nuclei and those in the ventrolateral medulla. However, there were quantitative differences in the labeling between the two groups. There was significantly more c-Fos expression in the trigeminal nucleus of the pseudoconditioned rabbits, but higher c-Fos levels were observed in the raphe nuclei of the conditioned animals.

Carrive et al. (1997a) investigated the changes in c-Fos immunoreactivity in the locus coeruleus (LC) after classical conditioning of the rabbit's nictitating membrane. Specifically, they compared unpaired versus paired presentations of a tone CS and a tactile US, near the eye. After two training sessions, only paired presentations resulted in acquisition of a conditioned response. This was associated with comparatively less LC c-Fos expression than with unpaired presentations. Similar observations have been reported for the ventrolateral medulla which is a major source of afferents to the LC.

3.2.4. Conditioned taste aversion

When consumption of a novel taste (e.g., saccharin or sucrose) is followed by exposure to a toxin, animals will avoid consumption of that taste in the future. This learned response, known as a conditioned taste aversion (CTA) can be demonstrated using a variety of a malaise-provoking drugs, such as lithium chloride (LiCl) or amphetamine. Thus CTA is a form of classical conditioning in which animals avoid a taste (CS) which has previously been paired with a treatment (US) that produces transient illness. Combining lesion, receptor antagonist, and enzyme inhibitor treatments with behavioral studies provided an ample evidence that number of brain structures, such as parabrachial nucleus (PBN), medial thalamus, gustatory (insular) cortex, and amygdala are essential for establishing the CTA memory trace.

This behavioral paradigm has been the most often used in studies on c-Fos expression in learning. To organize the data obtained by the various researchers, it might be worth stressing that there are at least three different stimuli, of various biological values to be considered: unconditioned stimulus, sensory stimulus, and the same sensory stimulus that have previously been employed as CS in conjunction with US. Noteworthy, the effects of the latter two treatments should also be considered in the context of studies described by Chaudhuri and Zangenhpour (Chapter V).

The c-Fos brain mapping in the animals exposed to LiCl administered intraperitoneally revealed labeling of a number of structures: the NTS, the PVN and supraoptic nuclei of the hypothalamus, amygdala (especially central), area postrema, PBN, and hypoglossal nucleus (Gu et al., 1993; Swank and Bernstein, 1994; Swank et al., 1995; Yamamoto et al., 1994a; Sakai and Yamamoto, 1997; Yamamoto et al., 1997; Thiele et al., 1996; Lamprecht and Dudai, 1995). Sakai and Yamamoto (1997) studied the relationship between the efficacy of inducing CTA to saccharin and c-Fos expression in the lower brainstem following administration of 13 different US in rats. The effective US were grouped into abdominal irritants, rewarding drugs and emetic agents. Regardless of the properties of US, good correlation was detected between the strength of CTA and c-Fos expression within the area postrema, caudal and intermediate subdivisions of NTS and the external lateral subnucleus of the PBN. Only hypertonic saline was exceptional because it induced strong c-Fos expression, but was not an effective US for CTA formation. Different dosages of the emetic LiCl induced CTA and c-Fos expression in a dose-dependent manner.

Sweet tasting substances, such as saccharin (most often investigated in this regard) or sucrose were capable of evoking c-Fos expression in PBN, amygdala as well as cingulate and parietal cortices even without pairing with the US (Yamamoto, 1993; Yamamoto et al., 1994a, 1997; Montag-Sallaz et al., 1999). However, the most remarkable result repeatedly studied in the context of CTA is the ability of the CS to evoke c-Fos accumulation in the NTS only after pairing with the US (Houpt et al., 1994, 1996b; Swank and Bernstein, 1994; Schafe et al., 1995; Swank et al., 1995; Thiele et al., 1996). Interestingly, Swank et al. (1995) compared the c-Fos response to the aversive saccharin CS to the response to quinine hydrochloride, which is innately aversive. Although behaviorally the animals' ingestive responses were quite similar, the saccharin CS induced significant elevations of c-Fos in NTS whereas the quinine did not. Furthermore, Houpt et al. (1996a,b) found that intraoral infusions of quinine were capable of inducing c-Fos in the NTS, however, only after acquisition of a CTA against quinine; quinine failed to evoke c-Fos expression in the NTS of unconditioned, non-contingently treated, or extinguished rats.

A number of further studies aimed at clarifying specific features of the CTA-related c-Fos activation. For instance, Swank (2000) reported increased c-Fos expression in the NTS of the rat in response to administration of a variety of pharmacologically diverse US. Before conditioning, the CS taste did not induce c-Fos in the NTS, but following pairing of the CS and US, subsequent CS presentation induced c-Fos in the NTS. In mice, post-conditioning c-Fos expression to the CS depended on contextual cues: when conditioning and testing occurred in a novel environment, CS saccharin caused an increase in c-Fos expression, and when conditioning and testing occurred in the home cage, CS saccharin produced a decrease in c-Fos expression relative to controls. Furthermore, merely placing an animal into a novel environment was sufficient to drive c-Fos expression in the NTS.

Schafe et al. (2000) evaluated c-Fos levels in the NTS in a context of two measures of conditioned fear: freezing and changes in mean arterial pressure and heart rate. Exposure to the taste CS resulted in a marked induction of c-Fos in the NTS, whereas exposure to a fear CS did not. Furthermore, exposure to a taste CS did not selectively lead to increases in mean arterial pressure or heart rate.

Yamamoto et al. (1997) examined c-Fos in the amygdala in rats after ingestion of taste solutions (either sucrose or saccharin), intragastric infusion of these solutions, or an intraperitoneal injection of LiCl. c-Fos-immunoreactive neurons were distributed most densely in the central nucleus of the amygdala in response to the LiCl injection, followed by the ingestion and intragastric infusion of sucrose. The intraoral infusion of sucrose, but not of saccharin, elicited intense c-Fos expression in the central nucleus after establishment of conditioned taste aversion to these taste stimuli.

Spray et al. (2000) examined whether a c-Fos response would occur during CTA when conditioning was accomplished by presenting the taste solution in a bottle instead of typically employed intraoral infusions. Intraoral and bottle methods generated aversions that were comparable, when judged by the behavioral response of solution rejection. However, elevations in c-Fos were seen only in animals conditioned with the intraoral method.

Swank et al. (1995) used c-Fos immunohistochemistry to define brain regions activated following either amphetamine or LiCl administration. Treatment with LiCl induced marked c-Fos accumulation in the NTS while amphetamine induced only light c-Fos staining in this area. Saccharin paired with amphetamine, however, resulted in enhanced c-Fos expression in NTS in a pattern quite similar to that seen to a LiCl-paired with the sweet taste.

Schafe et al. (1995) employed a variant of the chronic decerebrate rat preparation to explore whether circuitry intrinsic to the brainstem was sufficient for the induction of c-Fos in both as an unconditioned response to the LiCl and as a conditioned response to the saccharin. Using rats, which had a unilateral brain transection at the level of the superior colliculus, the authors found that the unconditioned c-Fos response to LiCl was unaltered by the transection, while the conditioned expression of c-Fos to the CS taste was evident only on the side of the NTS which retained neural connections with the forebrain.

Schafe and Bernstein (1996) combined electrolytic lesions of the amygdala with immunostaining for c-Fos to define neuroanatomical structures and pathways that contribute to induction of c-Fos in the NTS in CTA. Rats were given either unilateral or bilateral electrolytic lesions of the amygdala or sham operations. Following surgery 'paired' animals were given a single conditioning trial consisting of intraoral infusion of saccharin followed by injection with LiCl while 'unpaired' controls received a non-contingent saccharin–LiCl presentation. When tested, unilateral-lesioned rats displayed a CTA by rejecting the saccharin, but increases in c-Fos were evident only on the side of the NTS contralateral to the lesion. Rats with bilateral lesions showed no evidence of having acquired a CTA and no increase in c-Fos in the NTS relative to unpaired controls.

Following a similar approach Schafe and Bernstein (1998) as well as Cubero et al. (1999) employed electrolytic lesions of insular (gustatory) cortex, combined with immunostaining for c-Fos. Rats were given either unilateral or bilateral electrolytic lesions of insular cortex or sham operations. As in the previous study, 'paired' and 'unpaired' animals were distinguished. Rats with bilateral lesions showed no behavioral evidence of having acquired a CTA. Furthermore, bilateral electrolytic lesions of insular cortex blocked behavioral expression of a CTA, not only when lesions were placed prior to conditioning, but also when they were made after conditioning, but before testing. Increases in c-Fos in the NTS after the bilateral lesions were still evident, but reduced, relatively to 'sham' animals. Rats with unilateral lesions displayed a CTA by rejecting the saccharin, although increases in c-Fos on the side of the NTS ipsilateral to the lesion were reduced relative to that seen in 'sham' animals.

Houpt et al. (1997) attempted to determine if c-Fos expression in the NTS depended on subdiaphragmatic vagal afferent input to the NTS secondary to gastrointestinal symptoms during CTA expression (e.g., diarrhea). Thus, they quantified the induction of c-Fos in the NTS by sucrose infusions after total subdiaphragmatic vagotomy in rats with a previously acquired CTA against sucrose. Vagotomy had no apparent effect on the behavioral expression of the previously acquired CTA. There was also no significant difference between vagotomized and sham-vagotomized rats in the number of c-Fos-positive cells in the NTS after CTA expression.

3.3. MOTOR LEARNING

Kleim et al. (1996) measured the number of synapses per neuron and the percentage of c-Fos-positive neurons within layer II/III of the rat motor cortex after training on a complex motor learning task. Adult female rats were allocated randomly to either an acrobatic condition, a motor control condition, or an inactive control condition. Acrobatic condition animals were trained to traverse a complex series of obstacles, and each acrobatic condition animal was pair matched with a motor control animal that traversed an obstacle-free runway. Inactive control animals received no motor training. Acrobatic condition animals exhibited an overall increase in the number of synapses per neuron in comparison to motor and inactive control animals at later stages of training. Acrobatic condition animals also had a significantly higher overall percentage of c-Fos-positive cells in comparison to both controls, with a trend for the increase to be greater during the acquisition versus the maintenance phase.

3.4. SPATIAL AND RECOGNITION MEMORY

3.4.1. Spatial task

Vann et al. (2000) compared c-Fos levels during different spatial tasks in two experiments. In Experiment 1, matched groups of rats either ran a standard eight-arm radial maze task or were trained to run up and down just one arm of the maze; the number of runs and rewards was identical in both conditions. In Experiment 2, rats were trained on the eight-arm maze, but in different rooms. On the critical test day, both groups were run in the same room so that one group now performed with novel landmarks. All hippocampal subfields (dentate gyrus, CA3, CA1, dorsal, ventral, and caudal subiculum) showed a relative increase in c-Fos levels in the eight-arm (Experiment 1) and novel room (Experiment 2) conditions, the sole exception being the ventral subiculum in Experiment 2. Although increased c-Fos expression was found in both dorsal and ventral hippocampus, in Experiment 2 the relative increase was significantly greater in the dorsal hippocampus. Parahippocampal cortices responded heterogeneously: the perirhinal cortex failed to show increased c-Fos expression in both experiments, in contrast to the entorhinal and postrhinal cortices. Subsequent comparisons confirmed that the perirhinal and postrhinal cortices responded in qualitatively different ways, the perirhinal cortex differing from the rest of the hippocampal formation.

3.4.2. Visual recognition

Zhu et al. (1995) investigated expression of c-Fos in rat brain following exposing the animals to sets of either novel or highly familiar objects, or to the same pattern of illumination without objects being shown. Counts of stained nuclei were made in eight brain regions, where information about novel or familiar visual stimuli is likely to be processed or stored. The counts were relatively high in occipital visual association cortex and area TE of temporal cortex, intermediate in perirhinal cortex, entorhinal cortex, anterior cingulate cortex and the diagonal band of Broca, and low in the hippocampal formation and mediodorsal nucleus of the thalamus. The number of c-Fos-stained cells was significantly higher for the rats shown novel objects than for those shown familiar objects, in the perirhinal cortex, area TE, occipital cortex and anterior cingulate cortex. In the next study of Zhu et al. (1996), rats were also shown sets of novel and familiar objects; however, on each trial, two objects were shown simultaneously to a rat so that one eye saw a novel object while the other saw a familiar object. Thus novel and familiar objects were seen with the same conditions of alertness and eye movements. Familiar stimuli evoked c-Fos expression in significantly fewer neurons than novel stimuli in the perirhinal cortex and area TE of temporal cortex, and the ventral lateral geniculate nucleus of the thalamus, but not in the hippocampus or other areas sampled.

In one more study, Zhu et al. (1997) treated two groups of rats to individual novel visual objects. One group had been familiarized to the environmental context within which the objects were shown, the other experienced the situation for the first time. The c-Fos expression was determined in perirhinal cortex and the hippocampal formation. The ratio of counts in the hippocampal formation to that in perirhinal cortex was compared for the two groups and found to be significantly higher in the group experiencing the environment for the first time.

3.4.3. Social, maternal and sexual recognition, olfactory cues

Female mice form a memory for the pheromones of the male with which they mate. It has been proposed that the site of the synaptic changes underlying this memory is the accessory olfactory bulb, at the first level of the accessory olfactory system. Brennan et al. (1992) examined the c-Fos expression in the mitral and granule cells of the accessory olfactory bulb immediately after mating, during the period of memory formation finding transient increases in c-Fos expression in granule cells. The increase in the number of cells expressing c-Fos required the association of mating and pheromonal exposure, conditions also required for memory formation. Large increases in the number of mitral and granule cell nuclei expressing c-Fos were also observed following the infusion of the GABA antagonist bicuculline into the accessory olfactory bulb in the absence of mating. This procedure has previously been shown to result in the formation of a non-specific memory for male pheromones.

Fleming and Walsh (1994) mapped c-Fos expression in the functional pathways in the brain that mediate the onset and retention of maternal behavior. In the first two experiments, parturient rat dams were exposed to either pups or to other stimuli on Day 1 postpartum. Dams interacting with pups were either intact or sustained ventral somatosensory, olfactory, or combined desensitizations. Results showed that: (1) all intact pup-interacting dams showed elevated levels of c-Fos in the medial preoptic area (MPOA) and the medial and cortical amygdala as compared to control groups; and (2) olfactory and ventral somatosensory desensitization, either alone or in combination, did not decrease c-Fos expression in the MPOA. However, olfactory desensitizations did decrease c-Fos in the medial amygdala and the combined desensitizations significantly reduced c-Fos in both the basolateral and central amygdala. In the third study, dams were either exposed to pups or to other stimuli and were subsequently re-exposed to pups or to pup cues. Regardless of prior maternal experience, females who were able to interact with pups upon re-exposure showed increased c-Fos in the MPOA, the basolateral and central nuclei of the amygdala, and the nucleus accumbens when compared to females which did not interact with pups.

Fleming et al. (1994) examined the number of cells showing c-Fos immunoreactivity in the brains of hormonally primed parturient rat dams immediately following their first behavioral interactions with pups. Groups were exposed to newborn pups (pup), adult conspecifics (social), or a new food, or they were left alone in cages (control/isolate). Rats in the pup group had higher numbers of cells showing c-Fos within the MPOA nuclei than did the social, control/isolate, and, marginally, food groups and higher levels of c-Fos expression in a number of amygdaloid nucleusi (medial and cortical) and in cingulate and somatosensory cortices than did control/isolate or food groups. c-Fos levels in the amygdala did not differ between pup and social groups. There were also group differences in Fos-labeling in the olfactory bulbs, with the pup group showing the highest densities.

In sheep maternal behavior and the formation of the selective olfactory, ewe/lamb bond are induced by feedback to the brain from stimulation of the vagina and cervix during parturition. Da Costa et al. (1997) used in situ hybridization histochemistry to quantify changes in expression of c-*fos* related to the formation of the olfactory memory associated with bonding. Three different treatment groups were used. One group gave birth normally, became maternal and were allowed to interact with their lambs for 30 min. A second group received exogenous treatment with estradiol and progesterone to induce lactation and then received a 5-min period of artificial stimulation of the vagina and cervix (VCS), which reliably induces maternal behavior, but could not interact with lambs. A final control group received exogenous hormone treatment, but neither VCS nor interaction with lambs. Compared to the control group, post-partum animals and animals that had received VCS showed increased c-*fos* expression in a number of cortical regions (cingulate, entorhinal and somatosensory), the mediodorsal thalamic nucleus and the lateral habenula, the limbic system (bed nucleus of the stria terminalis, lateral septum, medial amygdala, dentate gyrus and the CA3 region of the hippocampus) and the hypothalamus (medial preoptic area, mediobasal hypothalamus, paraventricular nucleus, supraoptic nucleus and periventricular complex). The group that gave birth and had contact with their lambs for 30 min had significantly enhanced c-*fos* mRNA expression in the cingulate cortex compared to those receiving VCS and additionally showed significantly increased c-*fos* mRNA expression in olfactory processing regions (olfactory bulb, piriform cortex and orbitofrontal cortex).

3.4.4. Acquisition of copulatory reactions

Extensive analyses of behaviorally relevant c-*fos* mRNA and protein expression were carried out in a context of sexual behavior. Detailed review of these studies has been published elsewhere (Bialy and Kaczmarek, 1996). To summarize very briefly the conclusions, acquisition of copulatory reactions (see also Bialy et al., 1992, 2000) are most probably responsible for the reported increases in c-*fos* expression that were observed in such brain structures as medial preoptic area, bed nucleus of stria terminalis, medial amygdala, and central tegmental field as well parieto-occipital cortex.

3.5. VISUAL RECOGNITION AND PASSIVE AVOIDANCE IN THE CHICK

Anokhin et al. (1991) and Anokhin and Rose (1991) studied the expression of c-*fos* mRNA in the forebrains of young chicks subjected to different types of experience-dependent stimulation. Exposure to 1 h of activity in a rich visual environment produced a significant increase in forebrain c-*fos* mRNA in 1- or 2-day-old chicks; this increase was most marked in cerebellum and medial forebrain — especially intermediate medial hyperstriatum ventrale (IMHV) regions. Levels of c-*fos* mRNA were not elevated in birds which had been accommodated to the enriched environment for 2 days, suggesting that the novelty of presented stimuli play a critical role in the c-*fos* induction. Training chicks on a one-trial passive avoidance task (in which chicks learn to avoid pecking at a bitter-tasting bead) was followed by marked increases in c-*fos* mRNA in both left and right IMHV and lobus parolfactorius after the learning experience. Occlusion of one eye during training resulted in an asymmetrical expression of c-*fos* in the hemisphere contralateral to the open eye. Therefore induced c-*fos* expression in the chick forebrain during learning and its activation could not be attributed to 'stress', arousal or intensity of sensory stimulation alone. In a follow-up study, Freeman and Rose (1999) reported induction of c-Fos protein expression in 1-day-old chick forebrain after one-trial passive avoidance training.

3.6. IMPRINTING AND SONG LEARNING

McCabe and Horn (1994) investigated c-Fos expression in the chick brain after the imprinting experience. Chicks were either trained on an imprinting stimulus or dark-reared. Trained chicks were classified as good or poor learners by their preference score (a measure of the strength of imprinting). In the intermediate and medial part of the hyperstriatum ventrale (IMHV, a part of the chick forebrain critical for the learning process of imprinting that may be a site of information storage), significantly more c-Fos immunopositive nuclei were counted in good learners than in poor learners or dark-reared chicks. There was a positive correlation between counts of labeled nuclei and preference score that was not attributable to sensory activity per se, locomotor activity during training, or a predisposition to learn well; rather, the results indicated that the change in c-Fos immunoreactivity in the IMHV was related to learning. In the hyperstriatum accessorium, significantly fewer immunopositive nuclei were counted in good learners than in poor learners or in dark-reared chicks. In the dorsolateral hippocampal region, more immunopositive nuclei were counted in trained than in dark-reared chicks. No significant effects of training were found in the anterior hyperstriatum ventrale, lobus parolfactorius, neostriatum, medial hippocampal region, or ventrolateral hippocampal region.

c-Fos expression related to the bird-specific song learning was investigated repeatedly and it is reviewed elsewhere in this volume (Chapter IV).

3.7. SUMMARY OF LEARNING-RELATED c-*fos* BRAIN MAPPING

Results of the experiments described above clearly show that behavioral training activates c-*fos* expression in the brain. The long list of learning tasks that were found to involve this phenomenon (on either mRNA or protein level) includes recognition memory, motor learning, etc., as well as both classical and instrumental conditioning with either aversive or appetitive reinforcement with a number of modalities involved within the conditioning stimulus, such as visual, olfactory, gustatory, auditory, etc., and similarly wide range of unconditioned stimuli. Especially striking is a very high number of brain regions displaying elevated c-Fos expression in a context of behavioral experience. Notably, however, in many cases, induced c-Fos was observed in brain regions known to subserve the learning tasks under study. Furthermore, a number of studies with unilateral sensory occlusion or impairment of lateralized brain structures resulted in lateralized c-Fos response in downstream brain regions, clearly showing the c-Fos activation to depend on specific neuronal connections and it is not a widespread, non-specific response.

Interestingly, in the reviewed literature, an apparent controversy has emerged, whether the elevated c-*fos* expression correlates with either *learning* or *performing* the already acquired reaction. In particular, in a number of studies the authors claimed that c-*fos* expression could be detected following the testing session of the already learned behavior. However, very careful scrutiny of these reports reveals that in each case the testing session differed from the training ones (e.g., either in its duration or additional elements being included). Thus, it is clear that this so-called testing involved a learning component of novel aspects of the procedure! Notably, the answer to this issue has already been provided in an early report, showing increased c-*fos* mRNA accumulation after the single, but not multiple, training session of two-active avoidance reaction in rats (Nikolaev et al., 1992b).

Special considerations should also be given to the effects of post-training (in an absence of the US) exposure of the animals to the conditioned stimuli (e.g. sweet taste in the CTA paradigm or training environment in which footshocks were delivered during conditioned fear training). The c-*fos*/c-Fos expression was repeatedly found to be very pronounced after such experience. Most interestingly, in the CTA procedure, c-Fos labeling of the NTS could be revealed only after treatment with the conditioned taste, but not just after exposure to the same taste, not paired previously with the US. This clearly shows that learning modifies spatial pattern of the gene expression in the brain. However, as already pointed out by Tischmeyer and Grimm (1999), the gene activation in the re-exposure condition may be related to a number of phenomena, such as memory recall, expression of the learned behavior, relearning and extinction. The last possibility is especially important, as it is well established that the extinction process comprises a condition of a new learning experience, involving not the very same circuits that subserved original acquisition of the trace.

Further misconception surrounds the use of pseudoconditioning (random presentation of a number of CS and US, not allowing any association between them to be learned). This is a very good control procedure for acquisition of the CS–US association. However, it does not preclude *any learning* experience and indeed it has been shown that it results in learning of the *lack of association*. Existence of such learning can be proven by difficulties to train the previously pseudoconditioned animals an appropriate CS–US contingency (Rescorla, 1967). Thus, an elevated c-*fos* expression in the pseudotrained or yoked animals does not disprove its role in learning. On the other hand, the observation of a selective c-*fos* activation in a given brain region in animals exposed to appropriate CS–US contingency, but not in the pseudotrained ones may be indicative of specific involvement of this brain area in forming memory for the trained response.

Another critical issue is whether acquisition-related c-*fos* expression could be attributed to the learning process or to the stress accompanying this phenomenon. This is a very difficult issue to tackle. There is probably no learning without any component of stress, and by the same token, every stressful procedure results in learning (at least such as habituation). Indeed a number of studies have shown that c-*fos* expression accompanies the stress exposure (see, e.g., Campeau et al., 1991; Melia et al., 1994; Chen and Herbert, 1995; Cullinan et al., 1995; see also Kovacs, 1998). Interestingly, Melia et al. (1994) investigated the effects of acute and repeated restraint stress on c-*fos* expression in rat brain. The authors found that acute stress results in massive accumulation of c-*fos* mRNA in various brain regions. However, this response was completely habituated by the ninth restraint stress session. Furthermore, this habituation was stressor-specific as significant c-*fos* expression was observed in brains of the animals that were exposed to acute swim stress after they were treated to the repeated immobilization. Notably, adrenalectomy that completely abolished stress-related increases in plasma corticosterone levels did not affect c-*fos* activation following acute stress (Melia et al., 1994).

In conclusion, the c-Fos mapping studies provide paramount support for the existence of a brain-wide network to be involved in processing of information (or even underlying widespread plastic changes) in learning. However, at the same time, careful behavioral analyses into correlations between various animal reactions and c-Fos expression patterns reveal that specific aspects of information processing (plastic changes) are encoded by well-determined brain structures (such as emotions in the amygdala).

4. FUNCTIONAL ROLE FOR c-*fos* IN LEARNING AS INDICATED BY THE INTERVENTIVE APPROACHES

In aggregate, the aforementioned results and considerations raise the problem of the biological significance of c-Fos expression in a context of learning. One way to approach this issue is to continue the expression pattern studies, especially through extensive analyses of the complex animal behavior and thorough correlative analyses of the c-Fos expression in various brain regions (see Savonenko et al., 1999). Behavioral complexity of the task to be learned is especially important as simple procedures (such as, e.g., fear conditioning) do not allow for observation and score for multiple behavioral variables that may separately be linked to specific behavioral traits.

Another important approach is through functional alterations in c-Fos expression and/or activity. However, brain has proven to be a notoriously difficult organ to apply strategies aiming at altering gene expression in vivo. Use of c-*fos* knockout mice illustrates problems with this, especially recently heralded, approach. Paylor et al. (1994) carried out behavioral characterization of mice containing a mutant c-*fos* allele created via homologous recombination-based gene targeting. The majority of the tested c-*fos*-deficient animals were impaired in learning of the spatial version of the Morris water task. However, this poor performance in the spatial version of the task was highly correlated to their performance in the non-spatial version of the task which suggested that the animals displayed a behavioral impairment that interrupted their ability to perform adequately on both versions of the task with the same proficiency as wild-type and heterozygous litter mates. On the other hand, c-*fos* mutants were not impaired in a simple left/right discrimination in a T-maze. Furthermore, the mutants were also characterized by sensory, e.g., auditory deficiencies, what made interpretations of the observed behavioral deficits very difficult.

Recent developments with brain area-specific and inducible transgenes or gene knockouts may offer technological breakthroughs allowing the development-based deficiencies and their compensations to be overcome. However, the number of experiments employing these emerging technologies is still very limited. Moreover, they have not yet been applied successfully to investigate the role of c-*fos* expression in learning and memory formation. On the other hand, since the first anti-sense application for the brain (Chiasson et al., 1992), there was a number of studies involving c-*fos* antisense oligonucleotides towards addressing the c-*fos* function. However, one has to be aware of all of the limitations in interpreting the results of the experiments with antisense oligonucleotides. The works on cell cultures clearly show that there is a multitude of potential caveats and thus there is a necessity to use a plentitude of controls (Stein and Krieg, 1994; Stein, 1996; Neumann, 1997). Suffice to say that reliable experiment should employ several different oligonucleotides directed against various target mRNA regions, each of those oligonucleotides should be matched with missense, and nonsense counterparts, etc. In aggregate, this means that the number of animals tested for the function of each gene makes the learning studies almost impossible to perform and indeed apparently no such complete investigation has been described to date for c-*fos* (see Szklarczyk and Kaczmarek, 1999 for review). Nevertheless, the experiments described below, despite all the aforementioned reservations, lend support to a compelling suggestion that c-*fos* expression may indeed play a pivotal role in memory consolidation.Lamprecht and Dudai (1996) applied local microinjection into rat amygdala of phosphorothioate modified oligodeoxynucleotides (ODNs) antisense to c-*fos* mRNA several hours before CTA training and found that this treatment impaired taste aversion memory tested 3–5 days after conditioning. In contrast, injection of the antisense ODNs several days before training, or before testing, or into the basal ganglia, or injection of c-*fos* sense ODNs, had no effect on CTA memory. Notably, inhibition of translation by local microinjection of anisomycin into the amygdala shortly before as well as during CTA training, but not several days before training or shortly before testing, also impaired CTA memory.

Using a similar training paradigm in mice, Swank et al. (1996) found that blockade of US-induced c-Fos translation in another brain structure, the brainstem, by antisense oligonucleotides, specifically blocked both acquisition and extinction of a learned taste aversion, but did not impair sensory processing of either CS or US, suggesting that c-Fos antisense blocked associative events necessary to support taste aversion learning within this brain area.

Grimm et al. (1997) used an antisense phosphorothioate oligodeoxynucleotide to suppress, in vivo, the expression of c-Fos in rat brain evoked by training of a footshock-motivated brightness discrimination in a Y-maze. Intrahippocampal application of the oligodeoxynucleotide 10 and 2 h before starting a brightness discrimination training drastically reduced the induction of c-Fos immunoreactivity normally observed in limbic and cortical areas after the training session. Acquisition of the discrimination reaction was not affected by this treatment. In a re-learning test 24 h after the first training, retention of the discrimination reaction was specifically impaired compared with rats pretreated with control oligodeoxynucleotide or saline.

Morrow et al. (1999) examined inducible c-Fos immunoreactivity in subregions of the prefrontal cortex during a conditioned fear paradigm. During the acquisition phase, the rats were conditioned to fear a formerly neutral tone by pairing the tone with a footshock. The rats were then tested for fearful behavior by re-exposure to the tone without additional footshock. An antisense oligonucleotide directed against the c-*fos* mRNA and injected into the infralimbic/prelimbic cortex given 12 h, but not 72 h before the training suppressed c-Fos production without altering behavior. Three days after the acquisition session, rats were tested for fearful behavior as before and the authors found that the antisense oligonucleotide blockade of c-Fos production during acquisition was associated with a significantly less fearful response during the test session.

In addition to results obtained with rats and mice, c-*fos* oligonucleotides were also investigated in the chicken brain. Mileusnic et al. (1996) reported that suppression of c-Fos protein synthesis with antisense oligodeoxynucleotides prevented long-term retention of the passive avoidance memory when injected 10–11 h before training, but not if injections were made between 3 h pre- and 3 h post-training.

5. CONCLUDING REMARKS: CAN c-FOS PLAY A ROLE IN THE INTEGRATION OF INFORMATION IN LEARNING?

The aforementioned results and considerations strongly suggest that learning-related c-*fos*/c-Fos expression may play a functional role in the learning process. However, even accepting such a conclusion still allows for a number of interpretations as to the exact role of this protein in memory formation. Recently, several possibilities in this regard have been raised (Kaczmarek, 2000). One is replenishment. The concept of replenishment implies that learning- (or rather neuronal activity-, see below) evoked gene expression is involved in the metabolic recovery triggered by depletion of key cellular components during learning and thus serves to reinstate the same situation as before the training. Most learning conditions involve abrupt bursts of spiking activity — following relative quiescence periods — related to information processing. This may result in: (1) rapid depletion of neuronal components (e.g., synaptic release machinery, metabolic enzymes, etc.); and (2) the need to replace these exhausted elements (see Rainbow, 1979). Obviously, one would expect that a replenishment-related wave of gene expression should follow stimulation conditions, such as behavioral training. c-Fos expression pattern appears to be compatible with the concept of replenishment. Notably, replenishment hypothesis provides a good explanation for neuronal activity-driven induction of transcription factors. However, it does not explain the complete lack of gene activation in well-trained animals as discussed above.

Learning is thought to be based on reorganization of specific synaptic connections that cause a lasting change in neural functioning. Hence, another possible way to explain learning-evoked c-Fos expression is that it control genes that serve to produce proteins, whose function is to maintain the plastically reorganized neuronal connections. Most obviously, these proteins should be targeted to specific synapses to support their newly gained functions (these may include formation/loss as well as their strengthening/weakening of synapses).

The hypothesis of maintenance of plastic changes, as introduced above, can be extended further to suggest that learning-evoked gene expression may be central to the learning process. Gene regulatory regions always contain sites for binding multiple transcription factors and interactions between these factors are believed to be mandatory to drive gene expression (see Robertson et al., 1995). Thus, the regulatory regions of genes encoding proteins directly subserving synaptic reorganization ('effector' proteins in learning) may act as coincidence detectors, allowing a convergence of information provided by various transcription factors, activated by different signaling pathways of behavioral relevance, such as sensory information, arousal, motivation, reward, etc. (Kaczmarek, 1993a,b, 1995). There is a vast literature suggesting that there are separate neurotransmitter/receptor systems for conveying these inputs (see, e.g., Gold and Zornetzer, 1983; McGaugh, 1989; Decker and McGaugh, 1991). Next, they should activate specific second-messenger/kinase systems that deliver the signal to the nucleus to specific transcription factors. c-Fos may be one of transcription regulators activated in such a way. Interestingly, AP-1 already appears to act as a detector of inputs carried out by various kinase systems (see Kaminska et al., 1999). The molecular meaning of this transcriptional coincidence detection may be to trigger and elaborate long-term plastic changes.

In conclusion, it is still premature to resolve the functional role of learning-related c-Fos expression in the context of the aforementioned possibilities. Furthermore, approach to this issue appears to call for novel experimental strategies, allowing these options to be distinguished from one another. The answer to this problem may be pivotal in revealing the nature of the learning process, especially by shedding light on whether integration of information essential to learning may occur at the level of neuronal nuclei.

6. ABBREVIATIONS

AP-1	activatory protein 1, transcription factor
AMPA receptors	α-amino-3-hydroxy-5-methylisoxasole-4-propionic acid, class of glutamate receptors
cAMP	cyclic adenosine monophosphate, a second messenger
CRE	Ca^{2+}/cAMP responsive element, gene regulatory sequence
CREB	CRE binding proteins, transcription factors
CREM/ICER	cAMP responsive element modulator/inducible cAMP early repressor, transcription factor
CS	conditioned stimulus in behavioral training
CTA	conditioned taste aversion
2-DG	2-deoxyglucose
Elk-1	transcription factor
ERK	extracellularly regulated kinases, protein kinases of MAP kinase family
ETS	transcription factor
IMHV	intermediate medial hyperstriatum ventrale, avian brain structure
ITI	inter-trial intervals in behavioral training
ITR	inter-trial responses in behavioral training
LC	locus coeruleus

MAP kinases	mitogen-activated kinases, a class of protein kinases involved in signaling
MPOA	medial preoptic area
nbM	nucleus basalis of Meynert, brain structure
NFAT	nuclear factor of activated T-cells, transcription factor
NMDA receptors	*N*-methyl-D-aspartate receptors, class of glutamate receptors
NMR	nictitating membrane reflex
NTS	nuclei of the solitary tract (nucleus tractus solitarius)
ODN	oligodeoxynucleotide
PAG	periaqueductal gray
PBN	parabrachial nucleus
PND	postnatal days
PVN	paraventricular nucleus of the hypothalamus
SCN	suprachiasmatic nucleus, brain structure
SRE	serum responsive element, gene regulatory sequence
SRF	SRE-binding factors, transcription factors
TCF	ternary complex factors, transcription factors
TIMP-1	tissue inhibitor of metalloproteinases-1, a protein
US	unconditioned stimulus in behavioral training

7. ACKNOWLEDGEMENTS

Critical reading of the manuscript by Drs. R.K. Filipkowski, A. Savonenko, K. Zielinski is gratefully appreciated. The learning and memory studies in the author's laboratory have been supported by grants from KBN (State Committee for Scientific Research, Poland) and FNP (Foundation for Polish Science, Poland).

8. REFERENCES

Allingham K, Brennan PA, Distel H, Hudson R (1999): Expression of c-fos in the main olfactory bulb of neonatal rabbits in response to garlic as a novel and conditioned odour. *Behav Brain Res 104*:157–167.

Amir S, Stewart J (1996): Resetting of the circadian clock by a conditioned stimulus. *Nature 379*:542–545.

Amir S, Stewart J (1998a): Conditioned fear suppresses light-induced resetting of the circadian clock. *Neuroscience 86*:345–351.

Amir S, Stewart J (1998b): Conditioning in the circadian system. *Chronobiol Int 15*:447–456.

Amir S, Stewart J (1998c): Induction of Fos expression in the circadian system by unsignaled light is attenuated as a result of previous experience with signaled light: a role for Pavlovian conditioning. *Neuroscience 83*:657–661.

Amir S, Stewart J (1999): Conditioned and unconditioned aversive stimuli enhance light-induced fos expression in the primary visual cortex. *Neuroscience 89*:323–327.

Angel P, Karin M (1991): The role of Jun, Fos and the AP-1 complex in cell-proliferation and transformation. *Biochim Biophys Acta 1072*:129–157.

Anokhin KV (1997): Towards synthesis of systems and molecular genetics approaches to memory consolidation. *J Higher Nerv Act 47*:157–169.

Anokhin KV, Rose SPR (1991): Learning-induced increase of immediate early gene messenger RNA in the chick forebrain. *Eur J Neurosci 3*:162–167.

Anokhin KV, Mileusnic R, Shamakina IY, Rose SP (1991): Effects of early experience on c-fos gene expression in the chick forebrain. *Brain Res 544*:101–107.

Anokhin KV, Riabinin AE, Sudakov KV (2000): The expression of the c-fos gene in the brain of mice in the dynamic acquisition of defensive behavioral habits. *Zh Vyssh Nerv Deiat Pavlova 50*:88–94.

Beck CH, Fibiger HC (1995): Conditioned fear-induced changes in behavior and in the expression of the immediate early gene c-fos: with and without diazepam pretreatment. *J Neurosci 15*:709–720.

Berridge M (1986): Neurobiology: second messengers dualism in neuromodulation and memory. *Nature 323*:294–295.

Bertaina V, Destrade C (1995): Differential time courses of c-fos mRNA expression in hippocampal subfields following acquisition and recall testing in mice. *Cogn Brain Res 2*:269–275.

Bertaina-Anglade V, Tramu G, Destrade C (2000): Differential learning-stage dependent patterns of c-Fos protein expression in brain regions during the acquisition and memory consolidation of an operant task in mice. *Eur J Neurosci 12*:3803–3812.

Bialy M, Kaczmarek L (1996): *c-fos* expression as a tool to search for the neurobiological basis of the sexual behavior of males. *Acta Neurobiol Exp 56*:567–577.

Bialy M, Nikolaev E, Beck J, Kaczmarek L (1992): Delayed c-fos expression in sensory cortex following sexual learning in male rats. *Brain Res Mol Brain Res 14*:352–356.

Bialy M, Rydz M, Kaczmarek L (2000): Precontact 50-kHz vocalizations in male rats during acquisition of sexual experience. *Behav Neurosci 14*:983–990.

Brennan PA, Hancock D, Keverne EB (1992): The expression of the immediate-early genes c-fos, egr-1 and c-jun in the accessory olfactory bulb during the formation of an olfactory memory in mice. *Neuroscience 49*:277–284.

Bubser M, Deutch AY (1999): Stress induces Fos expression in neurons of the thalamic paraventricular nucleus that innvervate limbic forebrain sites. *Synapse 32*:13–22.

Cammarota M, Bevilaqua LR, Ardenghi P, Paratcha G, Levi de Stein M, Izquierdo I, Medina JH (2000): Learning-associated activation of nuclear MAPK, CREB and Elk-1, along with Fos production, in the rat hippocampus after a one-trial avoidance learning: abolition by NMDA receptor blockade. *Brain Res Mol Brain Res 76*:36–46.

Campeau S, Hayward MD, Hope BT, Rosen JB, Nestler EJ, Davis M (1991): Induction of the c-fos proto-oncogene in rat amygdala during unconditioned and conditioned fear. *Brain Res 565*:349–352.

Carretta D, Herve-Minvielle A, Bajo VM, Villa AE, Rouiller EM (1999a): c-Fos expression in the auditory pathways related to the significance of acoustic signals in rats performing a sensory-motor task. *Brain Res 841*:170–183.

Carretta D, Herve-Minvielle A, Bajo VM, Villa AE, Rouiller EM (1999b): Preferential induction of fos-like immunoreactivity in granule cells of the cochlear nucleus by acoustic stimulation in behaving rats. *Neurosci Lett 259*:123–126.

Carrive P, Kehoe EJ, Macrae M, Paxinos G (1997a): Fos-like immunoreactivity in locus coeruleus after classical conditioning of the rabbit's nictitating membrane response. *Neurosci Lett 223*:33–36.

Carrive P, Leung P, Harris J, Paxinos G (1997b): Conditioned fear to context is associated with increased Fos expression in the caudal ventrolateral region of the midbrain periaqueductal gray. *Neuroscience 78*:165–177.

Castro-Alamancos MA, Borrell J, Garcia-Segura LM (1992): Performance in an escape task induces fos-like immunoreactivity in a specific area of the motor cortex of the rat. *Neuroscience 49*:157–162.

Chen X, Herbert J (1995): Regional changes in c-*fos* expression in the basal forebrain and brainstem during adaptation to repeated stress: correlations with cardiovascular, hypothermic and endocrine responses. *Neuroscience 64*:675–685.

Chiasson BJ, Hooper ML, Murphy PR, Robertson HA (1992): Antisense oligonucleotide eliminates in vivo expression of c-fos in mammalian brain. *Eur J Pharmacol 227*:451–453.

Cubero I, Thiele TE, Bernstein IL (1999): Insular cortex lesions and taste aversion learning: effects of conditioning method and timing of lesion. *Brain Res 839*:323–330.

Cullinan WE, Herman JP, Battaglia DF, Akil H, Watson SJ (1995): Pattern and time of immediate early gene expression in rat brain following acute stress. *Neuroscience 64*:477–505.

Curran T, Morgan JI (1987): . *Mem fos BioEssays 7*:255–258.

Curran T, Miller AD, Zokas L, Verma IM (1984): Viral and cellular fos proteins: a comparative analysis. *Cell 36*:259–268.

Curran T, MacConnell WP, van Straaten F, Verma IM (1983): Structure of the FBJ murine osteosarcoma virus genome: molecular cloning of its associated helper virus and the cellular homolog of the v-fos gene from mouse and human cells. *Mol Cell Biol 3*:914–921.

Da Costa AP, Broad KD, Kendrick KM (1997): Olfactory memory and maternal behavior-induced changes in c-fos and zif/268 mRNA expression in the sheep brain. *Brain Res Mol Brain Res 46*:63–76.

Decker MW, McGaugh JL (1991): The role of interactions between the cholinergic system and other neuromodulatory systems in learning and memory. *Synapse 7*:151–168.

Diamond MI, Miner JN, Yoshinaga SK, Yamamoto KR (1990): Transcription factor interactions: selectors of positive or negative regulation from a single DNA element. *Science 249*:1266–1272.

Dragunow M (1996): A role for immediate-early transcription factors in learning and memory. *Behav Genet 26*:293–299.

Duncan GE, Knapp DJ, Breese GR (1996): Neuroanatomical characterization of Fos induction in rat behavioral models of anxiety. *Brain Res 713*:79–91.

Fleming AS, Walsh C (1994): Neuropsychology of maternal behavior in the rat: c-fos expression during mother–litter interactions. *Psychoneuroendocrinology 19*:429–443.

Fleming AS, Suh EJ, Korsmit M, Rusak B (1994): Activation of Fos-like immunoreactivity in the medial preoptic area and limbic structures by maternal and social interactions in rats. *Behav Neurosci 108*:724–734.

Foulkes NS, Borelli E, Sassone-Corsi P (1991): CREM gene: use of alternative DNA-binding domains generates multiple antagonists of cAMP-induced transcription. *Cell 64*:739–749.

Freeman FM, Rose SP (1999): Expression of Fos and Jun proteins following passive avoidance training in the day-old chick. *Learn Mem 6*:389–397.

Funk D, Amir S (2000): Enhanced fos expression within the primary olfactory and limbic pathways induced by an aversive conditioned odor stimulus. *Neuroscience 98*:403–406.

Gall CM, Hess US, Lynch G (1998): Mapping brain networks engaged by, and changed by, learning. *Neurobiol Learn Mem 70*:14–36.

Goelet P, Castelluci VF, Schacher S, Kandel ER (1986): The long and the short of long term memory — a molecular framework. *Nature 322*:419–423.

Gold PE, Zornetzer SF (1983): The mnemon and its juices: neuromodulation of memory processes. *Behav Neural Biol 38*:151–189.

Greenberg ME, Shyu AB, Belasco JG (1990): Deadenylation: a mechanism controlling c-fos mRNA decay. *Enzyme 44*:181–192.

Grimm R, Tischmeyer W (1997): Complex patterns of immediate early gene induction in rat brain following brightness discrimination training and pseudotraining. *Behav Brain Res 84*:109–116.

Grimm R, Schicknick H, Riede I, Gundelfinger ED, Herdegen T, Zuschratter W, Tischmeyer W (1997): Suppression of c-fos induction in rat brain impairs retention of a brightness discrimination reaction. *Learn Mem 3*:402–413.

Gu Y, Gonzalez MF, Chin DY, Deutsch JA (1993): Expression of c-fos in brain subcortical structures in response to nauseant lithium chloride and osmotic pressure in rats. *Neurosci Lett 157*:49–52.

Hall J, Thomas KL, Everitt BJ (2001): Fear memory retrieval induces CREB phosphorylation and Fos expression within the amygdala. *Eur J Neurosci 13*:1453–1458.

Herdegen T, Leah JD (1998): Inducible and constitutive transcription factors in the mammalian nervous system: control of gene expression by Jun, Fos and Krox, and CREB/ATF proteins. *Brain Res Rev 28*:370–490.

Hess US, Gall CM, Granger R, Lynch G (1997): Differential patterns of c-fos mRNA expression in amygdala during successive stages of odor discrimination learning. *Learn Mem 4*:262–283.

Hess US, Lynch G, Gall CM (1995a): Changes in c-fos mRNA expression in rat brain during odor discrimination learning: differential involvement of hippocampal subfields CA1 and CA3. *J Neurosci 15*:4786–4795.

Hess US, Lynch G, Gall CM (1995b): Regional patterns of c-fos mRNA expression in rat hippocampus following exploration of a novel environment versus performance of a well-learned discrimination. *J Neurosci 15*:7796–7809.

Heurteaux C, Messier C, Destrade C, Lazdunski M (1993): Memory processing and apamin induce immediate early gene expression in mouse brain. *Brain Res Mol Brain Res 18*:17–22.

Houpt TA, Philopena JM, Wessel TC, Joh TH, Smith GP (1994): Increased c-fos expression in nucleus of the solitary tract correlated with conditioned taste aversion to sucrose in rats. *Neurosci Lett 172*:1–5.

Houpt TA, Philopena JM, Joh TH, Smith GP (1996a): c-Fos induction in the rat nucleus of the solitary tract by intraoral quinine infusion depends on prior contingent pairing of quinine and lithium chloride. *Physiol Behav 60*:1535–1541.

Houpt TA, Philopena JM, Joh TH, Smith GP (1996b): c-Fos induction in the rat nucleus of the solitary tract correlates with the retention and forgetting of a conditioned taste aversion. *Learn Mem 3*:25–30.

Houpt TA, Berlin R, Smith GP (1997): Subdiaphragmatic vagotomy does not attenuate c-Fos induction in the nucleus of the solitary tract after conditioned taste aversion expression. *Brain Res 747*:85–91.

Hughes P, Dragunow M (1995): Induction of immediate-early genes and the control of neurotransmitter-regulated gene expression within the nervous system. *Pharmacol Rev 47*:133–178.

Irwin KB, Craig AD, Bracha V, Bloedel JR (1992): Distribution of c-fos expression in brainstem neurons associated with conditioning and pseudo-conditioning of the rabbit nictitating membrane reflex. *Neurosci Lett 148*:71–75.

Jain J, McCaffrey PG, Valge-Archer VE, Rao A (1992): Nuclear factor of activated T cells contains Fos and Jun. *Nature 356*:801–804.

Jaworski J, Biedermann IW, Lapinska J, Szklarczyk A, Figiel I, Konopka D, Nowicka D, Filipkowski RK, Hetman M, Kowalczyk A, Kaczmarek L (1999): Neuronal excitation-driven and AP-1-dependent activation of tissue inhibitor of metalloproteinases-1 gene expression in rodent hippocampus. *J Biol Chem 274*:28106–28112.

Johnson BA, Woo CC, Duong H, Nguyen V, Leon M (1995): A learned odor evokes an enhanced Fos-like glomerular response in the olfactory bulb of young rats. *Brain Res 699*:192–200.

Kaczmarek L (1986): Protooncogene expression during the cell cycle. *Lab Invest 54*:365–376.

Kaczmarek L (1990) Molecular biology of long lasting memory formation. *ESF Scientific Networks: Zeist 1-3 December, 1989.*

Kaczmarek L (1993a): Glutamate receptor-driven gene expression in learning. *Acta Neurobiol Exp 53*:187–196.

Kaczmarek L (1993b): Molecular biology of vertebrate learning: is c-fos a new beginning?. *J Neurosci Res 34*:377–381.

Kaczmarek L (1995): Towards understanding of the role of transcription factors in learning processes. *Acta Biochim Pol 42*:221–226.

Kaczmarek L (2000): Gene expression in learning processes. *Acta Neurobiol Exp 60*:419–424.

Kaczmarek L, Chaudhuri A (1997): Sensory regulation of immediate-early gene expression in mammalian visual cortex: implications for functional mapping and neural plasticity. *Brain Res Rev 23*:237–256.

Kaczmarek L, Kaminska B (1989): Molecular biology of cell activation. *Exp Cell Res 183*:24–35.

Kaczmarek L, Nikolajew E (1990): c-*fos* protooncogene expression and neuronal plasticity. *Acta Neurobiol Exp 50*:173–179.

Kaminska B, Kaczmarek L, Zangenehpour S, Chaudhuri A (1999): Rapid phosphorylation of Elk-1 transcription factor and activation of MAP kinase signal transduction pathways in response to visual stimulation. *Mol Cell Neurosci 13*:405–414.

Kaminska B, Pyrzynska B, Ciechomska I, Wisniewska M. (2000): Modulation of the composition of AP-1 complex and its impact on transcriptional activity. *Acta Neurobiol Exp 60*:395–402.

Kerppola TK, Luk D, Curran T (1993): Fos is a preferential target of glucocorticoid receptor inhibition of AP-1 activity in vitro. *Mol Cell Biol 13*:3782–3791.

Kleim JA, Lussnig E, Schwarz ER, Comery TA, Greenough WT (1996): Synaptogenesis and Fos expression in the motor cortex of the adult rat after motor skill learning. *J Neurosci 16*:4529–4535.

Konopka D, Szklarczyk AW, Filipkowski RK, Trauzold A, Nowicka D, Hetman M, Kaczmarek L (1998): Plasticity- and neurodegeneration-linked cyclic-AMP responsive element modulator/inducible cyclic-AMP early repressor messenger RNA expression in the rat brain. *Neuroscience 86*:499–510.

Kovacs KJ (1998): c-Fos as a transcription factor: a stressful (re)view from a functional map. *Neurochem Int 33*:287–297.

Lamprecht R, Dudai Y (1995): Differential modulation of brain immediate early genes by intraperitoneal LiCl. *NeuroReport 7*:289–293.

Lamprecht R, Dudai Y (1996): Transient expression of c-Fos in rat amygdala during training is required for encoding conditioned taste aversion memory. *Learn Mem 3*:31–41.

Li H-Y, Ericsson A, Sawchenko PE (1996): Distinct mechanisms underlie activation of hypothalamic neurosecretory neurons and their medullary catecholaminergic afferents in categorically different stress paradigms. *Proc Natl Acad Sci USA 93*:2359–2364.

Logan SK, Garabedian MJ, Campbell CE, Werb Z (1996): Synergistic transcriptional activation of the tissue inhibitor of metalloproteinases-1 promoter via functional interaction of AP-1 and Ets-1 transcription factors. *J Biol Chem 271*:774–782.

Lukasiuk K, Savonenko A, Nikolaev E, Rydz M, Kaczmarek L (1999): Defensive conditioning-related increase in AP-1 transcription factor in the rat cortex. *Mol Brain Res 67*:64–73.

Maleeva NE, Ivolgina GL, Anokhin KV, Limborskaia SA (1989): Analysis of the expression of the c-fos proto-oncogene in the rat cerebral cortex during learning. *Genetika 25*:1119–1121.

Maleeva NE, Bikbulatova LS, Ivolgina GL, Anokhin KV, Limborskaia SA, Kruglikov RI (1990): Activation of the c-fos proto-oncogene in different structures of the rat brain during training and pseudoconditioning. *Doklady Akad Nauk SSSR 314*:762–764.

McCabe BJ, Horn G (1994): Learning-related changes in Fos-like immunoreactivity in the chick forebrain after imprinting. *Proc Natl Acad Sci USA 91*:11417–11421.

McGaugh JL (1989): Involvement of hormonal and neuromodulatory systems in the regulation of memory storage. *Annu Rev Neurosci 12*:255–287.

Melia KR, Ryabinin AE, Schroeder R, Bloom FE, Wilson MC (1994): Induction and habituation of immediate early gene expression in rat brain by acute and repeated restraint stress. *J Neurosci 14*:5929–5938.

Milanovic S, Radulovic J, Laban O, Stiedl O, Henn F, Spiess J (1998): Production of the Fos protein after contextual fear conditioning of C57BL/6N mice. *Brain Res 784*:37–47.

Mileusnic R, Anokhin K, Rose SP (1996): Antisense oligodeoxynucleotides to c-fos are amnestic for passive avoidance in the chick. *NeuroReport 7*:1269–1272.

Miller AD, Curran T, Verma IM (1984): c-fos protein can induce cellular transformation: a novel mechanism of activation of a cellular oncogene. *Cell 36*:51–60.

Molina CA, Foulkes NS, Lalli E, Sassone-Corsi P (1993): Inducible and negative autoregulation of CREM: an alternative promoter directs the expression of ICER, an early response repressor. *Cell 75*:875–886.

Montag-Sallaz M, Welzl H, Kuhl D, Montag D, Schachner M (1999): Novelty-induced increased expression of immediate-early genes c-fos and arg 3.1 in the mouse brain. *J Neurobiol 38*:234–246.

Morrow BA, Elsworth JD, Inglis FM, Roth RH (1999): An antisense oligonucleotide reverses the footshock-induced expression of fos in the rat medial prefrontal cortex and the subsequent expression of conditioned fear-induced immobility. *J Neurosci 19*:5666–5673.

Neumann I (1997): Antisense oligonucleotides in neuroendocrinology: enthusiasm and frustration. *Neurochem Int 31*:363–378.

Nikolaev E, Kaminska B, Tischmeyer W, Matthies H, Kaczmarek L (1992a): Induction of expression of genes encoding transcription factors in the rat brain elicited by behavioral training. *Brain Res Bull 28*:479–484.

Nikolaev E, Werka T, Kaczmarek L (1992b): C-fos protooncogene expression in rat brain after long-term training of two-way active avoidance reaction. *Behav Brain Res 48*:91–94.

Papanikolaou NA, Sabban EI (2000): Ability of Egr-1 to activate tyrosine hydroxylase transcription in PC12 cells: cross-talk with AP-1 factors. *J Biol Chem 275*:26683–26689.

Paylor R, Johnson RS, Papaioannou V, Spiegelman BM, Wehner JM (1994): Behavioral assessment of c-fos mutant mice. *Brain Res 651*:275–282.

Pezzone MA, Lee WS, Hoffman GE, Rabin BS (1992): Induction of c-Fos immunoreactivity in the rat forebrain by conditioned and unconditioned aversive stimuli. *Brain Res 597*:41–50.

Pezzone MA, Lee WS, Hoffman GE, Pezzone KM, Rabin BS (1993): Activation of brainstem catecholaminergic neurons by conditioned and unconditioned aversive stimuli as revealed by c-Fos immunoreactivity. *Brain Res 608*:310–318.

Platenik J, Kuramoto N, Yoneda Y (2000): Molecular mechanisms associated with long-term consolidation of the NMDA signals. *Life Sci 67*:335–364.

Radulovic J, Kammermeier J, Spiess J (1998): Relationship between fos production and classical fear conditioning: effects of novelty, latent inhibition, and unconditioned stimulus preexposure. *J Neurosci 18*:7452–7461.

Rainbow TC (1979): Role of RNA and protein synthesis in memory formation. *Neurochem Res 4*:297–312.

Rescorla RA (1967): Pavlovian conditioning and its proper control procedures. *Psychol Rev 74*:71–80.

Robertson LM, Kerppola TK, Vendrell M, Luk D, Smeyne RJ, Bocchiaro C, Morgan JI, Curran T (1995): Regulation of c-fos expression in transgenic mice requires multiple interdependent transcription control elements. *Neuron 14*:241–252.

Rosen JB, Fanselow MS, Young SL, Sitcoske M, Maren S (1998): Immediate-early gene expression in the amygdala following footshock stress and contextual fear conditioning. *Brain Res 796*:132–142.

Sakai N, Yamamoto T (1997): Conditioned taste aversion and c-fos expression in the rat brainstem after administration of various USs. *NeuroReport 8*:2215–2220.

Sassone-Corsi P, Sisson JC, Verma IM (1988): Transcriptional autoregulation of the proto-oncogene fos. *Nature 334*:314–319.

Savonenko A, Filipkowski RK, Werka T, Zielinski K, Kaczmarek L (1999): Defensive conditioning-related functional heterogeneity among nuclei of the rat amygdala revealed by c-Fos mapping. *Neuroscience 94*:723–733.

Schafe GE, Bernstein IL (1996): Forebrain contribution to the induction of a brainstem correlate of conditioned taste aversion. I. The amygdala. *Brain Res 741*:109–116.

Schafe GE, Bernstein IL (1998): Forebrain contribution to the induction of a brainstem correlate of conditioned taste aversion. II. Insular (gustatory) cortex. *Brain Res 800*:40–47.

Schafe GE, Seeley RJ, Bernstein IL (1995): Forebrain contribution to the induction of a cellular correlate of conditioned taste aversion in the nucleus of the solitary tract. *J Neurosci 15*:6789–6796.

Schafe GE, Fitts DA, Thiele TE, LeDoux JE, Bernstein IL (2000): The induction of c-Fos in the NTS after taste aversion learning is not correlated with measures of conditioned fear. *Behav Neurosci 114*:99–106.

Schule R, Rangarajan P, Kliewer S, Ransone LJ, Bolado J, Yang N, Verma IM, Evans RM (1990): Functional antagonism between oncoprotein c-Jun and the glucocorticoid receptor. *Cell 62*:1217–1226.

Smith MA, Banerjee S, Gold PW, Glowa J (1992): Induction of c-fos mRNA in rat brain by conditioned and unconditioned stressors. *Brain Res 578*:135–141.

Sotty F, Sandner G, Gosselin O (1996): Latent inhibition in conditioned emotional response: c-fos immunolabelling evidence for brain areas involved in the rat. *Brain Res 737*:243–254.

Spray KJ, Halsell CB, Bernstein IL (2000): c-Fos induction in response to saccharin after taste aversion learning depends on conditioning method. *Brain Res 852*:225–227.

Stein CA (1996): Phosphorothioate antisense oligodeoxynucleotides: questions of specificity. *Trends Biotechnol 14*:147–149.

Stein CA, Krieg AM (1994): Problems in interpretation of data derived from in vitro and in vivo use of antisense oligodeoxynucleotides. *Antisense Res Drug Dev 4*:67–69.

Swank MW (2000): Conditioned c-Fos in mouse NTS during expression of a learned taste aversion depends on contextual cues. *Brain Res 862*:138–144.

Swank MW, Bernstein IL (1994): c-Fos induction in response to a conditioned stimulus after single trial taste aversion learning. *Brain Res 636*:202–208.

Swank MW, Schafe GE, Bernstein IL (1995): c-Fos induction in response to taste stimuli previously paired with amphetamine or LiCl during taste aversion learning. *Brain Res 673*:251–261.

Swank MW, Ellis AE, Cochran BN (1996): c-Fos antisense blocks acquisition and extinction of conditioned taste aversion in mice. *NeuroReport 7*:1866–1870.

Szabowski A, Maas-Szabowski N, Andrecht S, Kolbus A, Schorpp-Kistner M, Fusenig NE, Angel P (2000): c-Jun and JunB antagonistically control cytokine-regulated mesenchymal–epidermal interaction in skin. *Cell 103*:745–755.

Szklarczyk AW, Kaczmarek L (1999): Brain as a unique antisense environment. *Antisense Nucleic Acids Drug Dev 9*:105–116.

Thiele TE, Roitman MF, Bernstein IL (1996): c-Fos induction in rat brainstem in response to ethanol- and lithium chloride-induced conditioned taste aversions. *Alcohol Clin Exp Res 20*:1023–1028.

Tischmeyer W, Grimm R (1999): Activation of immediate early genes and memory formation. *Cell Mol Life Sci 55*:564–574.

Tischmeyer W, Kaczmarek L, Strauss M, Jork R, Matthies H (1990): Accumulation of c-fos mRNA in rat hippocampus during acquisition of a brightness discrimination. *Behav Neural Biol 54*:165–171.

Van Beveren C, van Straaten F, Curran T, Muller R, Verma IM (1983): Analysis of FBJ-MuSV provirus and c-fos (mouse) gene reveals that viral and cellular fos gene products have different carboxy termini. *Cell 32*:1241–1255.

Vann SD, Brown MW, Erichsen JT, Aggleton JP (2000): Fos imaging reveals differential patterns of hippocampal and parahippocampal subfield activation in rats in response to different spatial memory tests. *J Neurosci 20*:2711–2718.

Vosatka RJ, Hermanowski VA, Metz R, Ziff EB (1989): Dynamic interactions of c-fos protein in serum-stimulated 3T3 cells. *J Cell Physiol 138*:493–502.

Werlen G, Belin D, Conne B, Roche E, Lew DP, Prentki M (1993): Intracellular Ca2+ and the regulation of early response gene expression in HL-60 myeloid leukemia cells. *J Biol Chem 268*:16596–16601.

Woo CC, Oshita MH, Leon M (1996): A learned odor decreases the number of Fos-immunopositive granule cells in the olfactory bulb of young rats. *Brain Res 716*:149–156.

Yamamoto T (1993): Neural mechanisms of taste aversion learning. *Neurosci Res 16*:181–185.

Yamamoto T, Sako N, Sakai N, Iwafune A (1997): Gustatory and visceral inputs to the amygdala of the rat: conditioned taste aversion and induction of c-fos-like immunoreactivity. *Neurosci Lett 226*:127–130.

Yamamoto T, Shimura T, Sakai N, Ozaki N (1994a): Representation of hedonics and quality of taste stimuli in the parabrachial nucleus of the rat. *Physiol Behav 56*:1197–1202.

Yang-Yen HF, Chambard JC, Sun YL, Smeal T, Schmidt TJ, Drouin J, Karin M (1990): Transcriptional interference between c-Jun and the glucocorticoid receptor: mutual inhibition of DNA binding due to direct protein–protein interaction. *Cell 62*:1205–1215.

Zhang Y, Ji Y, Mei J (2000): Behavioral training-induced c-Fos expression in the rat nucleus basalis of Meynert during aging. *Brain Res 879*:156–162.

Zhu XO, Brown MW, McCabe BJ, Aggleton JP (1995): Effects of the novelty or familiarity of visual stimuli on the expression of the immediate early gene c-fos in rat brain. *Neuroscience 69*:821–829.

Zhu XO, McCabe BJ, Aggleton JP, Brown MW (1996): Mapping visual recognition memory through expression of the immediate early gene c-fos. *NeuroReport 7*:1871–1875.

Zhu XO, McCabe BJ, Aggleton JP, Brown MW (1997): Differential activation of the rat hippocampus and perirhinal cortex by novel visual stimuli and a novel environment. *Neurosci Lett 229*:141–143.

CHAPTER IX

Mapping neuropathology with inducible and constitutive transcription factors

ANDREE PEARSON AND MIKE DRAGUNOW

1. INTRODUCTION

Over the past few years, a wealth of data has been published to support the concept that nerve cell death during both development of the brain and during acute (e.g. stroke, head injury, status epilepticus) and chronic (e.g. Alzheimer's disease, Huntington's disease, amyolateral sclerosis, Creutzfeldt–Jakob disease, Parkinson's disease, etc.) neurodegeneration involves a programmed apoptotic cell death component (reviewed in Dragunow and Preston, 1995; Dragunow et al., 1997; Walton et al., 1999a; Walton and Dragunow, 2000). This concept first emerged in the early 1990s where mapping studies using immediate-early genes demonstrated that injured neurons expressed immediate-early genes, notably c-Jun, for prolonged periods before cell death (e.g. see Dragunow et al., 1993, and reviewed in Dragunow and Preston, 1995). These results suggested that nerve cell death during ischemia and status epilepticus, at least, involved a period of active gene expression analogous to previous studies showing that programmed cell death was involved in developmental nerve cell death. Since these initial pioneering studies, many reports have indicated that programmed nerve cell death occurs in the brain and spinal cord, and biochemical studies are now dissecting out the pathways responsible for this programmed nerve cell death. More recently, neuroanatomical mapping studies using antibodies to transcriptions factors and/or their activated states (phospho-specific antisera) have revealed biochemical pathways that may regulate nerve cell survival after brain injury (e.g. CRE response element binding protein (CREB) activation). Finally, certain transcription factors detect axonal damage (e.g. Jun and activating transcription factor (ATF-3) and may play a key role in activating an axonal regrowth program. These pathways are described in detail in the following sections of this chapter. In addition, we will review how the biochemical effectors of apoptosis (e.g. caspases, Fas) and anti-apoptosis (e.g. bcl-2) pathways in neurons may be regulated by transcription factors.

2. INDUCIBLE TRANSCRIPTION FACTORS

Inducible transcription factors (ITFs) are a subset of the immediate early gene family and are characterized by their fast induction of expression, usually within minutes of stimulation (reviewed by Hughes and Dragunow, 1995). Members of the ITFs include c-Jun and c-Fos, as well as JunB, JunD, FosB, ΔFosB, Fra-1, Fra-2 and Krox24. Expression of these proteins

Handbook of Chemical Neuroanatomy Vol. 19: Immediate Early Genes and Inducible Transcription Factors in Mapping of the Central Nervous System Function and Dysfunction
L. Kaczmarek and H.A. Robertson, editors

is controlled by constitutively expressed transcription factors (CTFs), and they are able to dimerize with members of both groups of proteins. Essentially ITFs are effectors of various signal transduction cascades which are switched on or off in response to stressful stimuli, such as trophic factor deprivation, exposure to hydrogen peroxide and free radicals, hypoxia, ischemia, heatshock and activation of death receptors, such as Fas and p75^{NTR} (further discussed below) (Ip and Davis, 1998; Mielke and Herdegen, 2000). The most studied of these transcription factors are c-Jun and c-Fos, and both have been linked to neuroprotective and neurodegenerative roles. Knockout mice of most of the ITFs have been reported, and with the exception of JunB and Fra-1, all give rise to quite different phenotypes indicating a diverse range of roles for this group of proteins (reviewed in Jochum et al., 2001).

2.1. c-Jun

Jun was first discovered as being the transforming oncogene of the avian sarcoma virus (Maki et al., 1987). Its cellular homologue is necessary for development, knockout mice lacking c-Jun die in gestation before day 16 postcoitum (Hilberg et al., 1993; Johnson et al., 1993) and at the same time, it was also discovered as being the major constituent of the AP-1 complex (Angel et al., 1988a). Phosphorylation of c-Jun enhances its ability to promote transcription from target genes, including c-*jun* itself (Angel et al., 1988b) and it is important in the regulation of the effects c-Jun elicits (Dragunow et al., 2000), particularly in response to stress, however, phosphorylation of c-Jun is not an essential requirement for it to be activated (Cruzalegui et al., 1999). c-Jun is phosphorylated by Jun N-terminal kinases (JNKs, also known as stress-activated protein kinases, SAPK), on Ser-63 and Ser-73, which is well characterized, and less well characterized, on Thr-91 and Thr-93, within the *trans*-activating domain in the N-terminus (Pulverer et al., 1991; Smeal et al., 1991; Papavassiliou et al., 1995). Phosphorylation of c-Jun has also been ascribed to extracellular related kinases (ERK) 1 and 2 activity in some cell types (Binetruy et al., 1991; Pulverer et al., 1991, 1993; Smeal et al., 1991) and more recently to p38 mitogen-activated protein (MAP) kinase (Yamagishi et al., 2001). In addition, c-Jun is also able to be phosphorylated on Tyr-170 by c-Abl, a tyrosine kinase which can operate within the nucleus (Barila et al., 2000).

Since c-Jun has been discovered, there has been a lot of evidence supporting a role for c-Jun in mediating both neuronal cell death and survival, and also as a marker for neurons surviving axotomy (also see Dragunow and Preston, 1995; Herdegen et al., 1997a; Ham et al., 2000). Indeed, studies involving cell lines from c-Jun knockout mice have shown c-Jun to be both pro- and anti-apoptotic (Bossy-Wetzel et al., 1997; Eferl et al., 1999; Wisdom et al., 1999; Kolbus et al., 2000; Shaulian et al., 2000).

2.1.1. c-Jun in neuronal apoptosis

Dragunow (1992) first showed that c-Jun was upregulated in medial septal neurons following axotomy and then went on to show that c-Jun immunoreactivity was upregulated in dying nerve cells after hypoxic–ischemic insult or following status epilepticus in rats (Dragunow et al., 1993, 1994). In addition to this, Estus et al. (1994) and Ham et al. (1995) found that nerve growth factor (NGF) withdrawal from rat sympathetic neurons caused apoptosis that could be prevented by the inhibition of c-Jun action using c-Jun antibodies or a dominant negative c-Jun mutant. c-Jun transiently increases upon removal of NGF and is expressed before the 'commitment point' of apoptosis (Estus et al., 1994). In PC12 cells, a rat pheochromocytoma cell line which responds to NGF by sprouting processes and a widely used neuronal cell culture model (Batistatou and Greene, 1991, 1993), treatment with a dominant negative c-Jun mutant also inhibited apoptosis induction after NGF withdrawal (Xia et al., 1995). Furthermore, studies in rat cerebellar granule cells show that serum deprivation and a lack of a depolarizing level of potassium give rise to apoptosis that is also preceded by a transient increase in c-Jun mRNA amidst a decrease of most other mRNA populations (Miller and Johnson, 1996; Tanabe et al., 1998). Phosphorylation of c-Jun protein on Ser-63 and Ser-73, which is associated with increased transcriptional activity, is also necessary for K^+/serum deprivation induced apoptosis in cerebellar granule cells (Watson et al., 1998) and in kainic acid-induced seizures in mice and apoptosis in murine hippocampal pyramidal cells (Behrens et al., 1999).

Since the initial demonstrations by Estus et al. (1994) and Ham et al. (1995) that c-Jun is required in sympathetic neurons for apoptosis, there have been many reports of the requirement for c-Jun in apoptosis of many neuronal cell types under various conditions. Exposure of murine cerebellar granule cells to kainate causes apoptotic cell death accompanied by a transient increase in c-Jun mRNA and protein levels (Cheung et al., 1998). c-Jun expression is also increased in rat cerebellar neurons undergoing apoptosis in response to ionizing radiation (Ferrer et al., 1996), and after okadaic acid, and hydrogen peroxide-induced apoptosis in PC12 cells and oligodendrocytes, respectively (Woodgate et al., 1999; Vollgraf et al., 1999). Upregulation of c-Jun expression has also been correlated with developmental apoptosis of neurons in both the substantia nigra pars compacta (SNpc) and substantia nigra pars reticulata (SNpr) in rats after lesion to the striatum using quinolinic acid (Oo et al., 1999).

How c-Jun induces apoptosis in neuronal cells is unknown; however, it was recently shown that c-Jun functions upstream of mitochondrial release of cytochrome *c*, since overexpression of a c-Jun dominant negative mutant inhibits cytochrome *c* release in response to NGF withdrawal from sympathetic neurons (Whitfield et al., 2001).

2.2. c-Fos

Like c-Jun, c-Fos has been linked to both death and survival roles. c-Fos was one of the first immediate early genes to be discovered, its viral counterpart was isolated from an osteogenic sarcoma virus (Curran and Teich, 1982). c-Fos appears to play an important role in development, especially in bone formation, since c-Fos knockout mice, whilst living as long as their wild-type littermates, suffer from severe osteopetrosis and altered hematopoiesis, do not have teeth and also have impaired fertility (Johnson et al., 1992; Wang et al., 1992). Furthermore, c-Fos was found to be upregulated in dying neuronal cells during normal cortical development of rats (Gonzalez-Martin et al., 1991, 1992). Generation of *fos-lacZ* transgenic mice and rats show that c-Fos is upregulated in terminally differentiated cells and prior to both developmental cell death and cell death induced by excitotoxic agents (Smeyne et al., 1992, 1993; Kasof et al., 1995). c-Fos is also upregulated in neuronal cell culture models of apoptosis (Woodgate et al., 1999), and in vulnerable neurons in the hippocampus following ischemic insults in rats (Wessel et al., 1991; An et al., 1993; Dragunow et al., 1994; Walton et al., 1998a). In contrast to this, Roffler-Tarlov et al. (1996), report that c-Fos knockout mice display normal neuronal cell death in the anterior horn of the spinal cord following sectioning of the sciatic nerve. Furthermore, following trophic withdrawal from PC12 cells, c-Fos is not upregulated until after the 'point of cellular commitment to death' (Estus et al., 1994; Mesner et al., 1995) and while c-Fos expression increases in the neurons of the substantia nigra following quinolinic acid lesion in rats, it is not temporally related to the onset of apoptosis (Oo et al., 1999). Oo et al. go on to suggest that c-Fos upregulation may not play an important role in the induction of apoptotic cell death, and that its increased expression may be related to stress effects. The observation that c-Fos expression increases following focal brain injury in mice supports this theory (Dragunow and Robertson, 1988; Dragunow et al., 1990).

2.3. OTHER ITFs

Less is known about the remaining Jun/Fos family members. All are found at higher levels in the developing rat brain than in the adult (Ferrer et al., 1996). FosB, Fra-1 and Fra-2 are upregulated following NGF withdrawal-induced apoptosis in sympathetic neurons (Estus et al., 1994), while JunB, JunD and Fra-2 are induced in PC12 cells in response to lead- or okadaic acid-induced apoptosis (Chakraborti et al., 1999; Woodgate et al., 1999). In PC12 cells, treatment with NGF, but not EGF, induces Fra-1 and Fra-2 levels and increases Fra-2/JunD binding to AP-1 and CRE elements, indicating a role for this dimer in PC12 cell differentiation (Boss et al., 2001). ERK, which undergoes sustained activation in response to NGF (Boss et al., 2001), is thought to be responsible for Fra-1 and Fra-2 phosphorylation and activation (Gruda et al., 1994; Murakami et al., 1997, 1999). Furthermore, JunB, which is phosphorylated in PC12 cells in response to NGF, is essential for NGF-induced neurite sprouting (Schlingensiepen et al., 1994). Fra-2 possibly has a neuroprotective role in rat hippocampus due to its upregulation in surviving neurons following middle cerebral arterial occlusion (Pennypacker et al., 2000a) and in regions of repair in mouse brain following neurotoxic insult (Pennypacker et al., 2000b).

3. CTFs: THE ATF FAMILY

In contrast to inducible transcription factors, the ATF family members (ATF-1, ATF-2, ATFa), along with CREB1 and CREB2, are constitutively expressed and the action of these proteins is controlled post-translationally, such as by phosphorylation. Like many ITFs, these proteins all contain within a DNA-binding domain, a group of basic amino acids and a leucine zipper region which enables dimerization. They are able to heterodimerize with either members within their own family or with inducible transcription factors to form activator protein-1 (AP-1) complexes which bind with high affinity to AP-1 sites, and can also bind as dimers to CRE sites within the promoter of target genes. ATF-3 is also a member of the ATF family by virtue of its homology with other proteins of this family; although its expression is inducible and as a homodimer is repressive; however, as a heterodimer with Jun family proteins, it is generally activating (Hsu et al., 1992; Wolfgang et al., 1997; Hai et al., 1999).

3.1. ATF-2 IN NERVE CELLS

ATF-2 is by far the most well-studied member of the ATF family, little is known about the roles of ATF-1, ATF-3 and ATFa in the brain. ATF-2 is ubiquitously expressed in the brain, and throughout the mammalian body (Maekawa et al., 1989; Takeda et al., 1991) and ATF-2 target genes include cyclin A (Shimizu et al., 1998; Beier et al., 2000), cyclin D1 (Beier et al., 1999), c-Jun (van Dam et al., 1993), tumor necrosis factor α (TNFα) (Tsai et al., 1996), transforming growth factor β2 (TGFβ2) (Kim et al., 1992) and E-selectin (Read et al., 1997). The activity of ATF-2 is regulated by phosphorylation on Thr-69 and Thr-71 and is targeted by both JNK and p38 MAP kinases (Gupta et al., 1995; van Dam et al., 1995; Raingeaud et al., 1996). It was recently reported that ATF-2 has intrinsic histone acetyltransferase activity which is phosphorylation dependent (Kawasaki et al., 2000). Two ATF-2 knockout mouse models with quite different phenotypes have been reported. The first by Reimold et al. (1996) describes mutant mice to have neurological abnormalities such as ataxia, loss of hearing, reduced Purkinje cell number in the cerebellum and enlarged ventricles as well as bone malformation. In contrast to this, null knockout mice generated by Maekawa et al. (1999) exhibited severe respiratory symptoms, similar to that of human meconium aspiration syndrome. Maekawa et al. (1999) go on to report that the knockout mice produced by Reimold et al. (1996) contain trace amounts of a splice form of ATF-2 which may compensate in some way for the lack of wild-type ATF-2 and cause a different phenotype to ATF-2 null mice. However, it is largely unknown why these two mice have such different phenotypes.

In vitro, ATF-2 has contradictory roles in nerve cell death and survival. Walton et al. (1998b) first showed that ATF-2 was phosphorylated (activated) in dying pyramidal CA1 neurons, but not in surviving dentate granule cells after hypoxia–ischemia,, suggesting that it might switch on nerve cell death. ATF-2 could perhaps heterodimerize with c-Jun to induce cell death, since c-Jun expression was also found to increase in the CA1 cells following hypoxia–ischemia (Walton et al., 1998b). Hu et al. (1999) also noted an increase in phosphorylated ATF-2 in CA1 cells in rat hippocampus during reperfusion following a period of cerebral ischemia. Furthermore, ATF-2 is activated in glioma cells during cyclosporin-induced apoptosis (Pyrzynska et al., 2000) and during cisplatin-induced apoptosis (Sanchez-Perez and Perona, 1999). ATF-2 also inhibits ERK1 activity and proliferation of a human squamous cell carcinoma line (Crowe and Shemirani, 2000). More recently, Leppa et al. (2001) reported overexpression of ATF-2 in PC12 cells caused apoptosis. Furthermore, overexpression of inactive ATF-2 in a late-stage melanoma cell line downregulates TNFα expression and increases susceptibility to UVC-induced apoptosis (Ivanov and Ronai, 1999). However, also in melanoma cell lines, activated ATF-2 plays an important role in regulating resistance to UV irradiation. In the presence of its kinase, p38, ATF-2 can increase TNFα expression and hence survival of melanoma cell lines (Ronai et al., 1998; Ivanov and Ronai, 2000). Phosphorylated ATF-2 was also found in high levels in malignant cell lines (Zoumpourlis et al., 2000).

Loss of ATF-2 expression is associated with axotomized neurons in rats (Robinson, 1996; Herdegen et al., 1997b; Martin-Villalba et al., 1998; Kreutz et al., 1999; Winter et al., 2000). In human postmortem tissue, increased ATF-2 expression has also been found in the cerebellar vermis from schizophrenia patients and in the cortex of Alzheimer's disease patients (Yamada et al., 1997; Kyosseva et al., 2000).

3.2. OTHER ATF FAMILY MEMBERS

The role of the other ATF members is unclear. ATF-1 dimerizes efficiently with CREB (Hurst et al., 1991; Rehfuss et al., 1991) and in cortical cultures, neuroprotection by iron chelator, cobalt chloride, is attributed to ATF-1/CREB activation (Zaman et al., 1999). ATF-1 also acts as a survival factor for melanoma cells (Jean et al., 2000). ATF-3 is induced by stress in many cell types (Chen et al., 1996) and more recently, it has been shown that axotomy induces ATF-3 (Takeda et al., 2000a; Tsujino et al., 2000). Interestingly, Jun can activate the ATF-3 promoter (Liang et al., 1996) suggesting that Jun and ATF3 may be co-regulated after axotomy and may control axonal regeneration. JNKs may be upstream regulators of c-Jun, and perhaps ATF-3, after axotomy, based upon their response to axotomy (Kenney and Kocsis,

1998), and activation in vitro during neurite sprouting (Heasley et al., 1996; Jho et al., 1997; Yao et al., 1997; Giasson et al., 1999).

4. UPSTREAM MEDIATORS OF ITFs AND CTFs

Upstream of the ITFs is the MAP kinase family, which can be divided into three groups: JNKs, ERKs and the p38 MAP kinases. The MAP kinase family are themselves activated by dual phosphorylation on threonine and tyrosine residues, within a tyr-x-thr motif (Davis, 1994). There are 10 known isoforms of JNK which are splice variants of proteins encoded by three genes: *jnk1*, *jnk2* and *jnk3* (Gupta et al., 1996). JNK1 and JNK2 are present throughout the body; however, JNK3 is mostly only found in the brain, heart and testis (Gupta et al., 1996). As well as c-Jun phosphorylation, JNKs are also responsible for phosphorylating ATF-2 on Thr-69 and Thr-71 (Gupta et al., 1995), as mentioned earlier, and Elk-1 on Ser-383 and Ser-389 (Cavigelli et al., 1995), both resulting in subsequent activation. JNK itself is activated by phosphorylation, JNK1 and JNK2 on Thr-183 and Tyr-185 and JNK3 on Thr-221 and Tyr-223 and upstream of JNK, activation of a cascade of MAP kinases are also required to be activated by phosphorylation (Ip and Davis, 1998).

Concurrent with the research into the role of c-Jun and other transcription factors in neuronal cell function has been the work on upstream regulators of these proteins. In 1995, Xia et al. reported a role for JNK and p38, and upstream regulator of these MAP kinases MEKK1, in apoptosis of differentiated PC12 cells after NGF withdrawal. Overexpression of constitutively activated MEKK1 also induced apoptosis in differentiated PC12 cells and this could be inhibited by co-expression with a dominant negative c-Jun mutant (Xia et al., 1995). Furthermore, NGF withdrawal was found to suppress ERK activity; however, overexpression of MKK1, which phosphorylates and activates ERK, promoted PC12 cell survival (Xia et al., 1995). Furthermore, JNK inhibition can rescue differentiated PC12 cells from apoptosis caused by UV irradiation, oxidative stress and NGF withdrawal (Maroney et al., 1998; Maroney et al., 1999). Conversely, Maroney et al. (1999) also reported that JNK activation is not required for naive PC12 cells induced to undergo apoptosis following serum withdrawal. In addition, Watson et al. (1998) found that while c-Jun is phosphorylated in response to deprivation of survival signals, JNK levels do not increase, suggesting another kinase is activating c-Jun. Eilers et al. (1998) also showed that JNK and c-Jun phosphorylation are upregulated in response to NGF withdrawal in sympathetic neurons from rats. Moreover apoptosis caused by expression of constitutively active form of MEKK1 could be prevented by co-expression of dominant negative SEK1 mutant, which in its wild-type form phosphorylates JNK, and is a target for MEKK1 phosphorylation (Eilers et al., 1998). In contrast, however, overexpression of the SEK1 dominant negative mutant alone did not inhibit apoptosis induced by NGF withdrawal, nor did its expression prevent c-Jun phosphorylation (Eilers et al., 1998). Interestingly, Coffey et al. (2000) report that JNK exists in three distinct pools within cerebellar granule neurons; a large cytoplasmic pool residing within neurites, a smaller pool which co-localizes with SEK1 and a third nuclear pool co-localizing with MKK7. In response to stress, JNK and SEK1 translocate to the nucleus, resulting in activation of c-Jun and an increase in c-Jun mRNA production (Coffey et al., 2000).

JNK3, the isoform selectively expressed in the brain, is required for kainate-induced apoptosis in vulnerable pyramidal cells in the hippocampus (Yang et al., 1997). JNK3 knockout mice are also less sensitive to kainate-induced seizures and neuronal apoptosis in the hippocampus than control animals as well as having reduced levels of AP-1 activity indicating an essential role for c-Jun phosphorylation in neuronal apoptosis (Yang et al., 1997). Other wise JNK3 knockout mice are viable and morphologically normal, as are JNK1 and JNK2 knockout mice (Dong et al., 1998; D.D. Yang et al., 1998). Interestingly, mice deficient in both *JNK1* and *JNK2* did not survive past embryonic day 12 and exhibited increased apoptosis in the forebrain, but less than control animals in the hindbrain (Kuan et al., 1999). However, knockout mice of other JNK combinations were phenotypically normal and viable thus suggesting a role for JNK1 and JNK2 together in regulation of regional neuronal apoptosis during development (Kuan et al., 1999). Different cell types express different isoforms of JNK. A mouse neuroblastoma cell line, Neuro2A, and a human neuroblastoma, SHSY5Y, express all three isoforms, whereas PC12 cells only express JNK1 (Mielke et al., 2000). More recently it has been reported that the JNK3 isoform, but not JNK1 or JNK2, is activated in apoptosis caused by sodium arsenite, an environmental toxin, in primary cortical neurons (Namgung and Xia, 2000).

Various cell culture models for Parkinson's disease support a role for JNKs in neuronal apoptosis. In rat neonatal striatal cells, dopamine induces apoptosis, which is thought to involve oxidation, in both a JNK- and SEK1-dependent manner (Luo et al., 1998). Treatment of a dopaminergic neuronal cell line with 6-OHDA induces apoptosis which is dependent on JNK activation, while treatment of the same cell line with MPTP induces necrotic cell death which is JNK independent (Choi et al., 1999). MPTP, in vivo, also induces c-Jun expression (Nishi, 1997) and activation of MKK4 and JNK (Saporito et al., 2000) in dying SNpc neurons. Furthermore, a JNK blocker, CEP 1347, inhibits MPTP-induced JNK phosphorylation and SNpc neuron loss (Saporito et al., 1999, 2000) and is currently in phase II clinical trials for the treatment of Parkinson's disease and Alzheimer's disease. The immunophilin ligand FK506 also inhibits SNpc death and c-Jun induction after medial forebrain bundle lesion (Winter et al., 2000). Parkinsonism neurological symptoms can also be caused by chronic manganese poisoning (Huang et al., 1989; Calne et al., 1994), and in PC12 cells, treatment with manganese causes apoptosis that requires SEK/JNK activation as well as c-Jun phosphorylation (Hirata et al., 1998). Furthermore, induction of apoptosis in dopaminergic neurons with bleomycin sulfate (BLM) or buthionine sulfoximine (BSO), both of which are models of oxidative stress, is also dependent on JNK activation, which goes on to induce transcription of AP-1 regulated genes. Protection against these two drugs by estradiol suppresses AP-1-driven transcription downstream of JNK and c-Jun activation (Sawada et al., 2000). JNK and SEK are also required for apoptosis-induced by overexpression of huntingtin containing an expanded polyglutamine tract in a hippocampal cell line (Liu, 1998), and JNK and p38 are upregulated in neuronal apoptosis induced by the antipsychotic drug haloperidol (Noh et al., 2000).

As well as a role in apoptosis, JNK activity, and phosphorylation of c-Jun by ERKs, has been reported to be required for NGF-induced differentiation of PC12 cells (Leppa et al., 1998; Giasson et al., 1999; Dragunow et al., 2000), thus the precise role of JNKs and JNK signaling in apoptosis versus differentiation remains to be fully elucidated. Activation of JNKs has also been associated with chemical preconditioning generating resistance to neuronal injury (i.e. a neuroprotective role, Sugino et al., 2000).

4.1. p38

As well as JNKs, there has been a lot of work on the role of p38 MAP kinases in neuronal death and survival. There are at least four p38 kinase isoforms, p38α, p38β, p38γ and p38δ which are derived from separate genes, as well as two splice variants of p38α (reviewed in

Mielke and Herdegen, 2000; Nebreda and Porras, 2000). p38 kinase isoforms are present in a variety of tissues, including the brain, and are activated by stress signals, especially TNFα and interleukin 1 (IL-1) (Whitmarsh and Davis, 1996). All isoforms are activated by phosphorylation on both Thr-180 and Tyr-182 by MKK3 or MKK6, which phosphorylate individual p38 kinases differentially (Enslen et al., 1998). Downstream targets of p38 kinase activity include ATF-2, Elk-1 (Raingeaud et al., 1996) and CHOP, a member of the C/EBP family which becomes activated in the cell in response to stress (Wang and Ron, 1996). More recently, a second p38β isoform has been cloned (p38β2), and was shown to be the predominant p38β isoform found in human brain, however while p38β2 was found to phosphorylate ATF-2, p38β did not (Enslen et al., 1998). This is in contrast to Jiang et al. (1996) who found p38β upregulated both ATF2 expression and phosphorylation. Furthermore, Enslen et al. (1998) found p38β2 to be specifically activated by MKK6 which is not found in the brain, suggesting the existence of an as yet unidentified MKK6 isoform. It is also unclear as to whether p38β2 is a splice variant or encoded by a different gene (Enslen et al., 1998).

4.1.1. p38 in nerve cells

p38 has been well documented to be upregulated in PC12 cells following NGF withdrawal (Xia et al., 1995; Kummer et al., 1997; Horstmann et al., 1998), in glutamate-induced apoptosis in cerebellar granule neurons (Kawasaki et al., 1997), after forebrain ischemia (Takagi et al., 2000), in arsenite-induced apoptosis in cortical neurons (Namgung and Xia, 2000) and in low potassium-induced apoptosis in cerebellar granule cells (Yamagishi et al., 2001). Inhibition of p38 kinase activity is also associated with the survival effect of insulin on fetal chick neurons (Heidenreich and Kummer, 1996). Furthermore, p38 kinase activity is upregulated in the CA1 and CA3 pyramidal neurons in the gerbil and rat hippocampus following transient forebrain ischemia (Ozawa et al., 1999; Sugino et al., 2000) and in rat brain cells with astrocyte-like morphology following middle cerebral artery occlusion implicating p38 in ischemic brain damage (Irving et al., 2000). However, the high basal level of p38 in the adult rat brain disappears after kainate injection (Mielke et al., 1999). It has also been reported that p38 is required for neurite outgrowth in PC12 cells (Morooka and Nishida, 1998; Iwasaki et al., 1999; Hansen et al., 2000) and Eilers et al. (1998) found that p38 is not activated following NGF withdrawal from rat sympathetic neurons. Thus, the role of p38 kinases in neuronal cells and the brain remain contradictory, different roles for p38 kinases could be attributed to distinct isoform activity. In support of this, overexpression of p38α in cardiomyocytes, HeLa and Jurkat cells induces apoptosis, while p38β is protective (Nemoto et al., 1998; Wang et al., 1998), and SB203580, a p38 inhibitor protects myocardium from ischemic death (Barancik et al., 2000). Apoptosis of T-cells (MacFarlane et al., 2000) and keratinocytes (Assefa et al., 2000) is also partly p38 mediated. In addition, axotomy-induced apoptosis of retinal ganglion cells is associated with increased phospho p39α and the p38 blocker SB203580 and the NMDA blocker MK801 both inhibited p38 phosphorylation and retinal ganglion cell apoptosis (Kikuchi et al., 2000). Castagne and Clarke (1999) also found that p38 blockers inhibited retinal ganglion cell apoptosis. Also, chlomethiazole may inhibit Jun induction and nerve cell death via p38 (Simi et al., 2000).

5. c-Jun/JNK/p38 AND CASPASES

Like c-Jun/JNK and related pathways, there is a wealth of literature on the role of caspases in apoptosis, and also much evidence to suggest the c-Jun/JNK/SEK/MEKK pathway is intertwined with caspase activation. While this chapter is focused on inducible and constitutive transcription factors and their upstream mediators involved in neuronal cell death, there is very little literature addressing how this signal transduction pathway interacts with caspase activation specifically in neurons. Caspases are cytoplasmic proteases which are expressed constitutively in all cells as inactive pro-enzymes. They are activated by cleavage and, as well as self-cleavage, they target a variety of cellular proteins which may ultimately contribute to the death of a cell (recently reviewed by Cryns and Yuan, 1998) including poly(ADP ribose)polymerase (PARP) whose cleavage by caspases has been well characterized (Lazebnik et al., 1994), cell cycle regulatory proteins, such as retinoblastoma protein (Rb) (An and Dou, 1996), and protein kinase C δ (PKCδ) (Emoto et al., 1995), as well as structural proteins α-fodrin (Cryns et al., 1996), actin (Mashima et al., 1995) and lamins (Orth et al., 1996; Takahashi et al., 1996). Of relevance to neuronal cells and possibly involved in the pathogenesis of neurodegenerative disorders, caspases have also been report to cleave the amyloid-β precursor protein (APP) (Barnes et al., 1998; Gervais et al., 1999; Weidemann et al., 1999), presenilins 1 and 2 (Kim et al., 1997; Loetscher et al., 1997; Walter et al., 1999), tau (Canu et al., 1998; Fasulo et al., 2000), and the polyglutamine-containing proteins: huntingtin (Goldberg et al., 1996; Wellington et al., 1998), ataxin-1 (Wellington et al., 1998), atrophin-1 (Ellerby et al., 1999a) and the androgen receptor (Ellerby et al., 1999b).

As discussed, JNK1 and JNK2 compound knockout mice have suggested a role for JNK1 and JNK2 in ensuring correct regional activation of caspase 3 (Kuan et al., 1999), thus, different cell types must have distinct survival and death pathways. In cerebellar granule neurons, c-Jun is phosphorylated upstream of caspase activation, since caspase inhibitors prevent apoptosis induced by decreases in K^+ concentration, but fail to have any effect on c-Jun induction or phosphorylation (Tanabe et al., 1998; Harada and Sugimoto, 1999). Furthermore, retinal ganglion cells undergoing apoptosis in response to transient ischemia express increased amounts of both c-Jun and caspase 2, but after rescue with brain-derived neurotrophic factor (BDNF) caspase 2 activity decreases, but c-Jun expression is unaffected (Kurokawa et al., 1999). This is also the case in PC12 cells deprived of NGF or treated with ceramide, in a dopaminergic cell line treated with staurosporine, and in sympathetic neurons deprived of NGF: JNK activation was not affected by treatment with general caspase inhibitor z-VAD·fmk or Boc-Asp·fmk (BAF) (Deshmukh et al., 1996; Park et al., 1996; Hartfield et al., 1998; Choi et al., 2000). Interestingly, JNK1, JNK2 and JNK3 were all elevated in response to BSO or BLM treatment of nigral dopaminergic neurons; however, while JNK2 and JNK3 activity was suppressed by caspase 3 inhibitor Ac-DMQD-CHO, JNK1 activity was not (Sawada et al., 2000). Furthermore, manganese-induced apoptosis in PC12 cells can be inhibited with p38 inhibitors after 24 h, but not after 48 h, and is independent of caspase 3 activity (Roth et al., 2000). Conversely, JNK and Jun activation in cortical cell cultures in response to treatment with lipid peroxidation product 4-hydroxy-2,3-nonenal occurs via a caspase 3-dependent pathway (Camandola et al., 2000). Moreover, in glioma cells, calphostin C, a PKC inhibitor, induces JNK/p38 activation and apoptotic nuclear damage in a caspase 3-dependent manner, but caspase inhibitors z-VAD·fmk and Ac-DEVD·cho fail to prevent calphostin C-induced cell membrane blebbing and death (Ozaki et al., 1999). However, inhibition of p38/JNK activity in calphostin C-treated cells did not prevent caspase 3 activation, but had similar effects as VAD·fmk and Ac-DEVD·cho did on apoptotic morphology: decreased nuclear condensation and cell contraction, but maintenance of membrane blebbing, indicating a further role for JNK/p38 downstream of caspase activation (Ozaki et al., 1999). Furthermore, caspase 3 activation in cortical neurons in response to calyculin A, a selective Ser/Thr phosphatase I and IIA inhibitor, was partially inhibited by p38 inhibitor, PD169316 (Ko et al., 2000). Further protection from calyculin A-induced apoptosis was offered by treatment with PD169316 and caspase inhibitor z-VAD·fmk (Ko et al., 2000). Arsenite-induced apoptosis in cortical neurons also induces JNK and p38 activation via a caspase–Bcl-2 dependent pathway (Namgung and Xia, 2000). Finally, apoptosis of endothelial cells involves a JNK1–caspase 3 pathway (Ho et al., 2000).

5.1. c-Jun/JNK/Bcl-2 INTERACTION

Bcl-2 and proteins of the Bcl-2-related family play an important role in regulation of caspase activation (reviewed by Merry and Korsmeyer, 1997). The discovery of Bcl-2 was the first of a large group now known as Bcl-2-related proteins, which can be divided into two groups: the anti-apoptotic members: Bcl-2, Bcl-X_L, Bcl-w, Bfl-1, Brag-1, Mcl-1 and A1, and the pro-apoptotic members Bax, Bak, Bcl-X_S, Bad, Bid, Bik and Hrk (Merry and Korsmeyer, 1997). Proapototic members of the Bcl-2 family are able to heterodimerize with anti-apoptotic members, and a rheostat model has been proposed whereby the cellular decision to die lies with the ratio of Bax homodimers, which lead the cell to apoptosis, to Bax:Bcl-2/Bcl-X_L heterodimers, whereby Bax is inactivated (Oltvai et al., 1993). Like caspase activation, Bcl-2 activity and the jun/JNK pathway are very much entwined. As discussed earlier, Xia et al. (1995) reported NGF withdrawal from PC12 cells caused JNK and p38 to become activated along with an increase in c-Jun phosphorylation concurrent with ERK downregulation. It was hypothesized that Bcl-2 could act upstream of JNK activity and subsequent reports have confirmed this. Ham et al. (1995) reported c-Jun to be upstream of Bcl-2 action, since overexpression of Bcl-2 in sympathetic neurons deprived of NGF did not have any effect on c-Jun levels. Over expression of Bcl-2 prevented apoptosis in naive PC12 cells after serum withdrawal (Park et al., 1996) and acting upstream of JNK, Bcl-2 prevented MEKK1 activation (Park et al., 1997). However, overexpression of JNK1 is able to override Bcl-2 function and induce apoptosis (Park et al., 1997).

Posttranslational phosphorylation appears to be important in regulation of the anti-apoptotic activity of Bcl-2. Bcl-2 phosphorylation on serine 70 by JNK1 has been reported (Srivastava et al., 1999; Deng et al., 2001), and overexpression of a Bcl-2 deletion mutant lacking a loop region including Ser-70 in a breast cancer cell line completely blocked apoptosis caused by common breast cancer treatment, paclitaxel. Overexpression of dominant negative SEK1 or of a MAP kinase specific phosphatase also blocked paclitaxel-induced apoptosis and Bcl-2 phosphorylation (Srivastava et al., 1999). Furthermore, overexpression of full-length Bcl-2 was unable to inhibit apoptosis in an immature B-cell line; however, when a Bcl-2 deletion mutant which was unable to be phosphorylated was overexpressed, apoptosis was blocked (Chang et al., 1997). Phosphorylation of Bcl-2 has also been reported by the brain specific JNK3 in the presence of Rac1, a moderate activator of JNK family members (Maundrell et al., 1997). It has also been reported that Bcl-2 phosphorylation causes cell cycle arrest rather than apoptosis, indeed Ling et al. (1998) found that when 82% of HeLa cells were undergoing mitosis, almost all Bcl-2 protein was phosphorylated, and furthermore, Yamamoto et al. (1999) found the ASK1/JNK1 pathway was responsible for Bcl-2 phosphorylation which attenuated the cell cycle at G_2/M. Conversely, others report that Bcl-2 phosphorylation is required for apoptosis (Ito et al., 1997; Ruvolo et al., 1998). However, the members of the JNK pathway have not been implicated in this phenomenon. Downstream of Bcl-2 action, JNK is also required for mitochondrial release of cytochrome C, which binds to Apaf1 causing procaspase 9 activation, which in turn activates caspase 3 and apoptosis (Tournier et al., 2000).

Bcl-X_L can be phosphorylated by JNK on Thr-47 and on Thr-115, following JNK translocation to the mitochondria in response to genotoxic stress (Kharbanda et al., 2000). Sequestration of JNK by a Bcl-X_L mutant within the mitochondria could provide a mechanism for inhibiting JNK activation of caspases downstream, thus causing apoptosis in some cell lines. In agreement with similar experiments with Bcl-2 mutants, overexpression of a Bcl-X_L mutant unable to be phosphorylated inhibited apoptosis more potently than wild-type Bcl-X_L, although there was no detectable difference in the ability of the phosphorylation mutant to heterodimerize with Bax (Kharbanda et al., 2000). In PC12 cells, Bcl-2 activity can also be regulated by insulin-like growth factor 1 (IGF-1), which prevents apoptosis induced by a variety of stimuli. This mechanism occurs via CREB activation which binds to a CRE site in the Bcl-2 promoter. CREB is activated via a specific p38 isoform, p38β, which activates MAPKAP-K3, which in turn phosphorylates CREB (Pugazhenthi et al., 1999). This appears to be an isoform-specific effect since p38α is required for apoptosis caused by Fas ligand and UV in HeLa and Jurkat cells (Nemoto et al., 1998; Wang et al., 1998).

5.2. Bax AND JNK/Jun PATHWAY

In contrast to Bcl-2, Bax is a proapoptotic member of the Bcl-2 family. Bax is required for apoptosis induced by NGF deprivation (Deckwerth et al., 1996) and Bax$^{-/-}$ knockout mice suffer from hyperplasia or hypoplasia depending on cell lineage (Knudson et al., 1995). While lymphocytes from knockout mice respond normally to death inducing stimuli (Knudson et al., 1995), sympathetic neurons remain atrophic unless NGF is supplied (Deckwerth et al., 1996). Bax is also required for apoptotic cell death in cerebellar granule cells caused by low K^+ concentrations, although not for excitotoxic cell death in the same cells in response to NMDA (Miller et al., 1997). Induction of apoptosis in cerebellar granule cells from Bax$^{-/-}$ mice increases c-Jun phosphorylation similarly to cells from wild-type mice, placing c-Jun activation upstream of Bax activity (Miller et al., 1997). However, c-Jun expression in Bax-deficient neurons deprived of NGF is maintained at an elevated level for much longer than neurons from control mice suggesting a role for Bax in c-Jun down regulation (Deckwerth et al., 1998). Interestingly, p38 plays a role in mediation of Bax activity in neurons after treatment with sodium nitroprusside, a nitric oxide (NO) donor. p38 stimulates Bax translocation from the cytosol to the mitochondria, causing activation of apoptosis, a phenomenon which can also occur in the absence of NO when endogenous p38 is activated by the overexpression of MKK3, an upstream activator of p38 (Ghatan et al., 2000).

6. c-Jun AND c-Fos IN ALZHEIMER'S DISEASE

Both cell culture studies and specific antibody staining of postmortem neurological tissue have implicated c-Jun and c-Fos in the etiology and/or pathogenesis of Alzheimer's disease (AD). Neurological hallmarks of AD, which is the most common cause of dementia, are the presence of neurofibrillary tangles containing hyperphosphorylated tau protein, and senile plaques consisting mostly of the amyloidogenic 42 amino acid peptide Aβ1–42. Neuronal loss in AD is thought to begin in the temporal lobe and then goes on to selectively affect other regions of the neocortex including the amygdala, nucleus basalis, anterior thalamus, as well as the locus ceruleus and raphe complex within the brainstem (Hopper and Vogel, 1976; Braak and Braak, 1991; Roberts et al., 1993). Most familial cases of AD are caused by mutations in the genes encoding presenilin 1 or 2, or amyloid precursor protein (APP), whose functions are largely unknown. Processing of APP by α, β- and then γ-secretases generates various forms of Aβ-amyloid (including Aβ1–40, Aβ1–42 and Aβ1–43), and in familial cases an excess of Aβ is thought to accumulate because of incorrect APP processing, although this could also occur in sporadic cases (see Chapman et al., 2001 for a recent review).

Evidence of apoptosis occurring in AD is controversial, however there is some compelling evidence to suggest it does play a role in the extensive loss of neurons characteristic of AD. Detection of apoptosis in the human brain has mostly relied on methods such as in situ end labeling of DNA (TUNEL) staining to detect DNA fragmentation, and several reports indicate AD brains have increased TUNEL staining in neurons, glia and astrocytes compared to control brains (Su et al., 1994; Dragunow et al., 1995; Lassmann et al., 1995; Smale et al., 1995; Li et al., 1997; Anderson et al., 2000). This method can also detect necrotic cell death, however many cells exhibiting positive TUNEL staining also display apoptotic morphology. Several studies have also reported increased activity of caspase 3 in various regions of postmortem AD brain tissue (Gervais et al., 1999; Selznick et al., 1999; Stadelmann et al., 1999). Furthermore, actin and fodrin, both targets for activated caspases, are found to be cleaved in AD brains (F. Yang et al., 1998; Rohn et al., 2001). Actin regulates DNase I which cleaves DNA into intranucleosomal bands during apoptosis (Kayalar et al., 1996), whilst fodrin, which was also found to be associated with neurofibrillary tangle formation in AD brains, is a major component of the membrane cytoskeleton in neurons (Goodman et al., 1987). Additionally, increased expression of the DNA repair enzyme PARP, which is upregulated in response to oxidative DNA damage, is increased in neurons and astrocytes in AD brains (Love et al., 1999).

An increase in c-Jun and c-Fos immunoreactivity has been described within the hippocampus and in the entorhinal cortex of Alzheimer's disease brains (Zhang et al., 1992; Anderson et al., 1994; MacGibbon et al., 1997a; Marcus et al., 1998). Furthermore, a correlation between neurofibrillary tangles and positive c-Jun and c-Fos antibody staining has been noted (Anderson et al., 1994). An increase in phosphorylated (activated) forms of both p38 and JNK activity has also been reported in AD brain tissue compared to controls (Hensley et al., 1999; Shoji et al., 2000; Zhu et al., 2001).

AP-1 subunits, such as c-Jun and c-Fos are further implicated in AD pathology because there is a putative AP-1-binding site in the promoter of the APP gene (Trejo et al., 1994). In addition to this, JNK has recently been shown to regulate APP activity by specifically phosphorylating APP at thr^{668} (Standen et al., 2001) and as well as p38 and ERK2, is also known to phosphorylate Tau at sites involved in AD pathology (Reynolds et al., 1997, 2000). Overexpression of presenilin 1 in HEK293 cells suppressed activation of JNK3 and its upstream mediators, SEK1 and MEKK1 (Kim et al., 2001). Moreover, biologically active mutants of presenilin 1 failed to inhibit JNK3 activity (Kim et al., 2001). In addition, presenilin 1 can suppress c-Jun homodimer activity through Jun-interacting factor 1 (Jif-1, Monteclaro and Vogt, 1993), however, presenilin 1 has no effect on Jun–Fos dimers (Imafuku et al., 1999).

In neuronal cell culture models, such as PC12 cells and several primary cell types, treatment with Aβ-amyloid induces apoptosis (Forloni et al., 1993; Loo et al., 1993) and this is accompanied by an increase in c-Jun and c-Fos expression (Anderson et al., 1995; Gillardon et al., 1996; Estus et al., 1997; Kihiko et al., 1999), and JNK activation (Shoji et al., 2000; Troy et al., 2001). Furthermore, treatment with the JNK inhibitor CEP-1347 or transfection with a dominant negative c-Jun plasmid gave protection from Aβ-amyloid-induced apoptosis (Bozyczko-Coyne et al., 2001; Troy et al., 2001). However, Gunn-Moore and Tavare (1998) found neither JNK or p38 were activated in primary cerebellar granule cells from rats in response to Aβ-amyloid treatment. Human amylin, the amyloidogenic peptide isolated from pancreatic islets of Langerhans in NIDDIM patients, is also known to be toxic to neuronal cells in culture (May et al., 1993; MacGibbon et al., 1997b). However, while induction of c-*jun* and c-*fos* as well as *junB* and *fosB* genes has been observed in human amylin-treated primary cortical neurons (Tucker et al., 1998), this effect is not seen in PC12 cells after exposure to human amylin (MacGibbon et al., 1997b).

6.1. ITFs IN OTHER NEUROLOGICAL DISORDERS

c-Jun and c-Fos have also been implicated in other neurological diseases, such as amyotrophic lateral sclerosis (ALS) and multiple sclerosis (MS). ALS is a progressive age-dependent disorder characterized by the loss of motoneurons in the brain, brainstem and spinal cord. Familial forms of ALS are caused by mutations in the gene encoding Cu/Zn superoxide dismutase (SOD-1) (Rosen et al., 1993) and transgenic mice expressing mutant SOD-1 exhibit a similar progressive loss of motoneurons to human ALS (Gurney et al., 1994). Induction of c-Jun is found in the motoneurons of the spinal cord and reticular formation of the brainstem of SOD1 transgenic mice (Jaarsma et al., 1996). However, although a strong induction of c-Jun mRNA has been observed in the spinal cord gray matter from ALS patients (Virgo and de Belleroche, 1995), Migheli et al. (1997) report most expression of c-Jun and JNK was found in astrocytes and glia, with no induction in motor neurons. Additionally, c-Fos induction has been found in the dorsal horn of the lumbar spinal cord in rats treated with cerebrospinal fluid from ALS patients and not from other neurological diseases (Manabe et al., 1999).

MS is a demyelinating disease of the CNS in which oligodendrocytes are thought to be targeted in an autoimmune response causing plaque formation. While there is loss of oligodendrocytes causing disruption of axonal function, there is rarely any neuronal cell loss and there is conflicting evidence as to whether death of oligodendrocytes in MS is by apoptosis or necrosis. However, oligodendrocytes do express 'death receptors' which are pro-apoptotic in vitro, and this is discussed later in the chapter. Martin et al. (1996) found c-Jun immunoreactivity was strong in areas surrounding plaques and neurons with axonal damage, but weak in uncompromised regions of the brain and in controls, and was accompanied by punctate JunB staining. In addition, Martin et al. (1996) found no increase in c-Fos, Fos-related antigens or JunD expression in MS brains. In contrast to this, Yu et al. (1991) reported an increased in c-Fos mRNA in white matter from MS patients compared to control tissue. More recently, Bonetti et al. (1999) found upregulation of c-Jun expression and JNK activity in oligodendrocytes within plaques from MS brains, without cell death. Furthermore, rats injected with demyelinating agent lysolecithin in the septohippocampal pathway and in the white matter of the cerebellum showed an increase in c-Jun expression in areas of demyelination (Lovas et al., 2000).

7. p75^{NTR} AND Fas IN NEURONAL CELL DEATH

Two 'death receptors' — Fas and p75^{NTR} — have been identified in neuronal cells and in glia, and they are implicated in neuronal death during development and in the adult brain. Both Fas and p75^{NTR} can induce cell death via transcriptional activation of c-Jun and ATF-2, and these transcription factors are also implicated in upregulating expression of the ligand for Fas: FasL. Fas and p75^{NTR} belong to the TNF superfamily of receptors (D'Souza et al., 1996) and other members include CD40, DR3, DR4 and DR5 (reviewed by Ashkenazi and Dixit, 1998). All these receptors recognize cell-surface-bound ligands, with the exception of $p75^{NTR}$, which is only known to bind soluble ligands, most notably NGF, whose binding, as well as being required for neuronal development, is also known to induce a death signal. The TNF family of receptors are transmembrane proteins which share a common motif in the extracellular domain and have weak homology within the intracellular domain, and it is this region that is classified as the 'death domain'.

7.1. Fas/FasL IN THE NERVOUS SYSTEM

Fas (also known as CD95 or Apo-1) was originally isolated as a cell surface receptor on various human cell lines that could transduce a signal for the cell to die after ligation with cytotoxic mouse-derived antibodies (Yonehara et al., 1989; Trauth et al., 1989). Fas is now known to bind a specific cytokine, FasL (Yonehara et al., 1989) and both Fas and its ligand play an important role in the regulation of the human immune system, removing surplus or harmful cells by apoptosis from the thymus during maturation (Nagata, 1997; Ashkenazi and Dixit, 1999). Similarly, in the developing nervous system, neurotrophic factors such as NGF and BDNF have been shown to play a role in removal/apoptosis of superfluous neurons via $p75^{NTR}$.

FasL expression can be controlled by both ITF and CTF activity because it contains an AP-1 site in its promoter region (Kasibhatla et al., 1998). Increased c-Jun phosphorylation and hence activation, via JNK, is associated with increased FasL expression in PC12 cells (Le-Niculescu et al., 1999). Also, Martin-Villalba et al. (1999) found that c-Jun is involved in the transcriptional control of the FasL during ischemia-induced apoptosis in neurons. Furthermore, in T-cells, Faris et al. (1998a) found that stress-induced FasL expression is mediated by Jun/ATF-2 binding to AP-1 sites within the FasL promoter in T-cells. Further to this, overexpression of constitutively activated MEKK1 also upregulates FasL expression (Faris et al., 1998b). In contrast, overexpression of p38, downregulates Fas expression in melanoma cells via NK-κB, causing increased resistance to apoptosis (Ivanov and Ronai, 2000). Interestingly, FasL has been shown to be upregulated following nerve fiber transection in rats, although not in neurons showing c-Jun immunoreactivity (Herdegen et al., 1998).

In the nervous system, Fas immunoreactivity is increased after spinal cord ischemia in rabbits prior to motoneuron cell death (Sakurai et al., 1998), and increased Fas expression has also been found in murine spinal cord following ischemia (Matsushita et al., 2000). Furthermore, treatment of motoneurons with Fas-FasL antagonist Fas-Fc, or caspase 8 inhibitor, IETD, inhibited about 75% of apoptosis caused by trophic factor deprivation (Raoul et al., 1999). In addition, activation of Fas by treatment with soluble FasL or Fas antibodies in the presence of trophic factors increased cell death by 40–50% (Raoul et al., 1999). FasL expression-induced Fas-mediated apoptosis in PC12 cells, as did KCl and trophic factor withdrawal from primary cerebellar granule neurons or neuronal PC12 cells, respectively (Le-Niculescu et al., 1999). Treatment of PC12 cells with a Fas antibody also induced apoptosis (Felderhoff-Mueser et al., 2000). In agreement with this, primary cerebellar cells from *gld* knockout mice (which have mutations in the *FasL* gene) were more resistant to apoptosis caused by low KCl levels (Le-Niculescu et al., 1999). Transiently increased levels of co-localized Fas and FasL immunoreactivity have also been found in apoptotic regions of the developing cerebral cortex in rats (Cheema et al., 1999). Felderhoff-Mueser et al. (2000) also found increased Fas expression in the dentate gyrus and CA1 and CA2 pyramidal cells in the hippocampus of rats after cerebral hypoxia–ischemia.

Oligodendrocytes from MS patients showed increased Fas expression in MS lesions in comparison to oligodendrocytes from control patients (D'Souza et al., 1996). However, D'Souza et al. (1996) also found that Fas-mediated death of oligodendrocytes in culture was not accompanied by DNA fragmentation or typical apoptotic morphology. In postmortem brains from MS sufferers, large numbers of glial cells displayed positive FasL immunoreactivity from acute and chronic lesions, compared with very little FasL staining found in non-inflammatory controls (Dowling et al., 1996). However, although FasL immunoreactivity was also seen in white matter from non-MS control brains, albeit less frequently, Fas was also found at an increased level in MS brains than control MS tissue. Furthermore, a high number of FasL-positive cells, most of which were oligodendrocytes, in MS lesions also had positive TUNEL staining, suggesting Fas/FasL could be involved in plaque pathogenesis in MS (Dowling et al., 1996).

Increased levels of soluble Fas have been found in the substantia nigra of parkinsonian brain tissue and in cerebrospinal fluid from Alzheimer's disease patients (Mogi et al., 1996; Martinez et al., 2000). An increase in Fas mRNA expression was also found in murine postischemic brains compared to neurological tissue from wild-type animals (Matsuyama et al., 1994, 1995). Moreover, reversible middle artery occlusion (MAO) in rats caused higher expression of FasL in postischemic brain tissue accompanied by co-localized increased TUNEL staining, and notably in *lpr* mice (which have a mutation in the *Fas* gene), reversible MAO-induced a significantly smaller infarct volume than in control rats (Martin-Villalba et al., 1999). In contrast, it was recently reported that neurons in the striatum from Huntington's disease (HD) brains and in the substantia nigra of Parkinson's disease (PD) brains showed less Fas and FasL expression than the same areas from control tissue (Ferrer et al., 2000).

7.2. HOW DOES Fas TRANSMIT ITS DEATH SIGNAL?

There is very little in the literature about the signal transduction cascade caused by Fas activation in neurons; however, in other cell types, binding of FasL to the Fas receptor causes trimerization of Fas and subsequent activation. It is proposed that ligation of Fas with FasL induces phosphorylation and enhanced binding of Jun and ATF-2 to AP-1 sites via activation of upstream regulators such as JNKs, suggesting that this is how the Fas/FasL pathway elicits apoptosis (Goillot et al., 1997; Juo et al., 1997; Toyoshima et al., 1997; Herr et al., 2000). Other reports, however, suggest that while JNK/Jun are activated during Fas-mediated apoptosis in non-neuronal cells, their activity is not required for cell death (Abreu-Martin et al., 1999; Herr et al., 1997).

8. NEURONAL SURVIVAL PATHWAYS

As well as an important role in cell death of various neuronal cell types under particular conditions, ITFs and CTFs can also regulate nerve cell survival and proliferative pathways. Again, c-Jun and c-Fos are the transcription factors best represented in the literature; however, there is increasing evidence that CREB plays an significant role in neuronal survival and it is less implicated in nerve cell apoptosis than other ITFs and CTFs.

8.1. c-Jun

Although c-Jun plays a central role in cell apoptosis in various circumstances, as discussed previously, c-Jun can also enhance cell survival. Indeed, over-expression of c-Jun in naive

PC12 cells elicits neurite sprouting and inhibits okadaic acid-induced apoptosis (Dragunow et al., 2000). Sprouting is regulated by the phosphorylation state of c-Jun on Ser-73 (Dragunow et al., 2000), and studies showing that TGFβ1 may protect cells from serum-deprivation-induced death by phosphorylating c-Jun on serine 73. Huang et al. (2000) suggest that Jun phosphorylation on serine 73 may also mediate neuroprotective effects of this transcription factor. Indeed, our working hypothesis is that c-Jun homodimers phosphorylated on Ser-73 mediate the neurite sprouting and neuroprotective effects of c-Jun, whereas phosphorylation on Ser-63 and heterodimerization with ATF-2 may mediate c-Jun's apoptotic effects (Dragunow et al., 2000). In support of this Leppa et al. (2001) showed that overexpression of c-Jun in naive cells affords protection from MEKK1-induced apoptosis and induces neurite sprouting and that ATF-2 overexpression induces apoptosis. However, Leppa et al. (2001) also report phosphorylation of c-Jun is not required for its protective effects in naive PC12 cells.

The apoptosis signal-regulating kinase 1 (ASK1), which is implicated in JNK and Jun-mediated apoptosis also regulates neurite sprouting and survival in PC12 cells (Takeda et al., 2000b) suggesting that it acts upstream of c-Jun in neuronal sprouting and survival. Other reports indicate c-Jun phosphorylation is important for regulation of survival effects: in a glioblastoma cell line, c-Jun phosphorylation is required for protection against DNA damaging agents; however, phosphorylation of c-Jun offers no protection against cytotoxic agents which do not cause DNA damage (Potapova et al., 2001). Also, phosphorylation of c-Jun protects fibroblasts from UV-induced apoptosis (Wisdom et al., 1999).

8.2. c-Jun IN AXOTOMY OF NEURONS

c-Jun is a sensitive marker of axonal injury. This was first demonstrated in the early 1990s in peripheral (Jenkins and Hunt, 1991; Leah et al., 1991) and central (Dragunow, 1992; Leah et al., 1993) neurons. The expression of c-Jun in the CNS appears to be associated with survival of the axotomized neurons (Butterworth and Dragunow, 1996).

Taken together with observations of c-Jun expression in regions of demyelination in MS neurological tissue (see Section 6.1), these results suggest that induction of c-Jun in axotomized or demyelinated neurons elicits an axonal regrowth program possibly by Jun/Jun homodimers (Dragunow et al., 2000).

8.3. c-Fos

As well as c-Jun being pleiotrophic, c-Fos seems to exhibit the same diversity in functionality. Studies using antisense c-Fos oligonucleotides have given compelling evidence for a role for c-Fos in neuroprotection. Rats treated with c-Fos antisense oligonucleotides suffered more injury in response to middle cerebral artery occlusion or following focal cerebral ischemia (Cui et al., 1999; Zhang et al., 1999). Furthermore, focal cerebral ischemia-induced expression of AP-1 target gene NGF was suppressed by c-Fos oligonucleotides in the rat hippocampus (Cui et al., 1999). Interestingly, the reverse effect also occurs, i.e. treatment of PC12 cells with NGF elicits a transient increase in c-Fos expression (Boss et al., 2001). In addition, reducing the expression of c-Fos in the amygdala in rats using antisense oligonucleotides leads to increased kindling seizures (Chiasson et al., 1998; Rocha and Kaufman, 1998). Also, treatment of rats with cerebral ischemia protective agent, *N*-acetyl-*O*-methyldopamine, increased expression of c-Fos in CA1 neurons in the hippocampus (Cho et al., 2001).

8.4. CREB1 AND CREB2

As detailed earlier in this review, ITFs, such as c-Jun and c-Fos, and CTF, ATF-2, are activated in neurons destined to undergo apoptosis after various brain insults, such as ischemia. In contrast, the CREB transcription factor, which is a CTF, is activated by phosphorylation (on Ser-133) only in neurons destined to survive similar injuries (Walton et al., 1996, and reviewed in Walton and Dragunow, 2000). These observations have been replicated by a number of researchers using different damage models (Tanaka et al., 1999, 2000; O'Dell et al., 2000; Zirpel et al., 2000). Activation of CREB in neurons may also mediate the neuroprotective effects of IGF-1 (Pugazhenthi et al., 1999, but see also Bonni et al., 1999), PACAP38 (Reglodi et al., 2000), NGF (Riccio et al., 1999) and BDNF (Bonni et al., 1999). Furthermore, overexpression of CREB promotes nerve cell survival (Walton et al., 1999b, but see also Saeki et al., 1999 who found that CREB overexpression-induced apoptosis). Downstream targets regulated by CREB include various members of the bcl-2 family, and they may mediate its neuroprotective effects. Also, c-Fos (He et al., 1998), Fra-2 (Pennypacker et al., 2000a,b), or JunB (Schenkel et al., 2000) could be involved.

8.5. UPSTREAM REGULATORS OF CREB

Possible upstream regulators of CREB phosphorylation in neurons destined to survive brain injury include ERK1/2 and Akt (Walton and Dragunow, 2000). Akt is likely to be an important upstream activator, although definitive in vivo studies of Akt phosphorylation in surviving neurons are lacking (Kitagawa et al., 1999; Jin et al., 2000). Additionally, a pathway for IGF-1 induction of phosphorylated CREB (pCREB) and bcl-2 via a novel p38β isozyme has also recently been reported (Pugazhenthi et al., 1999). ERK1/2 might also be involved as there is evidence for phosphorylation (activation) of ERK1/2 in neurons (dentate granule cells) destined to survive brain injuries (Hu et al., 2000; Sugino et al., 2000) in a pattern similar to that reported for pCREB. Other data supporting a neuroprotective role for ERK1/2 include studies showing that BDNF protects neonatal brain from ischemia via ERK1/2 (and not Akt, Han and Holtzman, 2000), and a study showing that PBN, a neuroprotective spin-trap agent, activates ERK1/2 in brain (Tsuji et al., 2000). Also, ischemic tolerance is associated with ERK phosphorylation and activity (Gonzalez-Zulueta et al., 2000; Gu et al., 2000; Hicks et al., 2000, but see also Sugino et al., 2000). Gonzalez-Zulueta et al. (2000) found that neuronal oxygen glucose preconditioning in neurons involves Ras-mediated ERK phosphorylation. Other studies linking Ras to cell survival in B-cells (Odajima et al., 2000) and fibroblasts (Wolfman and Wolfman, 2000), suggest a survival pathway of Ras–ERK–CREB. However, other reports implicate ERK1/2 in causing nerve cell death in HT22 cells after glutamate exposure (Stanciu et al., 2000; Satoh et al., 2000), in hippocampal neurons in vitro after prolonged seizures (Murray et al., 1998), and in mouse brain after ischemia (Alessandrini et al., 1999). Also, the MEK1 inhibitor PD98059 prevents ERK phosphorylation and neurite degeneration in hippocampal neurons evoked by Aβ-amyloid (Rapoport and Ferreira, 2000). Sugino et al. (2000) found that PD98059 failed to alter neuronal outcome after transient forebrain ischemia in gerbils. Thus, the roles of ERK1/2 in nerve cell survival/death are presently unclear, and contradictory. This may be due to different ERK isoforms performing different functions and also to the specificity of drugs used to block ERK activation.

8.6. Nf-κB AND NEURONAL SURVIVAL

The transcription factor Nf-κB is also a potent neuronal survival factor (reviewed in Mattson et al., 2000, but see also Schneider et al., 1999), perhaps acting via CREB (Delfino and Walker, 1999), or Ref-1 (Daily et al., 2001), which is known to be elevated in neurons resistant to brain injuries (Walton et al., 1997).

9. CONCLUSIONS

It is clear from the foregoing sections that much has been learnt about the biochemical pathways involved in nerve cell death, survival and repair during development and pathology (see Fig. 1). Brain mapping using inducible transcription factors, and more recently phosphorylation-specific antibodies to constitutive and inducible transcription factors, have provided the first clues about which pathways are responsible for nerve cell death, survival and axotomy in the brain after injury. The task ahead involves determining which inducible and constitutive transcription factors switch on the nerve cell death and life programs, how they themselves are activated (i.e. their upstream regulators) and which target genes they in turn regulate (i.e. downstream death and life effectors). In particular, determining how

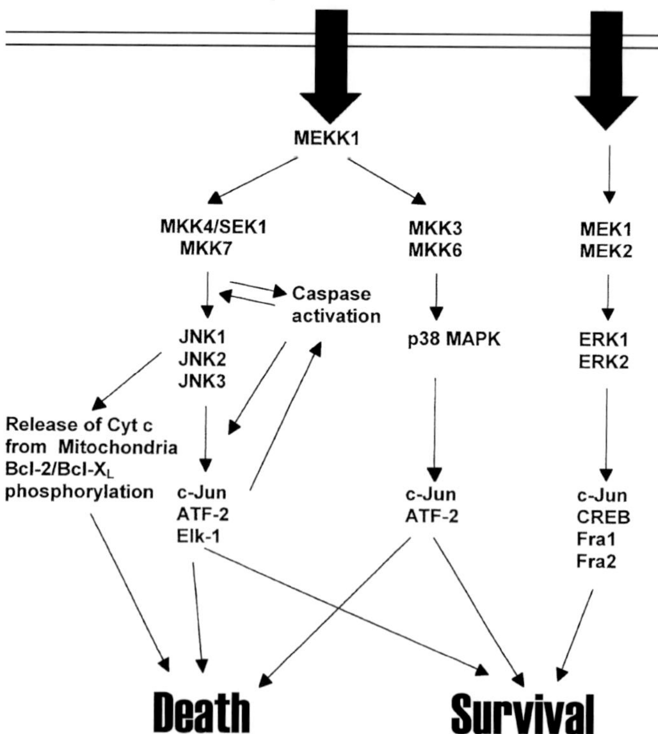

Fig. 1. Possible signal transduction cascades in neuronal cells in the mammalian brain leading to cell death or survival.

transcription factors interrelate with death effectors such as the caspases, as well as with other molecules involved in specific neurological disorders (e.g. synucleins, polyglutamines, amyloids, etc.) will yield a wealth of vital information into the control of nerve cell death, and the causes of neurodegeneration.

Of particular relevance are a number of recent studies showing that polyglutamines may interfere with transcription factor-mediated gene expression by sequestering polyglutamine-containing proteins such as CREB binding protein (CBP) (Kazantsev et al., 1999; McCampbell et al., 2000; Steffan et al., 2000), TATA binding protein (TBP) (Perez et al., 1998) and a TBP-associated factor, TAFII130 (Shimohata et al., 2000). The functional significance of these interactions was shown by studies where CBP or TAFII130 overexpression blocked polyglutamine-mediated neurotoxicity (McCampbell et al., 2000; Shimohata et al., 2000) and reversed suppression of CREB-mediated gene expression (Shimohata et al., 2000). These results further support the recent proposal that CREB switches on a neuronal survival program (as discussed earlier), which can be inhibited by oxidative stress (Zhang and Jope, 1999). Polyglutamine interference with transcription factor-mediated gene expression may also account for studies showing that gene expression is down-regulated in polyglutamine disorder brain models (Lin et al., 2000), for suppression of neurite sprouting and viability of PC12 cells (Li et al., 1999), and for the specific loss of Krox24 expression in striatal neurons in HD transgenic mouse brain (Silverman et al., 1998). Other reports show that proteins implicated in Alzheimer's disease can also interact with transcription factors. Apoe4 can enhance the transcriptional activity of CREB (Ohkubo et al., 2001), and presenilin 1 can suppress c-Jun homodimer function by accelerating the translocation of Jun-interacting factor (Jif-1, also known as QM, a negative regulator of c-Jun) from the cytoplasm to the nucleus (Imafuku et al., 1999), suggesting that both positive and negative regulation of transcription factor expression may contribute to neurodegeneration.

10. ABBREVIATIONS

AD	Alzheimer's disease
ALS	amyotrophic lateral sclerosis
AP-1	activatory protein 1, transcription factor
APP	amyloid precursor protein
ASK	apoptosis signal-regulating kinase
ATF	activating transcription factor
BDNF	brain-derived neurotrophic factor
CNS	central nervous system
CREB	cAMP and calcium responsive element binding protein
CTF	constitutive transcription factor
ERK	extracellular signal regulated kinase
HD	Huntington's disease
IGF-1	insulin-like growth factor 1
IL-1	interleukin 1
ITF	inducible transcription factor
JNK	c-Jun N-terminal kinase
MAPK	mitogen-activated protein kinases
MAO	middle artery occlusion
MEK	mitogen-activated protein kinase kinase

MKK	mitogen-activated protein kinase kinase kinase
MS	multiple sclerosis
NGF	nerve growth factor
NMDA	*N*-methyl-D-aspartate
NO	nitric oxide
PKC	protein kinase C
Rb	retinoblastoma protein
SEK	SAP (JNK) kinase
SNpc	substantia nigra pars compacta
SNpr	substantia nigra pars reticulata
TGF	transforming growth factor
TNF	tumor necrosis factor

11. ACKNOWLEDGEMENTS

We thank Emma Cropp for excellent secretarial assistance, and the New Zealand Health Research Council for supporting Professor Dragunow's research program.

12. REFERENCES

Abreu-Martin MT, Palladino AA, Faris M, Carramanzana NM, Nel AE, Targan SR (1999): Fas activates the JNK pathway in human colonic epithelial cells: lack of a direct role in apoptosis. *Am J Physiol 276*:G599–G605.

Alessandrini A, Namura S, Moskowitz MA, Bonventre JV (1999): MEK1 protein kinase inhibition protects against damage resulting from focal cerebral ischemia. *Proc Natl Acad Sci USA 96*:12866–12869.

An B, Dou QP (1996): Cleavage of retinoblastoma protein during apoptosis: an interleukin 1 beta-converting enzyme-like protease as candidate. *Cancer Res 56*:438–442.

An G, Lin TN, Liu JS, Xue JJ, He YY, Hsu CY (1993): Expression of c fos and *c-jun* family genes after focal cerebral ischemia. *Ann Neurol 33*:457–464.

Anderson AJ, Cummings BJ, Cotman CW (1994): Increased immunoreactivity for Jun- and Fos-related proteins in Alzheimer's disease: association with pathology. *Exp Neurol 125*:286–295.

Anderson AJ, Pike CJ, Cotman CW (1995): Differential induction of immediate early gene proteins in cultured neurons by beta-amyloid (A beta): association of c-Jun with A beta-induced apoptosis. *J Neurochem 65*:1487–1498.

Anderson AJ, Stoltzner S, Lai F, Su J, Nixon RA (2000): Morphological and biochemical assessment of DNA damage and apoptosis in Down syndrome and Alzheimer disease, and effect of postmortem tissue archival on TUNEL. *Neurobiol Aging 21*:511–524.

Angel P, Allegretto EA, Okino ST, Hattori K, Boyle WJ, Hunter T, Karin M (1988a): Oncogene jun encodes a sequence-specific trans-activator similar to AP-1. *Nature 332*:166–171.

Angel P, Hattori K, Smeal T, Karin M (1988b): The jun proto-oncogene is positively autoregulated by its product, Jun/AP-1. *Cell 55*:875–885.

Ashkenazi A, Dixit VM (1998): Death receptors: signaling and modulation. *Science 281*:1305–1308.

Ashkenazi A, Dixit VM (1999): Apoptosis control by death and decoy receptors. *Curr Opin Cell Biol 11*:255–260.

Assefa Z, Vantieghem A, Garmyn M, Declercq W, Vandenabeele P, Vandenheede JR, Bouillon R, Merlevede W, Agostinis P (2000): p38 Mitogen-activated protein kinase regulates a novel, caspase-independent pathway for the mitochondrial cytochrome c release in ultraviolet B radiation-induced apoptosis. *J Biol Chem 275*:21416–21421.

Barancik M, Htun P, Strohm C, Kilian S, Schaper W (2000): Inhibition of the cardiac p38-MAPK pathway by SB203580 delays ischemic cell death. *J Cardiovasc Pharmacol 35*:474–483.

Barila D, Mangano R, Gonfloni S, Kretzschmar J, Moro M, Bohmann D, Superti-Furga G (2000): A nuclear tyrosine phosphorylation circuit: c-Jun as an activator and substrate of c-Abl and JNK. *EMBO J 19*:273–281.

Barnes NY, Li L, Yoshikawa K, Schwartz LM, Oppenheim RW, Milligan CE (1998): Increased production of amyloid precursor protein provides a substrate for caspase-3 in dying motoneurons. *J Neurosci 18*:5869–5880.

Batistatou A, Greene LA (1991): Aurintricarboxylic acid rescues PC12 cells and sympathetic neurons from cell death caused by nerve growth factor deprivation: correlation with suppression of endonuclease activity. *J Cell Biol 115*:461–471.

Batistatou A, Greene LA (1993): Internucleosomal DNA cleavage and neuronal cell survival/death. *J Cell Biol 122*:523–532.

Behrens A, Sibilia M, Wagner EF (1999): Amino-terminal phosphorylation of c-Jun regulates stress-induced apoptosis and cellular proliferation. *Nat Genet 21*:326–329.

Beier F, Lee RJ, Taylor AC, Pestell RG, LuValle P (1999): Identification of the cyclin D1 gene as a target of activating transcription factor 2 in chondrocytes. *Proc Natl Acad Sci USA 96*:1433–1438.

Beier F, Taylor AC, LuValle P (2000): Activating transcription factor 2 is necessary for maximal activity and serum induction of the cyclin A promoter in chondrocytes. *J Biol Chem 275*:12948–12953.

Binetruy B, Smeal T, Karin M (1991): Ha-Ras augments c-Jun activity and stimulates phosphorylation of its activation domain. *Nature 351*:122–127.

Bonetti B, Stegagno C, Cannella B, Rizzuto N, Moretto G, Raine CS (1999): Activation of NF-kappaB and c-jun transcription factors in multiple sclerosis lesions. Implications for oligodendrocyte pathology. *Am J Pathol 155*:1433–1438.

Bonni A, Brunet A, West AE, Datta SR, Takasu MA, Greenberg ME (1999): Cell survival promoted by the Ras-MAPK signaling pathway by transcription-dependent and -independent mechanisms. *Science 286*:1358–1362.

Boss V, Roback JD, Young AN, Roback LJ, Weisenhorn DM, Medina-Flores R, Wainer BH (2001): Nerve growth factor, but not epidermal growth factor, increases Fra-2 expression and alters Fra-2/JunD binding to AP-1 and CREB binding elements in pheochromocytoma (PC12) cells. *J Neurosci 21*:18–26.

Bossy-Wetzel E, Bakiri L, Yaniv M (1997): Induction of apoptosis by the transcription factor c-Jun. *EMBO J 16*:1695–1709.

Bozyczko-Coyne D, O'Kane TM, Wu ZL, Dobrzanski P, Murthy S, Vaught JL, Scott RW (2001): CEP-1347/KT-7515, an inhibitor of SAPK/JNK pathway activation, promotes survival and blocks multiple events associated with Abeta-induced cortical neuron apoptosis. *J Neurochem 77*:849–863.

Braak H, Braak E (1991): Neuropathological staging of Alzheimer-related changes. *Acta Neuropathol 82*:239–259.

Butterworth NJ, Dragunow M (1996): Medial septal cholinergic neurons express c-Jun but do not undergo DNA fragmentation after fornix-fimbria transections. *Brain Res Mol Brain Res 43*:1–12.

Calne DB, Chu NS, Huang CC, Lu CS, Olanow W (1994): Manganism and idiopathic parkinsonism: similarities and differences. *Neurology 44*:1583–1586.

Camandola S, Poli G, Mattson MP (2000): The lipid peroxidation product 4-hydroxy-2,3-nonenal increases AP-1-binding activity through caspase activation in neurons. *J Neurochem 74*:159–168.

Canu N, Dus L, Barbato C, Ciotti MT, Brancolini C, Rinaldi AM, Novak M, Cattaneo A, Bradbury A, Calissano P (1998): Tau cleavage and dephosphorylation in cerebellar granule neurons undergoing apoptosis. *J Neurosci 18*:7061–7074.

Castagne V, Clarke PG (1999): Inhibitors of mitogen-activated protein kinases protect axotomized developing neurons. *Brain Res 842*:215–219.

Cavigelli M, Dolfi F, Claret FX, Karin M (1995): Induction of c-fos expression through JNK-mediated TCF/Elk-1 phosphorylation. *EMBO J 14*:5957–5964.

Chakraborti T, Kim KA, Goldstein GG, Bressler JP (1999): Increased AP-1 DNA binding activity in PC12 cells treated with lead. *J Neurochem 73*:187–194.

Chang BS, Minn AJ, Muchmore SW, Fesik SW, Thompson CB (1997): Identification of a novel regulatory domain in Bcl-X(L) and Bcl-2. *EMBO J 16*:968–977.

Chapman PF, Falinska AM, Knevett SG, Ramsay MF (2001): Genes, models and Alzheimer's disease. *Trends Genet 17*:254–261.

Cheema ZF, Wade SB, Sata M, Walsh K, Sohrabji F, Miranda RC (1999): Fas/Apo [apoptosis]-1 and associated proteins in the differentiating cerebral cortex: induction of caspase-dependent cell death and activation of NF-kappaB. *J Neurosci 19*:1754–1770.

Chen BPC, Wolfgang CD, Hai, T (1996): Analysis of ATF3, a transcription factor induced by physiological stresses and modulated by gadd153/Chop10. *Mol Cell Biol 16*:1157–1168.

Cheung NS, Carroll FY, Larm JA, Beart PM, Giardina SF (1998): Kainate-induced apoptosis correlates with c-Jun activation in cultured cerebellar granule cells. *J Neurosci Res 52*:69–82.

Chiasson BJ, Hong MG, Robertson HA (1998): Intra-amygdala infusion of an end-capped antisense oligodeoxynucleotide to c-fos accelerates amygdala kindling. *Brain Res Mol Brain Res 57*:248–256.

Cho S, Park EM, Kim Y, Liu N, Gal J, Volpe BT, Joh TH (2001): Early c-Fos induction after cerebral ischemia: a possible neuroprotective role. *J Cereb Blood Flow Metab 21*:550–556.

Choi WS, Yoon SY, Oh TH, Choi EJ, O'Malley KL, Oh YJ (1999): Two distinct mechanisms are involved in 6-hydroxydopamine- and MPP+-induced dopaminergic neuronal cell death: role of caspases, ROS, and JNK. *J Neurosci Res 57*:86–94.

Choi WS, Yoon SY, Chang II, Choi EJ, Rhim H, Jin BK, Oh TH, Krajewski S, Reed JC, Oh YJ (2000): Correlation between structure of Bcl-2 and its inhibitory function of JNK and caspase activity in dopaminergic neuronal apoptosis. *J Neurochem 74*:1621–1626.

Coffey ET, Hongisto V, Dickens M, Davis RJ, Courtney MJ (2000): Dual roles for c-Jun N-terminal kinase in developmental and stress responses in cerebellar granule neurons. *J Neurosci 20*:7602–7613.

Crowe DL, Shemirani B (2000): The transcription factor ATF-2 inhibits extracellular signal regulated kinase expression and proliferation of human cancer cells. *Anticancer Res 20*:2945–2949.

Cruzalegui FH, Hardingham GE, Bading H (1999): c-Jun functions as a calcium-regulated transcriptional activator in the absence of JNK/SAPK1 activation. *EMBO J 18*:1335–1344.

Cryns V, Yuan J (1998): Proteases to die for. *Genes Dev 12*:1551–1570.

Cryns VL, Bergeron L, Zhu H, Li H, Yuan J (1996): Specific cleavage of alpha-fodrin during Fas- and tumor necrosis factor-induced apoptosis is mediated by an interleukin-1beta-converting enzyme/Ced-3 protease distinct from the poly(ADP-ribose) polymerase protease. *J Biol Chem 271*:31277–31282.

Cui JK, Hsu CY, Liu PK (1999): Suppression of postischemic hippocampal nerve growth factor expression by a c-fos antisense oligodeoxynucleotide. *J Neurosci 19*:1335–1344.

Curran T, Teich NM (1982): Identification of a 39,000-dalton protein in cells transformed by the FBJ murine osteosarcoma virus. *Virology 116*:221–235.

Daily D, Vlamis-Gardikas A, Offen D, Mittelman L, Melamed E, Holmgren A, Barzilai A (2001): Glutaredoxin protects cerebellar granule neurons from dopamine-induced apoptosis by activating NF-kappa B via Ref-1. *J Biol Chem 276*:1335–1344.

Davis RJ (1994): MAPKs: new JNK expands the group. *Trends Biochem Sci 19*:470–473.

Deckwerth TL, Elliott JL, Knudson CM, Johnson EM, Snider WD, Korsmeyer SJ (1996): Bax is required for neuronal death after trophic factor deprivation and during development. *Neuron 17*:401–411.

Deckwerth TL, Easton RM, Knudson CM, Korsmeyer SJ, Johnson Jr. EM (1998): Placement of the BCL2 family member BAX in the death pathway of sympathetic neurons activated by trophic factor deprivation. *Exp Neurol 152*:150–162.

Delfino FJ, Walker WH (1999): NF-kappaB induces cAMP-response element-binding protein gene transcription in sertoli cells. *J Biol Chem 274*:35607–35613.

Deng X, Xiao L, Lang W, Gao F, Ruvolo P, May Jr. WS (2001): Novel role for jnk as a stress-activated bcl2 kinase. *J Biol Chem 276*:23681–23688.

Deshmukh M, Vasilakos J, Deckwerth TL, Lampe PA, Shivers BD, Johnson Jr. EM (1996): Genetic and metabolic status of NGF-deprived sympathetic neurons saved by an inhibitor of ICE family proteases. *J Cell Biol 135*:1341–1354.

Dong C, Yang DD, Wysk M, Whitmarsh AJ, Davis RJ, Flavell RA (1998): Defective T cell differentiation in the absence of Jnk1. *Science 282*:2092–2095.

Dowling P, Shang G, Raval S, Menonna J, Cook S, Husar W (1996): Involvement of the CD95 (APO-1/Fas) receptor/ligand system in multiple sclerosis brain. *J Exp Med 184*:1513–1518.

Dragunow M (1992): Axotomized medial septal-diagonal band neurons express Jun-like immunoreactivity. *Brain Res Mol Brain Res 15*:141–144.

Dragunow M, Preston K (1995): The role of inducible transcription factors in apoptotic nerve cell death. *Brain Res Brain Res Rev 21*:1–28.

Dragunow M, Robertson HA (1988): Brain injury induces c-fos protein(s) in nerve and glial-like cells in adult mammalian brain. *Brain Res 455*:295–299.

Dragunow M, de Castro D, Faull RL (1990): Induction of Fos in glia-like cells after focal brain injury but not during wallerian degeneration. *Brain Res 527*:41–54.

Dragunow M, Young D, Hughes P, MacGibbon G, Lawlor P, Singleton K, Sirimanne E, Beilharz E, Gluckman P (1993): Is c-Jun involved in nerve cell death following status epilepticus and hypoxic–ischaemic brain injury?. *Brain Res Mol Brain Res 18*:347–352.

Dragunow M, Beilharz E, Sirimanne E, Lawlor P, Williams C, Bravo R, Gluckman P (1994): Immediate-early gene protein expression in neurons undergoing delayed death, but not necrosis, following hypoxic–ischaemic injury to the young rat brain. *Brain Res Mol Brain Res 25*:19–33.

Dragunow M, Faull RL, Lawlor P, Beilharz EJ, Singleton K, Walker EB, Mee E (1995): In situ evidence for

DNA fragmentation in Huntington's disease striatum and Alzheimer's disease temporal lobes. *NeuroReport 6*:1053–1057.

Dragunow M, MacGibbon GA, Lawlor P, Butterworth N, Connor B, Henderson C, Walton M, Woodgate A, Hughes P, Faull RL (1997): Apoptosis, neurotrophic factors and neurodegeneration. *Rev Neurosci 8*:223–265.

Dragunow M, Xu R, Walton M, Woodgate A, Lawlor P, MacGibbon GA, Young D, Gibbons H, Lipski J, Muravlev A, Pearson A, During M (2000): c-Jun promotes neurite outgrowth and survival in PC12 cells. *Brain Res Mol Brain Res 83*:20–33.

D'Souza SD, Bonetti B, Balasingam V, Cashman NR, Barker PA, Troutt AB, Raine CS, Antel JP (1996): Multiple sclerosis: Fas signaling in oligodendrocyte cell death. *J Exp Med 184*:2361–2370.

Eferl R, Sibilia M, Hilberg F, Fuchsbichler A, Kufferath I, Guertl B, Zenz R, Wagner EF, Zatloukal K (1999): Functions of c-Jun in liver and heart development. *J Cell Biol 145*:1049–1061.

Eilers A, Whitfield J, Babij C, Rubin LL, Ham J (1998): Role of the Jun kinase pathway in the regulation of c-Jun expression and apoptosis in sympathetic neurons. *J Neurosci 18*:1713–1724.

Ellerby LM, Andrusiak RL, Wellington CL, Hackam AS, Propp SS, Wood JD, Sharp AH, Margolis RL, Ross CA, Salvesen GS, Hayden MR, Bredesen DE (1999a): Cleavage of atrophin-1 at caspase site aspartic acid 109 modulates cytotoxicity. *J Biol Chem 274*:8730–8736.

Ellerby LM, Hackam AS, Propp SS, Ellerby HM, Rabizadeh S, Cashman NR, Trifiro MA, Pinsky L, Wellington CL, Salvesen GS, Hayden MR, Bredesen DE (1999b): Kennedy's disease: caspase cleavage of the androgen receptor is a crucial event in cytotoxicity. *J Neurochem 72*:185–195.

Emoto Y, Manome Y, Meinhardt G, Kisaki H, Kharbanda S, Robertson M, Ghayur T, Wong WW, Kamen R, Weichselbaum R et al. (1995): Proteolytic activation of protein kinase C delta by an ICE-like protease in apoptotic cells. *EMBO J 14*:6148–6156.

Enslen H, Raingeaud J, Davis RJ (1998): Selective activation of p38 mitogen-activated protein (MAP) kinase isoforms by the MAP kinase kinases MKK3 and MKK6. *J Biol Chem 273*:1741–1748.

Estus S, Zaks WJ, Freeman RS, Gruda M, Bravo R, Johnson Jr. EM (1994): Altered gene expression in neurons during programmed cell death: identification of c-jun as necessary for neuronal apoptosis. *J Cell Biol 127*:1717–1727.

Estus S, Tucker HM, van Rooyen C, Wright S, Brigham EF, Wogulis M, Rydel RE (1997): Aggregated amyloid-beta protein induces cortical neuronal apoptosis and concomitant 'apoptotic' pattern of gene induction. *J Neurosci 17*:7736–7745.

Faris M, Latinis KM, Kempiak SJ, Koretzky GA, Nel A (1998a): Stress-induced Fas ligand expression in T cells is mediated through a MEK kinase 1-regulated response element in the Fas ligand promoter. *Mol Cell Biol 18*:5414–5424.

Faris M, Kokot N, Latinis K, Kasibhatla S, Green DR, Koretzky GA, Nel A (1998b): The c-Jun N-terminal kinase cascade plays a role in stress-induced apoptosis in Jurkat cells by up-regulating Fas ligand expression. *J Immunol 160*:134–144.

Fasulo L, Ugolini G, Visintin M, Bradbury A, Brancolini C, Verzillo V, Novak M, Cattaneo A (2000): The neuronal microtubule-associated protein tau is a substrate for caspase-3 and an effector of apoptosis. *J Neurochem 75*:624–633.

Felderhoff-Mueser U, Taylor DL, Greenwood K, Kozma M, Stibenz D, Joashi UC, Edwards AD, Mehmet H (2000): Fas/CD95/APO-1 can function as a death receptor for neuronal cells in vitro and in vivo and is upregulated following cerebral hypoxic–ischemic injury to the developing rat brain. *Brain Pathol 10*:17–29.

Ferrer I, Olive M, Ribera J, Planas AM (1996): Naturally occurring (programmed) and radiation-induced apoptosis are associated with selective c-Jun expression in the developing rat brain. *Eur J Neurosci 8*:1286–1298.

Ferrer I, Blanco R, Cutillas B, Ambrosio S (2000): Fas and Fas-L expression in Huntington's disease and Parkinson's disease. *Neuropathol Appl Neurobiol 26*:424–433.

Forloni G, Chiesa R, Smiroldo S, Verga L, Salmona M, Tagliavini F, Angeretti N (1993): Apoptosis mediated neurotoxicity induced by chronic application of beta amyloid fragment 25–35. *NeuroReport 4*:523–526.

Gervais FG, Xu D, Robertson GS, Vaillancourt JP, Zhu Y, Huang J, LeBlanc A, Smith D, Rigby M, Shearman MS, Clarke EE, Zheng H, Van Der Ploeg LH, Ruffolo SC, Thornberry NA, Xanthoudakis S, Zamboni RJ, Roy S, Nicholson DW (1999): Involvement of caspases in proteolytic cleavage of Alzheimer's amyloid-beta precursor protein and amyloidogenic A beta peptide formation. *Cell 97*:395–406.

Ghatan S, Larner S, Kinoshita Y, Hetman M, Patel L, Xia Z, Youle RJ, Morrison RS (2000): p38 MAP kinase mediates bax translocation in nitric oxide-induced apoptosis in neurons. *J Cell Biol 150*:335–347.

Giasson BI, Bruening W, Durham HD, Mushynski WE (1999): Activation of stress-activated protein kinases correlates with neurite outgrowth induced by protease inhibition in PC12 cells. *J Neurochem 72*:1081–1087.

Gillardon F, Skutella T, Uhlmann E, Holsboer F, Zimmermann M, Behl C (1996): Activation of c-Fos contributes to amyloid beta-peptide-induced neurotoxicity. *Brain Res 706*:169–172.

Goillot E, Raingeaud J, Ranger A, Tepper RI, Davis RJ, Harlow E, Sanchez I (1997): Mitogen-activated protein kinase-mediated Fas apoptotic signaling pathway. *Proc Natl Acad Sci USA 94*:3302–3307.

Goldberg YP, Nicholson DW, Rasper DM, Kalchman MA, Koide HB, Graham RK, Bromm M, Kazemi-Esfarjani P, Thornberry NA, Vaillancourt JP, Hayden MR (1996): Cleavage of huntingtin by apopain, a proapoptotic cysteine protease, is modulated by the polyglutamine tract. *Nat Genet 13*:442–449.

Gonzalez-Martin C, de Diego I, Fairen A, Mellstrom B, Naranjo JR (1991): Transient expression of c-fos during the development of the rat cerebral cortex. *Brain Res Dev Brain Res 59*:109–112.

Gonzalez-Martin C, de Diego I, Crespo D, Fairen A (1992): Transient c-fos expression accompanies naturally occurring cell death in the developing interhemispheric cortex of the rat. *Brain Res Dev Brain Res 68*:83–95.

Gonzalez-Zulueta M, Feldman AB, Klesse LJ, Kalb RG, Dillman JF, Parada LF, Dawson TM, Dawson VL (2000): Requirement for nitric oxide activation of p21(ras)/extracellular regulated kinase in neuronal ischemic preconditioning. *Proc Natl Acad Sci USA 97*:436–441.

Goodman SR, Zagon IS, Riederer BM (1987): Spectrin isoforms in mammalian brain. *Brain Res Bull 18*:787–792.

Gruda MC, Kovary K, Metz R, Bravo R (1994): Regulation of Fra-1 and Fra-2 phosphorylation differs during the cell cycle of fibroblasts and phosphorylation in vitro by MAP kinase affects DNA binding activity. *Oncogene 9*:2537–2547.

Gu Z, Jiang Q, Zhang G, Cui Z, Zhu Z (2000): Diphosphorylation of extracellular signal-regulated kinases and c-Jun N-terminal protein kinases in brain ischemic tolerance in rat. *Brain Res 860*:157–160.

Gunn-Moore FJ, Tavare JM (1998): Apoptosis of cerebellar granule cells induced by serum withdrawal, glutamate or beta-amyloid, is independent of Jun kinase or p38 mitogen activated protein kinase activation. *Neurosci Lett 250*:53–56.

Gupta S, Campbell D, Derijard B, Davis RJ (1995): Transcription factor ATF2 regulation by the JNK signal transduction pathway. *Science 267*:389–393.

Gupta S, Barrett T, Whitmarsh AJ, Cavanagh J, Sluss HK, Derijard B, Davis RJ (1996): Selective interaction of JNK protein kinase isoforms with transcription factors. *EMBO J 15*:2760–2770.

Gurney ME, Pu H, Chiu AY, Dal Canto MC, Polchow CY, Alexander DD, Caliendo J, Hentati A, Kwon YW, Deng HX et al. (1994): Motor neuron degeneration in mice that express a human Cu,Zn superoxide dismutase mutation. *Science 264*:1772–1775.

Hai T, Wolfgang CD, Marsee DK, Allen AE, Sivaprasad U (1999): ATF3 and stress responses. *Gene Expr 7*:321–335.

Ham J, Babij C, Whitfield J, Pfarr CM, Lallemand D, Yaniv M, Rubin LL (1995): A c-Jun dominant negative mutant protects sympathetic neurons against programmed cell death. *Neuron 14*:927–939.

Ham J, Eilers A, Whitfield J, Neame SJ, Shah B (2000): c-Jun and the transcriptional control of neuronal apoptosis. *Biochem Pharmacol 60*:1015–1021.

Han BH, Holtzman DM (2000): BDNF protects the neonatal brain from hypoxic–ischemic injury in vivo via the ERK pathway. *J Neurosci 20*:5775–5781.

Hansen T, Rehfeld JF, Nielsen FC (2000): Cyclic AMP-induced neuronal differentiation via activation of p38 mitogen-activated protein kinase. *J Neurochem 75*:1870–1877.

Harada J, Sugimoto M (1999): An inhibitor of p38 and JNK MAP kinases prevents activation of caspase and apoptosis of cultured cerebellar granule neurons. *Jpn J Pharmacol 79*:369–378.

Hartfield PJ, Bilney AJ, Murray AW (1998): Neurotrophic factors prevent ceramide-induced apoptosis downstream of c-Jun N-terminal kinase activation in PC12 cells. *J Neurochem 71*:161–169.

He H, Qi XM, Grossmann J, Distelhorst CW (1998): c-Fos degradation by the proteasome. An early, Bcl-2-regulated step in apoptosis. *J Biol Chem 273*:25015–25019.

Heasley LE, Storey B, Fanger GR, Butterfield L, Zamarripa J, Blumberg D, Maue RA (1996): GTPase-deficient G alpha 16 and G alpha q induce PC12 cell differentiation and persistent activation of cJun NH2-terminal kinases. *Mol Cell Biol 16*:648–656.

Heidenreich KA, Kummer JL (1996): Inhibition of p38 mitogen-activated protein kinase by insulin in cultured fetal neurons. *J Biol Chem 271*:9891–9894.

Hensley K, Floyd RA, Zheng NY, Nael R, Robinson KA, Nguyen X, Pye QN, Stewart CA, Geddes J, Markesbery WR, Patel E, Johnson GV, Bing G (1999): p38 kinase is activated in the Alzheimer's disease brain. *J Neurochem 72*:2053–2058.

Herdegen T, Skene P, Bahr M (1997a): The c-Jun transcription factor - bipotential mediator of neuronal death, survival and regeneration. *Trends Neurosci 20*:227–231.

Herdegen T, Blume A, Buschmann T, Georgakopoulos E, Winter C, Schmid W, Hsieh TF, Zimmermann M, Gass P

(1997b): Expression of activating transcription factor-2, serum response factor and cAMP/Ca response element binding protein in the adult rat brain following generalized seizures, nerve fibre lesion and ultraviolet irradiation. *Neuroscience 81*:199–212.

Herdegen T, Claret FX, Kallunki T, Martin-Villalba A, Winter C, Hunter T, Karin M (1998): Lasting N-terminal phosphorylation of c-Jun and activation of c-Jun N-terminal kinases after neuronal injury. *J Neurosci 18*:5124–5135.

Herr I, Wilhelm D, Bohler T, Angel PDK-M (1997): Activation of CD95 (Apo-1/Fas)signaling by ceramide mediates cancer therapy-induced apoptosis. *EMBO J 16*:6200–6208.

Herr I, Posovszky C, Di Marzio LD, Cifone MG, Boehler T, Debatin KM (2000): Autoamplification of apoptosis following ligation of CD95-L, TRAIL and TNF-alpha. *Oncogene 19*:4255–4262.

Hicks SD, Parmele KT, DeFranco DB, Klann E, Callaway CW (2000): Hypothermia differentially increases extracellular signal-regulated kinase and stress-activated protein kinase/c-Jun terminal kinase activation in the hippocampus during reperfusion after asphyxial cardiac arrest. *Neuroscience 98*:677–685.

Hilberg F, Aguzzi A, Howells N, Wagner EF (1993): c-jun is essential for normal mouse development and hepatogenesis. *Nature 365*:179–181.

Hirata Y, Adachi K, Kiuchi K (1998): Activation of JNK pathway and induction of apoptosis by manganese in PC12 cells. *J Neurochem 71*:1607–1615.

Ho FM, Liu SH, Liau CS, Huang PJ, Lin-Shiau SY (2000): High glucose-induced apoptosis in human endothelial cells is mediated by sequential activations of c-Jun NH(2)-terminal kinase and caspase-3. *Circulation 101*:2618–2624.

Hopper MW, Vogel FS (1976): The limbic system in Alzheimer's disease. A neuropathologic investigation. *Am J Pathol 85*:1–20.

Horstmann S, Kahle PJ, Borasio GD (1998): Inhibitors of p38 mitogen-activated protein kinase promote neuronal survival in vitro. *J Neurosci Res 52*:483–490.

Hsu JC, Bravo R, Taub R (1992): Interactions among LRF-1, JunB, c-Jun, and c-Fos define a regulatory program in the G1 phase of liver regeneration. *Mol Cell Biol 12*:4654–4665.

Hu BR, Fux CM, Martone ME, Zivin JA, Ellisman MH (1999): Persistent phosphorylation of cyclic AMP responsive element-binding protein and activating transcription factor-2 transcription factors following transient cerebral ischemia in rat brain. *Neuroscience 89*:437–452.

Hu BR, Liu CL, Park DJ (2000): Alteration of MAP kinase pathways after transient forebrain ischemia. *J Cereb Blood Flow Metab 20*:1089–1095.

Huang CC, Chu NS, Lu CS, Wang JD, Tsai JL, Tzeng JL, Wolters EC, Calne DB (1989): Chronic manganese intoxication. *Arch Neurol 46*:1104–1106.

Huang Y, Hutter D, Liu Y, Wang X, Sheikh MS, Chan AM, Holbrook NJ (2000): Transforming growth factor-beta 1 suppresses serum deprivation-induced death of A549 cells through differential effects on c-Jun and JNK activities. *J Biol Chem 275*:18234–18242.

Hughes P, Dragunow M (1995): Induction of immediate-early genes and the control of neurotransmitter-regulated gene expression within the nervous system. *Pharmacol Rev 47*:133–178.

Hurst HC, Totty NF, Jones NC (1991): Identification and functional characterisation of the cellular activating transcription factor 43 (ATF-43) protein. *Nucleic Acids Res 19*:4601–4609.

Imafuku I, Masaki T, Waragai M, Takeuchi S, Kawabata M, Hirai S, Ohno S, Nee LE, Lippa CF, Kanazawa I, Imagawa M, Okazawa H (1999): Presenilin 1 suppresses the function of c-Jun homodimers via interaction with QM/Jif-1. *J Cell Biol 147*:121–134.

Ip YT, Davis RJ (1998): Signal transduction by the c-Jun N-terminal kinase (JNK) - from inflammation to development. *Curr Opin Cell Biol 10*:205–219.

Irving EA, Barone FC, Reith AD, Hadingham SJ, Parsons AA (2000): Differential activation of MAPK/ERK and p38/SAPK in neurones and glia following focal cerebral ischaemia in the rat. *Brain Res Mol Brain Res 77*:65–75.

Ito T, Deng X, Carr B, May WS (1997): Bcl-2 phosphorylation required for anti-apoptosis function. *J Biol Chem 272*:11671–11673.

Ivanov VN, Ronai Z (1999): Down-regulation of tumor necrosis factor γ expression by activating transcription factor 2 increases UVC-induced apoptosis of late-stage melanoma cells. *J Biol Chem 274*:14079–14089.

Ivanov VN, Ronai Z (2000): p38 protects human melanoma cells from UV-induced apoptosis through down-regulation of NF-kappaB activity and Fas expression. *Oncogene 19*:3003–3012.

Iwasaki S, Iguchi M, Watanabe K, Hoshino R, Tsujimoto M, Kohno M (1999): Specific activation of the p38 mitogen-activated protein kinase signaling pathway and induction of neurite outgrowth in PC12 cells by bone morphogenetic protein-2. *J Biol Chem 274*:26503–26510.

Jaarsma D, Holstege JC, Troost D, Davis M, Kennis J, Haasdijk ED, de Jong VJ (1996): Induction of c-Jun immunoreactivity in spinal cord and brainstem neurons in a transgenic mouse model for amyotrophic lateral sclerosis. *Neurosci Lett 219*:179–182.

Jean D, Tellez C, Huang S, Davis DW, Bruns CJ, McConkey DJ, Hinrichs SH, Bar-Eli M (2000): Inhibition of tumor growth and metastasis of human melanoma by intracellular anti-ATF-1 single chain Fv fragment. *Oncogene 19*:2721–2730.

Jenkins R, Hunt SP (1991): Long-term increase in the levels of c-jun mRNA and jun protein-like immunoreactivity in motor and sensory neurons following axon damage. *Neurosci Lett 129*:107–110.

Jho EH, Davis RJ, Malbon CC (1997): c-Jun amino-terminal kinase is regulated by Galpha12/Galpha13 and obligate for differentiation of P19 embryonal carcinoma cells by retinoic acid. *J Biol Chem 272*:24468–24474.

Jiang Y, Chen C, Li Z, Guo W, Gegner JA, Lin S, Han J (1996): Characterization of the structure and function of a new mitogen-activated protein kinase (p38beta). *J Biol Chem 271*:17920–17926.

Jin KL, Mao XO, Nagayama T, Goldsmith PC, Greenberg DA (2000): Induction of vascular endothelial growth factor receptors and phosphatidylinositol 3′-kinase/Akt signaling by global cerebral ischemia in the rat. *Neuroscience 100*:713–717.

Jochum W, Passegue E, Wagner EF (2001): AP-1 in mouse development and tumorigenesis. *Oncogene 20*:2401–2412.

Johnson RS, Spiegelman BM, Papaioannou V (1992): Pleiotropic effects of a null mutation in the c-fos proto-oncogene. *Cell 71*:577–586.

Johnson RS, van Lingen B, Papaioannou VE, Spiegelman BM (1993): A null mutation at the c-jun locus causes embryonic lethality and retarded cell growth in culture. *Genes Dev 7*:1309–1317.

Juo P, Kuo CJ, Reynolds SE, Konz RF, Raingeaud J, Davis RJ, Biemann HP, Blenis J (1997): Fas activation of the p38 mitogen-activated protein kinase signalling pathway requires ICE/CED-3 family proteases. *Mol Cell Biol 17*:24–35.

Kasibhatla S, Brunner T, Genestier L, Echeverri F, Mahboubi A, Green DR (1998): DNA damaging agents induce expression of Fas ligand and subsequent apoptosis in T lymphocytes via the activation of NF-kappa B and AP-1. *Mol Cell 1*:543–551.

Kasof GM, Mandelzys A, Maika SD, Hammer RE, Curran T, Morgan JI (1995): Kainic acid-induced neuronal death is associated with DNA damage and a unique immediate-early gene response in c-fos-lacZ transgenic rats. *J Neurosci 15*:4238–4249.

Kawasaki H, Morooka T, Shimohama S, Kimura J, Hirano T, Gotoh Y, Nishida E (1997): Activation and involvement of p38 mitogen-activated protein kinase in glutamate-induced apoptosis in rat cerebellar granule cells. *J Biol Chem 272*:18518–18521.

Kawasaki H, Schiltz L, Chiu R, Itakura K, Taira K, Nakatani Y, Yokoyama KK (2000): ATF-2 has intrinsic histone acetyltransferase activity which is modulated by phosphorylation. *Nature 405*:195–200.

Kayalar C, Ord T, Testa MP, Zhong LT, Bredesen DE (1996): Cleavage of actin by interleukin 1 beta-converting enzyme to reverse DNase I inhibition. *Proc Natl Acad Sci USA 93*:2234–2238.

Kazantsev A, Preisinger E, Dranovsky A, Goldgaber D, Housman D (1999): Insoluble detergent-resistant aggregates form between pathological and nonpathological lengths of polyglutamine in mammalian cells. *Proc Natl Acad Sci USA 96*:11404–11409.

Kenney AM, Kocsis JD (1998): Peripheral axotomy induces long-term c-Jun amino-terminal kinase-1 activation and activator protein-1 binding activity by c-Jun and junD in adult rat dorsal root ganglia in vivo. *J Neurosci 18*:1318–1328.

Kharbanda S, Saxena S, Yoshida K, Pandey P, Kaneki M, Wang Q, Cheng K, Chen YN, Campbell A, Sudha T, Yuan ZM, Narula J, Weichselbaum R, Nalin C, Kufe D (2000): Translocation of SAPK/JNK to mitochondria and interaction with Bcl-x(L) in response to DNA damage. *J Biol Chem 275*:322–327.

Kihiko ME, Tucker HM, Rydel RE, Estus S (1999): c-Jun contributes to amyloid beta-induced neuronal apoptosis but is not necessary for amyloid beta-induced c-jun induction. *J Neurochem 73*:2609–2612.

Kikuchi M, Tenneti L, Lipton SA (2000): Role of p38 mitogen-activated protein kinase in axotomy-induced apoptosis of rat retinal ganglion cells. *J Neurosci 20*:5037–5044.

Kim JW, Chang TS, Lee JE, Huh SH, Yeon SW, Yang WS, Joe CO, Mook-Jung I, Tanzi RE, Kim TW, Choi EJ (2001): Negative regulation of the SAPK/JNK signaling pathway by presenilin 1. *J Cell Biol 153*:457–463.

Kim SJ, Wagner S, Liu F, O'Reilly MA, Robbins PD, Green MR (1992): Retinoblastoma gene product activates expression of the human TGF-beta 2 gene through transcription factor ATF-2. *Nature 358*:331–334.

Kim TW, Pettingell WH, Jung YK, Kovacs DM, Tanzi RE (1997): Alternative cleavage of Alzheimer-associated presenilins during apoptosis by a caspase-3 family protease. *Science 277*:373–376.

Kitagawa H, Warita H, Sasaki C, Zhang WR, Sakai K, Shiro Y, Mitsumoto Y, Mori T, Abe K (1999): Immunoreactive Akt, PI3-K and ERK protein kinase expression in ischemic rat brain. *Neurosci Lett 274*:45–48.

Knudson CM, Tung KSK, Tourtellotte WG, Brown GAJ, Korsmeyer SJ (1995): Bax-deficient mice with lymphoid hyperplasia and male germ cell death. *Science 270*:96–99.

Ko HW, Han KS, Kim EY, Ryu BR, Yoon WJ, Jung YK, Kim SU, Gwag BJ (2000): Synergetic activation of p38 mitogen-activated protein kinase and caspase-3-like proteases for execution of calyculin A-induced apoptosis but not N-methyl-D-aspartate-induced necrosis in mouse cortical neurons. *J Neurochem 74*:2455–2461.

Kolbus A, Herr I, Schreiber M, Debatin KM, Wagner EF, Angel P (2000): c-Jun-dependent CD95-L expression is a rate-limiting step in the induction of apoptosis by alkylating agents. *Mol Cell Biol 20*:575–582.

Kreutz MR, Bien A, Vorwerk CK, Bockers TM, Seidenbecher CI, Tischmeyer W, Sabel BA (1999): Co-expression of c-Jun and ATF-2 characterizes the surviving retinal ganglion cells which maintain axonal connections after partial optic nerve injury. *Brain Res Mol Brain Res 69*:232–241.

Kuan CY, Yang DD, Samanta Roy DR, Davis RJ, Rakic P, Flavell RA (1999): The Jnk1 and Jnk2 protein kinases are required for regional specific apoptosis during early brain development. *Neuron 22*:667–676.

Kummer JL, Rao PK, Heidenreich KA (1997): Apoptosis induced by withdrawal of trophic factors is mediated by p38 mitogen-activated protein kinase. *J Biol Chem 272*:20490–20494.

Kurokawa T, Katai N, Shibuki H, Kuroiwa S, Kurimoto Y, Nakayama C, Yoshimura N (1999): BDNF diminishes caspase-2 but not c-Jun immunoreactivity of neurons in retinal ganglion cell layer after transient ischemia. *Invest Ophthalmol Vis Sci 40*:3006–3011.

Kyosseva SV, Elbein AD, Hutton TL, Griffin WST, Mrak RE, Sturner WQ, Karson CN (2000): Increased levels of transcription factors Elk-1, cyclic adenosine monophosphate response element-binding protein, and activating transcription factor 2 in the cerebellar vermis of schizophrenic patients. *Arch Gen Psychiatry 57*:685–691.

Lassmann H, Bancher C, Breitschopf H, Wegiel J, Bobinski M, Jellinger K, Wisniewski HM (1995): Cell death in Alzheimer's disease evaluated by DNA fragmentation in situ. *Acta Neuropathol 89*:35–41.

Lazebnik YA, Kaufmann SH, Desnoyers S, Poirier GG, Earnshaw WC (1994): Cleavage of poly(ADP-ribose) polymerase by a proteinase with properties like ICE. *Nature 371*:346–347.

Leah JD, Herdegen T, Bravo R (1991): Selective expression of Jun proteins following axotomy and axonal transport block in peripheral nerves in the rat: evidence for a role in the regeneration process. *Brain Res 566*:198–207.

Leah JD, Herdegen T, Murashov A, Dragunow M, Bravo R (1993): Expression of immediate early gene proteins following axotomy and inhibition of axonal transport in the rat central nervous system. *Neuroscience 57*:53–66.

Le-Niculescu H, Bonfoco E, Kasuya Y, Claret FX, Green DR, Karin M (1999): Withdrawal of survival factors results in activation of the JNK pathway in neuronal cells leading to Fas ligand induction and cell death. *Mol Cell Biol 19*:751–763.

Leppa S, Saffrich R, Ansorge W, Bohmann D (1998): Differential regulation of c-Jun by ERK and JNK during PC12 cell differentiation. *EMBO J 17*:4404–4413.

Leppa S, Eriksson M, Saffrich R, Ansorge W, Bohmann D (2001): Complex functions of AP-1 transcription factors in differentiation and survival of PC12 cells. *Mol Cell Biol 21*:4369–4378.

Li SH, Cheng AL, Li H, Li XJ (1999): Cellular defects and altered gene expression in PC12 cells stably expressing mutant huntingtin. *J Neurosci 19*:5159–5172.

Li WP, Chan WY, Lai HW, Yew DT (1997): Terminal dUTP nick end labeling (TUNEL) positive cells in the different regions of the brain in normal aging and Alzheimer patients. *J Mol Neurosci 8*:75–82.

Liang G, Wolfgang CD, Chen BP, Chen TH, Hai T (1996): ATF3 gene. Genomic organization, promoter, and regulation. *J Biol Chem 271*:1695–1701.

Lin X, Antalffy B, Kang D, Orr HT, Zoghbi HY (2000): Polyglutamine expansion down-regulates specific neuronal genes before pathologic changes in SCA1. *Nat Neurosci 3*:157–163.

Ling YH, Tornos C, Perez-Soler R (1998): Phosphorylation of Bcl-2 is a marker of M phase events and not a determinant of apoptosis. *J Biol Chem 273*:18984–18991.

Liu YF (1998): Expression of polyglutamine-expanded Huntingtin activates the SEK1-JNK pathway and induces apoptosis in a hippocampal neuronal cell line. *J Biol Chem 273*:28873–28877.

Loetscher H, Deuschle U, Brockhaus M, Reinhardt D, Nelboeck P, Mous J, Grunberg J, Haass C, Jacobsen H (1997): Presenilins are processed by caspase-type proteases. *J Biol Chem 272*:20655–20659.

Loo DT, Copani A, Pike CJ, Whittemore ER, Walencewicz AJ, Cotman CW (1993): Apoptosis is induced by beta-amyloid in cultured central nervous system neurons. *Proc Natl Acad Sci USA 90*:7951–7955.

Lovas G, Palkovits M, Komoly S (2000): Increased c-Jun expression in neurons affected by lysolecithin-induced demyelination in rats. *Neurosci Lett 292*:71–74.

Love S, Barber R, Wilcock GK (1999): Increased poly(ADP-ribosyl)ation of nuclear proteins in Alzheimer's disease. *Brain 122*:247–253.

Luo Y, Umegaki H, Wang X, Abe R, Roth GS (1998): Dopamine induces apoptosis through an oxidation-involved SAPK/JNK activation pathway. *J Biol Chem 273*:3756–3764.

MacFarlane M, Cohen GM, Dickens M (2000): JNK (c-Jun N-terminal kinase) and p38 activation in receptor-mediated and chemically-induced apoptosis of T-cells: differential requirements for caspase activation. *Biochem J 348 (1)*:93–101.

MacGibbon GA, Lawlor PA, Walton M, Sirimanne E, Faull RL, Synek B, Mee E, Connor B, Dragunow M (1997a): Expression of Fos, Jun, and Krox family proteins in Alzheimer's disease. *Exp Neurol 147*:316–332.

MacGibbon GA, Cooper GJ, Dragunow M (1997b): Acute application of human amylin, unlike beta-amyloid peptides, kills undifferentiated PC12 cells by apoptosis. *NeuroReport 8*:3945–3950.

Maekawa T, Sakura H, Kanei-Ishii C, Sudo T, Yoshimura T, Fujisawa J, Yoshida M, Ishii S (1989): Leucine zipper structure of the protein CRE-BP1 binding to the cyclic AMP response element in brain. *EMBO J 8*:2023–2028.

Maekawa T, Bernier F, Sato M, Nomura S, Singh M, Inoue Y, Tokunaga T, Imai H, Yokoyama M, Reimold A, Glimcher LH, Ishii S (1999): Mouse ATF-2 null mutants display features of a severe type of meconium aspiration syndrome. *J Biol Chem 274*:17813–17819.

Maki Y, Bos TJ, Davis C, Starbuck M, Vogt PK (1987): Avian sarcoma virus 17 carries the jun oncogene. *Proc Natl Acad Sci USA 84*:2848–2852.

Manabe Y, Kashihara K, Shiro Y, Shohmori T, Abe K (1999): Enhanced Fos expression in rat lumbar spinal cord cultured with cerebrospinal fluid from patients with amyotrophic lateral sclerosis. *Neurol Res 21*:309–312.

Marcus DL, Strafaci JA, Miller DC, Masia S, Thomas CG, Rosman J, Hussain S, Freedman ML (1998): Quantitative neuronal c-fos and *c-jun* expression in Alzheimer's disease. *Neurobiol Aging 19*:393–400.

Maroney AC, Glicksman MA, Basma AN, Walton KM, Knight Jr. E, Murphy CA, Bartlett BA, Finn JP, Angeles T, Matsuda Y, Neff NT, Dionne CA (1998): Motoneuron apoptosis is blocked by CEP-1347 (KT 7515), a novel inhibitor of the JNK signaling pathway. *J Neurosci 18*:104–111.

Maroney AC, Finn JP, Bozyczko-Coyne D, O'Kane TM, Neff NT, Tolkovsky AM, Park DS, Yan CY, Troy CM, Greene LA (1999): CEP-1347 (KT7515), an inhibitor of JNK activation, rescues sympathetic neurons and neuronally differentiated PC12 cells from death evoked by three distinct insults. *J Neurochem 73*:1901–1912.

Martin G, Segui J, Diaz-Villoslada P, Montalban X, Planas AM, Ferrer I (1996): Jun expression is found in neurons located in the vicinity of subacute plaques in patients with multiple sclerosis. *Neurosci Lett 212*:95–98.

Martinez M, Fernandez-Vivancos E, Frank A, De la Fuente M, Hernanz A (2000): Increased cerebrospinal fluid fas (Apo-1) levels in Alzheimer's disease. Relationship with IL-6 concentrations. *Brain Res 869*:216–219.

Martin-Villalba A, Winter C, Brecht S, Buschmann T, Zimmermann M, Herdegen T (1998): Rapid and long-lasting suppression of the ATF-2 transcription factor is a common response to neuronal injury. *Brain Res Mol Brain Res 62*:158–166.

Martin-Villalba A, Herr I, Jeremias I, Hahne M, Brandt R, Vogel J, Schenkel J, Herdegen T, Debatin KM (1999): CD95 ligand (Fas-L/APO-1L) and tumor necrosis factor-related apoptosis-inducing ligand mediate ischemia-induced apoptosis in neurons. *J Neurosci 19*:3809–3817.

Mashima T, Naito M, Fujita N, Noguchi K, Tsuruo T (1995): Identification of actin as a substrate of ICE and an ICE-like protease and involvement of an ICE-like protease but not ICE in VP-16-induced U937 apoptosis. *Biochem Biophys Res Commun 217*:1185–1192.

Matsushita K, Wu Y, Qiu J, Lang-Lazdunski L, Hirt L, Waeber C, Hyman BT, Yuan J, Moskowitz MA (2000): Fas receptor and neuronal cell death after spinal cord ischemia. *J Neurosci 20*:6879–6887.

Matsuyama T, Hata R, Tagaya M, Yamamoto Y, Nakajima T, Furuyama J, Wanaka A, Sugita M (1994): Fas antigen mRNA induction in postischemic murine brain. *Brain Res 657*:342–346.

Matsuyama T, Hata R, Yamamoto Y, Tagaya M, Akita H, Uno H, Wanaka A, Furuyama J, Sugita M (1995): Localization of Fas antigen mRNA induced in postischemic murine forebrain by in situ hybridization. *Brain Res Mol Brain Res 34*:166–172.

Mattson MP, Culmsee C, Yu Z, Camandola S (2000): Roles of nuclear factor kappaB in neuronal survival and plasticity. *J Neurochem 74*:443–456.

Maundrell K, Antonsson B, Magnenat E, Camps M, Muda M, Chabert C, Gillieron C, Boschert U, Vial-Knecht E, Martinou JC, Arkinstall S (1997): Bcl-2 undergoes phosphorylation by c-Jun N-terminal kinase/stress-activated protein kinases in the presence of the constitutively active GTP-binding protein Rac1. *J Biol Chem 272*:25238–25242.

May PC, Boggs LN, Fuson KS (1993): Neurotoxicity of human amylin in rat primary hippocampal cultures: similarity to Alzheimer's disease amyloid-beta neurotoxicity. *J Neurochem 61*:2330–2333.

McCampbell A, Taylor JP, Taye AA, Robitschek J, Li M, Walcott J, Merry D, Chai Y, Paulson H, Sobue G, Fischbeck KH (2000): CREB-binding protein sequestration by expanded polyglutamine. *Hum Mol Genet 9*:2197–2202.

Merry DE, Korsmeyer SJ (1997): Bcl-2 gene family in the nervous system. *Annu Rev Neurosci 20*:245–267.

Mesner PW, Epting CL, Hegarty JL, Green SH (1995): A timetable of events during programmed cell death induced by trophic factor withdrawal from neuronal PC12 cells. *J Neurosci 15*:7357–7366.

Mielke K, Herdegen T (2000): JNK and p38 stress kinases — degenerative effectors of signal-transduction cascades in the nervous system. *Prog Neurobiol 61*:45–60.

Mielke K, Brecht S, Dorst A, Herdegen T (1999): Activity and expression of JNK1, p38 and ERK kinases, c-Jun N-terminal phosphorylation, and c-jun promoter binding in the adult rat brain following kainate-induced seizures. *Neuroscience 91*:471–483.

Mielke K, Damm A, Yang DD, Herdegen T (2000): Selective expression of JNK isoforms and stress-specific JNK activity in different neural cell lines. *Brain Res Mol Brain Res 75*:128–137.

Migheli A, Piva R, Atzori C, Troost D, Schiffer D (1997): c-Jun, JNK/SAPK kinases and transcription factor NF-kappa B are selectively activated in astrocytes, but not motor neurons, in amyotrophic lateral sclerosis. *J Neuropathol Exp Neurol 56*:1314–1322.

Miller TM, Johnson Jr. EM (1996): Metabolic and genetic analyses of apoptosis in potassium/serum-deprived rat cerebellar granule cells. *J Neurosci 16*:7487–7495.

Miller TM, Moulder KL, Knudson CM, Creedon DJ, Deshmukh M, Korsmeyer SJ, Johnson Jr. EM (1997): Bax deletion further orders the cell death pathway in cerebellar granule cells and suggests a caspase-independent pathway to cell death. *J Cell Biol 139*:205–217.

Mogi M, Harada M, Kondo T, Mizuno Y, Narabayashi H, Riederer P, Nagatsu T (1996): The soluble form of Fas molecule is elevated in parkinsonian brain tissues. *Neurosci Lett 220*:195–198.

Monteclaro FS, Vogt PK (1993): A Jun-binding protein related to a putative tumor suppressor. *Proc Natl Acad Sci USA 90*:6726–6730.

Morooka T, Nishida E (1998): Requirement of p38 mitogen-activated protein kinase for neuronal differentiation in PC12 cells. *J Biol Chem 273*:24285–24288.

Murakami M, Sonobe MH, Ui M, Kabuyama Y, Watanabe H, Wada T, Handa H, Iba H (1997): Phosphorylation and high level expression of Fra-2 in v-src transformed cells: a pathway of activation of endogenous AP-1. *Oncogene 14*:2435–2444.

Murakami M, Ui M, Iba H (1999): Fra-2-positive autoregulatory loop triggered by mitogen-activated protein kinase (MAPK) and Fra-2 phosphorylation sites by MAPK. *Cell Growth Differ 10*:333–342.

Murray B, Alessandrini A, Cole AJ, Yee AG, Furshpan EJ (1998): Inhibition of the p44/42 MAP kinase pathway protects hippocampal neurons in a cell-culture model of seizure activity. *Proc Natl Acad Sci USA 95*:11975–11980.

Nagata S (1997): Apoptosis by death factor. *Cell 88*:355–365.

Namgung U, Xia Z (2000): Arsenite-induced apoptosis in cortical neurons is mediated by c-Jun N-terminal protein kinase 3 and p38 mitogen-activated protein kinase. *J Neurosci 20*:6442–6451.

Nebreda AR, Porras A (2000): p38 MAP kinases: beyond the stress response. *Trends Biochem Sci 25*:257–260.

Nemoto S, Xiang J, Huang S, Lin A (1998): Induction of apoptosis by SB202190 through inhibition of p38beta mitogen-activated protein kinase. *J Biol Chem 273*:16415–16420.

Nishi K (1997): Expression of c-Jun in dopaminergic neurons of the substantia nigra in 1-methyl-4-phenyl-1,2,3,6-tetrahydropyridine (MPTP)-treated mice. *Brain Res 771*:133–141.

Noh JS, Kang HJ, Kim EY, Sohn S, Chung YK, Kim SU, Gwag BJ (2000): Haloperidol-induced neuronal apoptosis: role of p38 and c-Jun-NH(2)-terminal protein kinase. *J Neurochem 75*:2327–2334.

Odajima J, Matsumura I, Sonoyama J, Daino H, Kawasaki A, Tanaka H, Inohara N, Kitamura T, Downward J, Nakajima K, Hirano T, Kanakura Y (2000): Full oncogenic activities of v-Src are mediated by multiple signaling pathways. Ras as an essential mediator for cell survival. *J Biol Chem 275*:24096–24105.

O'Dell DM, Raghupathi R, Crino PB, Eberwine JH, McIntosh TK (2000): Traumatic brain injury alters the molecular fingerprint of TUNEL-positive cortical neurons in vivo: a single-cell analysis. *J Neurosci 20*:4821–4828.

Ohkubo N, Mitsuda N, Tamatani M, Yamaguchi A, Lee YD, Ogihara T, Vitek MP, Tohyama M (2001): Apolipoprotein E4 stimulates cAMP response element-binding protein transcriptional activity through the extracellular signal-regulated kinase pathway. *J Biol Chem 276*:3046–3053.

Oltvai ZN, Milliman CL, Korsmeyer SJ (1993): Bcl-2 heterodimerizes in vivo with a conserved homolog, Bax, that accelerates programmed cell death. *Cell 74*:609–619.

Oo TF, Henchcliffe C, James D, Burke RE (1999): Expression of c-fos, *c-jun*, and *c-jun* N-terminal kinase (JNK) in a developmental model of induced apoptotic death in neurons of the substantia nigra. *J Neurochem 72*:557–564.

Orth K, Chinnaiyan AM, Garg M, Froelich CJ, Dixit VM (1996): The CED-3/ICE-like protease Mch2 is activated during apoptosis and cleaves the death substrate lamin A. *J Biol Chem 271*:16443–16446.

Ozaki I, Tani E, Ikemoto H, Kitagawa H, Fujikawa H (1999): Activation of stress-activated protein kinase/c-Jun NH2-terminal kinase and p38 kinase in calphostin C-induced apoptosis requires caspase-3-like proteases but is dispensable for cell death. *J Biol Chem 274*:5310–5317.

Ozawa H, Shioda S, Dohi K, Matsumoto H, Mizushima H, Zhou CJ, Funahashi H, Nakai Y, Nakajo S, Matsumoto K (1999): Delayed neuronal cell death in the rat hippocampus is mediated by the mitogen-activated protein kinase signal transduction pathway. *Neurosci Lett 262*:57–60.

Papavassiliou AG, Treier M, Bohmann D (1995): Intramolecular signal transduction in c-Jun. *EMBO J 14*:2014–2019.

Park DS, Stefanis L, Yan CYI, Farinelli SE, Greene LA (1996): Ordering the cell death pathway. Differential effects of BCL2, an interleukin-1-converting enzyme family protease inhibitor, and other survival agents on JNK activation in serum/nerve growth factor-deprived PC12 cells. *J Biol Chem 271*:21898–21905.

Park J, Kim I, Oh YJ, Lee K, Han PL, Choi EJ (1997): Activation of c-Jun N-terminal kinase antagonizes an anti-apoptotic action of Bcl-2. *J Biol Chem 272*:16725–16728.

Pennypacker KR, Eidizadeh S, Kassed CA, O'Callaghan JP, Sanberg PR, Willing AE (2000a): Expression of fos-related antigen-2 in rat hippocampus after middle cerebral arterial occlusion. *Neurosci Lett 289*:1–4.

Pennypacker KR, Yang X, Gordon MN, Benkovic S, Miller D, O'Callaghan JP (2000b): Long-term induction of Fos-related antigen-2 after methamphetamine-, methylenedioxymethamphetamine-, 1-methyl-4-phenyl-1,2,3,6-tetrahydropyridine- and trimethyltin-induced brain injury. *Neuroscience 101*:913–919.

Perez MK, Paulson HL, Pendse SJ, Saionz SJ, Bonini NM, Pittman RN (1998): Recruitment and the role of nuclear localization in polyglutamine-mediated aggregation. *J Cell Biol 143*:1457–1470.

Potapova O, Basu S, Mercola D, Holbrook NJ (2001): Protective role for c-Jun in the cellular response to DNA damage. *J Biol Chem 276*:28546–28553.

Pugazhenthi S, Boras T, O'Connor D, Meintzer MK, Heidenreich KA, Reusch JE (1999): Insulin-like growth factor I-mediated activation of the transcription factor cAMP response element-binding protein in PC12 cells. Involvement of p38 mitogen-activated protein kinase-mediated pathway. *J Biol Chem 274*:2829–2837.

Pulverer BJ, Kyriakis JM, Avruch J, Nikolakaki E, Woodgett JR (1991): Phosphorylation of c-jun mediated by MAP kinases. *Nature 353*:670–674.

Pulverer BJ, Hughes K, Franklin CC, Kraft AS, Leevers SJ, Woodgett JR (1993): Co-purification of mitogen-activated protein kinases with phorbol ester-induced c-Jun kinase activity in U937 leukaemic cells. *Oncogene 8*:407–415.

Pyrzynska B, Mosieniak G, Kaminska B (2000): Changes of the trans-activating potential of AP-1 transcription factor during cyclosporin A-induced apoptosis of glioma cells are mediated by phosphorylation and alterations of AP-1 composition. *J Neurochem 74*:42–51.

Raingeaud J, Whitmarsh AJ, Barrett T, Derijard B, Davis RJ (1996): MKK3- and MKK6-regulated gene expression is mediated by the p38 mitogen-activated protein kinase signal transduction pathway. *Mol Cell Biol 16*:1247–1255.

Raoul C, Henderson CE, Pettmann B (1999): Programmed cell death of embryonic motoneurons triggered through the Fas death receptor. *J Cell Biol 147*:1049–1062.

Rapoport M, Ferreira A (2000): PD98059 prevents neurite degeneration induced by fibrillar beta-amyloid in mature hippocampal neurons. *J Neurochem 74*:125–133.

Read MA, Whitley MZ, Gupta S, Pierce JW, Best J, Davis RJ, Collins T (1997): Tumor necrosis factor alpha-induced E-selectin expression is activated by the nuclear factor-kappaB and c-JUN N-terminal kinase/p38 mitogen-activated protein kinase pathways. *J Biol Chem 272*:2753–2761.

Reglodi D, Somogyvari-Vigh A, Vigh S, Kozicz T, Arimura A (2000): Delayed systemic administration of PACAP38 is neuroprotective in transient middle cerebral artery occlusion in the rat. *Stroke 31*:1411–1417.

Rehfuss RP, Walton KM, Loriaux MM, Goodman RH (1991): The cAMP-regulated enhancer-binding protein ATF-1 activates transcription in response to cAMP-dependent protein kinase A. *J Biol Chem 266*:18431–18434.

Reimold AM, Grusby MJ, Kosaras B, Fries JW, Mori R, Maniwa S, Clauss IM, Collins T, Sidman RL, Glimcher MJ, Glimcher LH (1996): Chondrodysplasia and neurological abnormalities in ATF-2-deficient mice. *Nature 379*:262–265.

Reynolds CH, Utton MA, Gibb GM, Yates A, Anderton BH (1997): Stress-activated protein kinase/c-jun N-terminal kinase phosphorylates tau protein. *J Neurochem 68*:1736–1744.

Reynolds CH, Betts JC, Blackstock WP, Nebreda AR, Anderton BH (2000): Phosphorylation sites on tau identified by nanoelectrospray mass spectrometry: differences in vitro between the mitogen-activated protein kinases ERK2, c-Jun N-terminal kinase and P38, and glycogen synthase kinase-3beta. *J Neurochem 74*:1587–1595.

Riccio A, Ahn S, Davenport CM, Blendy JA, Ginty DD (1999): Mediation by a CREB family transcription factor of NGF-dependent survival of sympathetic neurons. *Science 286*:2358–2361.

Roberts GW, Nash M, Ince PG, Royston MC, Gentleman SM (1993): On the origin of Alzheimer's disease: a hypothesis. *NeuroReport 4*:7–9.

Robinson GA (1996): Changes in the expression of transcription factors ATF-2 and Fra-2 after axotomy and during regeneration in rat retinal ganglion cells. *Brain Res Mol Brain Res 41*:57–64.

Rocha L, Kaufman DL (1998): In vivo administration of c-Fos antisense oligonucleotides accelerates amygdala kindling. *Neurosci Lett 241*:111–114.

Roffler-Tarlov S, Brown JJ, Tarlov E, Stolarov J, Chapman DL, Alexiou M, Papaioannou VE (1996): Programmed cell death in the absence of c-Fos and c-Jun. *Development 122*:1–9.

Rohn TT, Head E, Su JH, Anderson AJ, Bahr BA, Cotman CW, Cribbs DH (2001): Correlation between caspase activation and neurofibrillary tangle formation in Alzheimer's disease. *Am J Pathol 158*:189–198.

Ronai Z, Yang YM, Fuchs SY, Adler V, Sardana M, Herlyn M (1998): ATF2 confers radiation resistance to human melanoma cells. *Oncogene 16*:523–531.

Rosen DR, Siddique T, Patterson D, Figlewicz DA, Sapp P, Hentati A, Donaldson D, Goto J, O'Regan JP, Deng HX et al. (1993): Mutations in Cu/Zn superoxide dismutase gene are associated with familial amyotrophic lateral sclerosis. *Nature 362*:59–62.

Roth JA, Feng L, Walowitz J, Browne RW (2000): Manganese-induced rat pheochromocytoma (PC12) cell death is independent of caspase activation. *J Neurosci Res 61*:162–171.

Ruvolo PP, Deng X, Carr BK, May WS (1998): A functional role for mitochondrial protein kinase Calpha in Bcl2 phosphorylation and suppression of apoptosis. *J Biol Chem 273*:25436–25442.

Saeki K, Yuo A, Suzuki E, Yazaki Y, Takaku F (1999): Aberrant expression of cAMP-response-element-binding protein ('CREB') induces apoptosis. *Biochem J 343 (1)*:249–255.

Sakurai M, Hayashi T, Abe K, Sadahiro M, Tabayashi K (1998): Delayed selective motor neuron death and fas antigen induction after spinal cord ischemia in rabbits. *Brain Res 797*:23–28.

Sanchez-Perez I, Perona R (1999): Lack of c-Jun activity increases survival to cisplatin. *FEBS Lett 453*:151–158.

Saporito MS, Brown EM, Miller MS, Carswell S (1999): CEP-1347/KT-7515, an inhibitor of c-jun N-terminal kinase activation, attenuates the 1-methyl-4-phenyl tetrahydropyridine-mediated loss of nigrostriatal dopaminergic neurons in vivo. *J Pharmacol Exp Ther 288*:421–427.

Saporito MS, Thomas BA, Scott RW (2000): MPTP activates c-Jun NH(2)-terminal kinase (JNK) and its upstream regulatory kinase MKK4 in nigrostriatal neurons in vivo. *J Neurochem 75*:1200–1208.

Satoh T, Nakatsuka D, Watanabe Y, Nagata I, Kikuchi H, Namura S (2000): Neuroprotection by MAPK/ERK kinase inhibition with U0126 against oxidative stress in a mouse neuronal cell line and rat primary cultured cortical neurons. *Neurosci Lett 288*:163–166.

Sawada H, Ibi M, Kihara T, Urushitani M, Honda K, Nakanishi M, Akaike A, Shimohama S (2000): Mechanisms of antiapoptotic effects of estrogens in nigral dopaminergic neurons. *FASEB J 14*:1202–1214.

Schenkel J, Winter C, Weiss C, Martin-Villalba A, Zimmermann M (2000): JUNB overexpression in transgenic mice results in protection against cell death of nigral neurons following axotomy. *Soc Neurosci Abstr 667.12*.

Schlingensiepen KH, Wollnik F, Kunst M, Schlingensiepen R, Herdegen T, Brysch W (1994): The role of Jun transcription factor expression and phosphorylation in neuronal differentiation, neuronal cell death, and plastic adaptations in vivo. *Cell Mol Neurobiol 14*:487–505.

Schneider A, Martin-Villalba A, Weih F, Vogel J, Wirth T, Schwaninger M (1999): NF-kappaB is activated and promotes cell death in focal cerebral ischemia. *Nat Med 5*:554–559.

Selznick LA, Holtzman DM, Han BH, Gokden M, Srinivasan AN, Johnson Jr. EM, Roth KA (1999): In situ immunodetection of neuronal caspase-3 activation in Alzheimer disease. *J Neuropathol Exp Neurol 58*:1020–1026.

Shaulian E, Schreiber M, Piu F, Beeche M, Wagner EF, Karin M (2000): The mammalian UV response: c-Jun induction is required for exit from p53-imposed growth arrest. *Cell 103*:897–907.

Shimizu M, Nomura Y, Suzuki H, Ichikawa E, Takeuchi A, Suzuki M, Nakamura T, Nakajima T, Oda K (1998): Activation of the rat cyclin A promoter by ATF2 and Jun family members and its suppression by ATF4. *Exp Cell Res 239*:93–103.

Shimohata T, Nakajima T, Yamada M, Uchida C, Onodera O, Naruse S, Kimura T, Koide R, Nozaki K, Sano Y, Ishiguro H, Sakoe K, Ooshima T, Sato A, Ikeuchi T, Oyake M, Sato T, Aoyagi Y, Hozumi I, Nagatsu T, Takiyama Y, Nishizawa M, Goto J, Kanazawa I, Davidson I, Tanese N, Takahashi H, Tsuji S (2000): Expanded polyglutamine stretches interact with TAFII130, interfering with CREB-dependent transcription. *Nat Genet 26*:29–36.

Shoji M, Iwakami N, Takeuchi S, Waragai M, Suzuki M, Kanazawa I, Lippa CF, Ono S, Okazawa H (2000): JNK activation is associated with intracellular beta-amyloid accumulation. *Brain Res Mol Brain Res 85*:221–233.

Silverman ES, Du J, Williams AJ, Wadgaonkar R, Drazen JM, Collins T (1998): cAMP-response-element-binding-protein-binding protein (CBP) and p300 are transcriptional co-activators of early growth response factor-1 (Egr-1). *Biochem J 336*:183–189.

Simi A, Ingelman-Sundberg M, Tindberg N (2000): Neuroprotective agent chlomethiazole attenuates c-fos, *c-jun*, and AP-1 activation through inhibition of p38 MAP kinase. *J Cereb Blood Flow Metab 20*:1077–1088.

Smale G, Nichols NR, Brady DR, Finch CE, Horton Jr. WE (1995): Evidence for apoptotic cell death in Alzheimer's disease. *Exp Neurol 133*:225–230.

Smeal T, Binetruy B, Mercola DA, Birrer M, Karin M (1991): Oncogenic and transcriptional cooperation with Ha-Ras requires phosphorylation of c-Jun on serines 63 and 73. *Nature 354*:494–496.

Smeyne RJ, Schilling K, Robertson L, Luk D, Oberdick J, Curran T, Morgan JI (1992): fos-lacZ transgenic mice: mapping sites of gene induction in the central nervous system. *Neuron 8*:13–23.

Smeyne RJ, Vendrell M, Hayward M, Baker SJ, Miao GG, Schilling K, Robertson LM, Curran T, Morgan JI (1993): Continuous c-fos expression precedes programmed cell death in vivo. *Nature 363*:166–169.

Srivastava RK, Mi QS, Hardwick JM, Longo DL (1999): Deletion of the loop region of Bcl-2 completely blocks paclitaxel-induced apoptosis. *Proc Natl Acad Sci USA 96*:3775–3780.

Stadelmann C, Deckwerth TL, Srinivasan A, Bancher C, Bruck W, Jellinger K, Lassmann H (1999): Activation of caspase-3 in single neurons and autophagic granules of granulovacuolar degeneration in Alzheimer's disease. Evidence for apoptotic cell death. *Am J Pathol 155*:1459–1466.

Stanciu M, Wang Y, Kentor R, Burke N, Watkins S, Kress G, Reynolds I, Klann E, Angiolieri MR, Johnson JW, DeFranco DB (2000): Persistent activation of ERK contributes to glutamate-induced oxidative toxicity in a neuronal cell line and primary cortical neuron cultures. *J Biol Chem 275*:12200–12206.

Standen CL, Brownlees J, Grierson AJ, Kesavapany S, Lau KF, McLoughlin DM, Miller CC (2001): Phosphorylation of thr(668) in the cytoplasmic domain of the Alzheimer's disease amyloid precursor protein by stress-activated protein kinase 1b (Jun N-terminal kinase-3). *J Neurochem 76*:316–320.

Steffan JS, Kazantsev A, Spasic-Boskovic O, Greenwald M, Zhu YZ, Gohler H, Wanker EE, Bates GP, Housman DE, Thompson LM (2000): The Huntington's disease protein interacts with p53 and CREB-binding protein and represses transcription. *Proc Natl Acad Sci USA 97*:6763–6768.

Su JH, Anderson AJ, Cummings BJ, Cotman CW (1994): Immunohistochemical evidence for apoptosis in Alzheimer's disease. *NeuroReport 5*:2529–2533.

Sugino T, Nozaki K, Takagi Y, Hattori I, Hashimoto N, Moriguchi T, Nishida E (2000): Activation of mitogen-activated protein kinases after transient forebrain ischemia in gerbil hippocampus. *J Neurosci 20*:4506–4514.

Takagi Y, Nozaki K, Sugino T, Hattori I, Hashimoto N (2000): Phosphorylation of c-Jun NH(2)-terminal kinase and p38 mitogen-activated protein kinase after transient forebrain ischemia in mice. *Neurosci Lett 294*:117–120.

Takahashi A, Alnemri ES, Lazebnik YA, Fernandes-Alnemri T, Litwack G, Moir RD, Goldman RD, Poirier GG, Kaufmann SH, Earnshaw WC (1996): Cleavage of lamin A by Mch2 alpha but not CPP32: multiple interleukin 1 beta-converting enzyme-related proteases with distinct substrate recognition properties are active in apoptosis. *Proc Natl Acad Sci USA 93*:8395–8400.

Takeda J, Maekawa T, Sudo T, Seino Y, Imura H, Saito N, Tanaka C, Ishii S (1991): Expression of the CRE-BP1 transcriptional regulator binding to the cyclic AMP response element in central nervous system, regenerating liver, and human tumors. *Oncogene 6*:1009–1014.

Takeda M, Kato H, Takamiya A, Yoshida A, Kiyama H (2000a): Injury-specific expression of activating transcription factor-3 in retinal ganglion cells and its colocalized expression with phosphorylated c-Jun. *Invest Ophthalmol Vis Sci 41*:2412–2421.

Takeda K, Hatai T, Hamazaki TS, Nishitoh H, Saitoh M, Ichijo H (2000b): Apoptosis signal-regulating kinase 1 (ASK1) induces neuronal differentiation and survival of PC12 cells. *J Biol Chem 275*:9805–9813.

Tanabe H, Eguchi Y, Shimizu S, Martinou JC, Tsujimoto Y (1998): Death-signalling cascade in mouse cerebellar granule neurons. *Eur J Neurosci 10*:1403–1411.

Tanaka K, Nogawa S, Nagata E, Suzuki S, Dembo T, Kosakai A, Fukuuchi Y (1999): Temporal profile of CREB phosphorylation after focal ischemia in rat brain. *NeuroReport 10*:2245–2250.

Tanaka K, Nogawa S, Ito D, Suzuki S, Dembo T, Kosakai A, Fukuuchi Y (2000): Activated phosphorylation of cyclic AMP response element binding protein is associated with preservation of striatal neurons after focal cerebral ischemia in the rat. *Neuroscience 100*:345–354.

Tournier C, Hess P, Yang DD, Xu J, Turner TK, Nimnual A, Bar-Sagi D, Jones SN, Flavell RA, Davis RJ (2000): Requirement of JNK for stress-induced activation of the cytochrome c-mediated death pathway. *Science 288*:870–874.

Toyoshima F, Moriguchi T, Nishida E (1997): Fas induces cytoplasmic apoptotic responses and activation of the MKK7-JNK/SAPK and MKK6-p38 pathways independent of CPP32-like proteases. *J Cell Biol 139*:1005–1015.

Trauth BC, Klas C, Peters AM, Matzku S, Moller P, Falk W, Debatin KM, Krammer PH (1989): Monoclonal antibody-mediated tumor regression by induction of apoptosis. *Science 245*:301–305.

Trejo J, Massamiri T, Deng T, Dewji NN, Bayney RM, Brown JH (1994): A direct role for protein kinase C and the transcription factor Jun/AP-1 in the regulation of the Alzheimer's beta-amyloid precursor protein gene. *J Biol Chem 269*:21682–21690.

Troy CM, Rabacchi SA, Xu Z, Maroney AC, Connors TJ, Shelanski ML, Greene LA (2001): beta-Amyloid-induced neuronal apoptosis requires c-Jun N-terminal kinase activation. *J Neurochem 77*:157–164.

Tsai EY, Jain J, Pesavento PA, Rao A, Goldfeld AE (1996): Tumor necrosis factor alpha gene regulation in activated T cells involves ATF-2/Jun and NFATp. *Mol Cell Biol 16*:459–467.

Tsuji M, Inanami O, Kuwabara M (2000): Neuroprotective effect of alpha-phenyl-N-tert-butylnitrone in gerbil hippocampus is mediated by the mitogen-activated protein kinase pathway and heat shock proteins. *Neurosci Lett 282*:41–44.

Tsujino H, Kondo E, Fukuoka T, Dai Y, Tokunaga A, Miki K, Yonenobu K, Ochi T, Noguchi K (2000): Activating transcription factor 3 (ATF3) induction by axotomy in sensory and motoneurons: a novel neuronal marker of nerve injury. *Mol Cell Neurosci 15*:170–182.

Tucker HM, Rydel RE, Wright S, Estus S (1998): Human amylin induces 'apoptotic' pattern of gene expression concomitant with cortical neuronal apoptosis. *J Neurochem 71*:506–516.

van Dam H, Duyndam M, Rottier R, Bosch A, de Vries-Smits L, Herrlich P, Zantema A, Angel P, van der Eb AJ (1993): Heterodimer formation of cJun and ATF-2 is responsible for induction of c-jun by the 243 amino acid adenovirus E1A protein. *EMBO J 12*:479–487.

van Dam H, Wilhelm D, Herr I, Steffen A, Herrlich P, Angel P (1995): ATF-2 is preferentially activated by stress-activated protein kinases to mediate c-jun induction in response to genotoxic agents. *EMBO J 14*:1798–1811.

Virgo L, de Belleroche J (1995): Induction of the immediate early gene c-jun in human spinal cord in amyotrophic lateral sclerosis with concomitant loss of NMDA receptor NR-1 and glycine transporter mRNA. *Brain Res 676*:196–204.

Vollgraf U, Wegner M, Richter-Landsberg C (1999): Activation of AP-1 and nuclear factor-kappaB transcription factors is involved in hydrogen peroxide-induced apoptotic cell death of oligodendrocytes. *J Neurochem 73*:2501–2509.

Walter J, Schindzielorz A, Grunberg J, Haass C (1999): Phosphorylation of presenilin-2 regulates its cleavage by caspases and retards progression of apoptosis. *Proc Natl Acad Sci USA 96*:1391–1396.

Walton M, Sirimanne E, Williams C, Gluckman P, Dragunow M (1996): The role of the cyclic AMP-responsive element binding protein (CREB) in hypoxic–ischemic brain damage and repair. *Brain Res Mol Brain Res 43*:21–29.

Walton M, Lawlor P, Sirimanne E, Williams C, Gluckman P, Dragunow M (1997): Loss of Ref-1 protein expression precedes DNA fragmentation in apoptotic neurons. *Mol Brain Res 44*:167–170.

Walton M, MacGibbon G, Young D, Sirimanne E, Williams C, Gluckman P, Dragunow M (1998a): Do c-Jun, c-Fos, and amyloid precursor protein play a role in neuronal death or survival?. *J Neurosci Res 53*:330–342.

Walton M, Woodgate AM, Sirimanne E, Gluckman P, Dragunow M (1998b): ATF-2 phosphorylation in apoptotic neuronal death. *Brain Res Mol Brain Res 63*:198–204.

Walton M, Connor B, Lawlor P, Young D, Sirimanne E, Gluckman P, Cole G, Dragunow M (1999a): Neuronal death and survival in two models of hypoxic–ischemic brain damage. *Brain Res Brain Res Rev 29*:137–168.

Walton M, Woodgate AM, Muravlev A, Xu R, During MJ, Dragunow M (1999b): CREB phosphorylation promotes nerve cell survival. *J Neurochem 73*:1836–1842.

Walton MR, Dragunow I (2000): Is CREB a key to neuronal survival?. *Trends Neurosci 23*:48–53.

Wang XZ, Ron D (1996): Stress-induced phosphorylation and activation of the transcription factor CHOP (GADD153) by p38 MAP kinase. *Science 272*:1347–1349.

Wang Y, Huang S, Sah VP, Ross Jr. J, Brown JH, Han J, Chien KR (1998): Cardiac muscle cell hypertrophy and apoptosis induced by distinct members of the p38 mitogen-activated protein kinase family. *J Biol Chem 273*:2161–2168.

Wang ZQ, Ovitt C, Grigoriadis AE, Mohle-Steinlein U, Ruther U, Wagner EF (1992): Bone and haematopoietic defects in mice lacking c-fos. *Nature 360*:741–745.

Watson A, Eilers A, Lallemand D, Kyriakis J, Rubin LL, Ham J (1998): Phosphorylation of c-Jun is necessary for apoptosis induced by survival signal withdrawal in cerebellar granule neurons. *J Neurosci 18*:751–762.

Weidemann A, Paliga KUDR, Reinhard FB, Schuckert O, Evin G, Masters CL (1999): Proteolytic processing of the Alzheimer's disease amyloid precursor protein within its cytoplasmic domain by caspase-like proteases. *J Biol Chem 274*:5823–5829.

Wellington CL, Ellerby LM, Hackam AS, Margolis RL, Trifiro MA, Singaraja R, McCutcheon K, Salvesen GS, Propp SS, Bromm M, Rowland KJ, Zhang T, Rasper D, Roy S, Thornberry N, Pinsky L, Kakizuka A, Ross CA, Nicholson DW, Bredesen DE, Hayden MR (1998): Caspase cleavage of gene products associated with triplet expansion disorders generates truncated fragments containing the polyglutamine tract. *J Biol Chem 273*:9158–9167.

Wessel TC, Joh TH, Volpe BT (1991): In situ hybridization analysis of c-fos and *c-jun* expression in the rat brain following transient forebrain ischemia. *Brain Res 567*:231–240.

Whitfield J, Neame SJ, Paquet L, Bernard O, Ham J (2001): Dominant-negative c-Jun promotes neuronal survival by reducing BIM expression and inhibiting mitochondrial cytochrome c release. *Neuron 29*:629–643.

Whitmarsh AJ, Davis RJ (1996): Transcription factor AP-1 regulation by mitogen-activated protein kinase signal transduction pathways. *J Mol Med 74*:589–607.

Winter C, Schenkel J, Burger E, Eickmeier C, Zimmermann M, Herdegen T (2000): The immunophilin ligand FK506, but not GPI-1046, protects against neuronal death and inhibits c-Jun expression in the substantia nigra pars compacta following transection of the rat medial forebrain bundle. *Neuroscience 95*:753–762.

Wisdom R, Johnson RS, Moore C (1999): c-Jun regulates cell cycle progression and apoptosis by distinct mechanisms. *EMBO J 18*:188–197.

Wolfgang CD, Chen BP, Martindale JL, Holbrook NJ, Hai T (1997): gadd153/Chop10, a potential target gene of the transcriptional repressor ATF3. *Mol Cell Biol 17*:6700–6707.

Wolfman JC, Wolfman A (2000): Endogenous c-N-Ras provides a steady-state anti-apoptotic signal. *J Biol Chem 275*:19315–19323.

Woodgate A, Walton M, MacGibbon GA, Dragunow M (1999): Inducible transcription factor expression in a cell culture model of apoptosis. *Brain Res Mol Brain Res 66*:211–216.

Xia Z, Dickens M, Raingeaud J, Davis RJ, Greenberg ME (1995): Opposing effects of ERK and JNK-p38 MAP kinases on apoptosis. *Science 270*:1326–1331.

Yamada T, Yoshiyama Y, Kawaguchi N (1997): Expression of activating transcription factor-2 (ATF-2), one of the cyclic AMP response element (CRE) binding proteins, in Alzheimer disease and non-neurological brain tissues. *Brain Res 749*:329–334.

Yamagishi S, Yamada M, Ishikawa Y, Matsumoto T, Ikeuchi T, Hatanaka H (2001): p38 mitogen-activated protein kinase regulates low potassium-induced c-Jun phosphorylation and apoptosis in cultured cerebellar granule neurons. *J Biol Chem 276*:5129–5133.

Yamamoto K, Ichijo H, Korsmeyer SJ (1999): BCL-2 is phosphorylated and inactivated by an ASK1/Jun N-terminal protein kinase pathway normally activated at G(2)/M. *Mol Cell Biol 19*:8469–8478.

Yang DD, Kuan CY, Whitmarsh AJ, Rincon M, Zheng TS, Davis RJ, Rakic P, Flavell RA (1997): Absence of excitotoxicity-induced apoptosis in the hippocampus of mice lacking the Jnk3 gene. *Nature 389*:865–870.

Yang DD, Conze D, Whitmarsh AJ, Barrett T, Davis RJ, Rincon M, Flavell RA (1998): Differentiation of $CD4^+$ T cells to Th1 cells requires MAP kinase JNK2. *Immunity 9*:575–585.

Yang F, Sun X, Beech W, Teter B, Wu S, Sigel J, Vinters HV, Frautschy SA, Cole GM (1998): Antibody to caspase-cleaved actin detects apoptosis in differentiated neuroblastoma and plaque-associated neurons and microglia in Alzheimer's disease. *Am J Pathol 152*:379–389.

Yao R, Yoshihara M, Osada H (1997): Specific activation of a c-Jun NH2-terminal kinase isoform and induction of neurite outgrowth in PC-12 cells by staurosporine. *J Biol Chem 272*:18261–18266.

Yonehara S, Ishii A, Yonehara M (1989): A cell-killing monoclonal antibody (anti-Fas) to a cell surface antigen co-downregulated with the receptor of tumor necrosis factor. *J Exp Med 169*:1747–1756.

Yu JS, Hayashi T, Seboun E, Sklar RM, Doolittle TH, Hauser SL (1991): Fos RNA accumulation in multiple sclerosis white matter tissue. *J Neurol Sci 103*:209–215.

Zaman K, Ryu H, Hall D, O'Donovan K, Lin KI, Miller MP, Marquis JC, Baraban JM, Semenza GL, Ratan RR (1999): Protection from oxidative stress-induced apoptosis in cortical neuronal cultures by iron chelators is associated with enhanced DNA binding of hypoxia-inducible factor-1 and ATF-1/CREB and increased expression of glycolytic enzymes, p21(waf1/cip1), and erythropoietin. *J Neurosci 19*:9821–9830.

Zhang L, Jope RS (1999): Oxidative stress differentially modulates phosphorylation of ERK, p38 and CREB induced by NGF or EGF in PC12 cells. *Neurobiol Aging 20*:271–278.

Zhang P, Hirsch EC, Damier P, Duyckaerts C, Javoy-Agid F (1992): c-fos protein-like immunoreactivity: distribution in the human brain and over-expression in the hippocampus of patients with Alzheimer's disease. *Neuroscience 46*:9–21.

Zhang Y, Widmayer MA, Zhang B, Cui JK, Baskin DS (1999): Suppression of post-ischemic-induced fos protein expression by an antisense oligonucleotide to c-fos mRNA leads to increased tissue damage. *Brain Res 832*:112–117.

Zhu X, Raina AK, Rottkamp CA, Aliev G, Perry G, Boux H, Smith MA (2001): Activation and redistribution of c-Jun N-terminal kinase/stress activated protein kinase in degenerating neurons in Alzheimer's disease. *J Neurochem 76*:435–441.

Zirpel L, Janowiak MA, Veltri CA, Parks TN (2000): AMPA receptor-mediated, calcium-dependent CREB phosphorylation in a subpopulation of auditory neurons surviving activity deprivation. *J Neurosci 20*:6267–6275.

Zoumpourlis V, Papassava P, Linardopoulos S, Gillespie D, Balmain A, Pintzas A (2000): High levels of phosphorylated c-Jun, Fra-1, Fra-2 and ATF-2 proteins correlate with malignant phenotypes in the multistage mouse skin carcinogenesis model. *Oncogene 19*:4011–4021.

CHAPTER X

c-Jun, JNK and p38: visualization of neuronal stress responses

THOMAS HERDEGEN AND STEPHAN BRECHT

1. INTRODUCTION

From the very beginning of the research of immediate-early genes (IEGs) in the brain, the expression of the c-Jun transcription factor (TF) was in the center of interest following neurodegenerative disorders. Already the first reports about its expression indicated a particular role of this AP-1 protein in neurodegenerative processes (Herdegen et al., 1991a,b; Leah et al., 1991) which might be only rivalled by JunD, but not of other AP-1 proteins (Herdegen et al., 1993). There are some basic properties of c-Jun upregulation which attribute this TF particular roles in neurobiological disorders: (i) the selective induction, e.g. following nerve fiber transection in the axotomized neurons (reviewed by Herdegen et al., 1997) or in those hippocampal neurons which are protected by conditioning following ischemia (Sommer et al., 1995); (ii) the lasting expression for weeks (Herdegen et al., 1993); (iii) the potential for auto-induction in neurons (Mielke et al., 1999); (iv) the role as nuclear substrate of the c-Jun N-terminal kinases (JNKs) which is supposed to act as major (apoptotic) degenerative effector in the nervous system (reviewed by Herdegen and Waetzig, 2001); moreover, c-Jun can act as docking molecule for JNKs with subsequent phosphorylation of other TFs such as JunD (Kallunki et al., 1996); (v) by functional interaction with ATF-2 (a putatively physiologic–protective TF), a nuclear substrate of p38 kinases, c-Jun is also involved as transcriptional endpoint into another cascade of stress kinases.

1.1. THE NEURONAL STRESS RESPONSE: THE SIGNAL TRANSDUCTION BY SO-CALLED STRESS KINASES

In neuronal and non-neuronal cells or tissues, the so-called stress response comprises more or less well defined enzymatic and genetic alterations following stressful stimuli such as deprivation of trophic factors, ionizing radiation, free radicals (H_2O_2 and peroxynitrite), hypoxia, ischemia, heat shock, production of lipid second messengers such as ceramide or activation of death domain receptors, e.g. by extracellular molecules such as TNFα or Fas-ligand (Karin, 1995; Pena et al., 1997; Ip and Davis, 1998). These potentially deleterious stimuli provoke intracellular reactions that either lead to (programmed) cell death or to defensive–protective adaptations. The bipartite nature of this response is not completely understood and some of the molecules involved, e.g. the c-Jun N-terminal kinases (JNK, also called stress-activated protein kinases, SAPK) and their substrates like the transcription

Handbook of Chemical Neuroanatomy Vol. 19: Immediate Early Genes and Inducible Transcription Factors in Mapping of the Central Nervous System Function and Dysfunction
L. Kaczmarek and H.A. Robertson, editors

factor c-Jun can be linked to both, neurodegeneration and neuroprotection (Herdegen et al., 1997). The protective and degenerative intracellular signal-transduction cascades are activated in parallel or branch from a common intermediate and the final outcome, death or survival with putatively functional recovery, is determined by the availability of trophic molecules, transmembranous stimulation or physico-chemical parameters that alter the balance of anti- and pro-degenerative intracellular programs. Branching points exist at various levels along the catalytic stream of the signal-transcription-coupling. (i) Already the formation of (soluble) monomers or oligomers can determine apoptotic or physiologic/protective intraneuronal reactions following receptor–ligand association as shown for the association of NGF=p75 receptor or TNFα=TNFα-receptor-I molecules (Dechant and Barde, 1997; Becher et al., 1998). (ii) Receptors are the second branching point. The intracellular domain of various receptors can associate with different adaptor molecules resulting in stimulation of specific cascades. For example, the TNFα-receptor-I (TNF-R1)/TRADD complex activates JNK directly through interaction with the noncytotoxic TNF receptor-associated factor protein 2 (TRAF2) whereas activation of NFκB through TNF-R1 requires a protein complex consisting of TRAF2/NIK or RIP. Independent of TRAF2, TNF-R1-induced apoptosis occurs via binding of a complex FADD/FELICE (Natoli et al., 1997, 1998). The Fas receptor engages two separate pathways to induce cell death: one pathway via DAXX that involves JNK activation is blocked by Bcl-2, and a second pathway via FADD/FELICE that is Bcl-2-insensitive (X. Yang et al., 1997). (iii) Interestingly, the same signal intermediate can exert different functions in dependence on the activatory upstream enzyme. For example, production of the lipid messenger ceramide by the acid sphingomyelinase leads to cell death including activation of JNK, whereas production of ceramide by the neutral sphingomyelinase is associated with physiological or reparative processes including activation of ERK (Pena et al., 1997). (iv) Information transfer by kinases is hierarchically organized by cascades of kinases (Fig. 1), and several of these cascades can be activated in parallel such as the cascades of JNKs and ERKs. In this case, kinases can propagate antagonistic programs and such a dualism has been found in starved PC12 cells with JNK and p38 stress kinases as mediators of apoptosis and the facilitating downregulation of ERK (Xia et al., 1995); similar, ionizing radiation or TNFα are strong activators of JNK but not of extracellular signal-regulated kinase (ERK), whereas H_2O_2, another cellular stressor, activates ERK but not JNK at lower concentrations (Hannun, 1996; Pena et al., 1997).

However, the idea of separately operating apoptotic or physiologic/reparative kinase systems is complicated by the possibility of reinforcing/antagonizing cross-talks at each level of the signal-transduction cascade. Thus, c-Jun N-terminal kinase kinase (JNKK) activates not only JNK, but also ERK with subsequent stimulation of Elk transcription factors and c-*fos* induction (Zinck et al., 1995; Minden and Karin, 1997). MEKK1, an upstream kinase of the JNK pathway, links the JNK and NFκB pathways by phosphorylation of the IKBα complex with subsequent dissociation of NFκB/IκB (Lee et al., 1997). (v) Finally, pre-existing and inducible transcription factors control in the nucleus the realization of different genetic programs in dependence on their phosphorylation and subsequent association with other transcription factors. Thus, following association with CBP or Fos proteins, the c-Jun transcription factor induces CRE- or AP-1-regulated target genes, respectively (Arias et al., 1994), and this operational range attributes to c-Jun the role of a nuclear branching point at the downstream level of gene transcription.

There is increasing evidence that neurons respond to stress with activation of signal-transcription cascades and gene expression that are well analyzed in non-neuronal cells such as immune cells, fibroblasts or tumor cells. These reactions comprise among others:

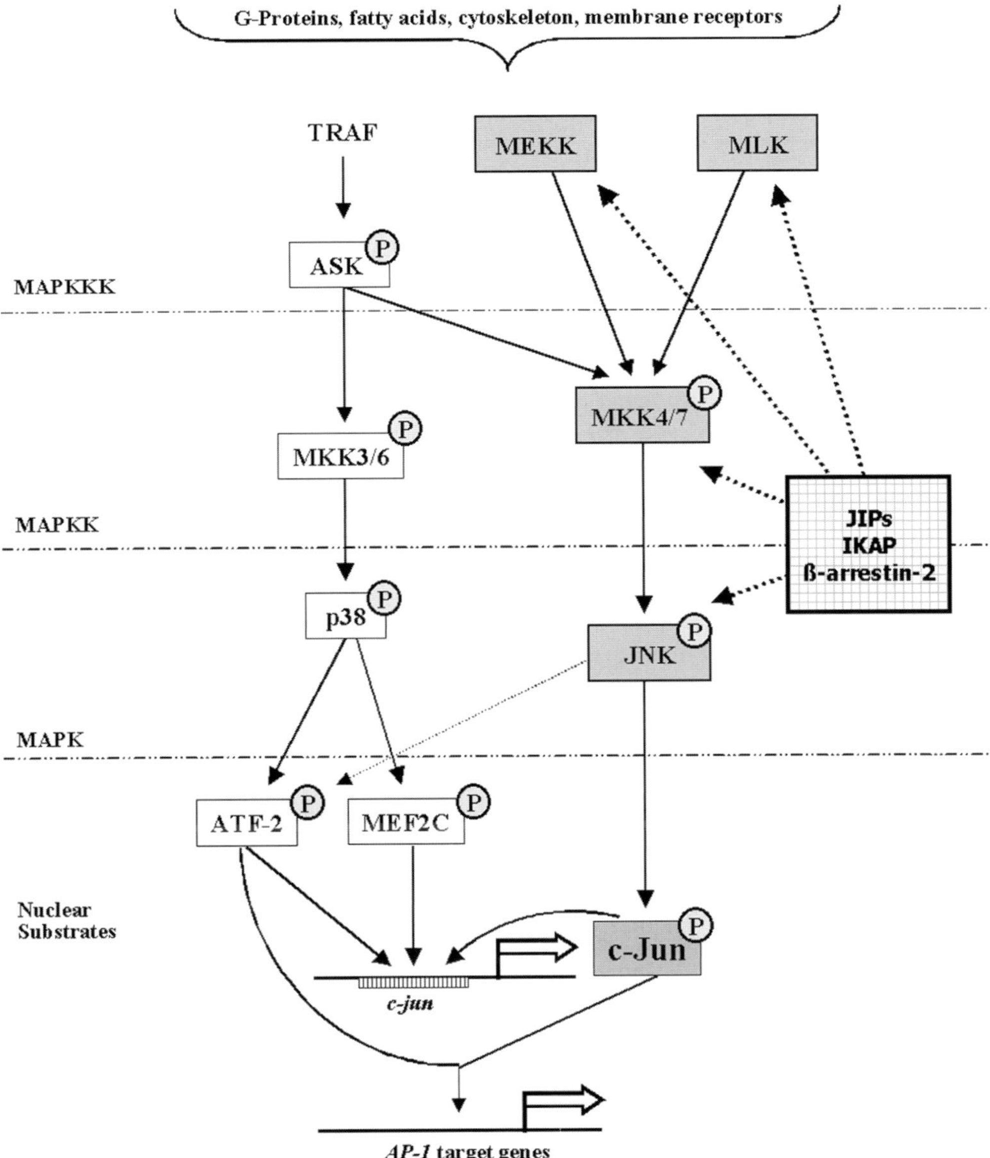

Fig. 1. The JNK-signal-transduction cascade. This graphic gives the different levels of the JNK-signal-transduction cascade comprising the most upstream MAP kinase kinase kinase (MAPKKK), the MAP kinase kinase (MAPKK), the MAP kinase (MAPK) and the nuclear substrates. The cytoplasmic substrates are not indicated. The scaffold proteins are set into the checked box which form scaffold complexes with the molecules framed by gray boxes and interconnected by dotted lines.

the expression and release of interleukins, TNFα or Fas-Ligand (CD95-ligand, APO-1), and their respective receptors (Matsuyama et al., 1995; Bruce et al., 1996); activation of proteases (Nicholson and Thornberry, 1997); and activation of stress kinases JNK or p38 (Xia et al., 1995; Herdegen et al., 1998; Mielke et al., 1999). Moreover, neurons are responsive to a variety of stressful stimuli that have been only recently considered as 'typical' neuronal stressors such as H_2O_2, ionizing radiation, or ceramide. The insight into these intraneuronal operations is meaningful for two reasons. Firstly, the regulated presence of these molecules enlarges our understanding of their functional repertoire. Secondly, the above-mentioned signal cascades are well defined in non-neuronal cells, and a variety of agonistic or antagonistic molecules has already been tested for analytic or therapeutic purposes (e.g. inhibitors of p38 or JNK) (Badger et al., 1996; Henry et al., 1998; Maroney et al., 1998). These findings raise the possibility to cut down apoptotic pathways or to enhance the protective programs of neurons.

1.2. TRANSCRIPTIONAL ACTIONS EXECUTED BY TRANSCRIPTION FACTORS

Knock-out of the c-*jun* gene is lethal for the developing embryo (Hilberg et al., 1993), whereas inhibition of the N-terminal phosphorylation by exchange of the serine residues 63 and 73 for alanin (c-JunAA) does not interfere with the development of an apparently intact and fertile phenotype (Behrens et al., 1999). Consequently, functional inactivation of c-Jun with preservation of its physical presence substantially differs from functional inactivation by physical absence: in contrast to the knock-out of c-*jun*, c-JunAA can still act as suppressor and antagonizing inhibitor of other transcription factors (TFs). Thus, for an understanding of the AP-1 functions, we have to consider their elaborate transcriptional actions (Chinenov and Kerppola, 2001). As with the majority of TFs, AP-1 proteins function in several ways.

(a) Induction of transcription. Regulated by their post-translational modifications (mainly by phosphorylation), their partners for the formation of homo- and/or heterodimers and scaffold proteins, AP-1 proteins bind to regulatory DNA sequences (AP-1 and ATF/CRE-like consensus motifs) and activate the basal transcription machinery (reviewed by Herdegen and Leah, 1998; Ferrell, 2000; Yaniv and Yisraeli, 2001). In the context of the transcriptional orchestration, the AP-1 proteins co-determine the time and the amount of gene transcription.

(b) Suppression of transcription. Transcriptional suppression can be achieved by transcription factor binding to regulatory DNA promoter sequences without activating of RNA-polymerase II complex. For the distinction of the transcriptional state (activating versus non-activating), it is mandatory to visualize the specific post-translational modifications which underlie the activation of the RNA-polymerase II by individual TFs. Thus, the mere presence of nuclear TF/AP-1 proteins does not allow conclusions about the transcriptional state of TFs.

(c) Modulation of transcription beyond the DNA-polymerase interface. AP-1 proteins can associate with other TFs such as steroid hormone receptor complexes, and this association prevents the DNA binding of both partners (Herrlich, 2001).

(d) Non-transcriptional effects. The presence of mRNA and/or proteins of TFs including AP-1 proteins in neurons outside the nucleus suggests functions of TFs beyond transcriptional control. For example, Fra-1 and the 55 kDa c-Fos isoform are present in synaptosomes and synaptic plasma membrane fractions from the rat cerebral cortex and hippocampus, from where they might be retransported into the nucleus (Paratcha et al., 2000). FosB immunoreactivities were found in dendrites following kainic acid-induced seizures (Gass et al., 1993) and in the nucleolus following axotomy (Robinson, 1995). ATF-2, which might be relevant for c-Jun-mediated transcription (Herdegen et al., 1997), is transported along the axon–target axis apparently serving as an information shuttle between perikaryon and axonal–synaptic events.

1.3. COUNTERACTING EFFECTS OF AP-1 PROTEINS IN DIFFERENT POPULATIONS OF CNS CELLS

Besides neurons, a variety of distinct cell populations mediate physiological functions and trigger pathological processes in the brain, such as microglia and astrocytes, oligodendrocytes and Schwann cells, ependymal cells and meningeal cells. It is important to realize that depending on the cell type similar actions of TFs might result in functionally opposite effects. Three examples will illustrate this critical issue.

(i) Neurons can be killed by apoptotic programs which include AP-1 activity (Behrens et al., 1999), but in parallel, apoptosis is also involved in the termination of microglial activity resembling the processes in immune cells (Raivich et al., 1999). In consequence, inhibition of apoptosis — one of the central aims of novel strategies for the treatment of neurodegenerative diseases — could also counteract the termination of microglial activity with pro-degenerative effects such as persistent phagocytosis and immune reactions.

(ii) AP-1 proteins are essential for cell cycle progression and are also involved in the differentiation of cells (Hilberg et al., 1993; Leppa and Bohmann, 1999). In consequence, expression of AP-1 proteins in non-neuronal cells such as astrocytes and meningeal cells in vivo might trigger fundamental physiological responses to injury which are inevitably blocked by AP-1 antagonization.

(iii) An important group of AP-1 target genes comprises the matrix metalloproteinases which are involved in inflammatory processes as well as in structural guidance depending on the expressing cell type (Xu et al., 2001).

The functional inhibition of AP-1 actions (e.g. by antisense- or decoy-oligonucleotides, inhibition of their upstream activatory kinases) as a strategy for the treatment of neurodegenerative disorders demands careful analysis of the physiological action of AP-1 proteins for the estimation of harmful side effects. Moreover, this analysis is the premise for answering the question to which extent neurodegenerative actions of AP-1 are pathologically enhanced physiological actions or due to qualitatively different properties such as binding of novel partners and the subsequent transcription of novel target genes. In addition, the inhibition of AP-1 expression (e.g. by antisense-oligonucleotides) negatively affects memory formation (Grimm et al., 1997) and physiological-adaptive processes such as synchronization of the endogenous clock (Wollnik et al., 1995). It remains a major challenge to develop strategies for AP-1 inhibition which selectively blocks degenerative actions without affecting physiological functions. Thus, null mutation of c-*fos* attenuates phenotypes of pathological activity such as kindling development as well as physiological functions such as reorganization of mossy fibers (Watanabe et al., 1996).

2. VISUALIZATION OF c-Jun/AP-1 PROTEINS IN THE ADULT BRAIN

2.1. THE TEMPOROSPATIAL EXPRESSION PATTERN

The AP-1 proteins substantially differ in their temporospatial expression patterns (Fig. 2). The analysis of their distribution alleviates the understanding of their putative functions since the combinatory variety is limited by the individual expression. In theory, more than 20 dimers can be formed by heterodimerization of Fos and Jun proteins or homodimerization of Jun proteins without considering the splice variants; this number is furthermore enhanced when we regard the heterodimers with non-AP-1 proteins (summarized in Herdegen and Leah,

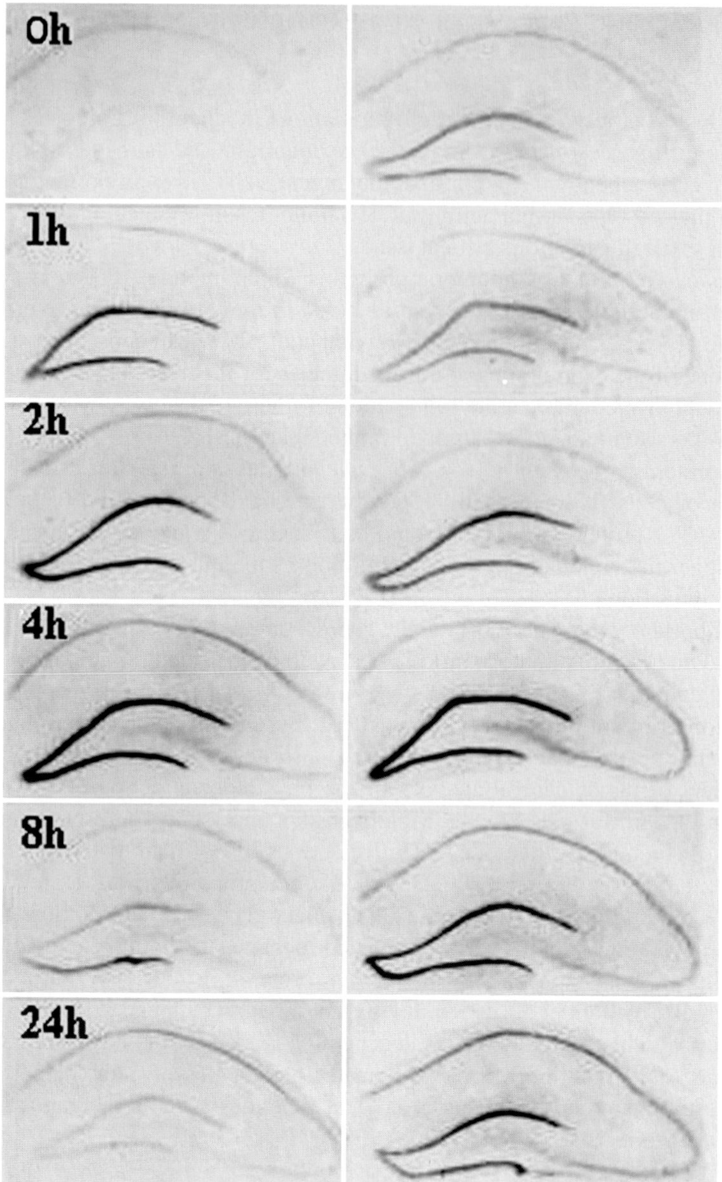

Fig. 2. Specific expression pattern of AP-1 proteins. Individual temporo-spatial patterns of expression of JunB (left hand side) and JunD (right hand side) in the rat hippocampus after systemic application of kainic acid (Gass et al., 1993).

1998). It may be helpful to roughly characterize the inducibility of the AP-1 proteins in the brain and their role in brain pathophysiology (for details see Herdegen and Leah, 1998).

(a) c-Jun. Basal expression is detected in various compartments including the dentate gyrus, motoneurons, peripheral neurons and nuclei of the autonomic nervous system. Following transynaptic stimulation, c-Jun shows a similar induction pattern as c-Fos and JunB. Stimulation by non-synaptic insults (i.e. transection of nerve fibers) selectively induces c-Jun and, to a lesser extent also JunD, but not other AP-1 proteins (Fig. 3) (Herdegen et al., 1997). This selective injury-triggered induction might be mediated through *cis*-activation of the distal AP-1 sequence of the c-*jun* promoter by c-Jun (Mielke et al., 1999).

(b) JunB. Almost absent under basal conditions, this Jun protein is mainly induced in neurons in a pattern and half-life very similar to the properties of c-Fos (Gass et al., 1993; Edelstein et al., 2000; Hata et al., 2000). Following injury, induction of JunB and c-Fos might primarily be mediated by transynaptic neuronal excitation; whether the inducibility of the *junB* promoter by IL-6 and CNTF (Coffer et al., 1995) is relevant for neuronal JunB expression, remains to be elucidated.

(c) JunD. Expressed in neurons and glial cells, JunD displays a wide basal expression in the majority of cells in the nervous system. Following transynaptic and non-synaptic stimulation, JunD expression is enhanced in a delayed but lasting fashion. The pattern of JunD expression following non-synaptic degenerative stimulation resembles that of c-Jun. Phosphorylation of JunD by JNKs, however, is only achieved following heterodimerization with c-Jun (Kallunki et al., 1996).

(d) c-Fos. Due to its fast expression, c-Fos becomes detectable within 30–35 min following the onset of an experimental stimulus from rather low or non-detectable basal expression. Its expression is rather transient, at least following termination of the primary inducing stimulus (Purkiss et al., 1993). In case of parallel induction of JunB, the c-Fos proteins might prefer this Jun isoform to c-Jun. The inducibility of c-Fos declines with progressing age (Salehi et al., 1999) and this observation supports the notion that c-Fos expression widely reflects the cellular responsiveness at the transcriptional level.

(e) FosB. The evident characteristic of FosB expression is its widespread absence in the brain under basal conditions and its delayed but protracted expression following intentional stimulation (Gass et al., 1993). In contrast to other AP-1 proteins, FosB immunoreactivity can display specific labeling of the nucleolus raising the question about its dimerization partner (Robinson, 1995). Importantly, splicing of the *fosB* gene yields truncated ΔFosB proteins with a comparatively persistent half-life which can confer robust inhibition of AP-1-dependent transcription (Chen et al., 1997). Knock-out of the *fosB* gene dramatically affects the maternal behavior towards newborn pubs, and FosB is involved in further physiological functions such as the circadian rhythm and the imprint of the addiction memory (Chen et al., 1997; Herdegen and Leah, 1998).

(f) Fra1, Fra2. The understanding of these AP-1 proteins is still handicapped by the insufficient characterization of the generation and structure of the proteins, and there is only recent evidence that Fra isoforms derive from the *fosB* gene (Chen et al., 1997). mRNA analysis demonstrated a lasting expression of these AP-1 members after transient stimuli indicating a particular role for memory formation. This notion is supported by presynaptic and dendritic localization of Fra-IR which changes following behavioral stimuli (Paratcha et al., 2000).

2.2. IMMUNOCYTOCHEMISTRY OF c-Jun

2.2.1. Specificity of the immunocytochemical signal

The individual expression of c-Jun can be visualized by a variety of specific antisera which most likely do not exhibit cross-reactivities with other Jun/AP-1 proteins as critically discussed elsewhere (Herdegen et al., 1995). The arguments for the specific visualization of c-Jun derives from the following observations: (i) in the untreated rat and mouse brain,

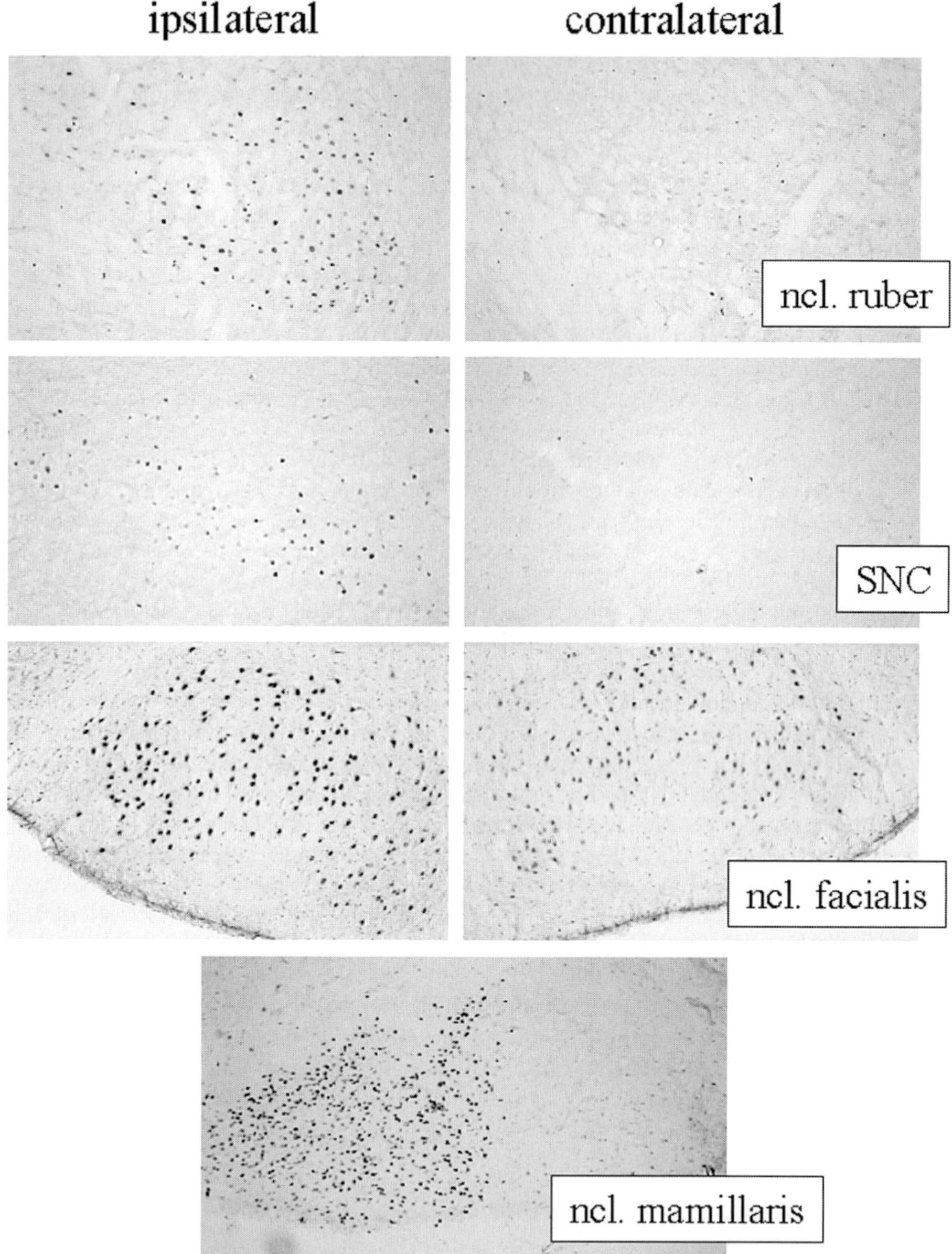

Fig. 3. c-Jun in axotomized neurons. Expression of c-Jun in axotomized neurons (left column) 3 days following transection of the corresponding nerve fiber tract, i.e. dorsal columns, medial forebrain bundle, facial nerve and mammillo-thalamic tract, respectively. SNC = substantia nigra pars compacta.

c-Jun is present in some compartments such as primary neurons, motoneurons or dentate gyrus whereas JunB is almost undetectable and JunD reveals a basal expression in apparently all neurons (Herdegen et al., 1991a,b, 1995); (ii) besides the basal expression pattern, the Jun antisera yielded different dynamics of induction of the individual Jun proteins: rapid induction of c-Jun and JunB following transynaptic excitation, e.g. noxious stimulation (Herdegen et al., 1991a,b; Tölle et al., 1994) or kainate excitotoxicity (Gass et al., 1993) with prolonged persistence of c-Jun, whereas JunD revealed a delayed inducibility; (iii) c-Jun immunoreactivity (IR), but not JunB-IR, rises in axotomized neurons, and this c-Jun-IR precedes the JunD-IR (Leah et al., 1991).

2.2.2. Cell types

The distinction of cell types expressing TFs such as AP-1 proteins has a twofold meaning. At first, it helps to understand the transcriptional regulators responsible for the realization of the individual cellular phenotype. Secondly, AP-1 proteins are blamed for their involvement in neurodegenerative–apoptotic processes, and in consequence, neuroprotective strategies aim at the antagonization of AP-1 actions. It became clear, however, that unspecific approaches affecting all (brain) cells might have negative side effects which include suppression of physiological, protective and/or anti-pathological actions (see Section 1.3).

2.2.3. The intracellular distribution of c-Jun

AP-1 proteins are supposed to act as TFs in the nucleus. In the last years, however, evidence has arisen that AP-1 are distributed outside the nucleus; the functional consequences of the underlying observations are entirely obscure (see Section 2.1). The extranuclear localization is not specific to AP-1 proteins, but affects also other TFs. For example, the constitutively expressed TF ATF-2 serves as axonal messenger retrogradely transported following axonal injury (Delcroix et al., 1999).

Extranuclear appearance could also be derived from blocking of nuclear translocation as reported for c-Fos following cellular starvation (Roux et al., 1990) but this result was never reproduced later on. The localization of c-Jun is almost exclusively intranuclear. The few observations on cytoplasmic c-Jun, e.g. in axotomized vagal motoneurons (Herdegen et al., 1991a,b), are rare (false positive?) examples which exclude a substantial role of c-Jun outside the nucleus.

3. THE RELATION OF AP-1 PROTEINS AND MAP KINASES AND THEIR PHYSIOLOGICAL FUNCTIONS

For the analysis of AP-1 functions, it is mandatory to understand the upstream system of kinases which is essential for the induction and/or activation of the transcription factors. The AP-1 dimers are part of the transcriptional effector system of the ERK, JNK and p38 MAP kinases, i.e. AP-1 proteins fulfill (some of) their functions in the context of MAP kinases. As with other kinase systems, the MAP kinases integrate intra- and extracellular signals and reinforce or antagonize the (parallel) information transfer of other signal-transduction pathways (Ichijo, 1999; Davis, 2000). Therefore, the understanding of the effector profile of MAP kinases provides crucial insight into the function of AP-1 proteins. We have to emphasize, however, that AP-1 proteins can modulate gene transcription independent of preceding MAP-kinase activation and, vice versa, MAP kinases have a wide range of action by post-translational modification of non-nuclear substrates.

3.1. THE c-Jun/JNK-AXIS

It is accepted knowledge that the transcriptional action of c-Jun strictly depends on its phosphorylation of the serine 63 and 73 residues (Gupta et al., 1996; Kallunki et al., 1996). Consequently, the study of JNKs in the brain enlightens the contextual understanding of their nuclear substrates c-Jun and (in c-Jun heterodimers) JunD (Kallunki et al., 1996).

However, data are also available on calcium-triggered c-Jun activation independent of both, the phosphorylation of N-terminal serines 63/73 and the activation of JNKs (Cruzalegui et al., 1999; Rizzo et al., 1999), and in PC12 cells, phosphorylation of the c-Jun serine residues 63 and 73 was found to be regulated by ERKs (Leppa et al., 1998). Besides these residues, other amino acids might be relevant for c-Jun activation since the exchange of threonine residues at position 91 and 93 to phosphate-mimicking aspartic acid residues enhances the 'gain of function' of the serines 63 and 73 (Musti et al., 1997). The presence of JNK-interacting protein (JIP), a positive as well as negative regulator of JNK-mediated c-Jun phosphorylation (Dickens et al., 1997), can channel JNK signaling away from c-Jun as shown in cerebellar granule cells where an increase in nuclear JNK activity parallels a decrease in c-Jun phosphorylation and expression (Coffey et al., 2000). Finally, primary hippocampal/cortical neurons from c-JunAA mutants are protected against kainic acid-induced apoptosis in spite of unaltered JNK activity (Behrens et al., 1999).

3.1.1. Physiological functions

In the adult mammalian brain, the three *JNK* genes are transcribed and are most likely present in their activated forms (Carletti et al., 1995; Martin et al., 1996; D.D. Yang et al., 1997; Mielke et al., 2000; Schauwecker, 2000) under normal conditions, i.e. during the absence of any intentional stimulation. One prominent feature of the c-Jun/JNK-axis is their involvement in neuronal plasticity and memory formation: (i) JNKs are largely residing in neurites in vitro (Coffey et al., 2000) and are retrogradely transported in neuronal axons in vivo (Fernyhough et al., 1999), or i.e. they act as signal molecules between the presynaptic terminal and the perikaryon; (ii) JNKs alter cytoskeletal functions by transcriptional regulation and phosphorylation of neurofilaments, tau proteins and microtubulins (Sadot et al., 1998; Reynolds et al., 2000); (iii) inversely, by inhibition of glucocorticoid receptor transcriptional activity, c-Jun and JNKs are effectors of specific cell programs initiated by depolymerization of the cytoskeleton after confluency (Oren et al., 1999); and, similarly, c-Jun and JNKs are activated by microtubular disarray (Wang et al., 1998; Yujiri et al., 1999); on the other hand, confluency of PC12 cells drastically inhibited c-Jun phosphorylation by JNKs (Lallemand et al., 1998); taken together, the c-Jun/JNK-axis participates in the sensoring for cytoskeletal alterations; (iv) JNK3 phosphorylates the threonine at position 668 of the Alzheimer's disease amyloid precursor protein (APP) (Standen et al., 2001) with physiological functions such as effects on neuronal differentiation; (v) JNKs receive transmembranous signals in microdomains (McDonald et al., 2000); and, finally (vi) JNKs initiate long-lasting genetic alterations by activation of TFs such as c-Jun, JunD or ATF-2 (Gupta et al., 1996; Kallunki et al., 1996). This plethora of properties renders JNKs as perfect candidate molecules for memory formation. In fact, visual stimulation, exposure to a novel environment and generation of electro-convulsive seizures (an experimental paradigm for neuronal plasticity) activates

JNK with N-terminal phosphorylation of c-Jun in the rodent brain (Xu et al., 1997; Brecht et al., 1999). Interestingly, repetition of electro-convulsive seizures dramatically reduces their effectiveness for c-Jun phosphorylation in the presence of ongoing JNK activation (Brecht et al., 1999). This habituation of the transcriptional processes indicates the presence of regulatory mechanisms for JNK-triggered c-Jun activation which are activated following lasting and/or repetitive stimulation underlying neuronal plasticity. It is important in this context that the glutamate receptor subunit 6 (GluR6, a member of the kainate receptor family) mediates seizure activity and spontaneous bursts (Telfeian et al., 2000) with reduced sensitivity to kainic acid in mice with GluR6 knock-out (Mulle et al., 1998). GluR6 transmission is coupled with JNK activation via PSD-95/SAP90 (a membrane-associated guanylate kinase), MLK2 and JIP (Yasuda et al., 1999; Sheng and Pak, 2000). In consequence, inhibition of JNKs or c-Jun might dissociate the essential axis between synaptic excitation and gene induction with resistance to seizure activity as seen following JNK3 knock-out and c-JunAA *knock-in*, but with the potential risk of inhibiting learning and memory formation.

Besides the adult nervous system, the c-Jun/JNK-axis has important functions for the maturation and differentiation of neurons in vivo and in vitro. In cerebellar neurons, nuclear JNK activity from MKK7 complexes increases during neuritogenesis (in parallel with reduced c-Jun activity) (Coffey et al., 2000); in PC12 cells, c-Jun and JNK contribute to neurite outgrowth (Kita et al., 1998; Leppa et al., 1998). During postnatal development, the three JNK isoforms decline at various degrees in the brain (Carboni et al., 1998), and these high perinatal levels indicate a role of JNKs during developmental neuroplasticity.

3.1.2. Variable activation of c-Jun and JNKs: questionable reliability

The expression and inducibility of individual JNK proteins, and in particular of their splice variants, however, under normal conditions and following patho-physiological stimuli are not yet resolved. This is true for in vivo and in vitro systems (Table 1). The data are also differing on the intracellular distribution of JNKs as well as on the N-terminal phosphorylation of c-Jun.

(i) There are studies reporting a weak and limited distribution of JNK1 in the brain which is also absent in the hippocampus (Carboni et al., 1998) while others describe a high JNK1-IR level which parallels that of JNK3 (Lee et al., 1999) and a substantial basal JNK1 activity in the hippocampus (Mielke et al., 1999). Inversely, information on strong and widely distributed JNK2 mRNA signals (Carboni et al., 1998) is in contrast to publications of low level JNK2-IR (Lee et al., 1999).

(ii) The intranuclear presence of JNKs under basal conditions and the nuclear translocation of JNKs following stress (with subsequent phosphorylation of c-Jun and/or JunD) are further issues of controversial observations. Under basal conditions, JNK3-IR was detected in the cytoplasm (Martin et al., 1996), whereas JNK1-IR was found in the cytoplasm and the nucleus (Schauwecker, 2000) or in the cytoplasm and the dendrites (Lee et al., 1999) or only in the nucleus (Oo et al., 1999). Visualization of phosphorylated JNK1 + JNK2 immunoreactivity (p-JNK-IR) yielded a complete absence in untreated gerbils, and a nuclear p-JNK-IR following ischemia in the hippocampus (Sugino et al., 2000). Differing observations exist on the absence (Herdegen et al., 1998) or presence (Schauwecker, 2000) of phosphorylated c-Jun in the dentate gyrus, an area with high constitutive expression of c-Jun. Following systemic application of adriamycin, JNK1 is translocated into the nucleus without alteration of the total JNK1 content in dorsal root ganglia, and in these cells, JNK1 and phosphorylated c-Jun are concentrated in nuclear domains enriched in splicing factors and perichromatin fibrils (Pena et

TABLE 1. *Expression and intracellular distribution of members of the JNK-MAP-cascade — overview of the expression and the activation of JNKs and their upstream regulators in the brain and in neuronal cells in vitro*

Isoform	Stimulus/disease	Area/cell type	Expression	Activity	Localization	Reference
JIP-1	MCA occlusion	cortex	↑		cytoplasm	Hayashi et al., 2000
MKK4	MPTP	striatum	=	↑	cytoplasm (only neurons)	Saporito et al., 2000
MKK4	MPTP	substantia nigra	=	↑	cytoplasm (only neurons)	Saporito et al., 2000
MKK4	anisomycin, serum deprivation	cerebellar granule neurons		↑	nucleus	Coffey et al., 2000
MKK7	postnatal development	forebrain, cerebellum		↑	nucleus	Coffey et al., 2000
Total JNK	striatal excitotoxic lesion	substantia nigra	↑		nucleus	Oo et al., 1999
Total JNK	anisomycin, serum deprivation	cerebellar granule neurons	↓	↑	nucleus (associated with MKK4)	Coffey et al., 2000
Total JNK	postnatal development	forebrain, cerebellum	=	↑	nucleus (associated with MKK7)	Coffey et al., 2000
Total JNK	MPTP	striatum	=	↑		Saporito et al., 2000
Total JNK	MPTP	substantia nigra	=	↑		Saporito et al., 2000
Total JNK	ALS	spinal cord		↑ (strong in astrocytes)		Migheli et al., 1997
Total JNK	Adriamycin	trigeminal ganglion neurons, rat	=	↑	increase in the nucleus	Pena et al., 2000
JNK1	NGF deprivation	differentiated PC12 cells		↑		Eilers et al., 1998
JNK1	neuritogenese (by protease inhibition)	PC12 cells		↑		Giasson et al., 1999
JNK1	postnatal development	brain (rat)	↓			Carboni et al., 1998
JNK1	embryonic development	CNS, PNS (mouse)	↑			Martin et al., 1996
JNK1	MCA occlusion	cortex	↑	↑	initially cytoplasm, later on nuclear	Hayashi et al., 2000
JNK1 (JNK3-activity)	transient hypoxia	fetal forebrain cultured cells, rat	↑ inhibited during hypoxia	↑		Chihab et al., 1998

TABLE 1. *(continued)*

Isoform	Stimulus/disease	Area/cell type	Expression	Activity	Localization	Reference
JNK1	ionizing radiation	brain homogenates rat	↑			Ferrer et al., 1997
JNK1	kainic acid	brain homogenates rat	↓			Ferrer et al., 1997
JNK1	kainic acid	hippocampus, mouse	↑ in susceptible cells	slight ↑ in resistant cells		Schauwecker, 2000
JNK1	sodium arsenite	cortical neurons	=	=		Namgung and Xia, 2000
JNK2	sodium arsenite	cortical neurons	=	=		Namgung and Xia, 2000
Phosphorylated JNK1 + 2	transient global ischemia	hippocampus (gerbil)	=	↑	nucleus	Sugino et al., 2000
JNK3	hypoxia	hippocampus			nucleus	Zhang et al., 1998
JNK3	postnatal development	brain (rat)	↓			Carboni et al., 1998
JNK3	transient hypoxia	fetal forebrain cultured cells, rat		↑		Chihab et al., 1998
JNK3	sodium arsenite	cortical neurons	=	↑		Namgung and Xia, 2000

↑ gives an increase, ↓ gives a decrease of expression and/or activity; = marks no significant alteration.

al., 2000). Kainic acid raised the JNK1-IR in the cytoplasm and the nucleus of hippocampal neurons based on a moderate increase in the 46-kDa-JNK fraction (Schauwecker, 2000), whereas MPTP provoked a moderate increase in the 54-kDa-JNK fraction in the substantia nigra (Saporito et al., 2000).

(iii) Substantial contradiction exists on the alteration of the expression of JNKs following neuronal stress. Downregulation of JNK1 expression (Chihab et al., 1998) and immunoreactivity (Ferrer et al., 1997) is opposed to enhanced JNK1-IR (Schauwecker, 2000) and an increase in JNK1 activity (Mielke et al., 1999) in the hippocampus following kainic acid excitotoxicity and fetal hypoxia. JNK1 expression was upregulated in the substantia nigra compacta following striatal lesion (Oo et al., 1999), whereas transcription of all three JNK isoform mRNAs did not change following severe stress in neural cell lines (Mielke et al., 2000).

(iv) Data based on the determination of JNK activity demand critical interpretation. Firstly, immunocomplex assays might display JNK autophosphorylation rather than JNK-triggered phosphorylation of the substrate (mainly GST-c-Jun) (Mielke et al., 2000). Secondly, analysis of the activity of individual JNK-isoforms are impeded by lack of specific antibodies. Thirdly, neuronal degeneration with the induction of one JNK isoform (e.g. JNK3) in the presence of high basal activity of other isoforms (e.g. JNK1 and JNK2) does not simply argue for a specific neurodegenerative role of the induced isoform (Namgung and Xia, 2000).

The reasons for these divergent findings at mRNA and protein levels as well as in immunocytochemistry are not clear. At least for immunocytochemistry, most reports had not been subjected to critical analysis of immunoreactivities. JNKs are highly conserved essential mediators of cellular information transfer and responses. Therefore, it is doubtful whether JNKs operate with such an individual cell-type-dependent and stimulus-dependent pattern of transcription, translation and intracellular distribution. More likely, activation of JNKs might be minutely modulated in 'microdomains' depending on the association with scaffold regulators and upstream kinases. This demands specific non-cross-reacting tools. Indeed, the recent analysis of intraneuronal JNK distribution and activation in cultured cerebellar granule cells has provided evidence like the existence of different JNK-pools defined by upstream kinases MKK4 and MKK7 and the JIP scaffold protein, with each pool differentially activated following stress and maturation (Coffey et al., 2000). Finally, the paradigm of JNK translocalization into the nucleus in dependence on its activation has been questioned by the observation that glucocorticoids inhibit the TNFα-induced phosphorylation of JNKs, but even enhance their nuclear translocalization (Gonzalez et al., 2000).

3.1.3. Another critical issue: the N-terminal phosphorylation of c-Jun

The persisting expressing of c-Jun under degenerative and regenerative conditions raises the question as to its role in neuronal injury. One major approach to this issue was the analysis of the N-terminal phosphorylation of c-Jun by JNKs, since this post-translational modification is the prerequisite of the activation of the polymerase II complex with subsequent transcription of c-Jun/AP-1-dependent genes (Kallunki et al., 1994). The first attractive hypothesis defined a link between neurodegeneration and c-Jun phosphorylation by JNKs; in this concept, c-Jun was supposed to execute the genetic programs and/or reinforces the degenerative actions of JNKs. Very soon, however, it became clear that the N-terminal phosphorylation of c-Jun is also part of the physiological function of c-Jun and that the actions of JNKs are important for the intact structural organization and functional plasticity of the nervous system (Kuan et al., 1999).

At present, the question for the specificity of antisera for the visualization of c-Jun in immunocytochemistry and Western blotting is still a very critical issue (Fig. 3). Unfortunately, the antisera against the phosphorylated N-terminus have been uncritically used ignoring the essential possibility that it is the phosphorylation of JunD which is co-recognized by the antisera. What underlies this cross-reactivity and what is the relevance of a clear distinction?

(i) The sequence of amino acids comprising the N-terminal serine residues (the substrates of the JNKs) is almost identical between c-Jun and JunD.

(ii) The c-Jun protein contains both the docking site and the phospho-acceptor site for JNKs. In contradistinction, JunB contains only the docking site and JunD contains only the phospho-acceptor site. In consequence, JunD can only be phosphorylated when c-Jun serves as docking site; in this context, JunB might serve as well as docking protein for JunD (Kallunki et al., 1996).

(iii) Due to its widespread basal expression, JunD is supposed to act as housekeeper for physiological functions. Moreover, JunD can function as anti-apoptotic player, e.g. by suppression of p53 (Weitzman et al., 2000). Thus, the immunocytochemical detection of the presumably apoptotic c-Jun might mark the activation of the anti-apoptotic JunD.

(iv) Once phosphorylated, JunD might dissociate from c-Jun or JunB with subsequent formation Fos/AP-1 heterodimers which are functionally different from c-Jun/AP-1 dimers. It remains to be elucidated whether JunD serves as substrate for MAP kinase phosphatases or whether phosphorylation of JunD is a lasting and stable post-translational modification.

(v) Provided that c-Jun can serve as docking molecule for JunD with JunD as major substrate (eventually depending on the JNK isoforms), the lack of N-terminal phosphorylation will change c-Jun into a transcriptional suppressor, and this action might be supported by JNK-dependent activation of the JDP-2, an endogenous nuclear inhibitor of c-Jun (Aronheim et al., 1997).

The relevance of these considerations are exemplified by the analysis of c-Jun phosphorylation in c-JunAA mice. In this mutant, the serine residues are exchanged into alanins by knock-in, and these amino acids cannot serve a substrate for JNKs (Behrens et al., 1999). The majority of the antisera against N-terminal phosphorylation of c-Jun generated an intense immunocytochemical reaction in the adult brain which was equal in c-JunAA and their control litter mutants. The inevitable consequence of these data is twofold: the anti-phospho-c-Jun antisera do either cross-react with JunD and/or recognize other phosphorylated serine residues of the N-terminal c-Jun. Antisera against the serine 63, however, failed to detect a substantial signal in c-JunAA knock-in mice which express normal levels of c-Jun and JunD (Fig. 4) (Behrens et al., 1999; S. Brecht and A. Fleischmann, unpublished observations).

3.1.4. c-Jun as suppressor — the meaning of antagonization and of absent phosphorylation

The absence of N-terminal phosphorylation does not simply mean lack of action. It is a trivial but nevertheless most essential fact (which is often ignored) that the cellular phenotype and its adaptive genetic alterations depend on the genes which are realized as de novo protein synthesis — regulated expression of genes immediately requires regulated suppression of genes since absence of proteins determines the cellular phenotype to the same extent as expression of proteins. Following axotomy, for example, neurons lose the transmittory phenotype by orchestrated suppression and restart the program for developmental formation of the neuron–target axis including (frustrane) attempts to reenter the cell cycle. Thus, suppression and induction of genetic programs go in parallel and have to be tightly controlled for successful repair and survival. Several data document the importance to differentiate between suppressive action versus mere functional inactivity of c-Jun.

(i) The knock-out of c-*jun* is lethal since the murine mutants do not undergo functional

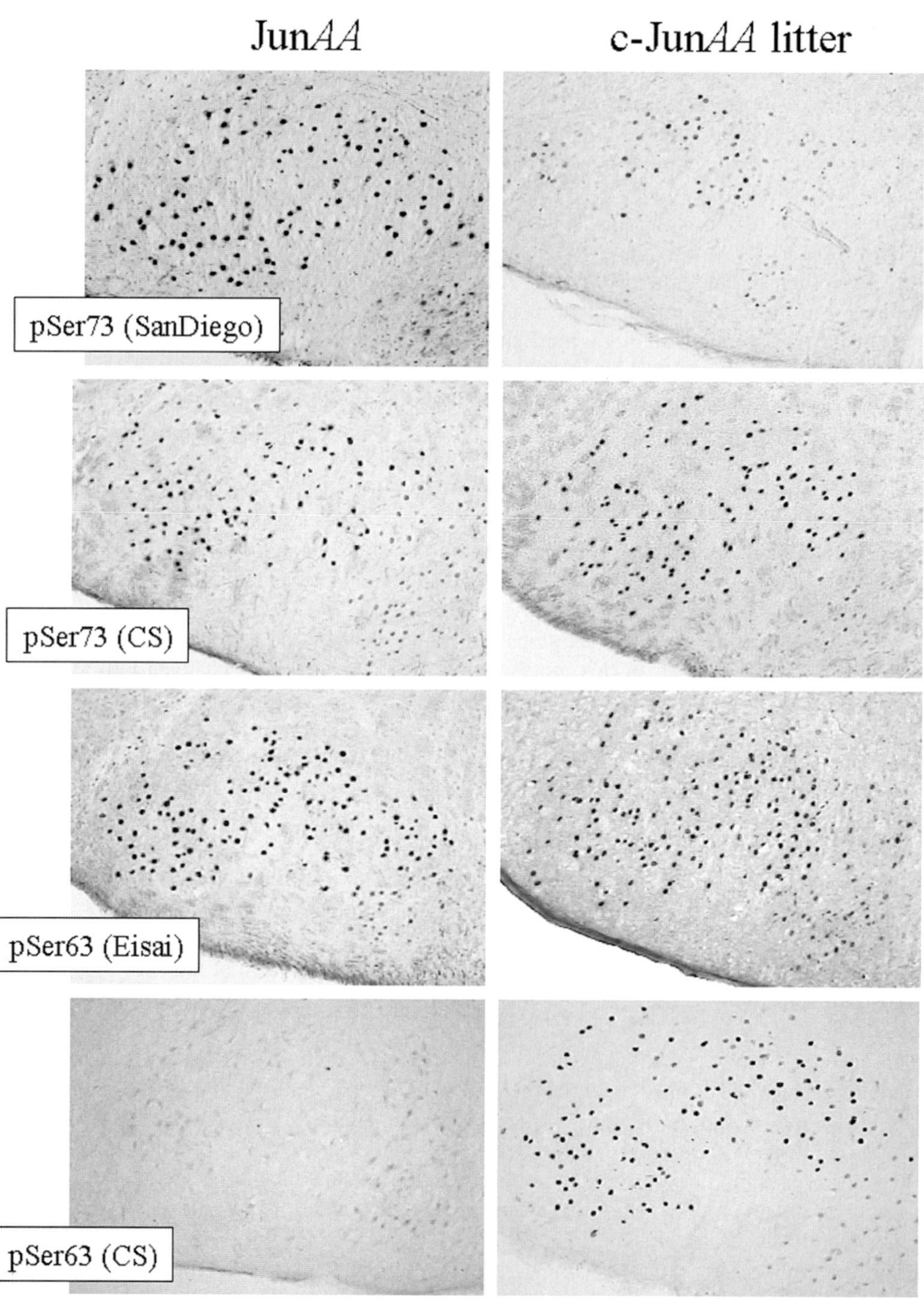

organogenesis of the liver (Hilberg et al., 1993). In contrast, c-JunAA mutant mice are fully viable and fertile (Behrens et al., 1999). Thus, the non-phosphorylated c-Jun provides essential (transcriptional?) actions, e.g. by suppression of genes and/or by inactivation of steroid hormone receptors-mediated transcription (Herrlich, 2001).

(ii) The transcription of c-Jun is antagonized by endogenous inhibitors such as Jun depressing proteins (JDPs). Trapped in JDP heterodimers, c-Jun can no longer activate the polymerase complex but most likely can still bind to and occupy regulatory promoter elements. Most interesting, JDPs are also substrates of JNKs and they enhance the pool of molecule which can compete as substrates for JNK similar as JunD or ATF-2 (Aronheim et al., 1997).

(iii) One argument for the specificity of the used anti-phospho-c-Jun antisera was its congruent distribution with the c-Jun immunoreactivity (Herdegen et al., 1998). Considering that c-Jun serves as docking adaptor for JNK with subsequent phosphorylation of JunD (in the presence of JDP-2), the phospho-c-Jun-IR is virtually congruent with that of c-Jun but derives from phosphorylated JunD (see Section 3.1.3).

(iv) The transcriptional suppression by c-Jun has been shown for several AP-1 targets.

4. PATHOLOGICAL RESPONSES AND AP-1 EXPRESSION FOLLOWING INTENTIONAL STIMULATION: THE PROBLEM OF LINKAGE

4.1. c-Jun

Strictly speaking, the visualization of AP-1 expression allows only the conclusion that an intentional stimulus (which usually starts a complex cascade of various pathophysiological reactions) causes the de novo synthesis of AP-1 proteins. The critical issue of the meaning of AP-1 expression for the cellular/neuronal reaction can be exemplified by the persistence of c-Jun in axotomized neurons — its lasting expression for weeks in surviving neurons or until the fragmentation of dying neurons allows a bipartite conclusion: (i) c-Jun is involved in the regenerative–protective efforts of injured neurons which last until realization of death by unbalanced degenerative programs or until the recovery with/without structural regeneration; (ii) c-Jun is involved in degenerative genetic programs which last until the 'successful' (apoptotic) termination or which come to an end with the onset of the neuronal recovery by protective–regenerative programs. And this observation is also true for both, the N-terminal phosphorylation of c-Jun and the activation of JNK (Herdegen et al., 1997, 1998).

←

Fig. 4. N-terminal phosphorylation of c-Jun. Specificity of antisera against the phosphorylated serine 63 and serine 73 residues (pSer63, pSer73) of the c-Jun protein in the murine facial ncl. 3 days following transection of the facial nerve fibers (A. Fleischmann and S. Brecht, in preparation). The left column gives the phospho-c-Jun-IR in c-Jun*AA* mice (Behrens et al., 1999); in these mice, Ser63 and Ser73 are mutated by knock-in into alanins which cannot be phosphorylated. The right column gives the phospho-c-Jun-IR of c-Jun*AA* litter controls. The following antisera are presented: pSer73 (San Diego), as described in Herdegen et al. (1998); pSer63 (Eisai), as described in Lallemand et al. (1998); pSer73 (CS) and pSer63 (CS) are commercially available from Cell Signaling (former New England Biolabs). These results indicate that antisera against pSer73 and pSer63 might cross-react with further epitopes such as serine residues from JunD (Kallunki et al., 1996); only one anti-pSer63 antisera (pSer63 CS) revealed a specific recognition of phosphorylated serine when used at higher dilution.

4.2. THE PROMISCUOUS EXPRESSION OF THE c-Fos PROTEIN, AND THE REMAINING AP-1 PROTEINS

The expression pattern of c-Fos, the main representative of the Fos family, is well characterized following numerous neurodegenerative events. Its putative pathogenic role, however, is still enigmatic. Null mutation of the *c-fos* gene attenuates the kindling development and kindling-induced structural plasticity (Watanabe et al., 1996), and this phenotype resembles the resistance to kainic acid seizures following JNK3 knock-out and c-JunAA mutation (D.D. Yang et al., 1997; Behrens et al., 1999). Thus, similar to the c-Jun/JNK-axis, the expression of c-Fos is involved in excitotoxicity (depending on dimerization with c-Jun?). In further comparison with c-Jun, the insight into functions of c-Fos is much less clarified. (i) There is still no marker available which indicates the transactive form of c-Fos corresponding to the N-terminal phosphorylation of c-Jun. (ii) Very often, pathological stimuli can be regarded as enhanced physiological stimuli; in consequence, pathological input should include the activation of both, physiological responses and pathological–degenerative programs. Thus, expression of c-Fos or JunB can represent the physiological and/or the degenerative part of the stimulus. (iii) Following some degenerative processes with selective induction of c-Jun (e.g. axotomy), the absence of c-Fos indicates an active suppression rather than the mere lack of the inducing input (reviewed by Herdegen et al., 1997) and the possibility of c-Fos degradation by c-Jun (Tsurumi et al., 1995). (iv) In contrast to JNK pathways, the upregulation of ERKs cannot be linked to neuronal degeneration, but rather to physiological programs and neuroprotection (Xia et al., 1995). Inhibition of ERK, however, does not worsen the ischemic progress in the gerbil (Sugino et al., 2000). On the other hand, push of calcium/cAMP pathways with activation of the CRE c-fos promoter site (with coincident activation of ERKs?) can be involved in pathological programs following calcium overload. (v) Degeneration can be regarded as a lasting/chronic harmful stimulus; the expression of c-Fos protein, however, can habituate to repetitive stimuli (Hsieh et al., 1998).

Similar as with c-Fos, the data on the expression of JunB, JunD, FosB and Fra proteins are more or less correlative, but nevertheless linked to processes involved in regeneration and degeneration: lasting FosB/JunD complexes reflected the adaptive neuronal responses in the denervated striatum following MPTP injection (Perez et al., 1998), Fra-2-IR persisted in neurons surviving an ischemic insult (Pennypacker et al., 2000). Differential effects of Jun proteins have been described for the regulation of hippocampal NGF expression following limbic seizures with participation of JunD in activation and of JunB in suppression (Elliott and Gall, 2000), and JunD might exert a major role in NGF production with variable AP-1 partners such as c-Fos or phosphorylated Fra-2 (Boss et al., 2001).

5. THE REGULATION OF PATHOLOGICAL–DEGENERATIVE ACTIONS OF THE c-Jun/JNK-AXIS IN THE BRAIN

The data on the role of c-Jun under physiological and pathological conditions provoke the essential question: which molecular mechanisms determine the tripartite functional transition of c-Jun from a physiological player to a neurodegenerative effector or to a modulator of neuroregeneration? JNKs form a central, but not exclusive, role for the c-Jun activation, which is also true for the nervous system. Therefore, we review the regulation of c-Jun with particular focus on its upstream N-terminal activators, JNKs. Several components can be defined for the regulation of the c-Jun/JNK-axis (Fig. 1).

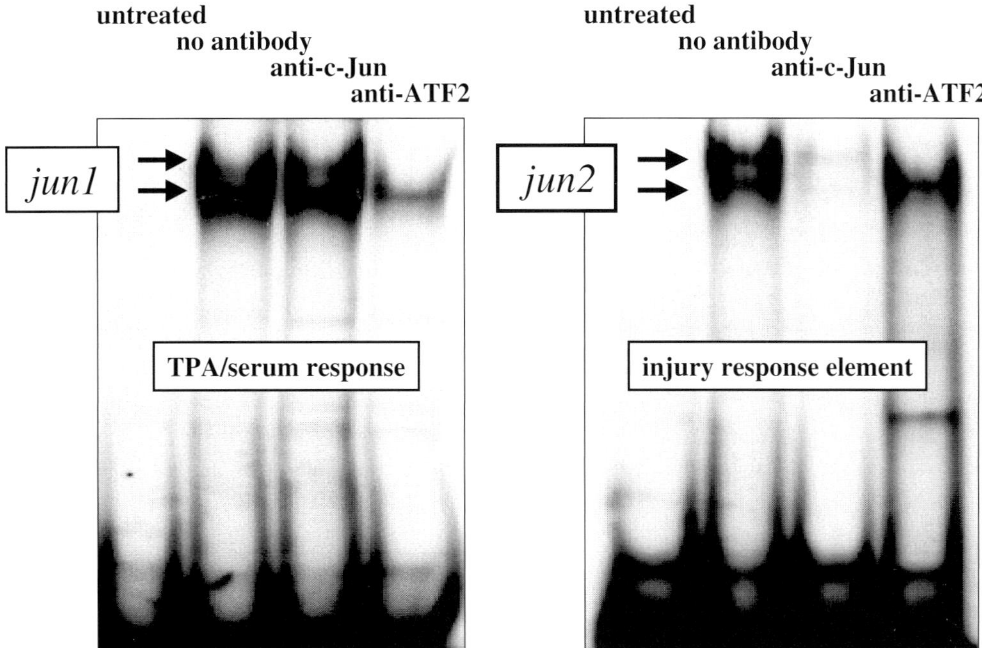

Fig. 5. Specific induction of the c-*jun* promoter – the *injury response element*. This electromobility-supershift assay (EMSA) demonstrates that the distal *jun2* promoter site of the c-*jun* promoter is mainly bound by c-Jun and to some extent by ATF-2, and this renders the *jun2* element as a converging point of JNK and p38 mediated signaling (Mielke et al., 1999).

(i) Induction. The classification of AP-1 proteins as immediate-early genes (IEGs) reflected their rapid de novo synthesis following serum deprivation or TPA stimulation (Almendral et al., 1988). The rather uniform expression is based on similar DNA response elements with the primarily functional designation of the TPA response element (TRE). This temporally staggered, but nevertheless congruent, AP-1 expression pattern is visible at the mRNA and protein level following transynaptic stimulation, e.g. following kainic acid or noxious stimulation (Beer et al., 1998; Herdegen and Leah, 1998). One essential particularity of c-Jun is its selective induction in the absence of other AP-1 proteins. Recently, we have defined the distal *jun2* (and to a lesser extent the proximal *jun1*) response element as regulatory elements for ATF-2 and c-Jun mediated c-*jun* induction in the brain (Fig. 5) (Mielke et al., 1999).

Besides ATF-2 and c-Jun, the p38 pathway with the transcription factor MEF2C has emerged as a novel regulator of the c-*jun* induction (Han et al., 1997; Marinissen et al., 1999); interestingly, MEF2C can mediate neuronal survival and neurite elongation during development (Okamoto et al., 2000). The down-regulation of ATF-2, a nuclear substrate of p38, is strictly coincident with c-Jun induction following neuronal injury (Figs. 6) and this might substantially change the competitive situation of p38 substrates, e.g. with a shift from ATF-2 to MEF2C phosphorylation with initiation of c-*jun* induction in the absence of ATF-2 and c-Jun.

(ii) Activation by N-terminal phosphorylation.

• *Scaffold proteins*. The N-terminal phosphorylation is the crucial step in the transcriptional activation of c-Jun, but the operative range of JNKs and ERKs and the c-Jun amino residues involved has still to be defined. Nevertheless, the activation of JNKs might be the most

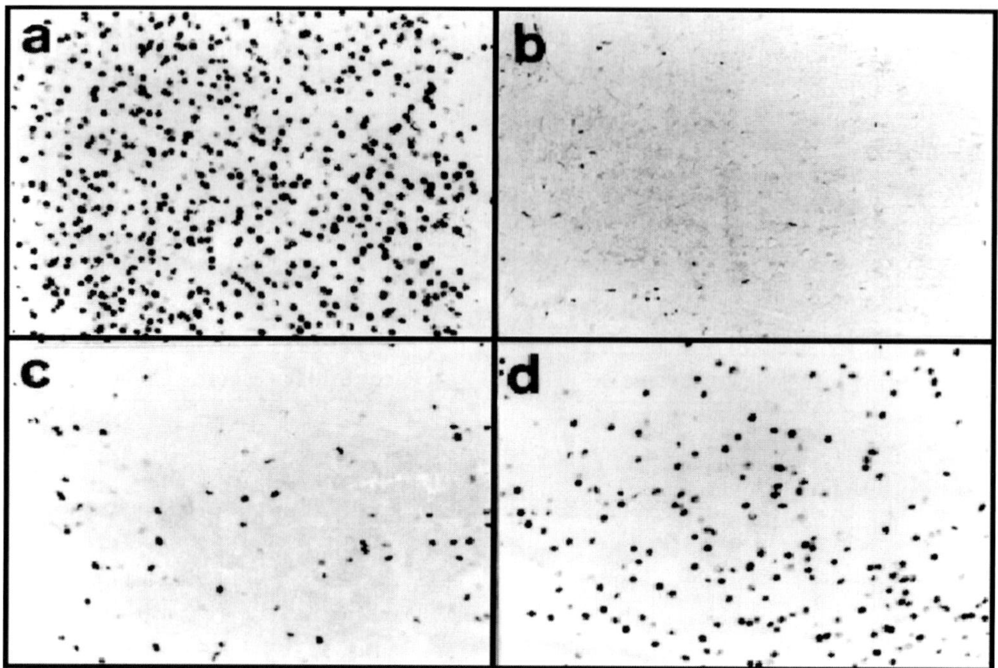

Fig. 6. Suppression of ATF-2. The constitutively expressed transcription factor is down-regulated following various degenerative stimuli compared to (*a*) basal expression, e.g. (*b*) 12 h, (*c*) 24 h, and (*d*) 72 h following transient occlusion of the middle cerebral artery in the rat entorhinal cortex (Martin-Villalba et al., 1998).

relevant step in the instrumentalization of c-Jun as a pro-degenerative effector. Therefore, the regulation of JNKs deserves particular attention. By complexation with distinct scaffold proteins, JNKs are coupled to individual signal-transduction units with different upstream activators. The group of JNK-interacting proteins (JIPs) selectively mediates signaling by the mixed-lineage kinase (MLK) > MAP kinase kinase 7 (MKK7) > JNK pathway, and their oligomeric complexes accumulate in peripheral cytoplasmic projections (Yasuda et al., 1999); interestingly, MLK2/3 can associate with the GluR6-PSD95 scaffolds (Liu YF, oral communication), and the activation of JNKs by MLK2 partially mediates apoptosis by polyglutamine-expanded huntingtin (Liu et al., 2000). Apart from renal *jip1* mRNA, the JIP1–3 proteins are highly expressed in the brain (Kim et al., 1999) with cytoplasmic and axonal–presynaptic distribution (Pellet et al., 2000), and complexes with MLK2 and activated JNK have been found along microtubules in fibroblasts (Nagata et al., 1998). Expression of JIPs can be enhanced at a transcriptional level following ischemia (Hayashi et al., 2000), neuronal denervation (Anguelova et al., 2000) or inhibition of JNK (Levresse et al., 2000) with an immediate impact on JNK activation since the overexpression of JIPs relative to JNKs blocks JNK functions (Ferrell, 2000). The JIP3/JSAP displays a preference for JNK3 (Ito et al., 1999) and antagonizes the activation of ERKs by inhibition of the Raf-1 mediated MEK1 phosphorylation (Kuboki et al., 2000). The β-arrestin 2 protein has recently been identified as novel scaffold for JNK3 in the murine brain which mediated the cytosolic retention of JNKs and the activation by ASK1 (McDonald et al., 2000).

In principle, scaffolds have the potential to facilitate signaling, but fluctuations of the pathway components will inevitably lead to a less efficient signal transfer. The facilitation of signal response by scaffolds ultimately leads to an ultrasensitivation with potential Hill coefficients of more than 10 (Ferrell, 2000). Thus, an increase in expression and activation of individual cascade members might be without relevance in the absence of concomitant stoichiometric alterations in scaffold proteins or decrease efficacy of signal transduction by disturbance of the balance of scaffold complexes.

• *Antagonistic phosphatases.* Besides inhibition of phosphorylation by scaffolds, activation of JNK can be terminated by the specific MAP kinase phosphatases 3/6 (MKP3/6) (Muda et al., 1996) and unspecifically by MKP1 or MKP2 (Keyse, 1998). Recently, we have shown an increase of nuclear MKP1 following transection of central nerve bundles in surviving mammillary neurons, but not in dying neurons of the substantia nigra compacta (Winter et al., 1998). On the other hand, downregulation of MKPs constitutes an effective step for the activation of JNKs as it has been reported following protein-damaging conditions (Meriin et al., 1999).

(iii) Parallel activation of agonistic or antagonistic signal cascades. Numerous signal-transduction cascades have the potential to activate JNKs, and — in the case of nuclear translocation — with subsequent activation of c-Jun. In the context of c-Jun/JNK actions during neurodegenerative processes, the influence of the three activating cascades p38-ATF2/MEF2C, ASK1-JNK and Akt-JNK deserves particular interest.

First of all, activation of p38 can enhance the induction of c-*jun* by phosphorylation of ATF-2 or MEF2C; furthermore, the phosphorylation of ATF-2 could interfere with the competition of c-Jun for binding to or being phosphorylated by JNKs (Gupta et al., 1996). Importantly, inhibition of p38-catalyzed substrate phosphorylation substantially protects against ischemic injury in the gerbil (Sugino et al., 2000).

Secondly, ASK1 was colocalized with activated JNK or activated p38 in apoptotic cells following spinal cord injury (Nakahara et al., 1999). Microtubular disarray, a putative feature of neuronal degeneration, induced c-Jun-dependent transcription via JNK and synergistic but parallel activation of ASK1 (Wang et al., 1998). ASK1 keeps the position of a master switch for activating both, p38 and, to a lesser extent, JNK with specific or synergistic functions. Thus, ASK1 triggers neurite extension in PC12 cells with downstream phosphorylation of JNK and p38; on the other hand, ASK1 is a crucial modulator of NGF withdrawal-induced apoptosis in PC12 cells and superior cervical ganglia with activated c-Jun/JNK-axis and p38 (Kanamoto et al., 2000). In contradistinction to ASK1, the Akt kinase (also called protein kinase B) which acts downstream of phosphatidylinositol-3-kinase (PI3-K), distinctly inhibits the activation of JNK and c-Jun and protects against neuronal apoptosis (Levresse et al., 2000).

The synergistic and antagonistic interplay of different signal cascades is nicely exemplified in trigeminal neurinoma cells (Marushige and Marushige, 1999): apoptosis was rapidly induced when the activation of JNK was coupled with the inhibition of PI3-K/Akt, and the induction was further enhanced by parallel inhibition of ERK; rapid apoptosis occurred when the JNK activation was coupled with p38 inhibition, and this was potentiated by inhibiting PI3-K/Akt; the concomitant inhibition of ERK and activation of Akt induced apoptosis without JNK activation, even though with a considerable delay (Marushige and Marushige, 1999).

6. p38 IN THE MAMMALIAN NERVOUS SYSTEM

6.1. FUNCTIONS

p38 kinases are widely distributed in mammalian tissues including brain, whereby only the p38α and p38β isoforms are present in the brain (H.H. Lee et al., 2000; Maruyama et al.,

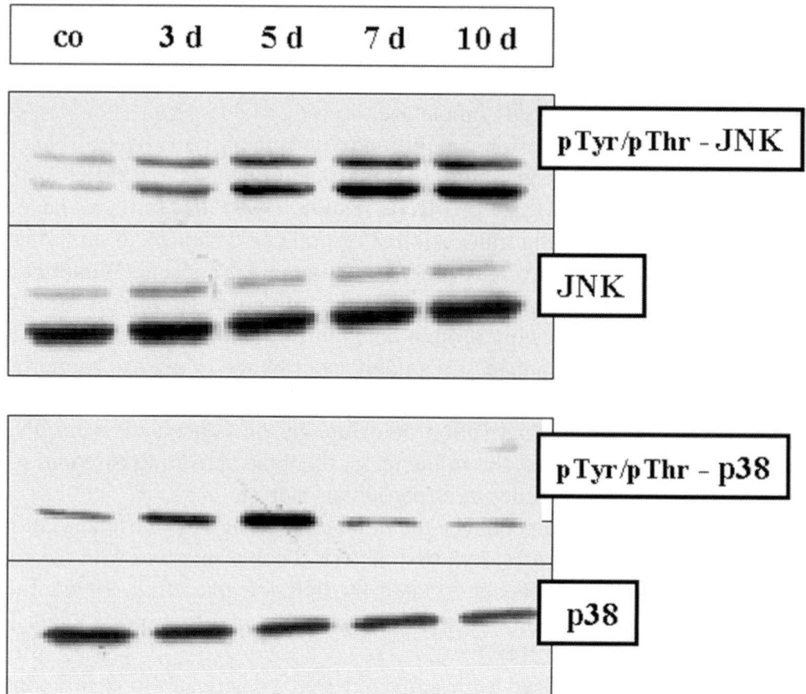

Fig. 7. Activation of JNK and p38 during differentiation of PC12 cells. Western blotting of PC12 homogenates demonstrates an ongoing increase of JNK activation and a transient increase of p38 activation during NGF mediated differentiation; the levels of JNK and p38 remain constant (V. Waetzig and T. Herdegen, unpublished observations).

2000). Comparable to JNK, p38 kinase is activated in PC12 cells following NGF-triggered differentiation without cell death (Fig. 7), and its activation is furthermore enhanced following NGF-withdrawal with subsequent apoptosis (Xia et al., 1995; Kummer et al., 1997; Eilers et al., 1998; K. Mielke et al., unpublished observations). In contrast, NGF-withdrawal in primary cultures of sympathetic neurons does not lead to activation of p38 (Eilers et al., 1998). p38α is phosphorylated in neurons and astrocytes up to 24 h following focal ischemia, whereas after 24 to 72 h, phospho-p38 was evident in activated microglia (Irving et al., 2000; Tian et al., 2000) and treatment with a second-generation p38 inhibitor, SB 239063, provides a significant reduction in infarct size, neurological deficits and inflammatory cytokine expression produced by focal ischemia (Barone and Parsons, 2000; Barone et al., 2001a,b) suggesting that the neuroprotective effect of p38 inhibitors may be indirect via inhibition of cytokine and other mediators involved in response of the brain to injury. In the cortical and hippocampal brain regions of patients with Alzheimer's disease (AD) p38α is exclusively localized in association with neurofibrillary pathology, tau and p38 was found to be physically associated and the majority of p38 was activated in AD neurons (Zhu et al., 2001). Because p38 is also implicated in mitotic arrest (Takenaka et al., 1998), these findings provide a link between the cell cycle re-entrant phenotype, cytoskeletal phosphorylation and oxidative stress in AD.

Recently, activation of the p38α kinase with subsequent activation of the transcription factor MEF2C was shown to prevent cell death during neuronal differentiation (Okamoto et al., 2000). Therefore, p38 activation per se does not refer to differentiation or apoptosis. The high basal activity of p38 in fetal chick neurons is decreased by treatment with insulin

(Heidenreich and Kummer, 1996). Because insulin has no significant effect on ERK and JNK activities, the inhibition of p38 activity is discussed as a mechanism by which insulin supports neuronal survival. Prevention of apoptosis by inhibition of p38 activity was also supported by the finding that pretreatment of PC12 cells or primary chick neurons with selective p38 inhibitors (SB203580, SB202190) promote neuronal survival after withdrawal of trophic factors (Horstmann et al., 1998). In contrary to the results in cell culture and chick neurons, a high basal activity of p38 was observed in the adult rat brain (S.H. Lee et al., 2000) that rapidly decreased after application of kainic acid (Mielke et al., 1999). The high basal activity in central neurons is in contradiction to the hypothesis that p38 acts as a stress kinase involved in apoptosis, but nothing so far is known about the function and activation of p38 in the central nervous system under physiological conditions. These findings are supported by data from mouse liver where constitutively active p38 is down-regulated by hypoxic–oxidative stress (Mendelson et al., 1996). Several findings indicate a role of p38 in the status of differentiation by arresting cells: (i) basal p38 activity in highly differentiated tissues (Mielke et al., 1999); (ii) the observation that p38 is activated in mammalian cultured cells (NIH 3T3) when the cells are arrested in the M phase; and (iii) injection of constitutively active p38 into cleaving *Xenopus* embryos induces mitotic arrest (Takenaka et al., 1998). There is some evidence that apoptosis in postmitotic neurons is associated with a frustrated attempt to reenter the mitotic cycle (Nuydens et al., 1998). Thus, down-regulation of p38 activity after stressful stimuli can be interpreted as part of neuronal purpose to reenter mitosis. Stabilization of postmitotic differentiation of neurons most likely is one physiological role of p38. The high variety of distinct substrates including transcription factors, protein kinases or cytoskeleton proteins argues for additional roles of p38 under physiological conditions.

The pool of present data provides evidence for p38 as a protein kinase whose activity and regulation depends on a variety of factors such as cell type, the status of cellular differentiation, and the external stimuli. Therefore, the meaning of the information on p38 that has been mainly provided by in vitro experiments is difficult for the nervous system and has to be dealt with care.

6.2. IMMUNOCYTOCHEMISTRY

Similar as with JNK-immunocytochemistry (IR), the visualization of p38 has provided conflicting data on the cellular and intracellular distribution of p38 in rodents (Table 2). The reports range from exclusively/predominantly neuronal or glial expression to nuclear or cytoplasmic distribution. In most cases, the careful analysis of the properties of the antiserum used is missing. In consequence, we cannot gain valid insights into the pathophysiology of p38 from the published data of the immunocytochemical localization.

6.3. EFFECT OF p38

Similar as with JNK, the family of p38 kinases is supposed to mediate pathophysiological-prodegenerative processes, but p38 are also involved in physiological and neuroprotective–regenerative events. Thus, inhibition of p38 reduced the extent of cell death and the infarction area (Barone and Parsons, 2000; Sugino et al., 2000; Legos et al., 2001), and p38 mediates neural cell death in vitro downstream of ASK1 (Ichijo et al., 1997).

On the other hand, p38 was linked with neurite outgrowth, and the transient increase of p38 (with activation of ATF-2?) is part of the initial cell body response and following NGF induced sprouting, respectively (Morooka and Nishida, 1998; Takeda et al., 2000;

TABLE 2. *Expression and intracellular distribution of p38 — verview of the expression and the activation of p38 in the brain and neural cells in vitro*

Stimulus/model	Cell type/area	Expression	Activity	Localization	Result	Reference
Untreated	mouse CNS	α, β p38		α: dendritic + cytoplasmatic β: nuclear	immunohistochemistry	S.H. Lee et al., 2000
Untreated	mouse CNS	basal level		mainly in white matter	immunohistochemistry	Maruyama et al., 2000
Ischemia	rat CNS		=		phospho-p38 in the hippocampus	Hicks et al., 2000
ATP	rat microglia		↑		TNFα mediated p38 activation	Hide et al., 2000
Ischemia	rat CNS	↑	↑	astrocytes	immunohistochemistry	Irving et al., 2000
Axotomy	Retinal ganglion cell	=	↑		immunohistochemistry	Kikuchi et al., 2000
NGF or serum withdrawal	Rat-1 and PC12		↑		p38 inhibition by insulin	Kummer et al., 1997
NGF	PC12 cells		↑		p38 inhibition reduces neurite growth	Morooka and Nishida, 1998
Ischemia	gerbil		↑	microglia	immunohistochemistry	Walton et al., 1998
Potassium	cerebellar granule cells		↑			Mao et al., 1999
Trauma	rat spinal cord		↑	in TUNEL pos. cells	immunohistochemistry	Nakahara et al., 1999
Ischemia	rat hippocampus		↑	neurons?	immunohistochemistry	Ozawa et al., 1999
Arsenite	rat cortical neurons		↑			Namgung and Xia, 2000
Ischemia	mouse forebrain	↑	↑	neurons	immunohistochemistry	Takagi et al., 2000
ASK1	PC12		↑		p38 inhibition reduces neurite outgrowth	Takeda et al., 2000
Angiostrongylus cantonensis	mouse		=		model for eosinophilic meningitis	H.H. Lee et al., 2000
Ischemia	gerbil	↑	↑	cytoplasmic	p38 inhibition reduces cell death	Sugino et al., 2000
Ischemia	rat		↑	neurons + microglia		Tian et al., 2000
Oncostatin M	human astrocytes		↑		p38 inhibition reduces IL-6 expression	Van et al., 2000
LTP and potassium chloride	rat hippocampus		↑		vitamin c and e reduces p38 activity	Vereker et al., 2000

TABLE 2. *(continued)*

Stimulus/model	Cell type/area	Expression	Activity	Localization	Result	Reference
Carbachol	cell line A2058		↑			Wood et al., 2000
Focal ischemia	rat		↑		p38 inhibition reduces stroke	Barone and Parsons, 2000, 2001a,b; Legos et al., 2001
LPS and IL-1β stimulation	astro-/micro-glia in culture				p38 inhibition reduces TNFα	Y.B. Lee et al., 2000
Calyculin A stimulation	mouse cortical neurons		↑		p38 inhibition reduces cell death	Ko et al., 2000
LTP and β-amyloid	rat hippocampal slice				p38 inhibition has no effect on LTP but counteracts β-amyloid	Saleshando and O'Connor, 2000
EGF	mouse cortical cultures				p38 inhibition does not prevent cell death	Cha et al., 2000
Kainate, NMDA	rat striatal slice				p38 inhibition does not prevent NFκB activation	Cruise et al., 2000
Substance P	U373 astrocytoma cell line		↑		p38 mediates IL-6 expression	Fiebich et al., 2000
NO	SH-SY5Y and cortical neurons				p38 inhibition blocked bax translocation	Ghatan et al., 2000
MEF-2 transfection	P9 precursor cells		↑		dominant negative p38 enhances apoptosis	Okamoto et al., 2000
Kainate	rat brain		↓		excitotoxicity lowers p38 activity	Mielke et al., 1999

↑ gives an increase, ↓ gives a decrease of expression and/or activity; = marks no significant alteration.

V. Waetzig and T. Herdegen, unpublished observations). In the adult rat brain, kainate-induced excitotoxicity resulted in lasting suppression of p38 activity (Mielke et al., 1999) arguing against a mere role of p38 in neuronal cell death.

The understanding of the actions of p38 is complicated by the observations that p38 triggers both neuronal and glial function. The release of microglial chemokines and cytokines can effectively be blocked by p38 inhibitors, and as discussed above for JNK, the pro-inflammatory effect of p38 might counteract the physiological functions in neurons.

Finally, to understand the functions of p38, a novel player has come into the focus of the neuroscientist: the transcription factor MEF2C (myocyte-specific enhancer binding factor 2). MEF2C is essential for the mitotic arrest of neurons and protects neurons against apoptosis during development (Okamoto et al., 2000). Increase in MEF2C expression in neurons and microglia was found in those areas which are resistant to death following global cerebral ischemia (Speliotes et al., 1996).

7. ABBREVIATIONS

AP-1	activator protein 1
ATF-2	activating transcription factor 2
CBP	CREB binding protein
ERK	extracellular regulated kinase
IR	immunoreactivity
JDP	Jun depressing protein
JIP	JNK interacting protein
JNK	c-Jun N-terminal kinase
JNKK	JNK kinase (= MKK = MAP kinase kinase)
MAPK	mitogen-activated kinase
MAPKAP	MAP kinase-activated protein kinase
MEKK	MAP kinase kinase kinase
MKP	mitogen-activated protein kinase phosphatase
MLK	mixed lineage kinase
NGF	nerve growth factor
NFκB	nuclear factor kappa B
SAPK	stress-activated protein kinase
SRF	serum response transcription factors
TNF	tumor necrosis factor

8. REFERENCES

Almendral JM, Sommer D, Macdonald-Bravo H, Burckhardt J, Perera J, Bravo R (1988): Complexity of the early genetic response to growth factors in mouse fibroblasts. *Mol Cell Biol 8*:2140–2148.

Anguelova E, Boularand S, Nowicki JP, Benavides J, Smirnova T (2000): Up-regulation of genes involved in cellular stress and apoptosis in a rat model of hippocampal degeneration. *J Neurosci Res 59*:209–217.

Arias J, Alberts AS, Brindle P, Claret FX, Smeal T, Karin M, Feramisco J, Montminy M (1994): Activation of cAMP and mitogen responsive genes relies on a common nuclear factor. *Nature 370*:226–229.

Aronheim A, Zandi E, Hennemann H, Elledge SJ, Karin M (1997): Isolation of an AP-1 repressor by a novel method for detecting protein–protein interactions. *Mol Cell Biol 17*:3094–3102.

Badger AM, Bradbeer JN, Votta B, Lee JC, Adams JL, Griswold DE (1996): Pharmacological profile of SB

203580, a selective inhibitor of cytokine suppressive binding protein/p38 kinase, in animal models of arthritis, bone resorption, endotoxin shock and immune function. *J Pharmacol Exp Ther 279*:1453–1461.

Barone FC, Parsons AA (2000): Therapeutic potential of anti-inflammatory drugs in focal stroke. *Expert Opin Invest Drugs 9*:2281–2306.

Barone FC, Irving EA, Ray AM, Lee JC, Kassis S, Kumar S, Badger AM, Legos JJ, Erhardt JA, Ohlstein EH, Hunter AJ, Harrison DC, Philpott K, Smith BR, Adams JL, Parsons AA (2001a): Inhibition of p38 mitogen-activated protein kinase provides neuroprotection in cerebral focal ischemia. *Med Res Rev 21*:129–145.

Barone FC, Irving EA, Ray AM, Lee JC, Kassis S, Kumar S, Badger AM, White RF, McVey MJ, Legos JJ, Erhardt JA, Nelson AH, Ohlstein EH, Hunter AJ, Ward K, Smith BR, Adams JL, Parsons AA (2001b): SB 239063, a second-generation p38 mitogen-activated protein kinase inhibitor, reduces brain injury and neurological deficits in cerebral focal ischemia. *J Pharmacol Exp Ther 296*:312–321.

Becher B, Barker PA, Owens T, Antel JP (1998): CD95-CD95L: can the brain learn from the immune system?. *Trends Neurosci 21*:114–117.

Beer J, Mielke K, Zipp M, Zimmermann M, Herdegen T (1998): Expression of c-jun, junB, c-fos, fra-1 and fra-2 mRNA in the rat brain following seizure activity and axotomy. *Brain Res 794*:255–266.

Behrens A, Sibilia M, Wagner EF (1999): Amino-terminal phosphorylation of c-Jun regulates stress-induced apoptosis and cellular proliferation. *Nat Genet 21*:326–329.

Boss V, Roback JD, Young AN, Roback LJ, Weisenhorn DM, Medina-Flores R, Wainer BH (2001): Nerve growth factor, but not epidermal growth factor, increases Fra-2 expression and alters Fra-2/JunD binding to AP-1 and CREB binding elements in pheochromocytoma (PC12) cells. *J Neurosci 21*:18–26.

Brecht S, Simler S, Vergnes M, Mielke K, Marescaux C, Herdegen T (1999): Repetitive electroconvulsive seizures induce activity of c-Jun N-terminal kinase and compartment-specific desensitization of c-Jun phosphorylation in the rat brain. *Brain Res Mol Brain Res 68*:101–108.

Bruce AJ, Boling W, Kindy MS, Peschon J, Kraemer PJ, Carpenter MK, Holtsberg FW, Mattson MP (1996): Altered neuronal and microglial responses to excitotoxic and ischemic brain injury in mice lacking TNF receptors. *Nature Medicine 2*:788–794.

Carboni L, Carletti R, Tacconi S, Corti C, Ferraguti F (1998): Differential expression of SAPK isoforms in the rat brain. An in situ hybridisation study in the adult rat brain and during post-natal development. *Brain Res Mol Brain Res 60*:57–68.

Carletti R, Tacconi S, Bettini E, Ferraguti F (1995): Stress-activated protein kinases, a novel family of mitogen-activated protein kinases, are heterogeneously expressed in the adult rat brain and differentially distributed from extracellular-signal-regulated protein kinases. *Neuroscience 69*:1103–1110.

Cha YK, Kim YH, Ahn YH, Koh JY (2000): Epidermal growth factor induces oxidative neuronal injury in cortical culture. *J Neurochem 75*:298–303.

Chen J, Kelz MB, Hope BT, Nakabeppu Y, Nestler EJ (1997): Chronic Fos-related antigens: stable variants of deltaFosB induced in brain by chronic treatments. *J Neurosci 17*:4933–4941.

Chihab R, Ferry C, Koziel V, Monin P, Daval JL (1998): Sequential activation of activator protein-1-related transcription factors and JNK protein kinases may contribute to apoptotic death induced by transient hypoxia in developing brain neurons. *Brain Res Mol Brain Res 63*:105–120.

Chinenov Y and Kerppola TK (2001): Close encounters of many kinds: Fos–Jun interactions that mediate transcription regulatory specificity. *Oncogene 20*:2438–2452.

Coffer P, Lutticken C, van Puijenbroek A, Klop-de Jonge M, Horn F, Kruijer W (1995): Transcriptional regulation of the junB promoter: analysis of STAT-mediated signal transduction. *Oncogene 10*:985–994.

Coffey ET, Hongisto V, Dickens M, Davis RJ, Courtney MJ (2000): Dual roles for c-Jun N-terminal kinase in developmental and stress responses in cerebellar granule neurons. *J Neurosci 20*:7602–7613.

Cruise L, Ho LK, Veitch K, Fuller G, Morris BJ (2000): Kainate receptors activate NF-kappaB via MAP kinase in striatal neurones. *Neuroreport 11*:395–398.

Cruzalegui FH, Hardingham GE, Bading H (1999): c-Jun functions as a calcium-regulated transcriptional activator in the absence of JNK/SAPK1 activation. *EMBO J 18*:1335–1344.

Davis RJ (2000): Signal transduction by the JNK group of MAP kinases. *Cell 103*:239–252.

Dechant G, Barde YA (1997): Signalling through the neurotrophin receptor p75NTR. *Curr Biol 7*:413–418.

Delcroix JD, Averill S, Fernandes K, Tomlinson DR, Priestley JV, Fernyhough P (1999): Axonal transport of activating transcription factor-2 is modulated by nerve growth factor in nociceptive neurons. *J Neurosci* (Online) *19*:RC24.

Dickens M, Rogers JS, Cavanagh J, Raitano A, Xia Z, Halpern JR, Greenberg ME, Sawyers CL, Davis RJ (1997): A cytoplasmic inhibitor of the JNK signal transduction pathway. *Science 277*:693–696.

Edelstein K, Beaule C, D'Abramo R, Amir S (2000): Expression profiles of JunB and c-Fos proteins in the rat circadian system. *Brain Res 870*:54–65.

Eilers A, Whitfield J, Babij C, Rubin LL, Ham J (1998): Role of the Jun kinase pathway in the regulation of c-Jun expression and apoptosis in sympathetic neurons. *J Neurosci 18*:1713–1724.

Elliott RC, Gall CM (2000): Changes in activating protein 1 (AP-1) composition correspond with the biphasic profile of nerve growth factor mRNA expression in rat hippocampus after hilus lesion-induced seizures. *J Neurosci 20*:2142–2149.

Fernyhough P, Gallagher A, Averill SA, Priestley JV, Hounsom L, Patel J, Tomlinson DR (1999): Aberrant neurofilament phosphorylation in sensory neurons of rats with diabetic neuropathy. *Diabetes 48*:881–889.

Ferrell JE (2000): What do scaffold proteins really do? *Science's STKE* 2000/*52*:1–3.

Ferrer I, Planas AM, Pozas E (1997): Radiation-induced apoptosis in developing rats and kainic acid-induced excitotoxicity in adult rats are associated with distinctive morphological and biochemical c-Jun/AP-1 (N) expression. *Neuroscience 80*:449–458.

Fiebich BL, Schleicher S, Butcher RD, Craig A, Lieb K (2000): The neuropeptide substance P activates p38 mitogen-activated protein kinase resulting in IL-6 expression independently from NF-kappaB. *J Immunol 165*:5606–5611.

Gass P, Herdegen T, Bravo R, Kiessling M (1993): Spatiotemporal induction of immediate early genes in the rat brain after limbic seizures: effects of NMDA receptor antagonist MK-801. *Eur J Neurosci 5*:933–943.

Ghatan S, Larner S, Kinoshita Y, Hetman M, Patel L, Xia Z, Youle RJ, Morrison RS (2000): p38 MAP kinase mediates bax translocation in nitric oxide-induced apoptosis in neurons. *J Cell Biol 150*:335–347.

Giasson BI, Bruening W, Durham HD, Mushynski WE (1999): Activation of stress-activated protein kinases correlates with neurite outgrowth induced by protease inhibition in PC12 cells. *J Neurochem 72*:1081–1087.

Gonzalez MV, Jimenez B, Berciano MT, Gonzalez-Sancho JM, Caelles C, Lafarga M, Munoz A (2000): Glucocorticoids antagonize AP-1 by inhibiting the activation/phosphorylation of JNK without affecting its subcellular distribution. *J Cell Biol 150*:1199–1208.

Grimm R, Schicknick H, Riede I, Gundelfinger ED, Herdegen T, Zuschratter W, Tischmeyer W (1997): Suppression of c-fos induction in rat brain impairs retention of a brightness discrimination reaction. *Learn Mem 3*:402–413.

Gupta S, Barrett T, Whitmarsh AJ, Cavanagh J, Sluss HK, Derijard B, Davis RJ (1996): Selective interaction of JNK protein kinase isoforms with transcription factors. *EMBO J 15*:2760–2770.

Han J, Jiang Y, Li Z, Kravchenko VV, Ulevitch RJ (1997): Activation of the transcription factor MEF2C by the MAP kinase p38 in inflammation. *Nature 386*:296–299.

Hannun YA (1996): Functions of ceramide in coordinating cellular responses to stress. *Science 274*:1855–1859.

Hata R, Maeda K, Hermann D, Mies G, Hossmann KA (2000): Dynamics of regional brain metabolism and gene expression after middle cerebral artery occlusion in mice. *J Cereb Blood Flow Metab 20*:306–315.

Hayashi T, Sakai K, Sasaki C, Zhang WR, Warita H, Abe K (2000): c-Jun N-terminal kinase (JNK) and JNK interacting protein response in rat brain after transient middle cerebral artery occlusion. *Neurosci Lett 284*:195–199.

Heidenreich KA, Kummer JL (1996): Inhibition of p38 mitogen-activated protein kinase by insulin in cultured fetal neurons. *J Biol Chem 271*:9891–9894.

Henry JR, Rupert KC, Dodd JH, Turchi IJ, Wadsworth SA, Cavender DE, Fahmy B, Olini GC, Davis JE, Pellegrino-Gensey JL, Schafer PH, Siekierka JJ (1998): 6-Amino-2-(4-fluorophenyl)-4-methoxy-3- (4-pyridyl)-1H-pyrrolo[2, 3-b]pyridine (RWJ 68354): a potent and selective p38 kinase inhibitor. *J Med Chem 41*:4196–4198.

Herdegen T, Leah JD (1998): Inducible and constitutive transcription factors in the mammalian nervous system: control of gene expression by Jun, Fos and Krox, and CREB/ATF proteins. *Brain Res Brain Res Rev 28*:370–490.

Herdegen T, Waetzig V (2001): AP-1 proteins in the brain: facts and fiction about effectors of neuroprotection and neurodegeneration. *Oncogene 21*:2424–2437 .

Herdegen T, Kovary K, Leah J, Bravo R (1991a): Specific temporal and spatial distribution of JUN, FOS, and KROX-24 proteins in spinal neurons following noxious transsynaptic stimulation. *J Comp Neurol 313*:178–191.

Herdegen T, Kummer W, Fiallos CE, Leah J, Bravo R (1991b): Expression of c-JUN, JUN B and JUN D proteins in rat nervous system following transection of vagus nerve and cervical sympathetic trunk. *Neuroscience 45*:413–422.

Herdegen T, Brecht S, Mayer B, Leah J, Kummer W, Bravo R, Zimmermann M (1993): Long-lasting expression of JUN and KROX transcription factors and nitric oxide synthase in intrinsic neurons of the rat brain following axotomy. *J Neurosci 13*:4130–4145.

Herdegen T, Kovary K, Buhl A, Bravo R, Zimmermann M, Gass P (1995): Basal expression of the inducible transcription factors c-Jun, JunB, JunD, c-Fos, FosB, and Krox-24 in the adult rat brain. *J Comp Neurol 354*:39–56.

Herdegen T, Skene P, Bahr M (1997): The c-Jun transcription factor — bipotential mediator of neuronal death, survival and regeneration. *Trends Neurosci 20*:227–231.

Herdegen T, Claret FX, Kallunki T, Martin-Villalba A, Winter C, Hunter T, Karin M (1998): Lasting N-terminal phosphorylation of c-Jun and activation of c-Jun N-terminal kinases after neuronal injury. *J Neurosci 18*:5124–5135.

Herrlich P (2001): Cross-talk between glucocorticoid receptor and AP-1. *Oncogene 20*:2465–2475.

Hicks SD, Parmele KT, DeFranco DB, Klann E, Callaway CW (2000): Hypothermia differentially increases extracellular signal-regulated kinase and stress-activated protein kinase/c-Jun terminal kinase activation in the hippocampus during reperfusion after asphyxial cardiac arrest. *Neuroscience 98*:677–685.

Hide I, Tanaka M, Inoue A, Nakajima K, Kohsaka S, Inoue K, Nakata Y (2000): Extracellular ATP triggers tumor necrosis factor-alpha release from rat microglia. *J Neurochem 75*:965–972.

Hilberg F, Aguzzi A, Howells N, Wagner EF (1993): c-jun is essential for normal mouse development and hepatogenesis. *Nature 365*:179–181.

Horstmann S, Kahle PJ, Borasio GD (1998): Inhibitors of p38 mitogen-activated protein kinase promote neuronal survival in vitro. *J Neurosci Res 52*:483–490.

Hsieh TF, Simler S, Vergnes M, Gass P, Marescaux C, Wiegand SJ, Zimmermann M, Herdegen T (1998): BDNF restores the expression of Jun and Fos inducible transcription factors in the rat brain following repetitive electroconvulsive seizures. *Exp Neurol 149*:161–174.

Ichijo H (1999): From receptors to stress-activated MAP kinases. *Oncogene 18*:6087–6093.

Ichijo H, Nishida E, Irie K, ten Dijke P, Saitoh M, Moriguchi T, Takagi M, Matsumoto K, Miyazono K, Gotoh Y (1997): Induction of apoptosis by ASK1, a mammalian MAPKKK that activates SAPK/JNK and p38 signaling pathways. *Science 275*:90–94.

Ip YT, Davis RJ (1998): Signal transduction by the c-Jun N-terminal kinase (JNK) — from inflammation to development. *Curr Opin Cell Biol 10*:205–219.

Irving EA, Barone FC, Reith AD, Hadingham SJ, Parsons AA (2000): Differential activation of MAPK/ERK and p38/SAPK in neurones and glia following focal cerebral ischaemia in the rat. *Brain Res Mol Brain Res 77*:65–75.

Ito M, Yoshioka K, Akechi M, Yamashita S, Takamatsu N, Sugiyama K, Hibi M, Nakabeppu Y, Shiba T, Yamamoto KI (1999): JSAP1, a novel jun N-terminal protein kinase (JNK)-binding protein that functions as a scaffold factor in the JNK signaling pathway. *Mol Cell Biol 19*:7539–7548.

Kallunki T, Su B, Tsigelny I, Sluss HK, Derijard B, Moore G, Davis R, Karin M (1994): JNK2 contains a specificity-determining region responsible for efficient c-Jun binding and phosphorylation. *Genes Dev 8*:2996–3007.

Kallunki T, Deng T, Hibi M, Karin M (1996): c-Jun can recruit JNK to phosphorylate dimerization partners via specific docking interactions. *Cell 87*:929–939.

Kanamoto T, Mota M, Takeda K, Rubin LL, Miyazono K, Ichijo H, Bazenet CE (2000): Role of apoptosis signal-regulating kinase in regulation of the c-Jun N-terminal kinase pathway and apoptosis in sympathetic neurons. *Mol Cell Biol 20*:196–204.

Karin M (1995): The regulation of AP-1 activity by mitogen-activated protein kinases. *J Biol Chem 270*:16483–16486.

Keyse SM (1998): Protein phosphatases and the regulation of MAP kinase activity. *Semin Cell Dev Biol 9*:143–152.

Kikuchi M, Tenneti L, Lipton SA (2000): Role of p38 mitogen-activated protein kinase in axotomy-induced apoptosis of rat retinal ganglion cells. *J Neurosci 20*:5037–5044.

Kim IJ, Lee KW, Park BY, Lee JK, Park J, Choi IY, Eom SJ, Chang TS, Kim MJ, Yeom YI, Chang SK, Lee YD, Choi EJ, Han PL (1999): Molecular cloning of multiple splicing variants of JIP-1 preferentially expressed in brain. *J Neurochem 72*:1335–1343.

Kita Y, Kimura KD, Kobayashi M, Ihara S, Kaibuchi K, Kuroda S, Ui M, Iba H, Konishi H, Kikkawa U, Nagata S, Fukui Y (1998): Microinjection of activated phosphatidylinositol-3 kinase induces process outgrowth in rat PC12 cells through the Rac-JNK signal transduction pathway. *J Cell Sci 111*:907–915.

Ko HW, Han KS, Kim EY, Ryu BR, Yoon WJ, Jung YK, Kim SU, Gwag BJ (2000): Synergetic activation of p38 mitogen-activated protein kinase and caspase-3-like proteases for execution of calyculin A-induced apoptosis but not *N*-methyl-D-aspartate-induced necrosis in mouse cortical neurons. *J Neurochem 74*:2455–2461.

Kuan CY, Yang DD, Samanta Roy DR, Davis RJ, Rakic P, Flavell RA (1999): The Jnk1 and Jnk2 protein kinases are required for regional specific apoptosis during early brain development. *Neuron 22*:667–676.

Kuboki Y, Ito M, Takamatsu N, Yamamoto K, Shiba T, Yoshioka K (2000): A scaffold protein in the c-Jun

NH2-terminal kinase signaling pathways suppresses the extracellular signal-regulated kinase signaling pathways [in process citation]. *J Biol Chem 275*:39815–39818.

Kummer JL, Rao PK, Heidenreich KA (1997): Apoptosis induced by withdrawal of trophic factors is mediated by p38 mitogen-activated protein kinase. *J Biol Chem 272*:20490–20494.

Lallemand D, Ham J, Garbay S, Bakiri L, Traincard F, Jeannequin O, Pfarr CM, Yaniv M (1998): Stress-activated protein kinases are negatively regulated by cell density. *EMBO J 17*:5615–5626.

Leah JD, Herdegen T, Bravo R (1991): Selective expression of Jun proteins following axotomy and axonal transport block in peripheral nerves in the rat: evidence for a role in the regeneration process. *Brain Res 566*:198–207.

Lee FS, Hagler J, Chen ZJ, Maniatis T (1997): Activation of the IkappaB alpha kinase complex by MEKK1, a kinase of the JNK pathway. *Cell 88*:213–222.

Lee JK, Park J, Lee YD, Lee SH, Han PL (1999): Distinct localization of SAPK isoforms in neurons of adult mouse brain implies multiple signaling modes of SAPK pathway. *Brain Res Mol Brain Res 70*:116–124.

Lee HH, Shiow SJ, Chung HC, Huang CY, Lin CL, Hsu JD, Shyu LY, Wang CJ (2000): Development of brain injury in mice by *Angiostrongylus cantonensis* infection is associated with the induction of transcription factor NF-kappaB, nuclear protooncogenes, and protein tyrosine phosphorylation. *Exp Parasitol 95*:202–208.

Lee SH, Park J, Che Y, Han PL, Lee JK (2000): Constitutive activity and differential localization of p38alpha and p38beta MAPKs in adult mouse brain. *J Neurosci Res 60*:623–631.

Lee YB, Schrader JW, Kim SU (2000): p38 map kinase regulates TNF-alpha production in human astrocytes and microglia by multiple mechanisms. *Cytokine 12*:874–880.

Legos JJ, Erhardt JA, White RF, Lenhard SC, Chandra S, Parsons AA, Tuma RF, Barone FC (2001): SB 239063, a novel p38 inhibitor, attenuates early neuronal injury following ischemia. *Brain Res 892*:70–77.

Leppa S, Bohmann D (1999): Diverse functions of JNK signaling and c-Jun in stress response and apoptosis. *Oncogene 18*:6158–6162.

Leppa S, Saffrich R, Ansorge W, Bohmann D (1998): Differential regulation of c-Jun by ERK and JNK during PC12 cell differentiation. *EMBO J 17*:4404–4413.

Levresse V, Butterfield L, Zentrich E, Heasley LE (2000): Akt negatively regulates the cJun N-terminal kinase pathway in PC12 cells [in process citation]. *J Neurosci Res 62*:799–808.

Liu YF, Dorow D, Marshall J (2000): Activation of MLK2-mediated signaling cascades by polyglutamine-expanded huntingtin. *J Biol Chem 275*:19035–19040.

Mao Z, Bonni A, Xia F, Nadal-Vicens M, Greenberg ME (1999): Neuronal activity-dependent cell survival mediated by transcription factor MEF2. *Science 286*:785–790.

Marinissen MJ, Chiariello M, Pallante M, Gutkind JS (1999): A network of mitogen-activated protein kinases links G protein-coupled receptors to the c-jun promoter: a role for c-Jun NH2-terminal kinase, p38s, and extracellular signal-regulated kinase 5. *Mol Cell Biol 19*:4289–4301.

Maroney AC, Glicksman MA, Basma AN, Walton KM, Knight Jr. E, Murphy CA, Bartlett BA, Finn JP, Angeles T, Matsuda Y, Neff NT, Dionne CA (1998): Motoneuron apoptosis is blocked by CEP-1347 (KT 7515), a novel inhibitor of the JNK signaling pathway. *J Neurosci 18*:104–111.

Martin JH, Mohit AA, Miller CA (1996): Developmental expression in the mouse nervous system of the p493F12 SAP kinase. *Brain Res Mol Brain Res 35*:47–57.

Martin-Villalba A, Winter C, Brecht S, Buschmann T, Zimmermann M, Herdegen T (1998): Rapid and long-lasting suppression of the ATF-2 transcription factor is a common response to neuronal injury. *Brain Res Mol Brain Res 62*:158–166.

Marushige K, Marushige Y (1999): Changes in the mitogen-activated protein kinase and phosphatidylinositol 3-kinase/Akt signaling associated with the induction of apoptosis. *Anticancer Res 19*:3865–3871.

Maruyama M, Sudo T, Kasuya Y, Shiga T, Hu B, Osada H (2000): Immunolocalization of p38 MAP kinase in mouse brain. *Brain Res 887*:350–358.

Matsuyama T, Hata R, Yamamoto Y, Tagaya M, Akita H, Uno H, Wanaka A, Furuyama J, Sugita M (1995): Localization of Fas antigen mRNA induced in postischemic murine forebrain by in situ hybridization. *Brain Res Mol Brian Res 34*:166–172.

McDonald PH, Chow CW, Miller WE, Laporte SA, Field ME, Lin FT, Davis RJ, Lefkowitz RJ (2000): Beta-arrestin 2: a receptor-regulated MAPK scaffold for the activation of JNK3. *Science 290*:1574–1577.

Mendelson KG, Contois LR, Tevosian SG, Davis RJ, Paulson KE (1996): Independent regulation of JNK/p38 mitogen-activated protein kinases by metabolic oxidative stress in the liver. *Proc Natl Acad Sci USA 93*:12908–12913.

Meriin AB, Yaglom JA, Gabai VL, Zon L, Ganiatsas S, Mosser DD, Sherman MY (1999): Protein-damaging stresses activate c-Jun N-terminal kinase via inhibition of its dephosphorylation: a novel pathway controlled by HSP72 [erratum appears in *Mol Cell Biol 19*(7):5235]. *Mol Cell Biol 19*:2547–2555.

Mielke K, Brecht S, Dorst A, Herdegen T (1999): Activity and expression of JNK1, p38 and ERK kinases, c-Jun N-terminal phosphorylation, and c-jun promoter binding in the adult rat brain following kainate-induced seizures. *Neuroscience 91*:471–483.

Mielke K, Damm A, Yang DD, Herdegen T (2000): Selective expression of JNK isoforms and stress-specific JNK activity in different neural cell lines. *Brain Res Mol Brain Res 75*:128–137.

Migheli A, Piva R, Atzori C, Troost D, Schiffer D (1997): c-Jun, JNK/SAPK kinases and transcription factor NF-kappa B are selectively activated in astrocytes, but not motor neurons, in amyotrophic lateral sclerosis. *J Neuropathol Exp Neurol 56*:1314–1322.

Minden A, Karin M (1997): Regulation and function of the JNK subgroup of MAP kinases. *Biochem Biophys Acta 1333*:F85–F104.

Morooka T, Nishida E (1998): Requirement of p38 mitogen-activated protein kinase for neuronal differentiation in PC12 cells. *J Biol Chem 273*:24285–24288.

Muda M, Theodosiou A, Rodrigues N, Boschert U, Camps M, Gillieron C, Davies K, Ashworth A, Arkinstall S (1996): The dual specificity phosphatases M3/6 and MKP-3 are highly selective for inactivation of distinct mitogen-activated protein kinases. *J Biol Chem 271*:27205–27208.

Mulle C, Sailer A, Perez-Otano I, Dickinson-Anson H, Castillo PE, Bureau I, Maron C, Gage FH, Mann JR, Bettler B, Heinemann SF (1998): Altered synaptic physiology and reduced susceptibility to kainate-induced seizures in GluR6-deficient mice. *Nature 392*:601–605.

Musti AM, Treier M, Bohmann D (1997): Reduced ubiquitin-dependent degradation of c-Jun after phosphorylation by MAP kinases. *Science 275*:400–402.

Nagata K, Puls A, Futter C, Aspenstrom P, Schaefer E, Nakata T, Hirokawa N, Hall A (1998): The MAP kinase kinase kinase MLK2 co-localizes with activated JNK along microtubules and associates with kinesin superfamily motor KIF3. *EMBO J 17*:149–158.

Nakahara S, Yone K, Sakou T, Wada S, Nagamine T, Niiyama T, Ichijo H (1999): Induction of apoptosis signal regulating kinase 1 (ASK1) after spinal cord injury in rats: possible involvement of ASK1-JNK and -p38 pathways in neuronal apoptosis. *J Neuropathol Exp Neurol 58*:442–450.

Namgung U, Xia Z (2000): Arsenite-induced apoptosis in cortical neurons is mediated by c-Jun N-terminal protein kinase 3 and p38 mitogen-activated protein kinase. *J Neurosci 20*:6442–6451.

Natoli G, Costanzo A, Ianni A, Templeton DJ, Woodgett JR, Balsano C, Levrero M (1997): Activation of SAPK/JNK by TNF receptor 1 through a noncytotoxic TRAF2-dependent pathway. *Science 275*:200–203.

Natoli G, Costanzo A, Guido F, Moretti F, Levrero M (1998): Apoptotic, non-apoptotic, and anti-apoptotic pathways of tumor necrosis factor signalling. *Biochem Pharmacol 56*:915–920.

Nicholson DW, Thornberry NA (1997): Caspases: killer proteases. *Trends Biochem Sci 22*:299–306.

Nuydens R, de Jong M, Van Den Kieboom G, Heers C, Dispersyn G, Cornelissen F, Nuyens R, Borgers M, Geerts H (1998): Okadaic acid-induced apoptosis in neuronal cells: evidence for an abortive mitotic attempt. *J Neurochem 70*:1124–1133.

Okamoto S, Krainc D, Sherman K, Lipton SA (2000): Antiapoptotic role of the p38 mitogen-activated protein kinase-myocyte enhancer factor 2 transcription factor pathway during neuronal differentiation. *Proc Natl Acad Sci USA 97*:7561–7566.

Oo TF, Henchcliffe C, James D, Burke RE (1999): Expression of c-fos, c-jun, and c-jun N-terminal kinase (JNK) in a developmental model of induced apoptotic death in neurons of the substantia nigra. *J Neurochem 72*:557–564.

Oren A, Herschkovitz A, Ben-Dror I, Holdengreber V, Ben-Shaul Y, Seger R, Vardimon L (1999): The cytoskeletal network controls c-Jun expression and glucocorticoid receptor transcriptional activity in an antagonistic and cell-type-specific manner. *Mol Cell Biol 19*:1742–1750.

Ozawa H, Shioda S, Dohi K, Matsumoto H, Mizushima H, Zhou CJ, Funahashi H, Nakai Y, Nakajo S, Matsumoto K (1999): Delayed neuronal cell death in the rat hippocampus is mediated by the mitogen-activated protein kinase signal transduction pathway. *Neurosci Lett 262*:57–60.

Paratcha G, de Stein ML, Szapiro G, Lopez M, Bevilaqua L, Cammarota M, de Iraldi AP, Izquierdo I, Medina JH (2000): Experience-dependent decrease in synaptically localized Fra-1. *Brain Res Mol Brain Res 78*:120–130.

Pellet JB, Haefliger JA, Staple JK, Widmann C, Welker E, Hirling H, Bonny C, Nicod P, Catsicas S, Waeber G, Riederer BM (2000): Spatial, temporal and subcellular localization of islet-brain 1 (IB1), a homologue of JIP-1, in mouse brain. *Eur J Neurosci 12*:621–632.

Pena E, Berciano MT, Fernandez R, Crespo P, Lafarga M (2000): Stress-induced activation of c-Jun N-terminal kinase in sensory ganglion neurons: accumulation in nuclear domains enriched in splicing factors and distribution in perichromatin fibrils. *Exp Cell Res 256*:179–191.

Pena LA, Fuks Z, Kolesnick R (1997): Stress-induced apoptosis and the sphingomyelin pathway. *Biochem Pharmacol 53*:615–621.

Pennypacker KR, Eidizadeh S, Kassed CA, O'Callaghan JP, Sanberg PR, Willing AE (2000): Expression of fos-related antigen-2 in rat hippocampus after middle cerebral arterial occlusion. *Neurosci Lett 289*:1–4.

Perez O, Ntilde OI, Mandelzys A, Morgan JI (1998): MPTP-Parkinsonism is accompanied by persistent expression of a Delta-FosB-like protein in dopaminergic pathways. *Brain Res Mol Brain Res 53*:41–52.

Purkiss RJ, Legg MD, Hunt SP, Davies SW (1993): Immediate early gene expression in the rat forebrain following striatal infusion of quinolinic acid. *Eur J Neurosci 5*:1653–1662.

Raivich G, Bohatschek M, Kloss CU, Werner A, Jones LL, Kreutzberg GW (1999): Neuroglial activation repertoire in the injured brain: graded response, molecular mechanisms and cues to physiological function. *Brain Res Brain Res Rev 30*:77–105.

Reynolds CH, Betts JC, Blackstock WP, Nebreda AR, Anderton BH (2000): Phosphorylation sites on tau identified by nanoelectrospray mass spectrometry: differences in vitro between the mitogen-activated protein kinases ERK2, c-Jun N-terminal kinase and P38, and glycogen synthase kinase-3beta. *J Neurochem 74*:1587–1595.

Rizzo MT, Leaver AH, Yu WM, Kovacs RJ (1999): Arachidonic acid induces mobilization of calcium stores and c-jun gene expression: evidence that intracellular calcium release is associated with c-jun activation. *Prostaglandins Leukot Essent Fatty Acids 60*:187–198.

Robinson GA (1995): Axotomy-induced regulation of c-Jun expression in regenerating rat retinal ganglion cells. *Brain Res Mol Brain Res 30*:61–69.

Roux P, Blanchard JM, Fernandez A, Lamb N, Jeanteur P, Piechaczyk M (1990): Nuclear localization of c-Fos, but not v-Fos proteins, is controlled by extracellular signals. *Cell 63*:341–351.

Sadot E, Jaaro H, Seger R, Ginzburg I (1998): Ras-signaling pathways: positive and negative regulation of tau expression in PC12 cells. *J Neurochem 70*:428–431.

Salehi M, Barron M, Merry BJ, Goyns MH (1999): Fluorescence in situ hybridization analysis of the fos/jun ratio in the ageing brain. *Mech Ageing Dev 107*:61–71.

Saleshando G, O'Connor JJ (2000): SB203580, the p38 mitogen-activated protein kinase inhibitor blocks the inhibitory effect of beta-amyloid on long-term potentiation in the rat hippocampus. *Neurosci Lett 288*:119–122.

Saporito MS, Thomas BA, Scott RW (2000): MPTP activates c-Jun NH(2)-terminal kinase (JNK) and its upstream regulatory kinase MKK4 in nigrostriatal neurons in vivo. *J Neurochem 75*:1200–1208.

Schauwecker PE (2000): Seizure-induced neuronal death is associated with induction of c-Jun N-terminal kinase and is dependent on genetic background [in process citation]. *Brain Res 884*:116–128.

Sheng M, Pak DT (2000): Ligand-gated ion channel interactions with cytoskeletal and signaling proteins. *Annu Rev Physiol 62*:755–778.

Sommer C, Gass P, Kiessling M (1995): Selective c-JUN expression in CA1 neurons of the gerbil hippocampus during and after acquisition of an ischemia-tolerant state. *Brain Pathol 5*:135–144.

Speliotes EK, Kowall NW, Shanti BF, Kosofsky B, Finklestein SP, Leifer D (1996): Myocyte-specific enhancer binding factor 2C expression in gerbil brain following global cerebral ischemia. *Neuroscience 70*:67–77.

Standen CL, Brownlees J, Grierson AJ, Kesavapany S, Lau KF, McLoughlin DM, Miller CC (2001): Phosphorylation of thr(668) in the cytoplasmic domain of the Alzheimer's disease amyloid precursor protein by stress-activated protein kinase 1b (Jun N-terminal kinase-3). *J Neurochem 76*:316–320.

Sugino T, Nozaki K, Takagi Y, Hattori I, Hashimoto N, Moriguchi T, Nishida E (2000): Activation of mitogen-activated protein kinases after transient forebrain ischemia in gerbil hippocampus. *J Neurosci 20*:4506–4514.

Takagi Y, Nozaki K, Sugino T, Hattori I, Hashimoto N (2000): Phosphorylation of c-Jun NH(2)-terminal kinase and p38 mitogen-activated protein kinase after transient forebrain ischemia in mice. *Neurosci Lett 294*:117–120.

Takeda K, Hatai T, Hamazaki TS, Nishitoh H, Saitoh M, Ichijo H (2000): Apoptosis signal-regulating kinase 1 (ASK1) induces neuronal differentiation and survival of PC12 cells. *J Biol Chem 275*:9805–9813.

Takenaka K, Moriguchi T, Nishida E (1998): Activation of the protein kinase p38 in the spindle assembly checkpoint and mitotic arrest. *Science 280*:599–602.

Telfeian AE, Federoff HJ, Leone P, During MJ, Williamson A (2000): Overexpression of GluR6 in rat hippocampus produces seizures and spontaneous nonsynaptic bursting in vitro. *Neurobiol Dis 7*:362–374.

Tian D, Litvak V, Lev S (2000): Cerebral ischemia and seizures induce tyrosine phosphorylation of PYK2 in neurons and microglial cells. *J Neurosci 20*:6478–6487.

Tölle TR, Herdegen T, Schadrack J, Bravo R, Zimmermann M, Ziegelgansberger W (1994): Application of morphine prior to noxious stimulation differentially modulates expression of Fos, Jun and Krox-24 proteins in rat spinal cord neurons. *Neuroscience 58*:305–321.

Tsurumi C, Ishida N, Tamura T, Kakizuka A, Nishida E, Okumura E, Kishimoto T, Inagaki M, Okazaki K, Sagata N (1995): Degradation of c-Fos by the 26S proteasome is accelerated by c-Jun and multiple protein kinases. *Mol Cell Biol 15*:5682–5687.

Van WN, Choi C, Repovic P, Benveniste EN (2000): Oncostatin M regulation of interleukin-6 expression in astrocytes: biphasic regulation involving the mitogen-activated protein kinases ERK1/2 and p38. *J Neurochem 75*:563–575.

Vereker E, O'Donnell E, Lynch MA (2000): The inhibitory effect of interleukin-1beta on long-term potentiation is coupled with increased activity of stress-activated protein kinases. *J Neurosci 20*:6811–6819.

Walton KM, DiRocco R, Bartlett BA, Koury E, Marcy VR, Jarvis B, Schaefer EM, Bhat RV (1998): Activation of p38MAPK in microglia after ischemia. *J Neurochem 70*:1764–1767.

Wang TH, Wang HS, Ichijo H, Giannakakou P, Foster JS, Fojo T, Wimalasena J (1998): Microtubule-interfering agents activate c-Jun N-terminal kinase/stress-activated protein kinase through both Ras and apoptosis signal-regulating kinase pathways. *J Biol Chem 273*:4928–4936.

Watanabe Y, Johnson RS, Butler LS, Binder DK, Spiegelman BM, Papaioannou VE, McNamara JO (1996): Null mutation of c-fos impairs structural and functional plasticities in the kindling model of epilepsy. *J Neurosci 16*:3827–3836.

Weitzman JB, Fiette L, Matsuo K, Yaniv M (2000): JunD protects cells from p53-dependent senescence and apoptosis. *Mol Cell 6*:1109–1119.

Winter C, Schenkel J, Zimmermann M, Herdegen T (1998): MAP kinase phosphatase 1 is expressed and enhanced by FK506 in surviving mamillary, but not degenerating nigral neurons following axotomy. *Brain Res 801*:198–205.

Wollnik F, Brysch W, Uhlmann E, Gillardon F, Bravo R, Zimmermann M, Schlingensiepen KH, Herdegen T (1995): Block of c-Fos and JunB expression by antisense oligonucleotides inhibits light-induced phase shifts of the mammalian circadian clock. *Eur J Neurosci 7*:388–393.

Wood MW, Segal JA, Mark RJ, Ogden AM, Felder CC (2000): Inflammatory cytokines enhance muscarinic-mediated arachidonic acid release through p38 mitogen-activated protein kinase in A2058 cells. *J Neurochem 74*:2033–2040.

Xia Z, Dickens M, Raingeaud J, Davis RJ, Greenberg ME (1995): Opposing effects of ERK and JNK-p38 MAP kinases on apoptosis. *Science 270*:1326–1331.

Xu J, Kim GM, Ahmed SH, Yan P, Xu XM, Hsu CY (2001): Glucocorticoid receptor-mediated suppression of activator protein-1 activation and matrix metalloproteinase expression after spinal cord injury. *J Neurosci 21*:92–97.

Xu X, Raber J, Yang D, Su B, Mucke L (1997): Dynamic regulation of c-Jun N-terminal kinase activity in mouse brain by environmental stimuli. *Proc Natl Acad Sci USA 94*:12655–12660.

Yang DD, Kuan CY, Whitmarsh AJ, Rincon M, Zheng TS, Davis RJ, Rakic P, Flavell RA (1997): Absence of excitotoxicity-induced apoptosis in the hippocampus of mice lacking the Jnk3 gene. *Nature 389*:865–870.

Yang X, Khosravi-Far R, Chang HY, Baltimore D (1997): Daxx, a novel Fas-binding protein that activates JNK and apoptosis. *Cell 89*:1067–1076.

Yaniv K, Yisraeli JK (2001): Defining cis-acting elements and trans-acting factors in RNA localization. *Int Rev Cytol 203*:521–539.

Yasuda J, Whitmarsh AJ, Cavanagh J, Sharma M, Davis RJ (1999): The JIP group of mitogen-activated protein kinase scaffold proteins. *Mol Cell Biol 19*:7245–7254.

Yujiri T, Fanger GR, Garrington TP, Schlesinger TK, Gibson S, Johnson GL (1999): MEK kinase 1 (MEKK1) transduces c-Jun NH2-terminal kinase activation in response to changes in the microtubule cytoskeleton. *J Biol Chem 274*:12605–12610.

Zhang Y, Zhou L, Miller CA (1998): A splicing variant of a death domain protein that is regulated by a mitogen-activated kinase is a substrate for c-Jun N-terminal kinase in the human central nervous system. *Proc Natl Acad Sci USA 95*:2586–2591.

Zinck R, Cahill MA, Kracht M, Sachsenmaier C, Hipskind RA, Nordheim A (1995): Protein synthesis inhibitors reveal differential regulation of mitogen-activated protein kinase and stress-activated protein kinase pathways that converge on Elk-1. *Mol Cell Biol 15*:4930–4938.

Zhu X, Raina AK, Rottkamp CA, Aliev G, Perry G, Boux H, Smith MA (2001): Activation and redistribution of c-Jun N-terminal kinase/stress activated protein kinase in degenerating neurons in Alzheimer's disease. *J Neurochem 76*:435–441.

CHAPTER XI

Elk-1: an important regulator of immediate early gene expression in the brain

PETER VANHOUTTE AND JOCELYNE CABOCHE

1. INTRODUCTION

In the central nervous system (CNS), neurotransmitter binding to its cognate receptor is associated with fast synaptic transmission along with short- and long-term synaptic plasticity. While phosphorylation of pre-existing synaptic proteins is required for short-term plasticity, long-lasting changes in neuronal efficacy depend critically on gene regulation and protein synthesis.

In considering how these events come about, it is important to understand the relationship between membrane events, that is the electrical activity of a neuron, and nuclear events, i.e. changes in the pattern of gene expression. A key intermediate in this process is the induction of signaling pathways that lead to the phosphorylation of constitutive transcription factors (CTFs) interacting with promoter regulatory elements. In this way, neurotransmitter release rapidly activates expression of a class of genes called immediate early genes (IEG) in post-synaptic neurons (for review, Sheng and Greenberg, 1990; Morgan and Curran, 1991). Most of these IEGs encode for inducible transcription factors (ITFs) such as c-*fos*, c-*jun*, *jun*B, or *zif268* that, in turn, will regulate expression of a later wave of genes coding for proteins directly involved in changes in synaptic efficacy or cellular morphology. These newly synthesized proteins mediate synaptic remodeling and modifications in neuronal structure essential for cognitive brain functions like learning and memory formation.

As stated above, CTF phosphorylation controls IEG transcription. Most attention has focused over the past few years on the nuclear CTF 'cAMP and calcium responsive element binding protein' (CREB), which appears to be a key component in memory formation and learning in various animal models (reviewed in Mayford and Kandel, 1999). More recently several groups, including ours, have focused on Elk-1, a CTF that seems to play a pivotal role between intracellular signaling events and IEG regulation in the adult CNS. In this review we will describe the molecular events leading to activation of Elk-1, as well as its mode of action via the serum response element (SRE), a DNA binding site located in the promoter of many IEGs. We also review data concerning its neuroanatomical distribution, its regulation in the CNS, and involvement in model systems of synaptic plasticity and long-term behavioral changes.

Handbook of Chemical Neuroanatomy Vol. 19: Immediate Early Genes and Inducible Transcription Factors in Mapping of the Central Nervous System Function and Dysfunction
L. Kaczmarek and H.A. Robertson, editors

2. THE TRANSCRIPTION FACTOR Elk-1

2.1. MOLECULAR PROPERTIES

Elk-1 belongs to the ternary complex factor (TCF) family, that also comprises SAP1 and SAP2/ERP/Net (Hipskind et al., 1991; Dalton and Treisman, 1992; Lopez et al., 1994; Price et al., 1995). These constitutive transcription factors utilize both protein/protein and protein/DNA contacts to bind the serum response element (SRE), a protein binding site initially described in the c-*fos* promoter to be required for its induction by serum (Treisman, 1986, 1987). This regulatory element, located at position −310 in the c-*fos* promoter, serves as the site of assembly of a multiprotein complex that contains a dimer of serum response factor (SRF) together with one molecule of TCF (Treisman, 1986, 1987; Schröter et al., 1987, 1990; Sharrocks et al., 1993; for review Treisman, 1992). Although SRF is rapidly phosphorylated in response to mitogenic stimulation and phosphorylation affects its DNA binding properties (Rivera et al., 1993), the major regulatory input received by the SRE can be attributed to TCF phosphorylation (Gille et al., 1992; Hill et al., 1993; Janknecht et al., 1993; Marais et al., 1993). In cell culture systems, TCFs represent major nuclear targets for MAPKs (mitogen-activated protein kinases) of the extracellular-signal regulated kinase (ERK) subfamily and the closely related stress-activated protein kinase (SAPK/JNK) and $p38^{MAPK}$ (for review Treisman, 1996) (Fig. 1). The activation of TCFs resulting from their phosphorylation by MAPKs, increases their ability to form a ternary complex with SRE and SRF in vitro and to strongly potentiate c-*fos* transcription (Zinck et al., 1993; Gille et al., 1995). SRE-driven transcriptional activation also involves recruitment of coactivators of the CBP/p300 family (for review Janknecht and Hunter, 1996; Goldman et al., 1997). CBP has been described to interact with the TCFs Elk-1 and SAP1-a (Janknecht and Nordheim, 1996a,b), as well as with SRF (Ramirez et al., 1997) and to act as a molecular bridge between

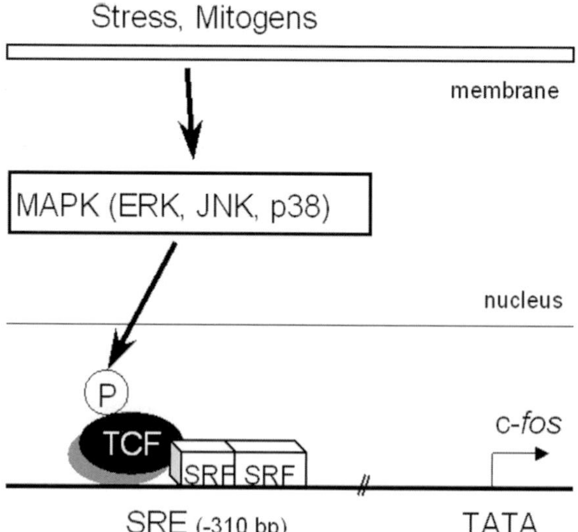

Fig. 1. Schematic presentation of TCF binding to the SRE through interaction with SRFs. The SRE serves as a site of assembly of a multiprotein complex that contains a dimer of SRFs, together with one molecule of TCF. The major regulatory input received by the SRE is attributed to TCF phosphorylation by MAPKs. This increases the ability of TCF to form a ternary complex with SRF and SRE and strongly potentiates c-*fos* transcription.

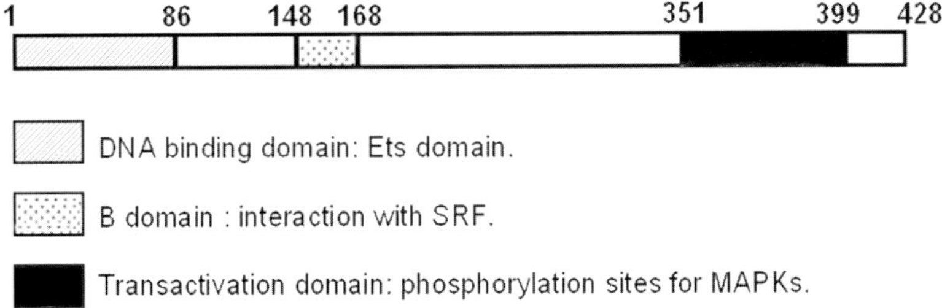

Fig. 2. Functional domains of Elk-1. Shown are the principal domains of Elk-1. In the N-terminal region: the DNA binding domain, which presents high homology with Ets domain. The B domain, encompassing amino acids 148–168, which mediate protein/protein interaction with SRF. The C-terminal region, which contains consensus phosphorylation sites for MAPKs necessary for conformational changes and increased affinity of Elk-1 for DNA.

transcription factors and components of the basal transcription machinery such as the RNA polymerase II (Kwok et al., 1994).

2.2. STRUCTURE OF Elk-1

Elk-1 represents the first identified TCF (Hipskind et al., 1991), and it contains four functional domains conserved with other members of the TCF family (Fig. 2): the N-term Ets-DNA binding domain (encompassing amino acids 1–85) (Shore and Sharrocks, 1995), a signature motif of 20 amino acids (the B domain), which mediates protein/protein interactions with SRF (Dalton and Treisman, 1992; Shore and Sharrocks, 1994; Ling et al., 1997), a combination of nuclear localization signal and MAPK docking site (the D domain) (Price et al., 1995; Yang et al., 1998a,b) and the C-terminal transactivation domain containing consensus phosphorylation sites for MAPKs (reviewed in Treisman, 1996). In this region, two residues, Ser^{383} and Ser^{389}, play a critical role in activation of Elk-1 by MAPKs. Mutating these residues to alanine, which cannot be phosphorylated, totally blocks SRE-driven gene induction (Janknecht et al., 1993; Marais et al., 1993). MAP kinase-dependent phosphorylation of Ser^{383} and Ser^{389} produces conformational changes of Elk-1 and thereby modulates its affinity for DNA (Yang et al., 1999). These events result in an increase in the ternary complex formation on the SRE and thereby in SRE-driven transactivation (Hipskind et al., 1991, 1994; Gille et al., 1995; Zinck et al., 1995).

2.3. SYNERGISM BETWEEN CRE AND SRE SITES IS REQUIRED FOR MAXIMAL ACTIVATION OF c-*fos*

Among the IEGs, perhaps the best studied is the c-*fos* proto-oncogene, which has provided a structural framework useful for studying how the activation of intracellular signaling cascades leads to changes in gene expression. This IEG is activated transcriptionnally by a wide plethora of extracellular stimuli (growth factors, cytokines, electrical or pharmacological stimuli, environmental stress), thus indicating that its promoter is a convergence point for multiple intracellular signaling networks.

In addition to the SRE site, the c-*fos* promoter contains at least one DNA response element that is essential for its activity-dependent regulation in the CNS: the cAMP and calcium response element (CRE), which lies just upstream of the c-*fos* TATA box. This element can be bound by members of the CREB/ATF family of bZIP transcription factors, in

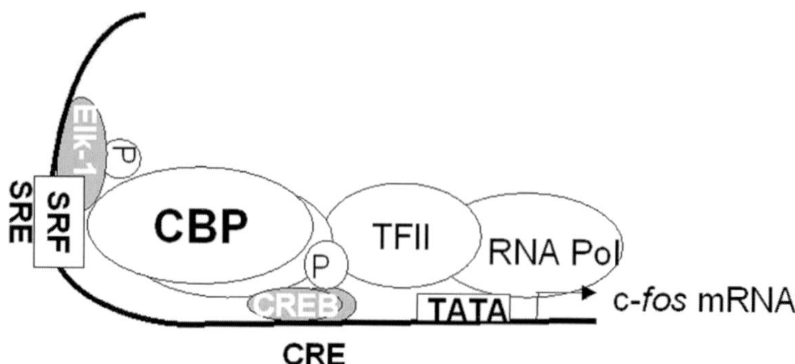

Fig. 3. Synergistic role of CREB and Elk-1 for c-*fos* transcriptional regulation. Phosphorylation of both Elk-1 and CREB recruits the coactivator CBP. By acting as a molecular bridge between these transcription factors and the basal transcriptional machinery, CBP potentiates c-*fos* transcription.

particular CREB (Montminy et al., 1986). The activation of CREB in response to increased intracellular levels of cAMP or Ca^{2+} involves the inducible phosphorylation of Ser^{133} by cAMP-dependent kinases (PKA) or by calcium/calmodulin dependent kinase (CamKII) (Gonzalez and Montminy, 1989; Sheng et al., 1990, 1991; Dash et al., 1991). CREB is also a target of the ERK and $p38^{MAPK}$ cascades, via MAPK activated protein kinases (MAPKAP-Ks) (for review Shaywitz and Greenberg, 1999). Once phosphorylated on Ser^{133}, CREB recruits CBP, which in turn promotes the assembly of the polymerase II transcription complex at the TATA box and leads to the initiation of transcription (Chrivia et al., 1993).

Increasing evidence now indicates that synergistic activation of CRE and SRE sites is necessary for full activation of c-*fos* (Fig. 3). The first, an elegant in vivo study from Robertson et al. (1995), showed that mutating either of these DNA sequences strongly impaired calcium-driven induction of a c-*fos* reporter gene in transgenic mice (Robertson et al., 1995). Similarly, transfection studies imply a role for these two elements in neurotrophin activation of IEGs (Bonni et al., 1995) and show that multiple Ca^{2+}-dependent pathways target them (Johnson et al., 1997). Finally, most in vivo studies in the CNS, aimed at analyzing intracellular mechanisms leading to c-*fos* induction in response to various extracellular stimuli, show a combined activation of Elk-1 and CREB. Thus, it is clear that considering how and where Elk-1 is activated in the brain is a critical issue for tracing intracellular events responsible for c-*fos* regulation in response to extracellular stimuli. Furthermore, Elk-1 is likely a key component in the transcriptional control of other genes such as *zif268*, *nur77*, *jun*B, that have one or multiple SREs in their promoters.

3. EXPRESSION OF Elk-1 IN THE BRAIN

3.1. Elk-1 mRNA EXPRESSION

The first evidence for Elk-1 mRNA expression in the brain was from global biochemical analyses, i.e. RNAse protection analysis. These studies showed high levels of Elk-1 mRNA expression in the brain in human, mouse and rat (Price et al., 1995). In situ hybridization studies from adult rat brain sections confirmed and considerably extended these observations. Elk-1 mRNA was found ubiquitously expressed with some regional differences in the rostrocaudal extension of the rat brain (Fig. 4A) (Sgambato et al., 1998a). Some regions expressed high levels of Elk-1 mRNA: olfactory tubercles, dentate gyrus of the hippocampus, cerebellum, while moderate levels were found in the thalamus, caudate-putamen, substantia nigra pars compacta, cerebral cortex; Elk-1 mRNA was undetectable in the external segment of globus pallidus (GPe) and substantia nigra pars reticulata (SNr). At the cellular level, Elk-1 mRNA hybridization signals were found mainly in cells of neuronal phenotype (Fig. 4B).

3.2. PROTEIN EXPRESSION

Elk-1 protein expression was studied by immunocytochemistry using an affinity-purified antibody directed against residues 407–428 of Elk-1 (Sgambato et al., 1998a). In brain sections this antibody detects Elk-1 protein in neuronal cells with both nuclear and cytoplasmic localization, including dendrites (Fig. 5A). A similar subcellular localization was found using antisera that localized Elk-1 in the nucleus in transiently transfected Hela cells (Pingoud et al., 1994) (Fig. 5B). Under the same experimental conditions, CREB was exclusively nuclear, matching its subcellular localization in cultured cells (Yamamoto et al., 1988). Thus, in contrast to cultured cell lines, where Elk-1 is localized to the nucleus (Janknecht et al., 1994), the scenario in neuronal cells in vivo is different, where Elk-1 is a potential substrate for MAPKs in the cytoplasm (see below).

Intriguingly, in some regions at least, Elk-1 protein was strongly expressed in the neuropil with very few immunoreactive cell bodies. For example, the GPe and SNr showed very low levels of mRNAs (see Fig. 4A) but strong immunoreactivity. This immunoreactivity localized to axon terminals of striatal neurons that project to these structures (Fig. 6). Accordingly, an unilateral kainic acid injection in the striatum that destroyed striatal neurons led to the corresponding loss of Elk-1 immunoreactivity in the projection field of lesioned striatal neurons, i.e. in the GPe and the SNr, ipsilaterally but not contralaterally to the striatal lesion (Fig. 6). Although the role of Elk-1 in axon terminals remains to be established, an attractive possibility is retrograde signaling from axon terminals to the nerve cell bodies. Interestingly, ERKs are also expressed in axon terminals (Ortiz et al., 1995) and can be transported retrogradely (Johanson et al., 1995; Martin et al., 1997). Furthermore, the transcription factors NFκ-B (Povelones et al., 1997) and ATF2 (Delcroix et al., 1999) show axonal transport. Thus, in the brain, CTFs are not necessarily nuclear proteins activated locally after nuclear translocation of kinases. It is now generally admitted that some of them, at least, can convey distant target-derived instructions from dendrites and/or axon terminals, to the nucleus where they can regulate gene expression.

3.3. sElk-1, A NEURONAL SPECIFIC ISOFORM OF Elk-1, DRIVES Elk-1 TO THE CYTOPLASM

In the process of investigating the functional significance of the unexpected subcellular localization of Elk-1 in the brain, we discovered a novel isoform sElk-1 (for short Elk-1), uniquely expressed in brain tissues and NGF-differentiated PC12 cells (Vanhoutte et al., 2001). Western blots of cerebral extracts probed with Elk-1 C-term-specific antibody (residus 407–428) described above showed not only Elk-1 at 52 kDa (the expected molecular weight for Elk-1), but also a 45 kDa form of Elk-1 (sElk-1). This smaller isoform was present in various cerebral structures, except for the SNr (Fig. 7A). Surprisingly, fractionation of cortical extracts showed that sElk-1 was exclusively nuclear while Elk-1 was both cytoplasmic and nuclear (Fig. 7B). The strictly nuclear localization of sElk-1 likely explains why it was absent in the SNr, a structure where Elk-1 protein was found in axon terminals but not cell bodies (see Fig. 6).

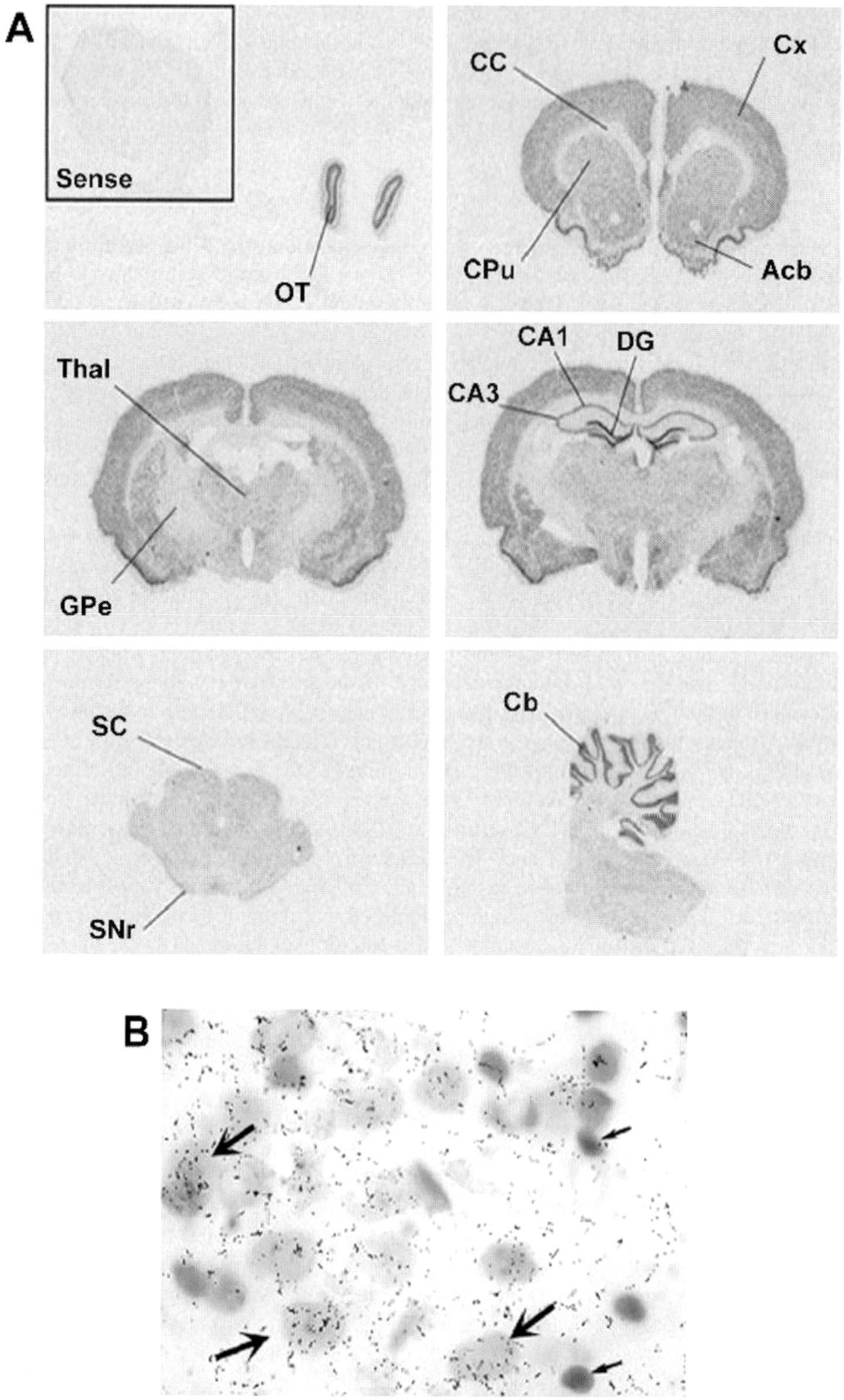

A
Sense
OT
CC
Cx
CPu
Acb
Thal
GPe
CA1
DG
CA3
SC
SNr
Cb

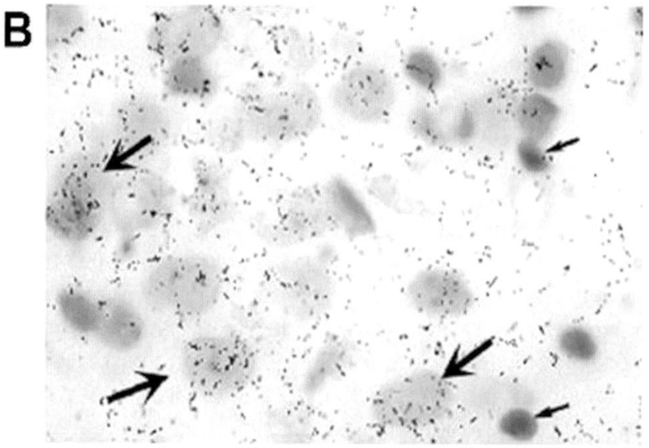
B

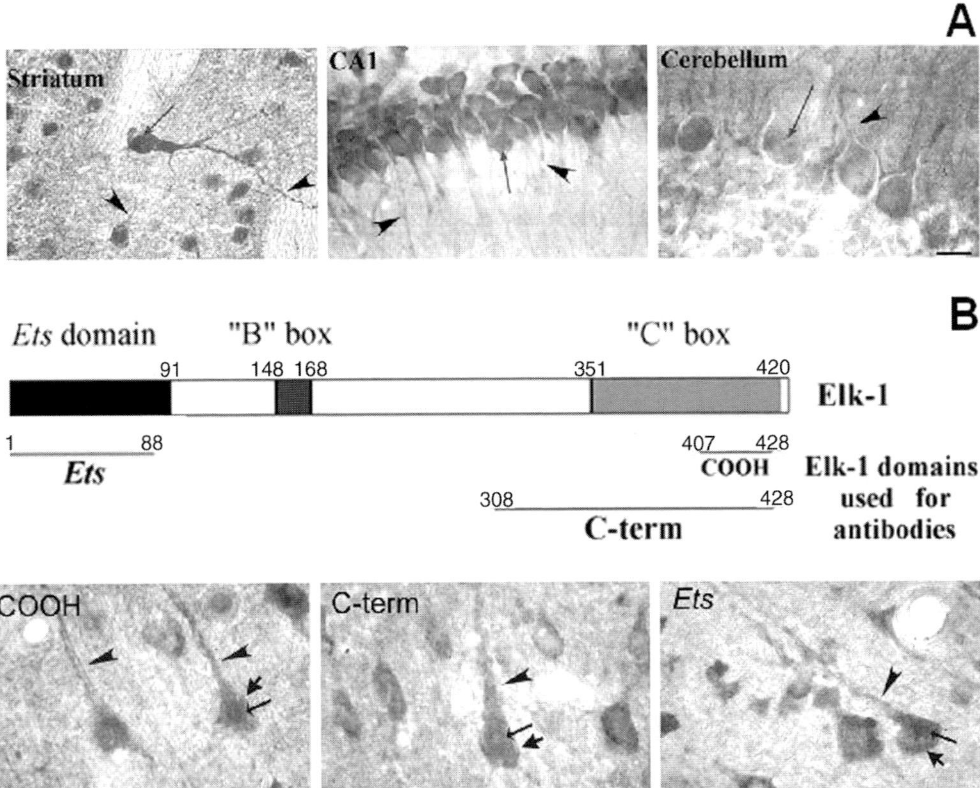

Fig. 5. Elk-1 is expressed in cytoplasmic and nuclear compartments in neuronal cells. (*A*) Immunocytochemical detection of Elk-1 using an affinity purified antibody directed against residues 407–428 or Elk-1. Shown are the striatum, the CA1 layer of hippocampus and Purkinje cell layer of the cerebellum. Note that in all three structures, the protein is present in the nucleus (thin-headed long arrows) and the cytosol, as well as dendritic processes (arrowheads). (*B*) Serial sections were processed for immunocytochemistry with three different Elk-1 antibodies (described in the diagram at the top): anti-COOH (407–428 residues), anti-C-term and anti-*Ets*. Note that all three antibodies show the protein in the nucleus (thin-headed long arrows), cytosol (short arrows) and also in dendritic processes (arrowheads) of neuronal cells (magnification ×500). Adapted from Sgambato et al., 1998a.

Further analyses demonstrated that sElk-1 arises from an internal translation start site in the Elk-1 sequence (Met 55) (Fig. 7C), which generated a protein lacking the first 54 amino acids (Fig. 7D). This deletion severely compromised the ability of sElk-1 to form complexes with SRE on the SRE, in vitro, and to activate SRE reporter genes. More pertinent to the neuronal-specific subcellular localization of Elk-1, we found that, in the presence of

←

Fig. 4. Expression of Elk-1 mRNA in rat brain. Rat cerebral sections were hybridized with P^{33}-cRNA-Elk-1 sense (insert) and antisense probes. The abbreviations are: OT, olfactory tubercles; Cx, cerebral cortex; Acb, accumbens nucleus; CC, corpus callosum; CPu, caudate-putamen; GPe, external globus pallidus; Thal, thalamus; CA, Amon Corn layer of the hippocampus; DG, dentate gyrus of the hippocampus; SC, superior colliculi; SNr, substantia nigra pars reticulata; Cb, cerebellum). (*B*) High magnification (×500) of Elk-1 mRNA hybridization: the cresyl violet counterstaining allows the characterization of glial (small arrows) and neuronal cells (large arrows) in the hippocampus. Note that hybridization signals are always observed on neuronal cells. Adapted from Sgambato et al., 1998a.

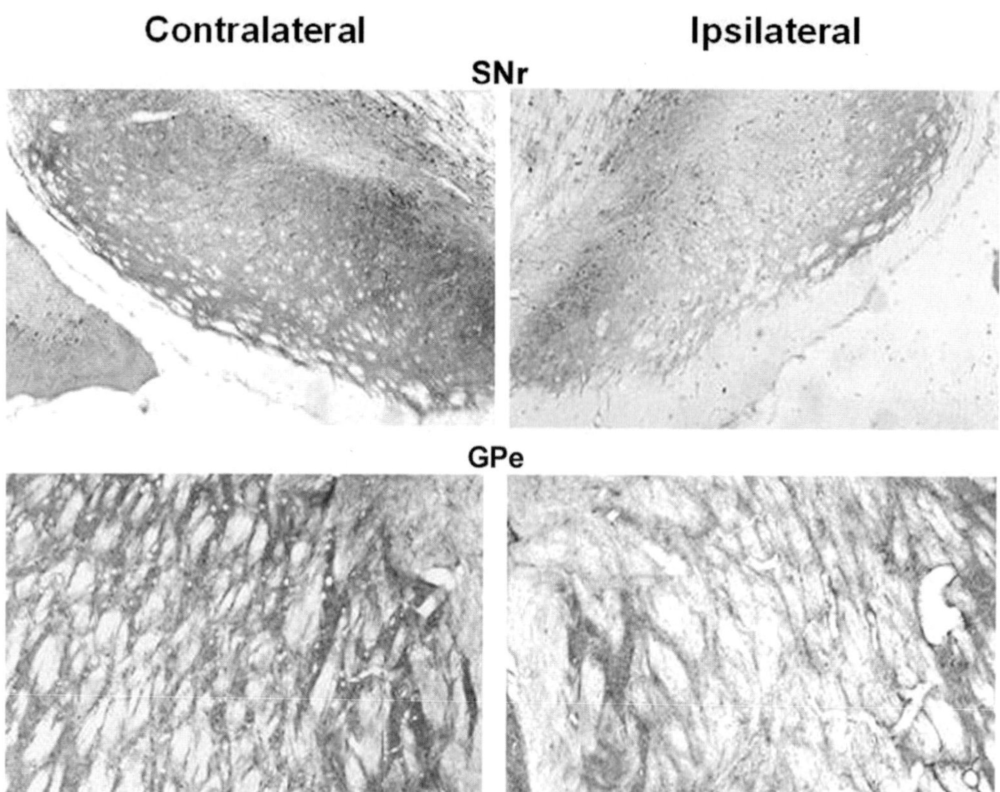

Fig. 6. Elk-1 is expressed in striato-pallidal and striato-nigral axon terminals. Elk-1 immunodetection in the substantia nigra pars reticulata (SNr) (upper panel) and the external segment of the globus pallidus (GPe) (lower panel) decreases ipsilaterally but not contralaterally to kainic acid lesion of striatal neurons. Adapted from Sgambato et al., 1998a.

sElk-1, Elk-1 loses its strictly nuclear localization, in NGF-treated PC12 cells and gains the nuclear/cytoplasmic pattern observed in the mature brain (Fig. 7E). Thus, the cytoplasmic localization of Elk-1 in neuronal cells seems to be related to sElk-1 expression. It is tempting to speculate that cytoplasmic Elk-1 might exercise a novel, specific function in the brain (see below).

4. REGULATION OF Elk-1 IN THE BRAIN

4.1. ERKs MEDIATE SIGNALING TO Elk-1 INDEPENDENTLY OF JNKs IN NEURONAL CELLS

On cell line models, in vitro, Elk-1 represents a point of convergence for ERKs, JNKs and $p38^{MAPK}$ (for review Treisman, 1995). ERKs are activated strongly by mitogens, and in the CNS, by neurotrophins and neurotransmitters (Grewal et al., 1999 for review), whereas SAPK/JNK and $p38^{MAPK}$ are induced by various cytokines and stresses, such as osmotic shock, UV irradiation and excitotoxicity (Cahill et al., 1996; Price et al., 1996). These proline-directed serine/threonine kinases are major effectors of signal transduction from the

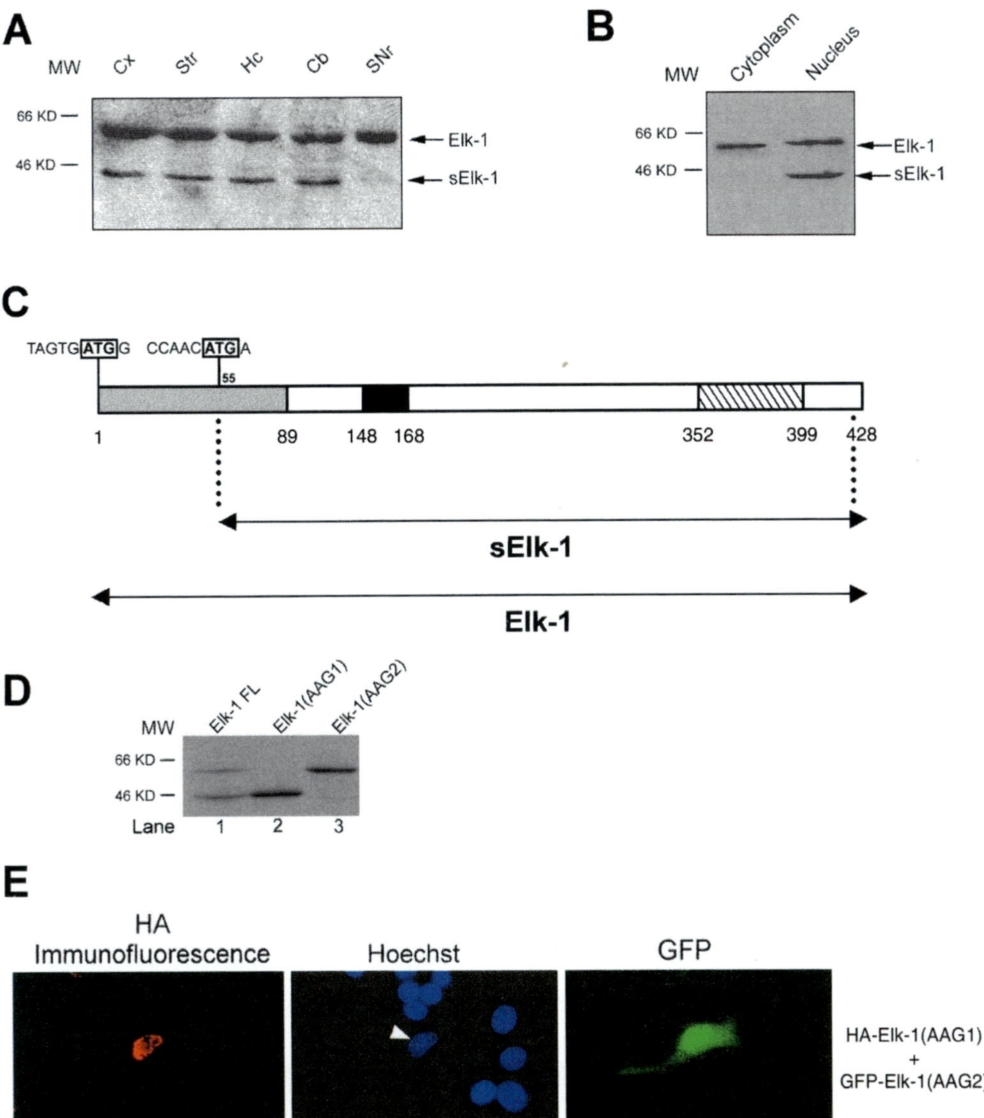

Fig. 7. sElk-1, a neuronal specific isoform of Elk-1, drives Elk-1 to the cytoplasm. (*A*) Western blot analysis of Elk-1 protein expression in extracts prepared from cerebral cortex (Cx), striatum (Str), hippocampus (Hc), cerebellum (Cb), and substantia nigra pars reticulata (SNr). The Elk-1 COOH antibody (407–428) reveals two bands, one at the apparent molecular weight of Elk-1 and the other at 45 kDa, corresponding to a small isoform of Elk-1 (s-Elk-1). (*B*) Cerebral extracts were fractionated into nucleus and cytoplasm and were analyzed by Western blotting with the Elk-1 antibody. Note the strictly nuclear expression of sElk-1, and nuclear/cytoplasmic expression of Elk-1. (*C*) The cDNA sequence for Elk-1 contains two in frame ATGs. The second one serves as an alternative translation start site and generates a protein lacking the 54 amino acids. (*D*) In vitro transcription translation from a Elk-1 cDNA wild type (WT, lane 1) containing the two ATGs, a cDNA mutated in its first ATG (Elk-1 (AAG1), lane 2) or in its second ATG (Elk-1 (AAG2), lane 3). Note that mutating the first ATG abrogates expression of Elk-1, while mutating the second one, inhibits expression of sElk-1. (*E*) Co-transfection of a HA-tagged Elk-1 (AAG1) (red) with GFP-tagged-Elk-1 (green) in PC12 cells, allowing expression of sElk-1 and Elk-1, respectively. An arrowhead in the Hoechst-staining panel (blue) points the nucleus of the transfected cell. Note the strictly nuclear expression of HA-sElk-1, and cytoplasmic localization of GFP-Elk-1 in the presence of an excess of HA-sElk-1. Adapted from Vanhoutte et al., 2000.

cell surface to the nucleus. While JNK and $p38^{MAPK}$ mediate the cellular response to stress and in many cases are involved in apoptosis (Davis, 2000; Mielke and Herdegen, 2000 for review), ERKs are implicated in cell growth and differentiation as well as in many functions in differentiated cells. In the adult CNS, where neurons are definitively post-mitotic, ERKs are abundant (Boulton et al., 1991; Fiore et al., 1993) and play important roles in the regulation of neuronal survival and synaptic plasticity (for review: Derkinderen et al., 1999; Grewal et al., 1999).

While all three MAPKs might signal to Elk-1 in neuronal cells, multiple studies indicate that activation of Elk-1 can be uncoupled from SAPK/JNK in vivo. Excitotoxic stimulation failed to induce hippocampal optosis in $JNK3^{-/-}$ mice but still caused a sustained activation of c-*fos* in this region (Yang et al., 1997). Using a semi in vivo system utilizing striatal slices, glutamate stimulation significantly increased levels of Elk-1 phosphorylation as measured by immunoblotting with an antibody specifically recognizing phospho-Ser^{383} Elk-1 (Vanhoutte et al., 1999). This correlated kinetically with increased levels of phospho-ERKs and phospho-JNKs, but not phospho-p38. Nevertheless, the increase in Elk-1 phosphorylation was completely blocked by preincubation with PD98059, an inhibitor of the ERK cascade via its action on the ERK activator MEK (Vanhoutte et al., 1999). Thus ERK and not JNK in glutamate-stimulated striatal slices target Elk-1. Similar results have been reported in in vivo model systems of neuronal plasticity. In a paradigm of conditioned taste aversive learning, Berman et al. (1998) found that both ERKs and JNKs were activated in the insular cortex, albeit with different time courses. While ERKs activation appeared about 30 min after drinking the taste solution, JNKs showed slower kinetics of induction. The time course of Elk-1 phosphorylation correlated with that of ERKs and not JNKs. Light stimulation increases the activity of ERKs and JNKs in the rat visual cortex (Kaminska et al., 1999), but with different kinetics. Once again phospho-Ser^{383} Elk-1 was detected concomitantly with the onset of ERK but not JNK activation in this structure. In contrast, the appearance of phospho-Ser^{63} Jun correlated temporally with the delayed induction of JNK. These authors concluded that light stimulation resulted in the differential expression of c-Fos and c-Jun, major components of the AP-1 transcriptional complex.

To account for the preferential targeting of Elk-1 by ERKs, and thereby explain the apparent discrepancy between in vitro data from cell lines and in vivo results in neurons, different hypotheses can be proposed: (1) Elk-1 may be localized intracellularly for activation by ERKs rather than JNKs (see below); (2) the major role of JNK activation may be to reorganize the cytoskeleton and gene regulation may be a secondary function of this pathway; (3) alternatively, activated JNKs may be preferentially targeted to c-Jun which is a stronger substrate of JNKs than Elk-1 in vitro (Gupta et al., 1996).

4.2. MAPK/ERKs PLAY A KEY ROLE IN NEURONAL FUNCTIONS

Studies predominantly carried out in non-neuronal cell lines have characterized what could be considered as the archetypal ERKs cascade. This involves the activation of the Ras family of GTPases via the translocation of guanine exchange factors to the membrane by various adapter proteins that bind specifically to phospho-tyrosine residues on the cytoplasmic tails of receptor tyrosine kinases. In neuronal cells, activation of Ras is mediated by a variety of receptor systems including receptor tyrosine kinase for peptide factors, G-protein-coupled serpentine receptors for neurotransmitters, and calcium influx through voltage-gated calcium channels (VSCCs) or NMDA receptors for glutamate (reviewed in Derkinderen et al., 1999; Grewal et al., 1999). Ras activates by as still-unknown mechanism the MAPK-kinase-kinases

(MAPKKKs) of the Raf family. Raf activates the MAPK-kinase (MAPKKs) or MEK1/2 by phosphorylation, which in turn activates ERKs by dual phosphorylation of their TEY motif.

ERK1 (44 kDa) and ERK2 (42 kDa), which are encoded by different genes, are expressed at high levels in the developing and adult CNS (Boulton et al., 1991; Flood et al., 1998). In neuronal cells, they are present in cytoplasmic compartments, including dendrites prior to their stimulation (Ortiz et al., 1995). Once activated, they can phosphorylate a large range of substrates in the cytoplasm, such as microtubule-associated protein 2 (MAP2), myelin basic protein T, synapsin, and SOS (a guanine exchange factor), as well as membrane-associated proteins, including epidermal growth factor receptor and phospholipase A2 (Seger and Krebs, 1995 for review). Activated ERKs also translocate to the nucleus (Chen et al., 1992) where they control gene expression, which may be relevant for their role in long-term neuronal adaptations.

A considerable body of literature implicates the Ras/ERK signaling cascade in synaptic plasticity and memory formation (Grewal et al., 1999; Orban et al., 1999 for review). Inhibiting this pathway using genetic or pharmacological approaches blocks activity-dependent strengthening of synaptic efficacy, also called long-term potentiation (LTP), in brain slices (English and Sweatt, 1997; Atkins et al., 1998) as well as in vivo (Davis et al., 2000; Rosenblum et al., 2000). Furthermore, ERK activation is also necessary for long-term behavioral changes, such as learning and memory (Atkins et al., 1998; Berman et al., 1998; Blum et al., 1999; Selcher et al., 1999; Walz et al., 1999a,b; Cammarota et al., 2000). This is well illustrated in RasGRF$^{-/-}$ mice lacking this neuronal-specific guanine exchange factor involved in Ras activation. Brambilla et al. (1997) failed to detect long-term memory for fear conditioning in these mice, while immediate learning and short-term memory were unchanged. Similarly, ERK play a key role in long-term and not short-term behavioral changes induced by addictive drugs (Pierce et al., 1999; Valjent et al., 2000). Thus it becomes clear that further elucidation of the components downstream of ERK signaling will help clarify the role of this cascade in neuronal adaptation as well as long-term behavioral changes.

4.3. ACTIVATION OF ERKs CORRELATES WITH IEGs INDUCTION IN THE CNS

Although ERKs can target numerous substrates within neuronal cells, their implication in long-term adaptive changes seems to be associated with IEG induction. This was illustrated in a model of electroconvulsive shocks (ECS), where Bhat et al. (1998) showed the temporal correlation between activation of ERKs, their nuclear translocation, phosphorylation of Elk-1 on Ser^{383} and c-Fos induction in the cerebral cortex. Phosphorylated ERK levels also increased in the cytoplasm, where they were associated with MAP2C phosphorylation. Intracortical injection of PD98059 blocked ECS-induced c-Fos expression in the nucleus and MAP2C phosphorylation in the cytoplasm. Thus, this model identified several different substrates of ERKs and indirectly implicated Elk-1 in c-*fos* induction.

This involvement was further illustrated in a model system of in vivo electrical stimulation of the cerebral cortex. In this model, c-Fos is induced in the striatum, the major target of cortical glutamatergic afferents, and this induction is topographically related to the stimulated cortical area (Sgambato et al., 1997). In particular, the stimulation of the sensorimotor cortex, which projects to the lateral part of the striatum (lSt), leads to increased levels of c-*fos* and *zif268* mRNA specifically in this region (Sgambato et al., 1998b) (Fig. 8A). Phospho-ERK immunolabeling showed increased levels in the lSt (Fig. 8B,C) that coincided spatially and temporally with c-*fos* and *zif268* mRNA induction (compare Fig. 8A with C). This

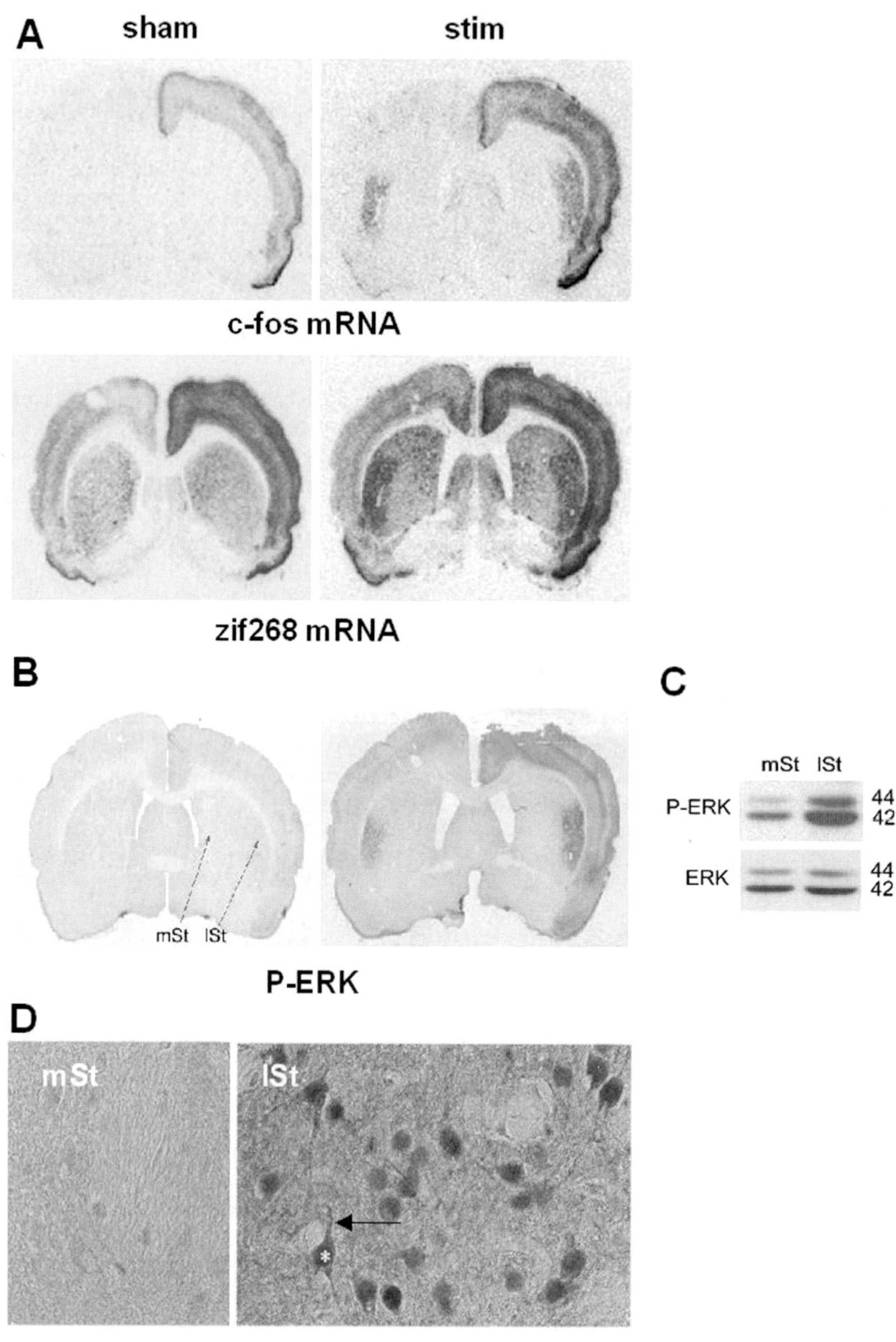
A
sham
stim
c-fos mRNA
zif268 mRNA
B
mSt
lSt
P-ERK
C
mSt
lSt
P-ERK
44
42
ERK
44
42
D
mSt
lSt
P-ERK

immunoreactivity was present in dendrites, soma and nuclei of striatal neurons specifically in the lSt of stimulated rats (Fig. 8D).

Western blotting and immunocytochemistry found a corresponding strong increase in phospho-Elk-1 immunoreactivity in the lSt. As with ERK, phospho-Elk-1 was found in dendritic, somatic and nuclear compartments (see Fig. 9B). Similarly, a strong hyperphosphorylation of CREB was found in the lSt, restricted however to the nucleus (see Fig. 9B).

The presence of activated ERKs, along with phospho-Elk-1 in cytoplasmic compartments, including dendrites, suggested the localized activation of these proteins near the point of calcium entry in the vicinity of glutamate receptors. Indeed, in the striatum, cortical glutamatergic afferences impinge precisely on dendrites (Smith and Bolam, 1990), where glutamate receptors of NMDA and AMPA subtypes are expressed (Bernard et al., 1997). Phospho-ERKs and phospho-Elk-1 were also detected at high levels in the nucleus, suggesting their nuclear translocation to effect IEG induction. To test this, PD98059 was injected into the striatum unilaterally. This abolished IEG induction, ipsilaterally to the PD98059 injection (Fig. 9A). PD98059 treatment blocked Elk-1 phosphorylation on Ser^{383} in the dendrites, soma and nuclei of striatal neurons (Fig. 9B), further supporting its strict dependence on ERKs. Moreover, this also inhibited CREB phosphorylation on Ser^{133}, indicating that CREB activation was also strictly dependent upon the ERK cascade (Fig. 9B). Although not analyzed in this study, a key intermediate between ERK and CREB phosphorylation could be the cytoplasmic CREB kinase *Rsk2*, a MAPKAP-K that plays a role in synaptic plasticity and memory formation (for review, Impey et al., 1999). To summarize, the ERK cascade controls glutamate-induced IEG induction in the striatum, through the combined activation of CREB and Elk-1.

Interestingly, ERK also controls IEG induction in an in vivo model of LTP, induced in the granule cells of dentate gyrus by tetanic stimulation of the perforant path. *zif268* mRNA was upregulated immediately after the end of the tetanus, in the dentate gyrus, ipsilaterally to the stimulation (Davis et al., 2000). This induction was specific of the tetanus since no upregulation of *zif268* mRNA was found in rats receiving low-frequency stimuli. ERK activation correlated spatio-temporally with *zif268* mRNA induction as well as Elk-1 and CREB hyperphosphorylation, specifically in tetanus-receiving rats. Once again, phospho-Elk-1 was found in cytoplasmic and nuclear compartments. Systemic injection of SL327, another inhibitor of MEK (Atkins et al., 1998; Selcher et al., 1999) blocked activation of both transcription factors (Fig. 10B,C) together with *zif268* mRNA induction (Fig. 10B) (Davis et al., 2000). This observation is interesting in light of the inhibitory role of SL327 on the maintenance of LTP observed in the granule cells of the dentate gyrus (Fig. 10A). However, the fact that, in the presence of SL327, LTP in the dentate gyrus decayed more rapidly than one would expect for a cascade involved in the protein synthesis-dependent phase of LTP,

---

←

Fig. 8. ERK activation coincides spatially and temporally with c-*fos* and *zif268* mRNA induction upon in vivo cortical stimulation. (*A*) Stimulation (stim) of the sensorimotor cortex, which projects bilaterally to the lateral part of the striatum (lSt), leads to increased levels of c-*fos* and *zif268* mRNA (analyzed by in situ hybridization) specifically in this region. Sham rats were implanted with stimulation electrodes in the cerebral cortex, but not stimulated. (*B*) Phospho-ERK (P-ERK) immunoreactivity processed on adjacent section showing a strict spatio-temporal coincidence with IEG mRNA induction. (*C*) Western blot analysis from striatal extracts of the lateral striatum (lSt) or medial striatum (mSt). Note that phospho-ERK1 and -ERK2 levels are increased in the lSt when compared to the mSt. Note also the lack of modification in total protein levels. (*D*) High magnification (×500) of P-ERK immunoreactivity in the lSt and mSt. Note the strong activation of ERK in both cytoplasmic (arrow) and nuclear (white asterisk) compartments of striatal neurons in the lSt. Adapted from Sgambato et al., 1998b.

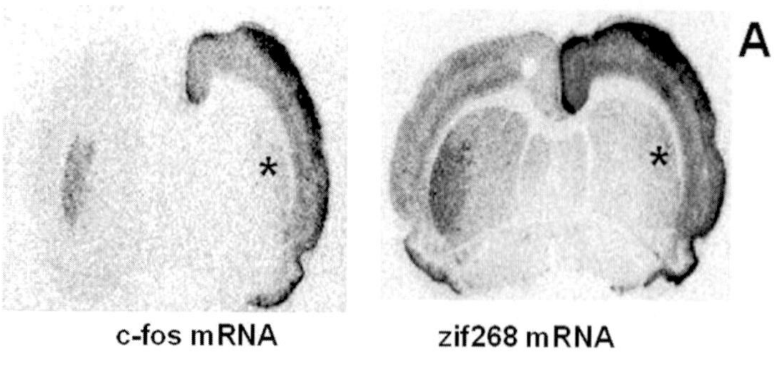

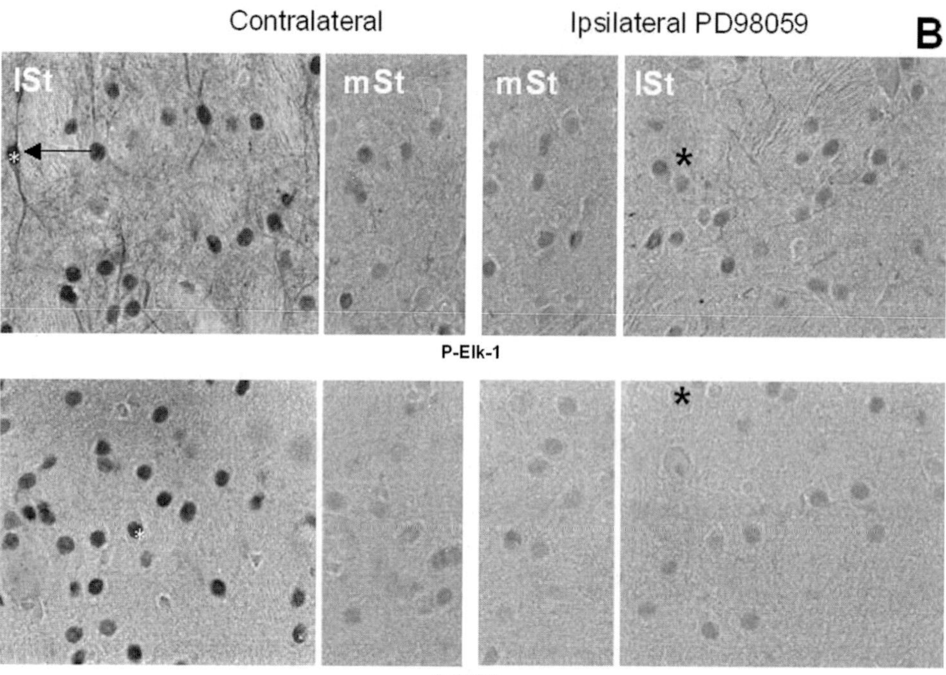

Fig. 9. Inhibition of cortico-striatal induced IEG induction along with Elk-1 and CREB after unilateral injection of PD98059: PD98059 was injected unilaterally (*) in the striatum prior to and during the cortico-striatal stimulation. (*A*) Inhibition of cortically driven IEG induction by PD98059. (*B*) Inhibition of cortically driven phosphorylation on Ser^{383}-Elk-1 (P-Elk-1) and phosphorylation on Ser^{133}-CREB (P-CREB) by PD98059. Note, in the contralateral side of PD98059 injection, the strong increase of P-Elk-1 and P-CREB in the lSt, specifically. Note also, that although P-CREB is restricted to the nucleus P-Elk-1 is found in cytoplasmic (including dendritic, arrow) and nuclear (white asterisk) compartments. Adapted from Sgambato et al., 1998b.

Fig. 10. LTP-induced IEG induction, Elk-1 and CREB activation are inhibited upon SL327 administration. ⟶ (*A*) LTP was induced in vivo in the dentate gyrus by unilateral tetanic stimulation of the perforant path (left panel). Systemic injection of SL327 but not its vehicle (DMSO) abolished LTP in the dentate gyrus (right panel). (*B*) Adjacent sections were processed for *zif268* mRNA in situ hybridization, P-ERK and P-Elk-1 immunoreactivity, from rats receiving the tetanus in the presence (right panels) or not (left panels) of SL327. Note the strong increase in *zif268* mRNA levels along with phosphorylation of ERK and Elk-1 ipsilaterally to the tetanus. Note also the total inhibition of *zif268* mRNA expression upon SL327 treatment. (*C*) CREB phosphorylation is shown at higher magnification from granule cells, ipsilaterally to the tetanus, in presence (right panel) or not (left panel) of SL327. Adapted from Davis et al., 2000.

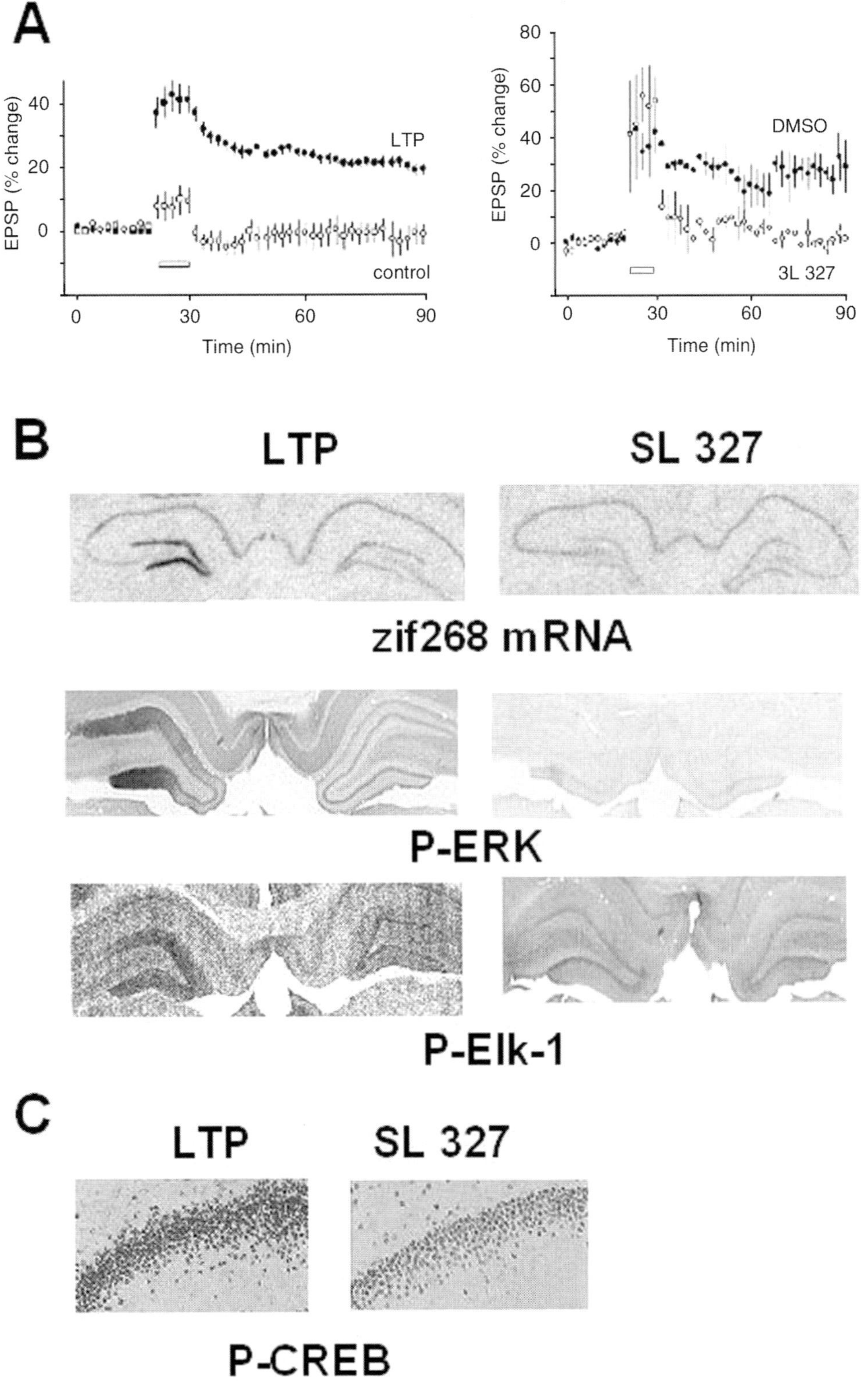
A
EPSP (% change)
40
20
0
LTP
control
0
30
60
90
Time (min)
80
60
40
20
0
DMSO
3L 327
Time (min)
B
LTP
SL 327
zif268 mRNA
P-ERK
P-Elk-1
C
LTP
SL 327
P-CREB

suggests that ERK may also contribute to an earlier phase of LTP and this may well involve one of the many cytoplasmic substrate of ERKs.

IEG expression is also a central feature of most drugs of abuse. IEGs are induced in discrete brain structures that are critically involved in behavioral responses related to addictive processes (Nestler and Aghajanian, 1997; Berke and Hyman, 2000 for review). In this way, acute psychostimulant administration has been reported to induce IEG expression in the striatum, a major cerebral target of dopamine (DA) inputs. Blocking DA receptors of the D1 subtype, abolishes IEG induction as well as behavioral responses. By elevating intracellular cAMP levels that activate PKA, DA–D1 receptors lead to CREB phosphorylation. However, recent evidences in PC12 and neuronal cells show that cAMP can also stimulate ERKs (Vossler et al., 1997; Impey et al., 1998; Yao et al., 1998; York et al., 1998; Roberson et al., 1999). Therefore, we investigated whether (1) ERK activation could occur upon acute cocaine administration, (2) this activation could be linked to IEG expression in the striatum, (3) blocking ERK activation modified rewarding properties related to chronic administration of cocaine (Valjent et al., 2000).

We were able to link a strong and transient cocaine-induced activation of ERKs in striatal neurons to D1 receptor stimulation. Upon cocaine treatment, phospho-ERK was found in the nucleus as was phospho-Ser^{383} Elk-1 immunoreactivity. Systemic administration of SL327 abolished both Elk-1 phosphorylation on Ser^{383} and c-Fos protein induction, at least in the dorsal part of the striatum. This was not the case ventrally, where c-Fos expression remained, although at lower levels. In this region, we suspect that c-fos regulation involves both ERK-dependent and -independent pathways, for instance direct targeting of CREB by PKA.

In conclusion, these data strongly support that ERKs control gene expression in various regions, in response to different extracellular stimuli, and this control depends on Elk-1 phosphorylation.

4.4. Elk-1 IS ACTIVATED IN MODEL SYSTEMS OF LEARNING AND MEMORY

Consolidation of long-term memory requires modulation of gene expression and synaptic resculpture. In a model system of conditioned taste aversive memory, Berman et al. (1998) searched for molecular mechanisms that might link the acquisition of taste information to long-lasting changes in the cortical circuits that are presumed to retain the representation of taste over time. They found an activation of ERKs and Elk-1 in the insular cortex of rats drinking the unfamiliar but not the familiar taste solution. Microinjection of PD98059 in the insular cortex before exposure to the novel taste attenuated long-term but not short-term memory of the sensory, motor and motivational faculties required to express long-term taste aversion memory. They thus concluded that ERK, by controlling Elk-1 phosphorylation and thereby gene expression, was involved in long-term storage of experience-dependent information in the cortex. More recently, Cammarota et al. (2000) detected phospho-Elk-1, along with phospho-ERK, phospho-CREB and c-Fos protein, in the nuclear fraction of hippocampal extracts in a model system of one-trial avoidance.

These findings fit with the hypothesis that ERK/Elk-1-induced modulation of gene expression is involved in long-term experience-dependent plasticity in the nervous system. This hypothesis is well illustrated also in the model system of cocaine administration where (1) the MEK inhibitor SL327 blocked Elk-1 phosphorylation along with c-Fos production in the striatum, (2) acute locomotor activity was poorly affected by SL327, and (3) cocaine-induced rewarding properties, evaluated in a place preference conditioning paradigm was totally prevented upon MEK inhibition (Valjent et al., 2000). We thus suggest that, by controlling the initial burst of gene expression, the ERK/Elk-1 signaling cassette is critically involved in neuronal adaptive processes underlying long-term behavioral responses. Further experiments, with genetically modified mice, will be required to further analyze the specific role of Elk-1 in synaptic plasticity and long-term memory formation.

We conclude with the recent suggestion that Elk-1-driven gene activation in transient transfection assays is also dependent upon Rsk2 (Brünning et al., 2000), the MAPKAP-K involved in ERK-dependent CREB phosphorylation. The gene encoding Rsk2 is mutated in the Coffin–Lowry syndrome in humans, a mental retardation disease. Thus by affecting the ability of Elk-1 to activate transcription via a still-unidentified mechanism, and by directly targeting CREB on Ser^{133}, this ERK-activated kinase could play a key role in genetic programs that are necessary for memory formation.

5. ABBREVIATIONS

CamK	calcium/calmodulin dependent kinase
CBP	CREB binding protein
CNS	central nervous system
CRE	cAMP and calcium responsive element
CREB	cAMP and calcium responsive element binding protein
CTF	constitutive transcription factor
ERK	extracellular-signal regulated kinase
IEG	immediate early gene
ITF	inducible transcription factor
LTP	long-term potentiation
MAPK	mitogen-activated protein kinase
MAPKAP-K	MAPK-activated protein kinase
PKA	cAMP-dependent kinase
SAPK/JNK	stress-activated protein kinase
SRE	serum response element
SRF	serum response factor
TCF	ternary complex factor

6. ACKNOWLEDGEMENTS

Acknowledgements to Robert Allan Hipskind for critical reading the manuscript.

7. REFERENCES

Atkins CM, Selcher JC, Petraitis JJ, Trzaskos JM, Sweatt JD (1998): The MAPK cascade is required for mammalian associative learning. *Nat Neurosci 1*:602–609.

Berke JD, Hyman SE (2000): Addiction, dopamine, and the molecular mechanisms of memory. *Neuron 25*:515–532.

Berman DE, Hazvi S, Rosenblum K, Seger R, Dudai Y (1998): Specific and differential activation of mitogen-activated protein kinase cascades by unfamiliar taste in the insular cortex of the behaving rat. *J Neurosci 18*:10037–10044.

Bernard V, Somogyi P, Bolam JP (1997): Cellular, subcellular, and subsynaptic distribution of AMPA-type glutamate receptor subunits in the neostriatum of the rat. *J Neurosci 17*:819–833.

Bhat RV, Engber TM, Finn JP, Koury EJ, Contreras PC, Miller MS, Dionne CA, Walton KM (1998): Region-specific targets of p42/p44MAPK signaling in rat brain. *J Neurochem 70*:558–571.

Blum S, Moore AN, Adams F, Dash PK (1999): A mitogen-activated protein kinase cascade in the CA1/CA2 subfield of the dorsal hippocampus is essential for long-term spatial memory. *J Neurosci 19*:3535–3544.

Bonni A, Ginty DD, Dudek H, Greenberg ME (1995): Serine 133-phosphorylated CREB induces transcription via a cooperative mechanism that may confer specificity to neurotrophin signals. *Mol Cell Neurosci 6*:168–183.

Boulton TG, Nye SH, Robbins DJ, Ip NY, Radziejewska E, Morgenbesser SD, DePinho RA, Panayotatos N, Cobb MH, Yancopoulos GD (1991): ERKs: a family of protein-serine/threonine kinases that are activated and tyrosine phosphorylated in response to insulin and NGF. *Cell 65*:663–675.

Brambilla R, Gnesutta N, Minichiello L, White G, Roylance AJ, Herron CE, Ramsey M, Wolfer DP, Cestari V, Rossi-Arnaud C, Grant SG, Chapman PF, Lipp HP, Sturani E, Klein R (1997): A role for the Ras signalling pathway in synaptic transmission and long-term memory. *Nature 390*:281–286.

Brünning JC, Gillette JA, Zhao Y, Bjorbaeck C, Kotzka J, Knebel B, Avci H, Hanstein B, Lingohr P, Moller DE, Krone W, Kahn CR, Muller-Wieland D (2000): Ribosomal subunit kinase-2 is required for growth factor-stimulated transcription of the c-Fos gene. *Proc Natl Acad Sci USA 97*:2462–2467.

Cahill MA, Janknecht R, Nordheim A (1996): Signalling pathways: jack of all cascades. *Curr Biol 6*:16–19.

Cammarota M, Bevilaqua LR, Ardenghi P, Paratcha G, Levi de Stein M, Izquierdo I, Medina JH (2000): Learning-associated activation of nuclear MAPK, CREB and Elk-1, along with Fos production, in the rat hippocampus after a one-trial avoidance learning: abolition by NMDA receptor blockade. *Brain Res Mol Brain Res 76*:36–46.

Chen RH, Sarnecki C, Blenis J (1992): Nuclear localization and regulation of erk- and rsk-encoded protein kinases. *Mol Cell Biol 12*:915–927.

Chrivia JC, Kwok RP, Lamb N, Hagiwara M, Montminy MR, Goodman RH (1993): Phosphorylated CREB binds specifically to the nuclear protein CBP. *Nature 365*:855–859.

Dalton S, Treisman R (1992): Characterization of SAP-1, a protein recruited by serum response factor to the c-fos serum response element. *Cell 68*:597–612.

Dash PK, Karl KA, Colicos MA, Prywes R, Kandel ER (1991): cAMP response element-binding protein is activated by Ca^{2+}/calmodulin- as well as cAMP-dependent protein kinase. *Proc Natl Acad Sci USA 88*:5061–5065.

Davis RJ (2000): Signal transduction by the JNK group of MAP kinases [in process citation]. *Cell 103*:239–252.

Davis S, Vanhoutte P, Pages C, Caboche J, Laroche S (2000): The MAPK/ERK cascade targets both Elk-1 and cAMP response element-binding protein to control long-term potentiation-dependent gene expression in the dentate gyrus in vivo. *J Neurosci 20*:4563–4572.

Delcroix JD, Averill S, Fernandes K, Tomlinson DR, Priestley JV, Fernyhough P (1999): Axonal transport of activating transcription factor-2 is modulated by nerve growth factor in nociceptive neurons. *J Neurosci (Online) 19*:RC24.

Derkinderen P, Enslen H, Girault JA (1999): The ERK/MAP-kinases cascade in the nervous system. *Neuroreport 10*:R24–34.

English JD, Sweatt JD (1997): A requirement for the mitogen-activated protein kinase cascade in hippocampal long term potentiation. *J Biol Chem 272*:19103–19106.

Fiore RS, Bayer VE, Pelech SL, Posada J, Cooper JA, Baraban JM (1993): p42 mitogen-activated protein kinase in brain: prominent localization in neuronal cell bodies and dendrites. *Neuroscience 55*:463–472.

Flood DG, Finn JP, Walton KM, Dionne CA, Contreras PC, Miller MS, Bhat RV (19): Immunolocalization of the mitogen-activated protein kinases p42MAPK and JNK1, and their regulatory kinases MEK1 and MEK4, in adult rat central nervous system. *J Comp Neurol 398*:373–392.

Gille H, Sharrocks AD, Shaw PE (1992): Phosphorylation of transcription factor p62TCF by MAP kinase stimulates ternary complex formation at c-fos promoter. *Nature 358*:414–417.

Gille H, Kortenjann M, Thomae O, Moomaw C, Slaughter C, Cobb MH, Shaw PE (1995): ERK phosphorylation potentiates Elk-1-mediated ternary complex formation and transactivation. *EMBO J 14*:951–962.

Goldman PS, Tran VK, Goodman RH (1997): The multifunctional role of the co-activator CBP in transcriptional regulation. *Recent Progr Hormon Res 52*:103–119.

Gonzalez GA, Montminy MR (1989): Cyclic AMP stimulates somatostatin gene transcription by phosphorylation of CREB at serine 133. *Cell 59*:675–680.

Grewal SS, York RD, Stork PJ (1999): Extracellular-signal-regulated kinase signalling in neurons. *Curr Opin Neurobiol 9*:544–553.

Gupta S, Barrett T, Whitmarsh AJ, Cavanagh J, Sluss HK, Derijard B, Davis RJ (1996): Selective interaction of JNK protein kinase isoforms with transcription factors. *EMBO J 15*:2760–2770.

Hill CS, Marais R, John S, Wynne J, Dalton S, Treisman R (1993): Functional analysis of a growth factor-responsive transcription factor complex. *Cell 73*:395–406.

Hipskind RA, Rao VN, Mueller CG, Reddy ES, Nordheim A (1991): Ets-related protein Elk-1 is homologous to the c-fos regulatory factor p62TCF. *Nature 354*:531–534.

Hipskind RA, Baccarini M, Nordheim A (1994): Transient activation of RAF-1, MEK, and ERK2 coincides kinetically with ternary complex factor phosphorylation and immediate-early gene promoter activity in vivo. *Mol Cell Biol 14*:6219–6231.

Impey S, Obrietan K, Wong ST, Poser S, Yano S, Wayman G, Deloulme JC, Chan G, Storm DR (1998): Cross talk between ERK and PKA is required for Ca^{2+} stimulation of CREB-dependent transcription and ERK nuclear translocation. *Neuron 21*:869–883.

Impey S, Obrietan K, Storm DR (1999): Making new connections: role of ERK/MAP kinase signaling in neuronal plasticity. *Neuron 23*:11–14.

Janknecht R, Hunter T (1996): Versatile molecular glue. Transcriptional control. *Curr Biol 6*:951–954.

Janknecht R, Nordheim A (1996a): MAP kinase-dependent transcriptional coactivation by Elk-1 and its cofactor CBP. *Biochem Biophys Res Commun 228*:831–837.

Janknecht R, Nordheim A (1996b): Regulation of the c-fos promoter by the ternary complex factor Sap-1a and its coactivator CBP. *Oncogene 12*:1961–1969.

Janknecht R, Ernst WH, Pingoud V, Nordheim A (1993): Activation of ternary complex factor Elk-1 by MAP kinases. *EMBO J 12*:5097–5104.

Janknecht R, Zinck R, Ernst WH, Nordheim A (1994): Functional dissection of the transcription factor Elk-1. *Oncogene 9*:1273–1278.

Johanson SO, Crouch MF, Hendry IA (1995): Retrograde axonal transport of signal transduction proteins in rat sciatic nerve. *Brain Res 690*:55–63.

Johnson CM, Hill CS, Chawla S, Treisman R, Bading H (1997): Calcium controls gene expression via three distinct pathways that can function independently of the Ras/mitogen-activated protein kinases (ERKs) signaling cascade. *J Neurosci 17*:6189–6202.

Kaminska B, Kaczmarek L, Zangenehpour S, Chaudhuri A (1999): Rapid phosphorylation of Elk-1 transcription factor and activation of MAP kinase signal transduction pathways in response to visual stimulation. *Mol Cell Neurosci 13*:405–414.

Kwok RP, Lundblad JR, Chrivia JC, Richards JP, Bachinger HP, Brennan RG, Roberts SG, Green MR, Goodman RH (1994): Nuclear protein CBP is a coactivator for the transcription factor CREB. *Nature 370*:223–226.

Ling Y, Lakey JH, Roberts CE, Sharrocks AD (1997): Molecular characterization of the B-box protein–protein interaction motif of the ETS-domain transcription factor Elk-1. *EMBO J 16*:2431–2440.

Lopez M, Oettgen P, Akbarali Y, Dendorfer U, Libermann TA (1994): ERP, a new member of the ets transcription factor/oncoprotein family: cloning, characterization, and differential expression during B-lymphocyte development. *Mol Cell Biol 14*:3292–3309.

Marais R, Wynne J, Treisman R (1993): The SRF accessory protein Elk-1 contains a growth factor-regulated transcriptional activation domain. *Cell 73*:381–393.

Martin KC, Michael D, Rose JC, Barad M, Casadio A, Zhu H, Kandel ER (1997): MAP kinase translocates into the nucleus of the presynaptic cell and is required for long-term facilitation in aplysia. *Neuron 18*:899–912.

Mayford M, Kandel ER (1999): Genetic approaches to memory storage. *Trends Genet 15*:463–470.

Mielke K, Herdegen T (2000): JNK and p38 stresskinase-degenerative effectors of signal-transduction-cascades in the nervous system. *Prog Neurobiol 61*:45–60.

Montminy MR, Sevarino KA, Wagner JA, Mandel G, Goodman RH (1986): Identification of a cyclic-AMP-responsive element within the rat somatostatin gene. *Proc Natl Acad Sci USA 83*:6682–6686.

Morgan JI, Curran T (1991): Proto-oncogene transcription factors and epilepsy. *Trends Pharmacol Sci 12*:343–349.

Nestler EJ, Aghajanian GK (1997): Molecular and cellular basis of addiction. *Science 278*:58–63.

Orban PC, Chapman PF, Brambilla R (1999): Is the Ras-MAPK signalling pathway necessary for long-term memory formation? [see comments]. *Trends Neurosci 22*:38–44.

Ortiz J, Harris HW, Guitart X, Terwilliger RZ, Haycock JW, Nestler EJ (1995): Extracellular signal-regulated protein kinases (ERKs) and ERK kinase (MEK) in brain: regional distribution and regulation by chronic morphine. *J Neurosci 15*:1285–1297.

Pierce RC, Pierce-Bancroft AF, Prasad BM (1999): Neurotrophin-3 contributes to the initiation of behavioral sensitization to cocaine by activating the Ras/Mitogen-activated protein kinase signal transduction cascade. *J Neurosci 19*:8685–8695.

Pingoud V, Zinck R, Hipskind RA, Janknecht R, Nordheim A (1994): Heterogeneity of ternary complex factors in HeLa cell nuclear extracts. *J Biol Chem 269*:23310–23317.

Povelones M, Tran K, Thanos D, Ambron RT (1997): An NF-kappaB-like transcription factor in axoplasm is rapidly inactivated after nerve injury in aplysia. *J Neurosci 17*:4915–4920.

Price MA, Rogers AE, Treisman R (1995): Comparative analysis of the ternary complex factors Elk-1, SAP-1a and SAP-2 (ERP/NET). *EMBO J 14*:2589–2601.

Price MA, Cruzalegui FH, Treisman R (1996): The p38 and ERK MAP kinase pathways cooperate to activate ternary complex factors and c-fos transcription in response to UV light. *EMBO J 15*:6552–6563.

Ramirez S, Ait-Si-Ali S, Robin P, Trouche D, Harel-Bellan A, Ait Si Ali S (1997): The CREB-binding protein (CBP) cooperates with the serum response factor for transactivation of the *c-fos* serum response element [erratum appears in *J Biol Chem 274*(25):18140]. *J Biol Chem 272*:31016–31021.

Rivera VM, Miranti CK, Misra RP, Ginty DD, Chen RH, Blenis J, Greenberg ME (1993): A growth factor-induced kinase phosphorylates the serum response factor at a site that regulates its DNA-binding activity. *Mol Cell Biol 13*:6260–6273.

Roberson ED, English JD, Adams JP, Selcher JC, Kondratick C, Sweatt JD (1999): The mitogen-activated protein kinase cascade couples PKA and PKC to cAMP response element binding protein phosphorylation in area CA1 of hippocampus. *J Neurosci 19*:4337–4348.

Robertson LM, Kerppola TK, Vendrell M, Luk D, Smeyne RJ, Bocchiaro C, Morgan JI, Curran T (1995): Regulation of c-fos expression in transgenic mice requires multiple interdependent transcription control elements. *Neuron 14*:241–252.

Rosenblum K, Futter M, Jones M, Hulme EC, Bliss TV (2000): ERKI/II regulation by the muscarinic acetylcholine receptors in neurons. *J Neurosci 20*:977–985.

Schröter H, Shaw PE, Nordheim A (1987): Purification of intercalator-released p67, a polypeptide that interacts specifically with the c-fos serum response element. *Nucleic Acids Res 15*:10145–10158.

Schröter H, Mueller CG, Meese K, Nordheim A (1990): Synergism in ternary complex formation between the dimeric glycoprotein p67SRF, polypeptide p62TCF and the c-fos serum response element. *EMBO J 9*:1123–1130.

Seger R, Krebs EG (1995): The MAPK signaling cascade. *FASEB J 9*:726–735.

Selcher JC, Atkins CM, Trzaskos JM, Paylor R, Sweatt JD (1999): A necessity for MAP kinase activation in mammalian spatial learning. *Learn Mem 6*:478–490.

Sgambato V, Abo V, Rogard M, Besson MJ, Deniau JM (1997): Effect of electrical stimulation of the cerebral cortex on the expression of the Fos protein in the basal ganglia. *Neuroscience 81*:93–112.

Sgambato V, Vanhoutte P, Pages C, Rogard M, Hipskind R, Besson MJ, Caboche J (1998a): In vivo expression and regulation of Elk-1, a target of the extracellular-regulated kinase signaling pathway, in the adult rat brain. *J Neurosci 18*:214–226.

Sgambato V, Pages C, Rogard M, Besson MJ, Caboche J (1998b): Extracellular signal-regulated kinase (ERK) controls immediate early gene induction on corticostriatal stimulation. *J Neurosci 18*:8814–8825.

Sharrocks AD, von Hesler F, Shaw PE (1993): The identification of elements determining the different DNA binding specificities of the MADS box proteins p67SRF and RSRFC4. *Nucleic Acids Res 21*:215–221.

Shaywitz AJ, Greenberg ME (1999): CREB: a stimulus-induced transcription factor activated by a diverse array of extracellular signals. *Annu Rev Biochem 68*:821–861.

Sheng M, Greenberg ME (1990): The regulation and function of c-fos and other immediate early genes in the nervous system. *Neuron 4*:477–485.

Sheng M, McFadden G, Greenberg ME (1990): Membrane depolarization and calcium induce c-fos transcription via phosphorylation of transcription factor CREB. *Neuron 4*:571–582.

Sheng M, Thompson MA, Greenberg ME (1991): CREB: a Ca^{2+}-regulated transcription factor phosphorylated by calmodulin-dependent kinases. *Science 252*:1427–1430.

Shore P, Sharrocks AD (1994): The transcription factors Elk-1 and serum response factor interact by direct protein–protein contacts mediated by a short region of Elk-1. *Mol Cell Biol 14*:3283–3291.

Shore P, Sharrocks AD (1995): The ETS-domain transcription factors Elk-1 and SAP-1 exhibit differential DNA binding specificities. *Nucleic Acids Res 23*:4698–4706.

Smith AD, Bolam JP (1990): The neural network of the basal ganglia as revealed by the study of synaptic connections of identified neurones. *Trends Neurosci 13*:259–265.

Treisman R (1986): Identification of a protein-binding site that mediates transcriptional response of the c-fos gene to serum factors. *Cell 46*:567–574.

Treisman R (1987): Identification and purification of a polypeptide that binds to the c-fos serum response element. *EMBO J 6*:2711–2717.

Treisman R (1992): The serum response element. *Trends Biochem Sci 17*:423–426.

Treisman R (1995): Journey to the surface of the cell: Fos regulation and the SRE. *EMBO J 14*:4905–4913.

Treisman R (1996): Regulation of transcription by MAP kinase cascades. *Curr Opin Cell Biol 8*:205–215.

Valjent E, Corvol J-C, Pagès C, Besson M-J, Maldonado R, Caboche J (2000): Involvement of the extracellular signal-regulated kinase cascade for cocaine-rewarding properties. *J Neurosci 20*:4563–4572.

Vanhoutte P, Barnier JV, Guibert B, Pages C, Besson MJ, Hipskind RA, Caboche J (1999): Glutamate induces phosphorylation of Elk-1 and CREB, along with c-fos activation, via an extracellular signal-regulated kinase-dependent pathway in brain slices. *Mol Cell Biol 19*:136–146.

Vanhoutte P, Nissen J, Brugg B, Della Gaspera B, Besson MJ, Hipskind RA, Caboche J (2000): Opposing roles of Elk-1 and its brain specific isoform, sElk-1, in NGF-induced PC12 differentiation. *J Biol Chem 276*:5189–5196.

Vossler MR, Yao H, York RD, Pan MG, Rim CS, Stork PJ (1997): cAMP activates MAP kinase and Elk-1 through a B-Raf- and Rap1-dependent pathway. *Cell 89*:73–82.

Walz R, Roesler R, Barros DM, de Souza MM, Rodrigues C, Sant'Anna MK, Quevedo J, Choi HK, Neto WP, DeDavid e Silva TL, Medina JH, Izquierdo I (1999a): Effects of post-training infusions of a mitogen-activated protein kinase kinase inhibitor into the hippocampus or entorhinal cortex on short- and long-term retention of inhibitory avoidance. *Behav Pharmacol 10*:723–730.

Walz R, Roesler R, Quevedo J, Rockenbach IC, Amaral OB, Vianna MR, Lenz G, Medina JH, Izquierdo I (1999b): Dose-dependent impairment of inhibitory avoidance retention in rats by immediate post-training infusion of a mitogen-activated protein kinase kinase inhibitor into cortical structures. *Behav Brain Res 105*:219–223.

Yamamoto KK, Gonzalez GA, Biggs WHD, Montminy MR (1988): Phosphorylation-induced binding and transcriptional efficacy of nuclear factor CREB. *Nature 334*:494–498.

Yang DD, Kuan CY, Whitmarsh AJ, Rincon M, Zheng TS, Davis RJ, Rakic P, Flavell RA (1997): Absence of excitotoxicity-induced apoptosis in the hippocampus of mice lacking the Jnk3 gene. *Nature 389*:865–870.

Yang SH, Whitmarsh AJ, Davis RJ, Sharrock AD (1998a): Differential targeting of MAP kinases to the ETS-domain transcription factor Elk-1. *EMBO J 17*:1740–1749.

Yang SH, Yates PR, Whitmarsh AJ, Davis RJ, Sharrocks AD (1998b): The Elk-1 ETS-domain transcription factor contains a mitogen-activated protein kinase targeting motif. *Mol Cell Biol 18*:710–720.

Yang SH, Shore P, Willingham N, Lakey JH, Sharrocks AD (1999): The mechanism of phosphorylation-inducible activation of the ETS-domain transcription factor Elk-1. *EMBO J 18*:5666–5674.

Yao H, York RD, Misra-Press A, Carr DW, Stork PJ (1998): The cyclic adenosine monophosphate-dependent protein kinase (PKA) is required for the sustained activation of mitogen-activated kinases and gene expression by nerve growth factor. *J Biol Chem 273*:8240–8247.

York RD, Yao H, Dillon T, Ellig CL, Eckert SP, McCleskey EW, Stork PJ (1998): Rap1 mediates sustained MAP kinase activation induced by nerve growth factor. *Nature 392*:622–626.

Zinck R, Hipskind RA, Pingoud V, Nordheim A (1993): c-fos transcriptional activation and repression correlate temporally with the phosphorylation status of TCF. *EMBO J 12*:2377–2387.

Zinck R, Cahill MA, Kracht M, Sachsenmaier C, Hipskind RA, Nordheim A (1995): Protein synthesis inhibitors reveal differential regulation of mitogen-activated protein kinase and stress-activated protein kinase pathways that converge on Elk-1. *Mol Cell Biol 15*:4930–4938.

CHAPTER XII

The Egr transcription factors and their utility in mapping brain functioning

JOHN LEAH AND PETER A. WILCE

1. STRUCTURE OF THE Egr GENES AND PROTEINS, AND THEIR DNA BINDING

The defining feature of the Egr family of proteins is their three zinc finger motifs, which recognize the same nine-base-pair segment of DNA. Once bound to the promoters of target genes the proteins act in concert with other transcription factors to regulate the expression of these genes. The precise mechanisms by which they do this are yet to be clearly defined.

1.1. NAMING OF THE Egrs

During the course of their discovery, different members of the Egr family of transcription factors have been given different names: Egr-1 (NGFI-A, Krox-24, *zif268*, TIS8, ZENK), Egr-2 (*krox-20*), Egr-3 (Pilot) and Egr-4 (NGFI-C, pAT133). There is now growing acceptance of the Egr nomenclature, an acronym for 'early growth response', since NGFI-B is unrelated to the others (Beckmann and Wilce, 1997; Herdegen and Leah, 1998); and most importantly there is now clearly a family of at least four of these factors (O'Donovan et al., 1999).

1.2. STRUCTURE OF THE GENES

This has previously been reviewed in detail (Beckmann and Wilce, 1997; Herdegen and Leah, 1998; Hughes et al., 1999; O'Donovan et al., 1999). The genes encoding Egr-1 to -4 are each present as single copies with a single exon that contains the zinc finger sequence. *egr*-1 has two exons and one intron (Tsai-Morris et al., 1988). *egr*-2 consists of two exons and one intron that encode a single transcript (Chavrier et al., 1988). The human EGR-3 is organized in two exons (Holst et al., 1993).

The promoters of all the Egr genes have not been fully characterized. Thus far the human *EGR*-1 is known to contain two Sp1 elements (through which the COUP transcription factor acts (Pipaon et al., 1999), one AP-1, two CRE (calcium/cAMP response element), one ERE (Egr response element), five SRE (serum response elements) with their associated Ets binding sites (Schwachtgen et al., 2000), an element similar to that for NFκB (Aicher et al., 1999a), as well as a possible estrogen response element (Slade and Carter, 2000). The *egr*-2 promoter contains four CRE-like elements but apparently, and surprisingly, no SREs (Cochran, 1993). Since they all contain an ERE, all *Egr* genes have the potential to auto-regulate their

Handbook of Chemical Neuroanatomy Vol. 19: Immediate Early Genes and Inducible Transcription Factors in Mapping of the Central Nervous System Function and Dysfunction
L. Kaczmarek and H.A. Robertson, editors

expression. For example, *Egr-4* is able to down-regulate its own expression, as well as that of c-*fos* in vitro (Zipfel et al., 1997).

1.3. PROTEIN STRUCTURE AND DNA BINDING

The strongest similarity amongst the Egr proteins is in their highly conserved DNA-binding domain (Swirnoff and Milbrandt, 1995). This recognizes the DNA sequence GCGG/TGGGCG. The Egr proteins belong to the Cys2His2 class, which share extensive homology throughout the zinc finger DNA-binding domain. The zinc fingers are present as tandemly repeated units of 28–30 amino acids containing two cysteines and two histidines at invariant positions within the consensus sequence. The invariant cysteine and histidine residues direct folding around a tetrahedrally coordinated zinc ion forming a finger-like projection which then binds DNA in a sequence-specific manner with each finger spanning three nucleotides. (Miller et al., 1985; Berg, 1988). Egr-1, -2 and -3 also have a repressive domain rich in basic amino acids.

The Cys2His2 zinc fingers are located at their C-termini. Three of the four zinc fingers of Egr-1 and Egr-2 share a 92% homology, and at the nucleotide level they have an overall homology of 87%. All have a nuclear translocation signal located between amino acids 315 and 430, i.e. within the zinc finger region. Truncation of this region maintains the proteins in the cytoplasm. Three zinc fingers are necessary and sufficient for nuclear localization.

Each Egr protein has distinct DNA-binding properties and transcriptional regulatory activities. Individual members have distinct binding affinities to the ERE, which may provide numerous combinatorial possibilities for transactivation of target sequences. DNA-binding sites for the Egr family and Sp1 (also a zinc finger transcription factor) overlap and, in some cases at least, this renders the binding of the Egrs and Sp1 mutually exclusive. However, some Sp1 sites may also be Egr-binding sites. Other transcription factor proteins binding to an ERE include the GC-rich binding factor, ETF, as well as AP-2 and AP-4.

In the brain, Egr-3 exists as several full-length and truncated isoforms generated by different translation initiation sites (which is unusual in eukaryotes, Sachs et al., 1997). Proteins produced from these initiation sites differ in their ability to instigate transcription, and production of different amounts of the isoforms may allow for differential regulation of Egr target genes containing multiple vs single EREs (O'Donovan and Baraban, 1999; O'Donovan et al., 2000).

2. EXPRESSION OF THE Egrs IN CELL CULTURE AND IN THE BRAIN

The Egrs, like the other ITFs, have a basal expression and can be induced by numerous extracellular stimuli in many types of neuronal and non-neuronal cells as detailed previously (Beckmann and Wilce, 1997; Herdegen and Leah, 1998; Hughes et al., 1999).

2.1. Egrs IN CULTURE

In glial cells, all Egrs can be activated via the M1 acetylcholine receptor (von der Kammer et al., 1998). Egr-1 is inducible in cultured astrocytes via metabotropic glutamate receptors (Condorelli et al., 1993) and ATP (Hung et al., 2000), and in C6 glioma cells by sphingosine 1-phosphate via the MAPK pathway (Sato et al., 1999). NGF induces expression of *egr*-2, but not *egr*-1, mRNA in PC12 cells (Cosgaya et al., 1998). Treatment of cultured sensory ganglion neurons with NGF and other growth factors increases the expression of both *egr*-1 and -2 (Kendall et al., 1994). The *egr*-1 gene is rapidly induced in PC12 cells by TPA, NGF, EGF and depolarization, and is superinduced by cycloheximide (Milbrandt, 1987; Sukhatme et al., 1988; Waters et al., 1990). *egr*-1 is also strongly up-regulated by incubating PC12 cells with the calcineurin inhibitors FK-506 or cyclosporine (Enslen and Soderling, 1994). *egr*-2 is induced in these cells by serum and TPA (DeFranco et al., 1993). *egr*-1 is more inducible than is *egr*-2 since it cannot be induced in neuronal PC12 cells by cAMP (Joseph et al., 1988), or by cAMP in general (Mechta et al., 1989).

2.2. Egrs IN THE BRAIN

2.2.1. Basal expression

In the brain, all four *egr* genes have a basal expression, i.e one that is maintained by normal ongoing synaptic or neurohormonal activity and not by any intrinsic genetic program (Beckmann and Wilce, 1997; Herdegen and Leah, 1998; Shirayama et al., 1999). In the cortex, there is a strong basal expression of Egr-1, -3 and -4 in layers II and VII of the cerebral cortex, whereas Egr-2 occurs mostly in layers II and III. In the cortex, Egr-1 and -2 may have some constitutive expression maintained by processes entirely of intracellular origin since their expression is insensitive to blockade of synaptic transmission by NMDA antagonists. In addition, the NMDA antagonist, MK-801 does not block kainate-induced Egr-1 expression in the cortex, and has no effect on the stimulus-induced or basal expressions of Egr-2. However, MK-801 and tetrodotoxin (TTX) abolish the persisting expressions of *egr*-1 and its protein in the visual and piriform cortices, and hippocampus (Worley et al., 1991; Gass et al., 1993), indicating they result from natural on-going synaptic activity and are basal expressions (Worley et al., 1993). All Egrs are present in the striatum and hippocampus, especially in the CA1–3 regions.

2.2.2. Induced expression

As detailed previously, numerous physiological and pharmacological stimuli will induce expression of the Egrs in particular areas of the brain (Beckmann and Wilce, 1997; Herdegen and Leah, 1998; Hughes et al., 1999; O'Donovan et al., 1999). More recently, the expression of the Egrs has been used to characterize some additional and important systems. *egr*-1 and -3 mRNAs were shown to be expressed after kainate-induced seizures particularly in the hippocampus, cortex and amygdala, showing the typical peak at 30 min and returning to basal levels after several hours (Honkaniemi and Sharp, 1999). In the CA1 and CA3 regions their expression persisted for at least 24 h, most likely due to the toxic effects of kainate on these particular neurons. *egr*-3 appears in the ventral suprachiasmatic nucleus after light stimulation (Morris et al., 1998). L-DOPA and amitriptyline induce *egr*-1 mRNA in the denervated striatum, and hippocampus and cortex, respectively (Johansson et al., 1998; Khan et al., 1999). With repeated administration, the levels remain elevated, i.e. there is no habituation of the inducibility. The cocaine-induced expression of *egr*-1 in the caudate putamen and shell of the nucleus accumbens requires 5-HT_3 receptors (Humblot et al., 1998). A vasopressin V1 agonist induces *egr*-1 in a subpopulation of cultured hippocampal glia, but not in the neurons (Brinton et al., 1998). *egr*-1 is induced in the striatum by the metabotropic glutamate agonist ACPD and this is only partially dependent upon NMDA receptor functioning (Wang, 1998).

Noxious stimulation causes a rapid increase in Egr-1 in the CA1 region of the hippocampus suggesting that these neurons mediate pain-associated memory (Wei et al., 2000).

Further recent examples of the induction of Egrs are by granulocyte colony-stimulating factor (acting via just one SRE-1) (Mora-Garcia and Sakamoto, 2000), EGF (Tsai et al., 2000), nitric oxide (Chiu et al., 1999), insulin (Barroso and Stantisteban, 1999), growth hormone (Hodge et al., 1998), zinc (Park and Koh, 1999), pheromones (Brennan et al., 1999), in striatal neurons after cortical stimulation (Sgambato et al., 1998) or after the neurotensin-mediated synergistic effect of D1 and D2 agonists (Alonso et al., 1999), and in arthritic synovial fibroblasts (Aicher et al., 1999a).

3. SIGNAL TRANSDUCTION PATHWAYS INVOLVED IN Egr EXPRESSION

3.1. THE NMDA RECEPTOR AND Egr EXPRESSION

Previous experiments showed that virtually all the expressions of c-Fos induced by activation of many different types of receptors also requires calcium entry into the neuron via the NMDA receptor (Beckmann and Wilce, 1997; Herdegen and Leah, 1998). Inhibition of the NMDA receptor by blockade of glutamate-binding with 2-AP5 or CPP, or its ion channel with MK-801, abolishes most basal and stimulus-induced expressions of c-Fos. This presumably occurs because all receptors can be linked to the NMDA receptor and affect its functioning. Thus, dopamine receptors must phosphorylate NMDA receptor subunits by activating PKA and/or PKC pathways before dopamine can induce c-Fos in the striatum (Leveque et al., 2000).

NMDA receptors are also important for expression of the Egrs. Thus i.p. injection of NMDA induces Egr-1 in many areas of the rat brain (Beckmann and Wilce, 1997). However, the basal expression of Egr-1 in cortical layers V and VI is unaffected by MK-801 (Gass et al., 1993). This lack of effect of MK-801 is also seen with induced expressions. For example, in the hippocampus, pretreatment with MK-801 (5 mg/kg) 30 min beforehand strongly decreased the expression of Egr-2 induced by brief focal seizures, but was without effect on Egr-1 (Hughes et al., 1998). MK-801 is also unable to block the expression of Egr-2 induced by spreading depression and seizures (Herdegen et al., 1993). Finally, NMDA receptor-independent expression of Egr-1 and -2 is seen in non-neuronal cells surrounding a focal injury in the hippocampus (Dragunow and Hughes, 1993).

NMDA antagonists can also *increase* Egr-1 expression in several brain areas (Gass et al., 1993). This may not be a 'toxic' effect of MK-801, as is proposed for the MK-801-induced expression of c-Fos in some brain areas, such as the retrosplenial cortex. Instead, it may indicate a sub-class of NMDA receptors that *represses* Egr-1 expression. The requirement for NMDA receptor activity in induction of the other Egrs remains to be studied.

3.2. THE MITOGEN-ACTIVATED PATHWAY AND Egr EXPRESSION

In all the recent experiments described at the end of Section 2.2.2. above, expression of the Egrs was inhibited by blocking the MAP kinase pathway using novel, specific antagonists. These results suggest that this pathway is more important in regulating Egr expression than those activated by the NMDA receptor. However, the *egr*-1 appearing after induction of hippocampal LTP is induced by MAPK/Elk-1 phosphorylation of SRF as well as by MAPK phosphorylation of CREB (Davis et al., 2000). In PC12 cells, half of the NGF-induced induction of *egr*-1 mRNA is mediated via MAPK while the remainder is via PI 3-kinase, with a possible involvement of JNK (Jun N-terminal kinase) (Kumahara et al., 1999). *egr*-1 is induced in these cells by activation of any combination of two SREs or AP-1-like elements in its promoter (DeFranco et al., 1993), whereas activation of just one SRE by the MAP kinase pathway is sufficient for its induction (Schwachtgen et al., 2000). Thus, different elements are responsible for induction of the *egr* genes in different cell types, but generally the *egr* genes are mostly induced via activation of the most proximal SRE (Aicher et al., 1999b; Mora-Garcia and Sakamoto, 2000). Occupation of the AP-1 and EREs seems to have only a small effect (Aicher et al., 1999b), and in some cell types activating the CRE *inhibits* Egr-1 induction (Aicher et al., 1999a). The situation is likely to be more complex because the serum response factor gene itself contains an Egr-1 binding site as well as two SREs (Spencer and Misra, 1999). Overall, current evidence indicates that the MAPK pathway is particularly important in inducing the zinc-finger-containing Egr family, whereas the leucine zipper family of ITFs is more dependent upon pathways activated by the NMDA receptor.

4. USING Egr EXPRESSIONS TO MAP ACTIVATED NEURONS IN THE BRAIN

Initial experiments using ITF expression to identify activated neurons used mostly c-Fos, but with the advent of specific antibodies, the Egrs are becoming a popular adjunct. This is especially the case in the sensory systems. Non-noxious tactile stimulation causes a weak expression of ITFs in the somatosensory cortex (Kenneth and Mack, 1992), and stimulation of vibrissae evokes a weak Egr-1 expression in the contralateral somatosensory cortex (Mack et al., 1992, 1995). However, in conscious rats when the whiskers were remotely stimulated there was a stronger expression in many, but not all neurons in the corresponding cortical barrels, mostly in layer IV stellate neurons (Melzer and Steiner, 1997).

Noxious sensory stimuli, e.g. caused by colorectal distension or i.p. injection of acetic acid, induced Egr-1 expression in the spinal cord: in laminae I–II, the neck of the dorsal horn, lamina X, the dorsal grey commissure and parasympathetic preganglionic neurons, principally in the lumbosacral and upper lumbar segments (Traub et al., 1992; Lanteri-Minet et al., 1993, 1995). Excitation by intraperitoneal acetic acid evoked their expression in laminae I–II, with a more mediolateral spread, but most labelling occurs in thoracolumbar segments where visceral afferent terminations are more dense (Presley et al., 1990; Lanteri-Minet et al., 1993). Chemical stimulation of the viscera, or electrical stimulation of a visceral nerve, also induces Egr-1 bilaterally in many brain areas including the nucleus of the solitary tract, the lateral parabrachial area (mostly in the rostrodorsal part) and in the Edinger–Westphal nucleus. In contrast, noxious mechanical stimulation causes expression mostly in the most caudal part of the medulla oblongata (McKitrick et al., 1992; Lanteri-Minet et al., 1995).

In the visual sensory system, *egr*-1 is expressed in ocular dominance columns in monkeys monocularly exposed to 2 or 5 h light after 24 h of darkness. Accordingly, when light stimulation of one eye is suppressed, or TTX is injected intravitreally, the basal expression of *egr*-1 in the visual cortex decreases, revealing the ocular dominance columns. This columnar pattern is similar to that revealed by cytochrome oxidase staining of the visual cortex from long-term monocularly deprived monkeys (Chaudhuri and Cynader, 1993; Chaudhuri et al., 1995). Complete deafferentation of one visual cortex by sectioning an optic nerve, the corpus callosum as well as the anterior commissure results in a decrease in *egr*-1 in the affected cortex (Zhang et al., 1995).

Recent experiments illustrate the utility of Egr expression in mapping activated neurons in

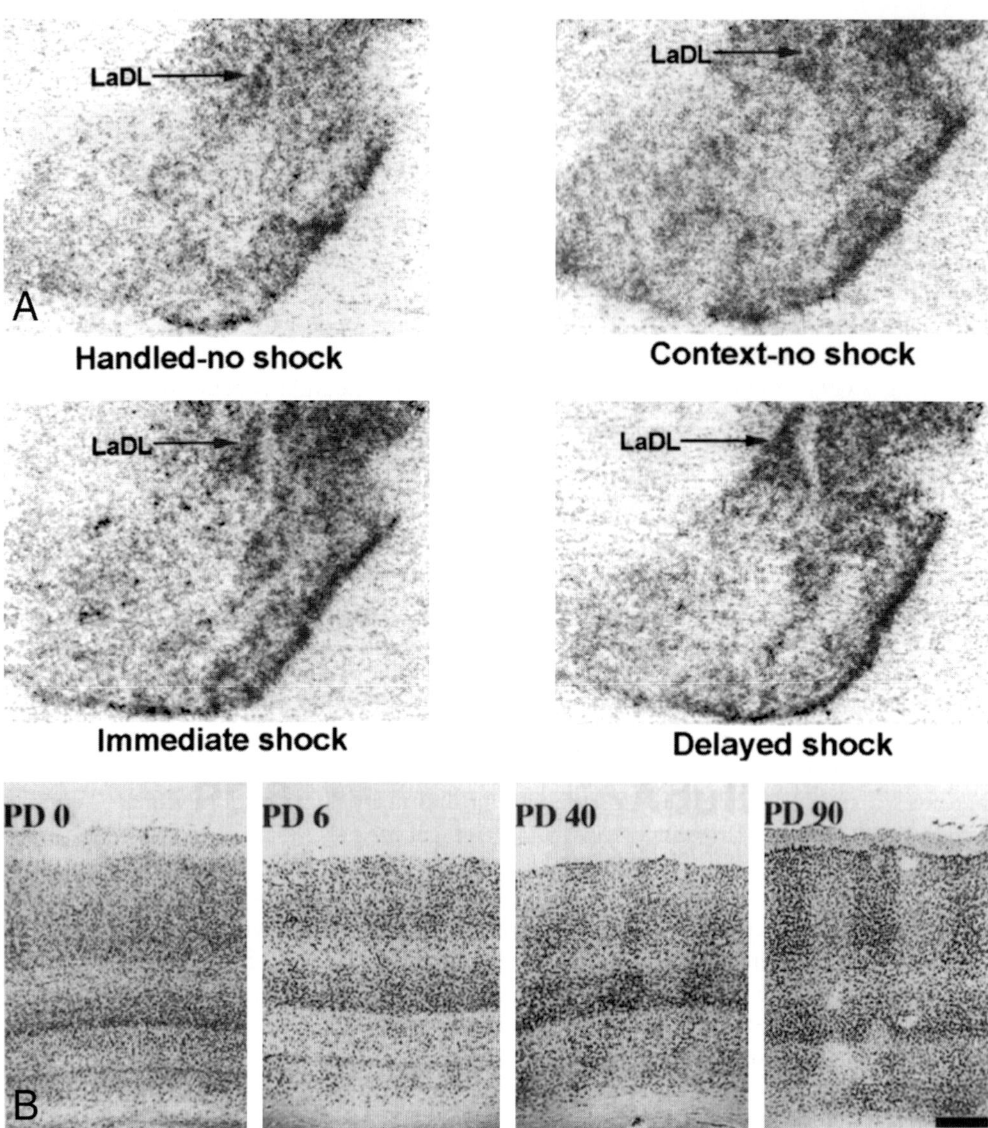

Fig. 1. Differing spatial patterns of expression of the Egrs in the brain after conditioning and during development. (*A*) The amygdala of the rat. *egr*-1 mRNA is expressed selectively in the dorsolateral area of the lateral nucleus of the amygdala (LaDL) after one-trial context fear conditioning. None of the other *egr* mRNAs are expressed with this stimulation, but both Egr-1 and -2 are expressed in this nucleus after C-fiber stimulation (see Fig. 2). (From Malkani and Rosen, 2000). (*B*) The visual cortex of the monkey. After monocular deprivation followed by light stimulation, Egr-1 is expressed in ocular dominance columns during development of the visual cortex. (From Kaczmarek et al., 1999.)

the brain. In the visual cortex, Kaczmarek et al. (1999) showed that, in primates, monocular deprivation followed by light stimulation induces Egr-1 during the critical period of visual cortical development. The light-induced Egr-1 labeling appeared as well-delineated ocular dominance columns that could be evoked from the first postnatal day throughout the critical period (Fig. 1B). Many previous experiments which used c-Fos and other ITFs as markers had failed to reveal these columns.

A further striking example is mapping brain nuclei involved in singing in birds. Thus, when zebra finches and canaries hear playbacks of, and acquire songs of their species, Egr-1 is expressed specifically in the caudomedial neostriatum (NCM) of the telencephalon. This occurs only during a critical period during development (Jin and Clayton, 1997). The adult birds that heard the song but did not respond with singing showed expression of *egr*-1 in the NCM and other auditory areas, whereas birds that sang in response showed additional expression in the brain nuclei required to produce the songs (Mello and Ribeiro, 1998). Most importantly, further experiments (Jarvis et al., 1998) also showed that single whistles (syllables) caused labeling in cells in specific locations within the NCM, and that the *patterns* of Egr-1 staining can be related to the organization of the syllables into families and their intrafamily relationships (Ribeiro et al., 1998). Artificial notes of the same frequency produced synthetically or on a guitar failed to produce the same labeling patterns. Surprisingly, the Egr-1 labeling patterns produced by complex or simple combinations of syllables were not the patterns that would be expected to be formed by simply adding the expressions produced by each single syllable (Ribeiro et al., 1998). In budgerigars (parrots), hearing of conspecific warbles also evokes Egr-1 expression in large areas of the caudal medial forebrain, whereas singing evokes it in nine, more discrete, areas: seven in the lateral and anterior telencephalon, one in the thalamus and one in the midbrain (Jarvis and Mello, 2000). This pattern was found to be very similar in songbirds, hummingbirds and parrots, three species in which singing is thought to have evolved independently (Jarvis et al., 2000). Song learning in finches was also found to activate Egr-1 in brain regions other than the known song-control nuclei (Bolhuis et al., 2000).

Expression of Egr-1 has also allowed detailed maps of brain neurons activated in response to the noxious inputs. In a series of carefully controlled experiments the CNS nuclei activated by noxious somatic compared to visceral pain (caused by ankle inflammation and cyclophosphamide-induced cystitis, respectively), have been mapped (Lanteri-Minet et al., 1995; Bon et al., 1998). Visceral noxious stimulation-produced labeled cells were the dorsal vagal complex, paratrigeminal nucleus, ventrocaudal bulbar reticular formation, the dorsal region of the bed nucleus of the stria terminalis and the caudal region of the central nucleus of the amygdala. It is proposed that these regions form an 'anatomo-functional network' that processes cystitis-related noxious inputs. Somatic noxious stimulation activated the ventrocaudal bulbar reticular formation, the lateral parabrachial area, nucleus cuneiformis and the central gray, but not the dorsal vagal complex, bed nucleus of the stria terminalis, or the amygdala.

There are several recent examples associating the Egrs with conditioning, learning, and the establishment of memory. After two-way active avoidance learning where light and sound, or light alone, are paired with an electric foot shock there are increased levels of *egr*-1 in the hippocampus and visual cortex (Nikolaev et al., 1992). Continued performance results in diminution of the training-induced *egr*-1 expression. Also, the levels of *egr*-1 expressed are probably related to the efficacy of learning because DBA mice, which have a lower learning ability than do C57 mice, also show lower basal levels of *egr*-1 in the hippocampus (Fordyce et al., 1994). Female mice form a memory of the male's pheromones during the 4 h following mating when *egr*-1 levels increase in the granule and mitral cells of the accessory olfactory bulbs (Brennan et al., 1999). Lastly, in macaques trained to recognize a cue stimulus and then recall its paired associate for a reward, Egr-1 appears in patches in area 36 of the inferiotemporal cortex, which is well known to be involved in memory. In contrast, no expression occurs in monkeys simply learning to discriminate a rewarded from a non-rewarded stimulus (Okuno and Miyashita, 1996).

Egr expression has also been utilized in studies of sleep and its deprivation. In rats, 6 to 12 h after sleep deprivation during the day there is a strong expression of *egr*-1 in many cortical areas and the caudate putamen, but a decrease in the thalamus and hippocampus. In contrast, sleep deprivation during the night elicited its expression in only a few cortical areas (Pompeiano et al., 1997). In rats deprived for 24 h the *egr*-1 expression is hardly different to that in control animals (Pompeiano et al., 1997). This is related to activity in the locus coeruleus so that the inducibility of the *egr*-1 is probably modulated by noradrenalin. In normal rats there is a generalized decrease in Egr-1 during both slow wave and REM sleep, whereas in rats exposed to a stimulating environment during the day, there is a marked *increase* of Egr-1 expression in the hippocampus and cerebral cortex during the subsequent REM sleep (Ribeiro et al., 1999).

5. CONSIDERATIONS WHEN USING Egr EXPRESSION TO LOCATE ACTIVATED NEURONS

Here we discuss several aspects of the induction of the Egrs that must be considered when using their expression to map those neurons in the brain activated by intentional stimulation.

5.1. TEMPORAL EXPRESSION OF THE Egrs

Expression of ITFs in the brain is typically examined 1 h after stimulation. However, their expression can be highly dependent upon the duration of the stimulus, as well as the time after the stimulus at which it is examined (Herdegen and Leah, 1998). For example, 30 to 60 min of light exposure evokes *egr*-1 expression in the suprachiasmatic nucleus, whereas 2 h of stimulation reduces its expression in this nucleus (Tanaka et al., 1999). A single brief (5 min) stimulation of sensory nerve C-fibers induces complex temporal patterns of the Egrs in different brain regions. In the hippocampus, Egr-1 shows a fluctuating expression over the subsequent 36 h, Egr-2 does not appear, and Egr-3 has a stable expression (Fig. 2A). In contrast in the hindlimb cortex and amygdala, both Egr-1 and -2 are expressed, and both show the classical peak expression at one hour and subsequent decline, with Egr-1 levels increasing again at later times (Fig. 2B). Also, after electroconvulsive-induced seizure, Egr-1 shows the classic transient expression peaking at 1 h and falling to control levels by 4 h. Egr-3 synthesis is delayed by 2 h, remains elevated for several hours and returns to baseline only after 24 h. The DNA-binding of the Egr-1 and -3 parallel these expression patterns (O'Donovan et al., 1998). Further *egr*-1 and -2 are expressed after induction of LTP in the hippocampus of awake rats (Williams et al., 2000). Both mRNAs are degraded within 20 min and the Egr-1 protein is gone by the typical 4 h. The Egr-2 protein remains elevated for at least 36 h (Demmer et al., 1993), suggesting it has a role in the stabilization of LTP (see Section 6).

Following 15 min stimulation of the motor cortex, there is a parallel phosphorylation of ERK and consequential induction of *egr*-1 mRNA in the corresponding region of the striatum. However, the phosphorylated ERK immunostaining then decreases (probably due to MAPK phosphatase-1 which is also induced), whereas Egr-1 levels remain elevated for at least one hour (Sgambato et al., 1998).

Noxious stimulation has been reported to induce Egr-1 selectively in CA1 neurons (Wei et al., 2000). In this case, the expression of Egr-1 was examined within 1 h of the stimulation, but several hours later it also appears in all other hippocampal regions (Pearse and Leah, 2001). Therefore, it is imperative that the expression of an Egr be examined over a time course of

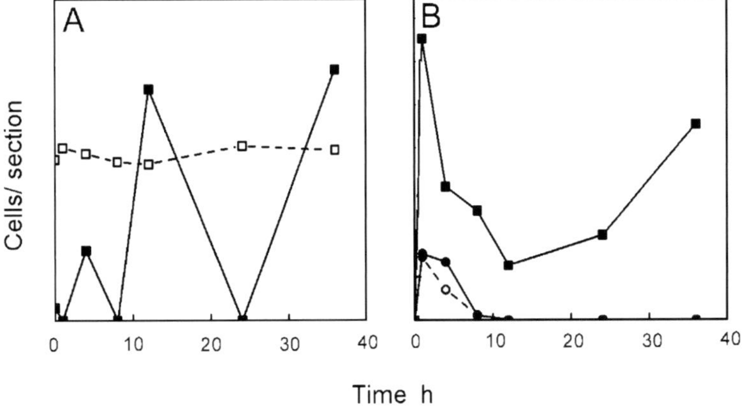

Fig. 2. Differing temporal patterns of expression of the Egr proteins, as determined by immunocytochemistry, in different areas of the rat brain after electrical stimulation of C-fibers in a sciatic nerve. (*A*) The CA1 region of the hippocampus. Egr-1 (■) shows a fluctuating expression whereas the numbers of Egr-3-labeled neurons (□) do not change. Egr-2 is not expressed in the hippocampus after such stimulation. (*B*) Layer V of the hindlimb area of the cortex. Both Egr-1 (■) and Egr-2 (•) are expressed in this area after C-fiber stimulation, although the numbers of Egr-2-labeled neurons are less than Egr-1. Egr-1 (not shown) and Egr-2 (○) are also expressed in the LaDL region of the amygdala after C-fiber stimulation. (Redrawn from Pearse and Leah, 2001; and Pearse et al., unpublished.)

several hours after stimulation before concluding that it is not induced, and that the particular neurons are not activated by the stimulus.

5.2. DIFFERENTIAL EXPRESSION OF THE Egrs AND THEIR DNA BINDING

In various cell types in vitro, stimulation can cause a co-expression or a differential expression of the four members of the Egr family (Kharbanda et al., 1991). This also occurs in the brain after stimulation. Malkani and Rosen (2000) best exemplify this by showing that, of the four *egr* mRNAs, only *egr*-1 is expressed upon contextual fear conditioning. Moreover, it is expressed uniquely in the dorsolateral part of the amygdaloid complex, and it is expressed only after the conditioning and not after a retention test given 24 h later. This experiment shows that the different members of the Egr family can have surprisingly specific regional expression patterns, and that the expression of all four Egrs should be examined when mapping activated neurons.

All four Egrs are expressed in the hippocampus after electroconvulsant seizure, although only Egr-1 and Egr-3 are found in DNA-binding complexes (O'Donovan et al., 1998). In contrast, chronic environmental enrichment leads to increased expression of *egr*-1 in the CA2 region of the hippocampus, but a decreased expression of *egr*-2 (Kim et al., 1994). Induction of LTP in awake rats induces expression of Egr-1 and Egr-2 in the hippocampus, but only Egr-1 participates in the subsequent DNA-binding activity (Williams et al., 2000). These results indicate that the other Egr proteins are translated at very low levels, and/or that their DNA binding is blocked. Chronic ethanol exposure selectively affects Egr activity in several brain areas. Discrepancies in DNA-binding activity, protein expression and the number of immunoreactive cells suggested that factors other than simply the level of protein expression were important in the ethanol effect. Changes in the activity of NAB1 and NAB2 the protein inhibitors of Egr-1 and Egr-2 may account for some alteration in Egr DNA-binding and similar proteins may modulate the DNA-binding activity of Egr-3 (Depaz et al., 2000). The results also illustrate that expression of an ITF does not necessarily mean that it affects the activity of genes whose promoters contain Egr-binding elements. However, there is probably partial functional redundancy amongst the Egr proteins because in Egr-1 knockout mice Egr-4 can substitute for Egr-1 in regulating luteinizing hormone (Tourtellotte et al., 2000).

5.3. STIMULATION PARADIGMS

Expression of Egrs can be highly dependent upon the characteristics of an inducing stimulus, as well as the *pattern* of stimulation when more than one stimulus is applied. Thus 1 to 2 h after 5 Hz burst, but not continuous, regular stimulation of the medial forebrain bundle, *egr*-1 appears in the medial striatum. With 8 Hz stimulation its expression expands to cover the medial, central and lateral striatal areas, as well as the nucleus accumbens and lateral septum (Chergui et al., 1997). Although this is due simply to increased dopamine release at the higher frequency, the experiment does emphasize how expression of the Egr can be dependent upon the characteristics of a simple stimulus.

After the intense input caused by electrical stimulation of all the C-fibers in one sciatic nerve in the rat, Egr-1, but not Egr-2, appears in all areas of the hippocampus (Pearse and Leah, 2001). However, when a weak noxious cutaneous stimulus, which itself does not induce expression of the Egrs, is applied simultaneously to each hindpaw, Egr-2 is also expressed in the hippocampus. Thus weak coincident inputs can induce expression of an Egr when a single intense stimulus does not.

5.4. CONTEXT

While expression of the Egrs can be induced by numerous simple physiological and pharmacological stimuli, this expression can be affected by the context in which the stimulus is given. As noted above, the pattern of expression of *egr*-1 caused by sleep deprivation depends upon whether it occurs during the day or during the night (Pompeiano et al., 1997). This aspect is further exemplified by the work of Jarvis et al. (1998). When male zebra finches direct their singing to females during their courtship dances Egr-1 appears in quite different forebrain nuclei than when their song is not so directed. Further, non-contingent infusion of cocaine into the brains of mice elicits *egr*-1 expression in the nucleus accumbens, medial caudate putamen and frontal cortex, whereas self-administration of the drug does not cause any such *egr*-1 expression (Kuzmin and Johansson, 1999).

5.5. EXPRESSION OUTSIDE AREAS KNOWN TO BE EXCITED BY A STIMULUS

In the mapping of singing-associated nuclei in budgerigars described above (Jarvis and Mello, 2000) the areas in which Egr-1 is induced are much more extensive and different in shape than would be expected from anatomical studies. Additionally, song learning in finches activates Egr-1 in brain regions other than the known song-control nuclei (Bolhuis et al., 2000). The same occurs in the rat brain after a noxious stimulus where many nuclei unrelated to pain processing, such as the visual lateral geniculate nucleus and the auditory inferior colliculus, show changes in Egr-1 expression (D.D. Pearse et al., unpublished).

5.6. mRNA AND PROTEIN LEVELS

Finally, discrepancies between the presence of the Egr mRNAs and their proteins occur. For example, phencyclidine induces *egr*-1 mRNA in layers III and IV of the retrosplenial cortex, but the protein levels in layer IV are unaffected, and in the outer layers, it is reduced (Shirayama et al., 1999). As described below, after induction of LTP both *egr*-1 and -2 mRNAs are transiently expressed. The Egr-1 protein has gone by 8 h, but Egr-2 remains for more than 30 h. Thus ITF proteins can have very long expression times *independent* of their mRNAs' half-lives such that in situ hybridization and immunocytochemical labeling could give very different results.

6. ROLES

Several possible roles for the Egrs have been discussed previously (Beckmann and Wilce, 1997; Herdegen and Leah, 1998; Hughes et al., 1999; O'Donovan et al., 1999). Egr-1 binds to the promoters and initiates transcription of the genes for phenylethanolamine *N*-methyl transferase, neurofilament, synapsin I, synaptophysin I and II, apolipoprotein A1, platelet-derived growth factor B, and *nur77*, and it represses the adenosine deaminase gene. Additionally, the *egr*-1 gene contains an ERE and is thus itself a target of Egr-1 proteins. Egr-2 regulates some Hox genes to cause peripheral nerve myelination and hindbrain segmentation during development. The JunB and JunD promoters also contain one or more EREs so their activity might be under some control by the Egrs.

6.1. NEURONAL DEVELOPMENT

In PC12 cells, neurite extension is accompanied by increased Egr-1 expression, which suggested that Egrs might have a role in axonal growth in the developing and/or adult brain. However, over-expression of Egr-1 *reduces* NGF-induced neurite outgrowth in these cells (Matsumoto et al., 2000). In contradistinction, over-expression of Egr-1 in N2A neuroblastoma cells causes them to increase synthesis of the microtubule-associated protein MAP1B and to extend long neurites (Pignatelli et al., 1999). Treatment of these cells with *egr*-1 antisense oligonucleotides prevents morphological differentiation induced by serum withdrawal. Finally, in the brains of the mutant *stargazer* mice, which show inherent and extensive sprouting of hippocampal mossy fibers, there is no expression of Egr-1 (Nahm and Noebels, 1998). A single electrically induced stimulus in hippocampal neurons induces expression of *egr*-1 and its protein, and of Egr-2 protein in dentate granule neurons. MK-801 potently inhibits the expression of Egr-2, but not that of *egr*-1 or its protein (Hughes et al., 1998). MK-801 also reduces the mossy fiber sprouting induced by this type of stimulation (kindling), and it is thus proposed that *egr*-2 but not *egr*-1 is involved in this sprouting (Hughes et al., 1998). Thus, the role of Egr-1 in axonal growth remains to be delineated. In mice in which the *egr*-2 gene is inactivated during development, a population of neurons in the lateral reticular nucleus appears to be missing in the adult (Borday et al., 1997). In addition to its developmental role in Schwann cells during peripheral nerve myelination, Egr-2 is induced in vivo in these glial cells by glutamate and ATP (Stevens and Fields, 2000).

6.2. TARGET GENES

In addition to those listed previously, Egr-1 up-regulates the nicotinic acetylcholine receptor (Carrasco-Serrano et al., 2000) and the acetylcholinesterase gene (von der Kammer et al., 1998). It transactivates the p75 NGF receptor after nerve injury (Nikam et al., 1995), the TGF-beta1 that is probably involved in sympathetic neuron differentiation (Kim et al., 1994), and neuropeptide Y (Wernerson et al., 1998). Egr-1 *inhibits* production of neuroserpin, a protease inhibitor that modulates the activity of tissue plasminogen activator (tPA) and thereby may affect synaptic activity (Berger et al., 1999).

6.3. NEURAL PLASTICITY AND LTP

Activation of Egr-1 in non-neuronal cells results in production of TGF-beta1, plasminogen activator inhibitor-1 (PAI-1), fibroblast growth factor-2 (Jin et al., 2000), fibronectin (Liu et al., 2000), FasL (Xiao et al., 1999), and E-cadherin (Hosono et al., 2000). Egr-2 also causes a strong expression of FasL (Rengarajan et al., 2000). These proteins can all increase cell adhesion and probably neuronal connectivity. They may also increase interaction with the extracellular matrix, such as would be required during visual cortical development (Kaczmarek et al., 1999) and learning (Miyashita et al., 1998). In addition, Egr-1 is up-regulated by the transcription factor COUP-TF that is expressed mainly in neural tissue, regulates cell–cell interactions, and whose deletion causes nervous system defects (Pipaon et al., 1999).

The Egr proteins are crucial in long-term potentiation. After induction of LTP in awake rats both *egr*-1 and -2 mRNAs appear in hippocampal neurons. Both mRNAs disappear before 2 h and the Egr-1 protein before 8 h. However, the Egr-2 protein remains for more than 30 h, suggesting it has a role in the maintenance of LTP (Demmer et al., 1993). Egr-1 knock-out mice do not show synaptic potentiation in the hippocampus (Wei et al., 2000).

6.4. INTERACTION WITH OTHER TRANSCRIPTION FACTORS

Both the JunB and JunD promoters contain two or more EREs, and Egr-4 can repress c-*fos* (Zipfel et al., 1997). Egr-1 also binds to the CREB-binding protein CBP/p300 (Silverman et al., 1998). Egr-1 inhibits NF-κB activity, apparently by sequestering it from its DNA-binding site. It also inhibits TNF-alpha activity (Di Battista and Martel-Pelletier, 1999). These actions may contribute to its role in apoptosis (Chapman and Perkins, 2000). Egr-2 plays a crucial role in segmentation of the hindbrain. In doing so, it directly regulates the homeobox transcription factors, Hoxa-2, Hoxb-2 and Hoxb-3 (Seitanidou et al., 1997).

6.5. COINCIDENCE DETECTORS

Compared to Egr-1, and to members of the other ITF families, the ‘threshold’ for stimulus-induced expression of Egr-2 in the brain is high (Herdegen et al., 1993). For example, neither the intense excitation produced by bicuculline-induced seizures, nor stimulation of all the C-fibers in the sciatic nerve are able to elicit its expression in the hippocampus (Herdegen et al., 1993; Pearse and Leah, 2001), even though several other ITFs are induced by these stimuli. However, Egr-2 can be induced by certain *patterns* of stimulation. Thus, in the rat hippocampus a single or repeated noxious stimulus to the hindpaws induces Egr-1 but not Egr-2. But when a noxious stimulus is given simultaneously to each hindpaw, Egr-2 appears in all hippocampal areas (Pearse and Leah, 2001). In this case, Egr-2 may thus have a role as a ‘genetic coincidence detector’.

6.6. ROLES OUTSIDE THE NUCLEUS

There are clear instances of the ITFs, c-Fos and the Fos-related antigens (Fras) and the constitutive transcription factors CREB and ATF-2, being present in the cytoplasm and axoplasm of some types of neurons (Herdegen et al., 1993; Herdegen and Leah, 1998). Egr-2, but not Egr-1, also shows a distinct cytoplasmic immunoreactivity in motor and sensory ganglion neurons (Herdegen et al., 1993). A truncated form of Egr-1 occurs exclusively in the cytoplasm (Matheny et al., 1994), but it remains to be determined whether Egr-3 and -4 can also occur here. We should not dismiss the possibility that Egrs have functions in the cytoplasm that complement their roles in the nucleus, or they are in the cytoplasm 'on standby' so that in these neurons the Egr-responsive genes can be activated more quickly in response to inputs. More significantly, and in relation to the latter possibility Egr-1, like CREB, has been identified in hippocampal dendrites, and it was proposed that it translocates directly from the synapse to the nucleus to more rapidly alter gene expression (Crino et al., 1998).

7. HISTOLOGICAL MAPPING AND ELECTROPHYSIOLOGICAL EXCITATION

Probably because their expression in the central nervous system was first observed after seizure (Morgan et al., 1987) or after stimulation of sensory nerves (Hunt et al., 1987), we have since held in our minds an association between electrophysiological activity of neurons and ITF expression. However, there are numerous examples where this is not the case. This uncertain relationship and its implication for using c-Fos in mapping studies have previously been considered by several authors (Gass et al., 1997). Here we examine the problem with respect to the Egrs.

There are several instances where the numbers of c-Fos-labeled cells clearly *does* correlate with the strength of a sensory input and the degree of electrophysiological excitation of the affected neurons. In the pain system, the number of neurons expressing c-Fos in the spinal dorsal horn and in the brainstem parabrachial nucleus is highly correlated with the strength and/or duration of a peripheral noxious stimulus (Herdegen and Leah, 1998). In monocularly deprived rats the electrophysiological responses of neurons in the different layers of the cortex is also strongly correlated with their expression of Egr-1 in the same layers after stimulation of the deprived eye (Caleo et al., 1999).

However, there are many counter-examples. The most compelling are those in which the neurons' electrical activity was monitored, or was known to occur in response to the same stimulus that induced the Egr expression. Stimulation of the motor cortex excites the neurons and elicits *egr*-1 mRNA transcription in the subthalamic nucleus (Sgambato et al., 1998, 1999). Pharmacological block of the corticostriatal pathway strongly reduced this excitation of the subthalamic neurons, but there was still a strong expression of *egr*-1. However, after lesion block of the corticostriatal pathway, the subthalamic neurons were again excited by cortical stimulation, but *egr*-1 remained absent. Lastly, systemic injection of cholecystokinin causes neurons in the nucleus of the solitary tract and the magnocellular neurons of the hypothalamus to become electrophysiologically active, but *egr*-1 was not induced (Luckman, 1997).

Application of NMDA agonists to one side of the dorsal surface of the trigeminal nucleus caudalis to excite laminae I–II neurons induces both c-Fos and Egr-1 on that side, but only Egr-1 was expressed contralaterally. When substance P was used as the excitant, both c-Fos and Egr-1 were again expressed ipsilaterally but only Egr-1 was expressed contralaterally (Koganemaru et al., 2000). Is it Fos or Egr that maps the excited neurons? A second example is the different expressions of Egr-1 -2 and -3 in the hippocampus after peripheral noxious stimulation described above. From these results, one could infer that the noxious stimulus either caused an electrical activity in the neurons that fluctuated over the subsequent 30 h, or was without any effect.

A further set of examples that indicate dissociation between the electronic and genetic activities of neurons comes from experiments where particular drugs are given singly or repeatedly. For example, repeated doses of caffeine reduce both its behavioral (i.e. its electrophysiological) effects and its ability to decrease *egr*-1 mRNA levels in the nucleus accumbens (Svenningsson et al., 1999). In contradistinction, repeated administration of D2 receptor antagonists reduces both the behavioral effects and induction of *egr*-1 in striatopallidal neurons (Steiner and Gerfen, 1999).

These recent results re-emphasize that electrical activity per se does not necessarily induce expression of any particular ITF in the neuron. Thus it is imperative to use a 'bank' of Egrs and other ITFs when mapping activated neurons, and this is increasingly being done. However, it may be that in some instances electrical activity does not induce expression of *any* ITF in the neuron, but this has not been specifically looked for.

In addition, a neuron may receive inputs that evoke EPSPs that lead to calcium influx, activation of signal transduction pathways, ITF expression, and consequential changes in its genetic activity. However, it may be that these EPSPs do not reach action potential threshold and affect the neuron's electronic activity. ITF expression is the result of the activation, or co-activation of signal transduction pathways in response to a particular afferent input, or pattern of inputs. It may even be that these inputs cause *no* changes in the neuron's electronic status, but only activate its signal transduction pathways and thereby causes ITF expression. This 'silent transmission' in the brain would be distinct from electrical transmission, and may have functions that remain to be revealed.

8. ABBREVIATIONS

2-APV	2-amino-phosphono-valeric acid
AP-1	activatory protein-1
cAMP	cyclic adenosinomonophosphate
CRE	Ca^{2+}/cAMP responsive element, gene regulatory sequence
EGF	epidermal growth factor
EPSP	excitatory postsynaptic potential
ERE	Egr response element
ETS	transcription factor
FasL	Fas ligand
ITF	inducible transcription factor
JNK	c-Jun N-terminal kinase
LTP	long-term potentiation
MAPK	mitogen-activated protein kinases
NCM	caudomedial neostriatum of the telencephalon
NF-κB	nuclear factor κ-B, transcription factor
NGF	nerve growth factor
NMDA	*N*-methyl-D-aspartate
PKA	protein kinase A
SRE	serum responsive element
TNF	tumor necrosis factor
TPA	tetradecyl phorbol 13-acetate
TTX	tetrodotoxin

9. REFERENCES

Aicher WK, Dinkel A, Grimbacher B, Haas C, Seydlitz-Kurzbach EV, Peter HH, Eibel H (1999a): Serum response elements activate and cAMP responsive elements inhibit expression of transcription factor Egr-1 in synovial fibroblasts of rheumatoid arthritis patients. *Int Immunol 11*:47–61.

Aicher WK, Sakamoto KM, Hack A, Eibel H (1999b): Analysis of functional elements in the human Egr-1 gene promoter. *Rheumatol Int 18*:207–214.

Alonso R, Gnanadicom H, Frechin N, Fournier M, Le Fur G, Soubrie P (1999): Blockade of neurotensin receptors suppresses the dopamine D1/D2 synergism on immediate early gene expression in the rat brain. *Eur J Neurosci 11*:967–974.

Barroso I, Stantisteban P (1999): Insulin-induced early growth response gene (Egr-1): mediates a short term repression of rat malic enzyme gene transcription. *J Biol Chem 274*:17997–18004.

Beckmann AM, Wilce PA (1997): Egr transcription factors in the nervous system. *Neurochem Int 31*:477–510.

Berg JM (1988): Proposed structure for the zinc-binding domains from transcription factor IIIA and related proteins. *Proc Natl Acad Sci USA 85*:99–102.

Berger P, Kozlov SV, Cinelli P, Kruger SR, Vogt L, Sonderegger P (1999): Neuronal depolarization enhances the transcription of the neuronal serine protease inhibitor neuroserpin. *Mol Cell Neurosci 14*:455–467.

Bolhuis JJ, Zijlstra GG, den Boer-Visser AM, Van Der Zee EA (2000): Localized neuronal activation in the zebra finch brain is related to the strength of song learning. *Proc Natl Acad Sci USA 97*:2282–2285.

Bon K, Lanteri-Minet M, Michiels JF, Menetrey D (1998): Cyclophosphamide cystitis as a model of visceral pain in rats: A c-fos and Krox-24 study at telencephalic levels, with a note on pituitary adenylate cyclase activating polypeptide (PACAP). *Exp Brain Res 122*:165–174.

Borday V, Kato F, Champagnat J (1997): A ventral pontine pathway promotes rhythmic activity in the medulla of neonatal mice. *Neuroreport 8*:3679–3683.

Brennan PA, Schellinck HM, Keverne EB (1999): Patterns of expression of the immediate-early gene egr-1 in the accessory olfactory bulb of female mice exposed to pheromonal constituents of male urine. *Neuroscience 90*:1460–1463.

Brinton RD, Yamazaki R, Gonzalez CM, O'Neill K, Schreiber SS (1998): Vasopressin-induction of the immediate early gene, NGFI-A, in cultured hippocampal glial cells. *Mol Brain Res 57*:73–85.

Caleo M, Lodovichi C, Pizzorusso T, Maffei L (1999): Expression of the transcription factor Zif268 in the visual cortex of monocularly deprived rats, effects of nerve growth factor. *Neuroscience 91*:1017–1026.

Carrasco-Serrano C, Viniegra S, Ballesta JJ, Criado M (2000): Phorbol ester activation of the nicotinic acetylcholine receptor alpha7 subunit gene, involvement of transcription factor Egr-1. *J Neurochem 74*:932–939.

Chapman NR, Perkins ND (2000): Inhibition of the RelA(p65): NF-kappaB subunit by Egr-1. *J Biol Chem 275*:4719–4725.

Chaudhuri A, Cynader MS (1993): Activity-dependent expression of the transcription factor Zif268 reveals ocular dominance columns in monkey visual cortex. *Brain Res 605*:349–353.

Chaudhuri A, Matsubara JA, Cynader MS (1995): Neuronal activity in primate visual cortex assessed by immunostaining for the transcription factor Zif268. *Vis Neurosci 12*:35–45.

Chavrier P, Janssen-Timmen U, Mattei MG, Zerial M, Bravo R, Charnay P (1988): Nucleotide structure, chromosome location, and expression of the mouse zinc finger gene Krox-20: multiple gene products and coregulation with the proto-oncogene c-fos. *Mol Cell Biol 9*:787–797.

Chergui K, Svenningsson P, Nomikos GG, Gondon F, Fredholm BB, Svennson TH (1997): Increased expression of NGFI-A mRNA in the rat striatum following stimulation of the medial forebrain bundle. *Eur J Neurosci 9*:2370–2382.

Chiu JJ, Wung BS, Hsieh HJ, Lo LW, Wang DL (1999): Nitric oxide regulates shear stress-induced early growth response-1 expression via the extracellular signal-regulated kinase pathway in endothelial cells. *Circ Res 85*:238–246.

Cochran BH (1993): Regulation of immediate early gene expression. *NIDA Res Monogr 125*:3–24.

Condorelli DF, Dell'Albani P, Amico C, Kaczmarek L, Nicoletti F, Lukasiuk K, Stella AM (1993): Induction of primary response genes by excitatory amino acid receptor agonists in primary astroglial cultures. *J Neurochem 60*:877–885.

Cosgaya JM, Perez-Juste G, Aranda A (1998): Retinoic acid regulates selectively the expression of immediate early response genes in PC12 cells. *FEBS Lett 429*:254–258.

Crino P, Khodakhah K, Becker K, Ginsberg S, Hemby S, Erbwine J (1998): Presence and phosphorylation of transcription factors in developing dendrites. *Proc Natl Acad Sci USA 95*:2313–2318.

Davis S, Vanhoutte P, Pages C, Caboche J, Laroche S (2000): The MAPK/ERK cascade targets both Elk-1 and cAMP response element-binding protein to control long-term potentiation-dependent gene expression in the dentate gyrus in vivo. *J Neurosci 20*:4563–4572.

DeFranco C, Damon DH, Endoh M, Wagner JA (1993): Nerve growth factor induces transcription of NGFI-A through complex regulatory elements that are also sensitive to serum and phorbol 12-myristate 13-acetate. *Mol Endocrinol 7*:365–379.

Demmer J, Dragunow M, Lawlor PA, Mason SE, Leah JD, Abraham WC, Tate WP (1993): Differential expression of immediate early genes after hippocampal long-term potentiation in awake rats. *Mol Brain Res 17*:279–286.

Depaz IM, Goodenough S, Wilce PA (2000): Chronic ethanol has region-selective effects on Egr-1 and Egr-3 DNA-binding activity and protein expression in the rat brain. *Neurochem Int 37*:473–482.

Dragunow M, Hughes P (1993): Differential expression of immediate-early proteins in non-nerve cells after focal brain injury. *Int J Neurosci 11*:249–255.

Di Battista JA, Martel-Pelletier J (1999): Suppression of tumor necrosis factor (TNF-alpha): gene expression by prostaglandin E(2): role of early growth response protein-1 (Egr-1). *Osteoarthritis Cartilage 7*:395–398.

Enslen H, Soderling TR (1994): Roles of calmodulin-dependent protein kinases and phosphatase in calcium-dependent transcription of immediate early genes. *J Biol Chem 269*:20872–20877.

Fordyce DE, Bhat RV, Baraban JM, Wehner JM (1994): Genetic and activity-dependent regulation of *zif*268 expression: association with spatial learning. *Hippocampus 4*:559–568.

Gass P, Herdegen T, Bravo R, Kiessling M (1993): Induction and suppression of immediate early genes (IEGs) in specific rat brain regions by the non-competitive NMDA receptor antagonist MK-801. *Neuroscience 53*:749–758.

Gass P, Bruehl C, Herdegen T, Kiessling M, Lutzenburg M, Witte OW (1997): Induction of FOS and JUN proteins during focal epilepsy: congruencies with and differences to [^{14}C]deoxyglucose metabolism. *Mol Brain Res 46*:177–184.

Herdegen T, Leah JD (1998): Inducible and constitutive transcription factors in the mammalian nervous system, control of gene expression by Jun Fos and Krox and CREB/ATF proteins. *Brain Res Rev 28*:370–490.

Herdegen T, Kiessling M, Bele S, Bravo R, Zimmermann M, Gass P (1993): The KROX-20 transcription factor in the rat central and peripheral nervous systems, novel expression pattern of an immediate early gene-encoded protein. *Neuroscience 57*:41–52.

Hodge C, Liao J, Stofega M, Guan K, Carter-Su C, Schwartz J (1998): Growth hormone stimulates phosphorylation and activation of elk-1 and expression of c-fos egr-1 and junB through activation of extracellular signal-regulated kinases 1 and 2. *J Biol Chem 273*:31327–31336.

Holst C, Skerka C, Lichter P, Bialonski A, Zipfel PF (1993): Genomic organization, chromosomal localization and promoter function of the human zinc-finger gene pAT133. *Hum Mol Genet 2*:367–372.

Honkaniemi J, Sharp FR (1999): Prolonged expression of zinc finger immediate-early gene mRNAs and decreased protein synthesis following kainic acid induced seizures. *Eur J Neurosci 11*:10–17.

Hosono S, Gross I, English MA, Hajra KM, Fearon ER, Licht JD (2000): E-cadherin is a WT1 target gene. *J Biol Chem 275*:10943–10953.

Hughes PE, Young D, Preston KM, Yan Q, Dragunow M (1998): Differential regulation by MK801 of immediate-early genes, brain-derived neurotrophic factor and trk receptor mRNA induced by a kindling after-discharge. *Mol Brain Res 53*:138–151.

Hughes PE, Alexi T, Walton M, Williams CE, Dragunow M, Clarke RG, Gluckman PD (1999): Activity and injury-dependent expression of inducible transcription factors, growth factors and apoptosis-related genes within the central nervous system. *Prog Neurobiol 57*:421–450.

Humblot N, Thiriet N, Gobaille S, Aunis D, Zwiller J (1998): The serotonergic system modulates the cocaine-induced expression of the immediate early genes egr-1 and c-fos in rat brain. *Ann NY Acad Sci 844*:7–20.

Hung AC, Huang HM, Tsay HJ, Lin TN, Kuo JS, Sun SH (2000): ATP-stimulated c-fos and zif268 mRNA expression is inhibited by chemical hypoxia in a rat brain-derived type 2 astrocyte cell line RBA-2. *J Cell Biochem 77*:323–332.

Hunt SP, Pini A, Evan G (1987): Induction of c-fos-like protein in spinal neurons following sensory stimulation. *Nature 328*:632–634.

Jarvis ED, Mello CV (2000): Molecular mapping of brain areas involved in parrot vocal communication. *J Comp Neurol 419*:1–31.

Jarvis ED, Schaff C, Grossman MR, Ramos JA, Nottebohm F (1998): For whom the bird sings, context-dependent gene expression. *Neuron 21*:775–788.

Jarvis ED, Ribeiro S, da Silva ML, Ventura D, Vielliard J, Mello CV (2000): Behaviourally driven gene expression reveals song nuclei in hummingbird brain. *Nature 406*:628–632.

Jin H, Clayton DF (1997): Localized changes in immediate-early gene regulation during sensory and motor learning in zebra finches. *Neuron 19*:1049–1059.

Jin Y, Sheikh F, Detillieux KA, Cattini PA (2000): Role for early growth response-1 protein in alpha(1)-adrenergic stimulation of fibroblast growth factor-2 promoter activity in cardiac myocytes. *Mol Pharmacol 57*:984–990.

Johansson IM, Bjartmar L, Marcusson J, Ross SB, Seckl JR, Olsson T (1998): Chronic amitriptyline treatment induces hippocampal NGFI-A glucocorticoid receptor and mineralocorticoid receptor mRNA expression in rats. *Mol Brain Res 62*:92–95.

Joseph LJ, Le Beau MM, Jamieson GA, Acharya S, Shows TB, Rowley JD, Sukhatme VP (1988): Molecular cloning, sequencing, and mapping of EGR2, a human early growth response gene encoding a protein with zinc-finger binding structure. *Proc Natl Acad Sci USA 85*:7164–7168.

Kaczmarek L, Zangenehpour S, Chaudhuri A (1999): Sensory regulation of immediate-early genes c-fos and zif268 in monkey visual cortex at birth and throughout the critical period. *Cereb Cortex 9*:179–187.

Kendall G, Ensor E, Brar-Rai A, Winter J, Latchman DS (1994): Nerve growth factor induces expression of immediate-early genes NGFI-A (Egr-1) and NGFI-B (nur77) in adult dorsal root ganglion neurons. *Mol Brain Res 25*:73–79.

Kenneth JM, Mack PA (1992): Induction of transcription factors in somatosensory cortex after tactile stimulation. *Mol Brain Res 12*:141–147.

Khan SM, Smith TS, Bennett JP (1999): Effects of single and multiple treatments with L-dihydroxyphenylalanine (L-DOPA): on dopamine receptor-G protein interactions and supersensitive immediate early gene responses in striata of rats after reserpine treatment or with unilateral nigrostriatal lesions. *J Neurosci Res 55*:71–79.

Kharbanda S, Nakamura T, Stone R, Hass R, Bernstein S, Datta Sukhatme VP, Kufe D (1991): Expression of the early growth response 1 and 2 zinc finger genes during induction of monocytic differentiation. *J Clin Invest 88*:571–577.

Kim SJ, Park K, Rudkin BB, Dey BR, Sporn MB, Roberts AB (1994): Nerve growth factor induces transcription of transforming growth factor-beta1 through a specific promoter element in PC12 cells. *J Biol Chem 269*:3739–3744.

Koganemaru M, Takasaki M, Nishimori T (2000): Simultaneous activation of *N*-methyl-D-aspartate and neurokinin-1 receptors modulates c-Fos and Zif/268 expression in the rat trigeminal nucleus caudalis. *Neuroscience 98*:317–323.

Kumahara E, Ebihara T, Saffen D (1999): Nerve growth factor induces zif268 gene expression via MAPK-dependent and -independent pathways in PC12 cells. *J Biochem (Tokyo) 125*:541–553.

Kuzmin A, Johansson B (1999): Expression of c-fos, NGFI-A and secretogranin II mRNA in brain regions during initiation of cocaine self-administration in mice. *Eur J Neurosci 11*:3694–3700.

Lanteri-Minet M, Isnardon P, De Pommery J, Menetrey D (1993): Spinal and hindbrain structures involved in visceroception and visceronociception as revealed by the expression of Fos, Jun and Krox-24 proteins. *Neuroscience 55*:737–753.

Lanteri-Minet M, Bon K, de Pommery J, Michiels JF, Menetrey D (1995): Cyclophosphamide cystitis as a model of visceral pain in rats: model elaboration and spinal structures involved as revealed by the expression of c-Fos and Krox-24 proteins. *Exp Brain Res 105*:220–232.

Leveque JC, Macias W, Rajadhyaksha A, Carlson RR, Barczak A, Kang S, Li XM, Coyle JT, Huganir RL, Heckers S, Konradi C (2000): Intracellular modulation of NMDA receptor function by antipsychotic drugs. *J Neurosci 20*:4011–4020.

Liu C, Yao J, Mercola D, Adamson E (2000): The transcription factor EGR-1 directly transactivates the fibronectin gene and enhances attachment of human glioblastoma cell line U251. *J Biol Chem 257*:20315–20323.

Luckman SM (1997): Comparison of the expression of c-fos, nur77 and egr1 mRNA in rat hypothalamic magnocellular neurons and their putative afferent projection neurons: cell- and stimulus-specific induction. *Eur J Neurosci 9*:2443–2451.

Mack KJ, Cortner J, Mack P, Farnham PJ (1992): Krox-20 messenger RNA and protein expression in the adult central nervous system. *Mol Brain Res 14*:117–123.

Mack KJ, Yi SD, Chang S, Millan N, Mack P (1995): NGF1-C expression is affected by physiological stimulation and seizures in the somatosensory cortex. *Mol Brain Res 29*:140–146.

Malkani S, Rosen JB (2000): Specific induction of early growth response gene 1 in the lateral nucleus of the amygdala following contextual fear conditioning in rats. *Neuroscience 97*:693–702.

Matsumoto T, Akiama N, Kizaka-Kondoh S, Noda M (2000): Transient over-expression of NGFI-A gene suppresses NGF-induced neurite outgrowth in PC12 cells. *Neuroreport 11*:1001–1005.

Matheny C, Day ML, Milbrandt J (1994): The nuclear localization signal of NGFI-A is located within the zinc finger DNA binding domain. *J Biol Chem 269*:8176–8181.

McKitrick DJ, Krukoff TL, Calaresu FR (1992): Expression of c-fos protein in rat brain after electrical stimulation of the aortic depressor nerve. *Brain Res 599*:215–222.

Mechta F, Piette J, Hirai S-I, Yaniv M (1989): Stimulation of protein kinase C or protein kinase A mediated signal transduction pathways shows three modes of response among serum inducible genes. *New Biol 3*:297–304.

Mello CV, Ribeiro S (1998): ZENK protein regulation by song in the brain of songbirds. *J Comp Neurol 393*:426–438.

Melzer P, Steiner H (1997): Stimulus-dependent expression of immediate-early genes in rat somatosensory cortex. *J Comp Neurol 380*:145–153.

Milbrandt J (1987): A nerve growth factor-induced gene encodes a possible transcriptional regulatory factor. *Science 238*:797–799.

Miller J, McLachlan AD, Klug A (1985): Repetitive zinc-binding domains in the protein transcription factor IIIA from *Xenopus* oocytes. *EMBO J 4*:1609–1614.

Miyashita Y, Morita M, Naya Y, Yoshida M, Tomita H (1998): Backward signal from medial temporal lobe in neural circuit reorganization of primate inferotemporal cortex. *Czech Repub Acad Sci III 321*:185–192.

Mora-Garcia P, Sakamoto KM (2000): Granulocyte colony-stimulating factor induces Egr-1 up-regulation through interaction of serum response element-binding proteins. *J Biol Chem 275*:22418–22426.

Morgan JI, Cohen DR, Hempstead JL, Curran T (1987): Mapping patterns of c-fos expression in the central nervous system after seizure. *Science 237*:192–197.

Morris ME, Viswanathan N, Kuhlman S, Davis FC, Weitz CJ (1998): A screen for genes induced in the suprachiasmatic nucleus by light. *Science 279*:1544–1547.

Nahm WK, Noebels JL (1998): Non obligate role of early or sustained expression of immediate-early gene proteins c-fos, c-jun, and zif/268 in hippocampal mossy fiber sprouting. *J Neurosci 18*:9245–9255.

Nikam SS, Tennekoon GI, Christy BA, Yoshino JE, Rutkowski JL (1995): The zinc finger transcription factor Zif268/Egr-1 is essential for Schwann cell expression of the p75 NGF receptor. *Mol Cell Biol 6*:337–348.

Nikolaev E, Werka T, Kaczmarek L (1992): c-fos protooncogene expression in rat brain after long-term training of two-way active avoidance reaction. *Behav Brain Res 48*:91–94.

O'Donovan KJ, Baraban JM (1999): Major Egr-3 isoforms are generated via alternate translation start sites and differ in their abilities to activate transcription. *Mol Cell Biol 19*:4711–4718.

O'Donovan KJ, Wilkens EP, Baraban JM (1998): Sequential expression of Egr-1 and Egr-3 in hippocampal granule cells following electroconvulsive stimulation. *J Neurochem 70*:1241–1248.

O'Donovan KJ, Tourtellotte WG, Milbrandt J, Barbaran JM (1999): The EGR family of transcription-regulatory factors: progress at the interface of molecular and systems neuroscience. *Neurosci 22*:167–173.

O'Donovan KJ, Levkovitz Y, Ahn D, Baraban JM (2000): Functional comparison of egr-3 transcription factor isoforms, identification of an activation domain in the N-terminal segment absent from egr3beta, a major isoform expressed in brain. *J Neurochem 75*:1352–1357.

Okuno H, Miyashita Y (1996): Expression of the transcription factor zif268 in the temporal cortex of monkeys during visual paired associate learning. *Eur J Neurosci 8*:2118–2128.

Park JA, Koh JY (1999): Induction of an immediate early gene egr-1 by zinc through extracellular signal-regulated kinase activation in cortical culture, its role in zinc-induced neuronal death. *J Neurochem 73*:450–456.

Pearse DD, Leah JD (2001): Jun Fos and Krox in the hippocampus after C-fiber stimulation: simultaneous-input-dependent expression and nuclear speckling. *Brain Res 894*:193–208.

Pignatelli M, Cortes-Canteli M, Santos A, Perez-Castillo A (1999): Involvement of the NGFI-A gene in the differentiation of neuroblastoma cells. *FEBS Lett 461*:37–42.

Pipaon C, Tsai SY, Tsai MJ (1999): COUP-TF upregulates NGFI-A gene expression through an Sp1 binding site. *Mol Cell Biol 19*:2734–2745.

Pompeiano M, Cirelli C, Ronca-Testoni S, Tononi G (1997): NGFI-A expression in the rat brain after sleep deprivation. *Mol Brain Res 46*:143–153.

Presley RW, Menetrey D, Levine JD, Basbaum AI (1990): Systemic morphine suppresses noxious stimulus-evoked Fos protein-like immunoreactivity in the rat spinal cord. *J Neurosci 10*:323–335.

Rengarajan J, Mittelstad PR, Mages HW, Gerth AJ, Kroczek RA, Ashwell JD, Glimmer LH (2000): Sequential involvement of NFAT and Egr transcription factors in FasL regulation. *Immunity 12*:293–300.

Ribeiro S, Cecchi GA, Magnasco MO, Mello CV (1998): Toward a song code: evidence for a syllabic representation in the canary brain. *Neuron 21*:359–371.

Ribeiro S, Goyal V, Mello CV, Pavlides C (1999): Brain gene expression during REM sleep depends on prior waking experience. *Learn Mem 6*:500–508.

Sachs AB, Sarnow P, Hentze MW (1997): Starting at the beginning, middle and end, translation initiation in eukaryotes. *Cell 89*:831–838.

Sato K, Ishikawa K, Ui M, Okajima F (1999): Sphingosine 1-phosphate induces expression of early growth response-1 and fibroblast growth factor-2 through mechanism involving extracellular signal-regulated kinase in astroglial cells. *Mol Brain Res 74*:182–189.

Schwachtgen JL, Campbell CJ, Braddock M (2000): Full promoter sequence of human early growth response factor-1 (Egr-1), demonstration of a fifth functional serum response element. *DNA Seq 10*:429–432.

Seitanidou T, Schneider-Maunoury S, Desmarquet C, Wilkinson DG, Charnay P (1997): Krox-20 is a key regulator of rhombomere-specific gene expression in the developing hindbrain. *Mech Dev 65*:31–42.

Sgambato V, Pages C, Rogard M, Besson JM, Caboche J (1998): Extracellular signal-regulated kinase (ERK): controls immediate early gene induction on corticostriatal stimulation. *J Neurosci 18*:8814–8825.

Sgambato V, Maurice N, Besson JM, Thierry AM, Deniau JM (1999): Effect of functional impairment of corticostriatal transmission on a cortically evoked expression of c-Fos and zif268 in the rat basal ganglia. *Neuroscience 93*:1313–1321.

Shirayama Y, Hashimoto K, Matsuki H, Tsunashima K, Iyo M, Higuchi T, Minabe Y (1999): Increased expression of zif268 mRNA in rat retrosplenial cortex following administration of phencyclidine. *Brain Res 839*:180–185.

Silverman ES, Du J, Williams AJ, Wadgaonkar R, Drazen JM, Collins T (1998): cAMP-responsive-element-binding-protein binding protein (CBP) and p300 are transcriptional co-activators of early growth response factor-1 (Egr-1). *Biochem J 336*:183–189.

Slade JP, Carter DA (2000): Cyclic expression of egr-1/NGFI-A in the rat anterior pituitary, a molecular signal for ovulation?. *J Neuroendocrinol 12*:671–676.

Spencer JA, Misra RP (1999): Expression of the SRF gene occurs through a Ras/Sp/SRF-mediated-mechanism in response to serum growth signals. *Oncogene 18*:7319–7327.

Steiner H, Gerfen CR (1999): Enkephalin regulates acute dopamine receptor antagonist-induced immediate-early gene expression in striatal neurons. *Neuroscience 88*:795–810.

Stevens B, Fields RD (2000): Response of Schwann cells to action potentials in development. *Science 287*:2267–2271.

Svenningsson P, Nomikos GG, Fredholm BB (1999): The stimulatory action and the development of tolerance to caffeine is associated with alterations in gene expression in specific brain regions. *J Neurosci 19*:4011–4022.

Sukhatme VP, Cao X, Chang LC, Tsai-Morris C-H, Stamenkovich D, Ferreira PCP, Cohen DR, Edwards SA, Shows TB, Curran T, Le Beau MM, Adamson ED (1988): A zinc finger-encoding gene coregulated with c-fos during growth and differentiation, and after cellular depolarization. *Cell 53*:37–43.

Swirnoff AH, Milbrandt J (1995): DNA-binding specificity of NGFI-A and related zinc finger transcription factors. *Mol Cell Biol 15*:2275–2287.

Tanaka M, Iijima N, Amaya F, Tamada Y, Ibata Y (1999): NGFI-A gene expression induced in the rat suprachiasmatic nucleus by photic stimulation: spread into hypothalamic periventricular somatostatin neurons and GABA receptor involvement. *Eur J Neurosci 11*:3178–3184.

Tourtellotte WG, Nagarajan R, Bartke A, Milbrandt J (2000): Functional compensation by Egr-4 in Egr-1-dependent luteinizing hormone regulation and Leydig cell steroidogenesis. *Mol Cell Biol 20*:5261–5268.

Traub RJ, Herdegen T, Gebhard GF (1992): Differential expression of c-FOS and c-JUN in two regions of the rat spinal cord following noxious distension. *Neurosci Lett 160*:121–125.

Tsai JC, Liu L, Guan J, Aird WC (2000): The Egr-1 gene is induced by epidermal growth factor in ECV304 cells and primary endothelial cells. *Am J Physiol 279*:1414–1424.

Tsai-Morris CH, Cao XM, Sukhatme VP (1988): Nucleotide 5′ flanking sequence and genomic structure of Egr-1, a murine mitogen inducible zinc finger encoding gene. *Nucleic Acids Res 16*:8835–8846.

Von der Kammer H, Mayhaus M, Albrecht C, Enderich J, Wegner M, Nitsch RM (1998): Muscarinic acetylcholine receptors activate expression of the EGR gene family of transcription factors. *J Biol Chem 273*:14538–14544.

Wang JQ (1998): Regulation of immediate early gene c-fos and zif/268 mRNA expression in rat striatum by metabotropic glutamate receptor. *Mol Brain Res 57*:46–53.

Waters CM, Hancock DC, Evan GI (1990): Identification and characterisation of the *egr*-1 gene product as an inducible, short-lived, nuclear phosphoprotein. *Oncogene 5*:669–674.

Wei F, Xu ZC, Qu Z, Milbrandt J, Zhuo M (2000): Role of EGR1 in hippocampal synaptic enhancement induced by tetanic stimulation and amputation. *J Cell Biol 149*:1325–1334.

Wernerson J, Johansson I, Larsson U, Minth-Worby C, Pahlman S, Andersson G (1998): Activated transcription of the human neuropeptide Y gene in differentiating SH-SY5Y neuroblastoma cells is dependent on transcription factors AP-1, AP-2alpha, and NGFI. *J Neurochem 70*:1887–1897.

Williams JM, Beckmann AM, Mason-Parker SE, Abraham WC, Wilce PA, Tate WP (2000): Sequential increase in Egr-1 and AP-1 DNA binding activity in the dentate gyrus following the induction of long-term potentiation. *Mol Brain Res 77*:258–266.

Worley PF, Christy BA, Nakabeppu Y, Bhat RV, Cole AJ, Baraban JM (1991): Constitutive expression of zif268 in neocortex is regulated by synaptic activity. *Proc Natl Acad Sci USA 88*:5106–5110.

Worley PF, Bhat RV, Baraban JM, Erickson CA, McNaughton BL, Barnes CA (1993): Thresholds for synaptic activation of transcription factors in hippocampus: correlation with long-term enhancement. *J Neurosci 13*:4776–4786.

Xiao S, Matsui K, Fine A, Zhu B, Marshak-Rothstein A, Widom RL, Ju ST (1999): FasL promoter activation by IL-2 through SP1 and NFAT but not Egr-2 and Egr-3. *Eur J Immunol 29*:3456–3465.

Zhang F, Vanduffel W, Schiffmann SN, Mailleux P, Arckens L, Vandesande F, Orban GA, Vanderhaeghen J-J (1995): Decrease of zif-268 and c-fos and increase of c-jun mRNA in the cat areas 17, 18 and 19 following complete visual deafferentation. *Eur J Neurosci 7*:1292–1296.

Zipfel PF, Decker EL, Holst C, Sherka C (1997): The human zinc finger protein EGR-4 acts as autoregulatory transcriptional repressor. *Biochim Biophys Acta 1354*:134–144.

CHAPTER XIII

CREB, plasticity and memory

SHEENA A. JOSSELYN, SATOSHI KIDA, SANDRA PEÑA DE ORTIZ AND ALCINO J. SILVA

1. CREB AND TRANSCRIPTION

1.1. THE MULTIGENE CREB FAMILY

cAMP Responsive Element Binding Protein (CREB) is a member of a family (CREB/ATF) of structurally related transcription factors that bind to promoter CRE sites. In mammals, at least three genes encode the CREB-like proteins, *CREB*, *CREM* (cAMP Response Element Modulator) and *ATF-1* (Activating Transcription Factor) (Hoeffler et al., 1988; Rehfuss et al., 1991; Foulkes et al., 1992). These three genes share a high degree of sequence homology and are well conserved throughout evolution. The CREB-like proteins share many structural and functional polymorphisms, and are expressed in a wide range of tissues and cell types.

1.2. STRUCTURAL FEATURES

Transcription factors from the CREB family generally contain three key domains that mediate transcriptional activation, DNA binding and dimerization. The transcription activation domain is located on the N-terminal part of CREB and CREM and contains glutamine-rich domains (Q domain) that flank a cluster of phosphorylation sites (P-box, also referred to as the kinase-inducible domain (KID)) by which various kinases regulate the transactivational potential of CREB. DNA binding is mediated by a region rich in arginine and lysine residues (basic region), while dimerization is mediated by an adjacent leucine zipper domain (bZIP) (Kerppola and Curran, 1995). Most of the sequence homology amongst the different members of the CREB family of proteins is restricted to the bZIP region (Hai et al., 1989). However, not all of the CREB-like proteins contain these three key domains. For example, an isoform of the CREM gene, the Inducible cAMP Early Repressor (ICER), contains only the bZIP and the DNA binding domains (Molina et al., 1993).

1.3. CREB/CREM ISOFORMS

The human and mouse CREB gene is composed of 11 exons (Hoeffler et al., 1990; Waeber et al., 1991; Cole et al., 1992). Three main transcriptional activators, α (Gonzalez et al., 1989), β (Blendy et al., 1996) and δ (Yamamoto et al., 1990) are generated from the CREB gene by alternative splicing. In addition to these transcriptional activators, the CREB family also includes repressors of transcription. For example, the CREM gene codes at least four

Handbook of Chemical Neuroanatomy Vol. 19: Immediate Early Genes and Inducible Transcription Factors in Mapping of the Central Nervous System Function and Dysfunction
L. Kaczmarek and H.A. Robertson, editors

isoforms that repress CRE-dependent transcription: the CREM α, β and γ proteins as well as ICER (Foulkes et al., 1991; Molina et al., 1993). The diversity of CREB isoforms is produced by alternative splicing, start sites (e.g., CREBβ) and even alternate promoters (e.g., ICER) (Molina et al., 1993).

In *Drosophila* and *Aplysia*, the CREB gene produces several isoforms by alternative splicing, some of which act as transcriptional activators and others as transcriptional repressors. Similar to the mammalian CREM gene, the *Drosophila* CREB gene expresses both transcriptional activators and repressors. The activator (dCREB2-a) contains the three key protein domains described above, while the repressor (dCREB2-b) contains a P-box but lacks the glutamine-rich region, which is required for the transactivation of CREB (Yin et al., 1994). The *Aplysia* CREB1 gene expresses three isoforms: ApCREB1a, ApCREB1b and ApCREB1c (Bartsch et al., 1995). ApCREB1a functions as a transcriptional activator, similar to mammalian CREB. ApCREB1b contains only the bZIP domain and lacks the N-terminal transcription activation domain of ApCREB1a, and functions as a repressor. ApCREB1c is a truncated protein that lacks a nuclear localization signal.

1.4. DNA BINDING

The CRE site in the somatostatin promoter (5′-TGACGTCA-3′) was the first CREB-binding site described (Montminy and Bilezikjian, 1987). These palindromic consensus CRE sequences are typically located 100 nucleotides upstream from the TATA box in the promoter regions of the target genes (Comb et al., 1986; Montminy et al., 1986; Short et al., 1986). CREB binds as a dimer to the CRE sites with an affinity of approximately 1 to 2 nM (Richards et al., 1996). Promoters may include more than one copy of the CRE sequence, and there may be considerable sequence variability between functional CRE sites. In fact, the exact nucleotide sequence of these CRE sites may affect CREB binding. Adding to this variability, and providing a further mechanism of transcriptional regulation, are the different transcriptional efficiencies of the various members of the CREB/ATF family (these vary by 10–20 fold).

There are two categories of factors that recognize CRE sites: those that form dimers with CREB and those that do not (Kerppola and Curran, 1995; Shaywitz and Greenberg, 1999). The first group of factors that are able to dimerize with CREB includes CREB, CREM and ATF-1. CREM and ATF-1 recognize, and bind to, CRE sites as homodimers or as heterodimers with CREB (Hai and Curran, 1991; Loriaux et al., 1994a), perhaps owing to a highly conserved leucine zipper region (Foulkes et al., 1991; Hoeffler et al., 1991; Hurst et al., 1991). Factors that cannot dimerize with CREB include c-Jun, some of the other ATFs, such as ATF-4, and members of the CAAT/enhancer binding protein (C/EBP) gene family (Yun et al., 1990; Hai and Curran, 1991; Hummler et al., 1994). Heterodimerization may regulate CREB function by changing the affinity of these factors for the DNA sequences to which they normally bind. Moreover, this heterodimerization may cross-link distinct intracellular signaling systems, allowing a rich and complex regulation of gene function.

1.5. TRANSCRIPTIONAL ACTIVATION

The transcriptional activity of CREB is regulated by phosphorylation of Ser133 in the P-box or KID of the protein (Gonzalez and Montminy, 1989). This domain includes several consensus phosphorylation sites for a variety of kinases, including protein kinase A (PKA), protein kinase C (PKC), casein kinases (CKII), calmodulin kinases (CaMK), glycogen synthase kinase-3 (GSK-3), p34^{cdc2}, p70^{s6k}, mitogen-activated p90 rsk, MAP kinase/extracellular signal-regulated kinase (ERK1/2), stress-regulated mitogen-activated protein kinase-2 and Akt/PKB (Montminy et al., 1986; Gonzalez and Montminy, 1989; Gonzalez et al., 1989; Yamamoto et al., 1990; Dash et al., 1991; Sheng et al., 1991; Fiol et al., 1994; Tan et al., 1996; Xing et al., 1996; Bullock and Habener, 1998; De Cesare et al., 1998; Deak et al., 1998) that may either increase or decrease the transcriptional activity of CREB (Brindle and Montminy, 1992; Sassone-Corsi, 1995). Thus, although CREB was initially identified as a mediator of the cAMP pathway, the KID domain of CREB may be phosphorylated by kinases in the calcium/calmodulin and/or growth factor pathways. An illustrative example of each of these distinct pathways that converge on CREB, namely cAMP, calcium/calmodulin and growth factors, will be discussed briefly.

Increases in the level of intracellular cAMP produced by forskolin administration (Deisseroth et al., 1998; Impey et al., 1998a; Sheng et al., 1990) or by activation of adenylyl cyclase by transmembrane receptors, such as the D_1 dopaminergic receptor (Liu and Graybiel, 1996), for example, activates PKA by dissociating the regulatory (R) from the catalytic (C) subunits. The C subunits of PKA passively translocate to the nucleus where they may phosphorylate CREB at Ser133 to induce transcription (Bacskai et al., 1993; Hagiwara et al., 1993).

A second pathway that leads to CREB activation is mediated by increases in intracellular calcium (Ca^{2+}). Driven by the activation of synaptic *N*-methyl-D-aspartate receptors (NMDARs) and L-type Ca^{2+} channels, the increase in intracellular Ca^{2+} may lead to phosphorylation of CREB via the CaMK family of serine/threonine kinases. Increases in Ca^{2+} in the nucleus do not seem to be sufficient for CREB phosphorylation in response to synaptic signals (Deisseroth et al., 1996). Instead, increased Ca^{2+} levels located within 1–2 μm of the cell membrane seem to be crucial, suggesting that a calcium sensor (e.g., calmodulin (CaM)) associated with synaptic membranes may trigger a cascade of events leading to the activation of CREB in the nucleus (Deisseroth et al., 1996). Thus, increases in intracellular Ca^{2+} may increase the concentration of Ca^{2+}/CaM complex, which binds to, and activates, CaMKs that, in turn, phosphorylate CREB.

Phosphopeptide mapping indicates that CaMKI, CaMKII and CaMKIV all phosphorylate CREB at Ser133 in vitro (Dash et al., 1991; Sheng et al., 1991; Enslen et al., 1994; Matthews et al., 1994; Sun et al., 1994). However, CaMKII, in addition to phosphorylating CREB at Ser133, also phosphorylates CREB at Ser142, which inactivates the transcriptional activating properties of CREB by preventing dimerization but not DNA binding (Parker et al., 1998; Wu and McMurray, 2001). Furthermore, CaMKII-induced inhibition of CREB transcriptional activity overshadows the stimulatory effects of PKA (Gonzalez and Montminy, 1989) and CaMKIV (Sheng and Greenberg, 1990; Sun et al., 1996) on CREB.

CaMKIV may phosphorylate CREB at Ser133 following membrane depolarization in neuronal cells (Bito et al., 1996). Cotransfection of constitutively active CaMKIV drives CREB-dependent gene expression (Enslen et al., 1994; Matthews et al., 1994; Sun et al., 1994) while interfering with CAMKIV function inhibits membrane depolarization-induced Ser133 phosphorylation (Bito et al., 1996). Furthermore, CaMKIV-deficient mice show impairments in inducible CREB phosphorylation and decreases in the expression of c-Fos, an immediate-early gene that has CRE sequences in the promoter region (Ho et al., 2000). Therefore, it may be that the Ca^{2+} signal generated at the synapse is conveyed to the nucleus by transport of calmodulin across the nuclear membrane and activation of CREB by CaMKs (Deisseroth et al., 1996).

However, a recent study suggests that CREB may be phosphorylated in the nucleus even when protein import into the nucleus is blocked with wheat-germ agglutinin (Hardingham et al., 2001). This suggests that Ca^{2+} alone may carry the signal from dendrites to the nucleus. Therefore, waves of Ca^{2+} from the site of entry at the synapse may be propagated and amplified by intracellular stores of Ca^{2+} into the nucleus to activate CREB (Hardingham et al., 1997, 2001).

A third pathway by which CREB may be activated is via a cascade of kinase activity initiated by nerve growth factor (NGF). NGF stimulation activates NGF receptors (tyrosine kinase receptor, Trk receptors) that stimulate guanine-nucleotide exchange factors (GEF) to activate Ras, a small G protein. Activated Ras, in turn, stimulates the serine/threonine kinase, Raf, that triggers activation of MEK, and its targets, the ERK1/2 members of the MAPK family (Blenis et al., 1991). One downstream substrate of the Ras/ERK pathway is a 90 kDa ribosomal S-6 kinase-2 (RSK-2). Upon activation, both ERKs and RSKs translocate to the nucleus where they may phosphorylate CREB at Ser133 (Chen et al., 1992; Xing et al., 1996; Finkbeiner et al., 1997).

The functional importance of these different signaling cascades that culminate in the phosphorylation of CREB at Ser133 is not entirely clear. The complexity of the pathways upstream from CREB may allow for tight, fine-tuned regulation of CRE-mediated transcription. There is evidence for substantial cross-talk between the pathways that converge on CREB. For instance, CaMKIV produces a wave of CREB phosphorylation with a rapid onset and a rapid offset whereas the Ras–ERK–RSK2 pathway promotes a slow phase of CREB phosphorylation (Wu et al., 2001a). The distinct kinetic properties of the upstream pathways may allow CREB to compute information regarding the exact nature of synaptic stimuli. Perhaps this complexity allows for specific stimuli to be translated into specific patterns of gene expression.

Phosphorylation at Ser133 is essential for the activation of CREB (Yamamoto et al., 1990; Lee et al., 1993). Although CREB proteins bind to CRE sites as dimers, phosphorylation of both partners does not seem to be required for transcriptional activation. Nevertheless, dimers in which both partners are phosphorylated (at Ser133) are more active than hemiphosphorylated dimers (Loriaux et al., 1994b). Thus, the degree of phosphorylation may be another avenue to control CREB-dependent transcription.

Phosphorylation of CREB at Ser133 does not have an appreciable impact on the structure of CREB (Richards et al., 1996), but promotes the phosphorylation-dependent interaction with the KIX domain of the CREB Binding Protein (CBP) or its close relative p300 (Chrivia et al., 1993). CBP and p300 are ubiquitously expressed coactivator proteins that may physically link phosphorylated CREB with the basal transcriptional complex (Kwok et al., 1994; Nakajima et al., 1997a,b). Overexpression of CBP enhances stimulus-induced transcription of a CRE-reporter gene, an effect that depends on the phosphorylation of CREB at Ser133 (Kwok et al., 1994). Conversely, inhibiting the function of CBP, or the formation of a CREB–CBP complex, blocks CREB-mediated transcription (Arias et al., 1994; Hu et al., 1999). CBP may facilitate transcription by recruiting RNA polymerase II to the transcription machinery through an interaction with RNA helicase A (Kwok et al., 1994; Swope et al., 1996; Nakajima et al., 1997a,b). In addition, both CBP and p300 may also possess intrinsic histone acetyltransferase (HAT) activity that may facilitate access of the basal transcription factors to the core promoter region by decondensing the chromatin (Bannister and Kouzarides, 1995, 1996; Ogryzko et al., 1996; Kouzarides, 1999).

Although binding of phosphorylated CREB to CBP is required for transcription, the recruitment of CBP does not appear to be sufficient for transcription. In order for transcription to be initiated, it seems that CBP, itself, must also be activated (Chawla et al., 1998; Hu et al., 1999). CBP may be phosphorylated via the Ca^{2+} signal transduction pathway, possibly through a CaMKIV mechanism (Chrivia et al., 1993; Kwok et al., 1994). However, the ERK1/2 cascade does not seem to phosphorylate CBP as activation of the ERK1/2 pathway alone (by electrical activity in the presence of inhibitors of CaM kinases, selective stimulation of ERK1/2 with growth factors, or by genetic means) leads to CREB phosphorylation at Ser133 without appreciable induction of CREB-dependent transcription (Sheng et al., 1988; Bonni et al., 1995; Chawla et al., 1998; Hardingham et al., 1999; Hu et al., 1999). It seems, therefore, that CREB-mediated transcription requires two activation events (CREB and CBP).

CBP and p300 are global coactivators that interact with a number of other transcription factors, including AP-1, NF-κB, c-Fos, c-Jun, and transcriptional activator sites such as phorbol ester elements (TREs) and Serum-Responsive Elements (SREs) (Arias et al., 1994; Nordheim, 1994; Kamei et al., 1996). Thus, it may be that competition for the availability of CBP may also regulate the transcriptional responses mediated by CREB (Kamei et al., 1996).

Surprisingly, a recent report indicates that the activation of CREM is not always phosphorylation-dependent. ACT (activator of CREM in testis, found exclusively in the male germ cell) is a tissue-specific coactivator that interacts with CREM independent of its phosphorylation state (Fimia et al., 1999). This finding suggests that there may be another pathway in the regulation of CRE-mediated transcription, at least in this tissue.

1.6. TRANSCRIPTIONAL REPRESSION

Just as phosphorylation of Ser133 seems to be critical for activation of CREB, dephosphorylation of this residue is important for inactivation of CREB. As with all other phosphoproteins, therefore, the level of CREB phosphorylation at Ser133 reflects a balance between the oppositional actions of kinases and phosphatases, such as protein phosphatase 1 (PP-1 and PP-2) (Hagiwara et al., 1992). For example, dephosphorylation of CREB at Ser133 may be initiated by the activation of calcineurin (PP-2B) by the Ca^{2+}–CaM pathway. Calcineurin may then activate the nuclear phosphatase PP-1 that goes on to dephosphorylate CREB (Bito et al., 1996).

In addition to dephosphorylation, the transcriptional activity of CREB may also be actively suppressed by transcriptional repressors. For example, the CREM α, β and γ repressors lack the glutamine-rich transactivating domains, but seem to bind CRE sites normally (Foulkes et al., 1991; Laoide et al., 1993). Phosphorylated CREB/CREMα complexes activate transcription, thus showing that repression by these factors is not mediated by heterodimerization with activators (Loriaux et al., 1994a). Instead, repressor dimers may compete with CREB activators for CRE sites. As the repressors are unable to interact with the basal transcription machinery (they lack Q-domains; but see above (Fimia et al., 1999)), the result may be a silencing of CRE-containing promoters.

Furthermore, the 3′ end of the CREM gene contains an alternative promoter that codes for ICER, a strong CREB repressor. The ICER promoter contains CRE sites, and, therefore, is induced by CREB/ATF activation (Molina et al., 1993). It is possible that activation of CREB results in the transcription of repressor isoforms, such as ICER. The accumulation of repressors could, in turn, lead to the eventual repression of CREB-dependent transcription. This regulatory feedback may have an impact on a number of biological functions, including plasticity.

Thus, the transcriptional activity of CREB seems to be determined by complex interactions between kinases and phosphatases and active transcriptional repression systems. These competing systems, furthermore, may allow for fine-tuning of CREB-mediated transcriptional responses.

2. PLASTICITY AND MEMORY

2.1. MEMORY AND PROTEIN SYNTHESIS

There is extensive evidence from a variety of animal model systems showing that protein synthesis during, or shortly after, training is essential for the formation of long-term memory (LTM) (Davis and Squire, 1984; Matthies, 1989). For example, systemic administration of the protein synthesis inhibitor anisomycin, before or immediately after training, blocks LTM (typically measured 24 h following training) but not short-term memory (STM; typically measured 30 min to 2 h following training) for conditioned fear (Abel et al., 1997; Bourtchouladze et al., 1998). A similar pattern of results is observed following infusion of anisomycin icv (Schafe et al., 1999) or directly into the amygdala (Gewirtz et al., 1998; Schafe and LeDoux, 2000), a neural structure known to be critical for the acquisition of fear conditioning (Davis, 1992; LeDoux, 1992; Fanselow and Kim, 1994; Fanselow and LeDoux, 1999).

Several lines of evidence show that CREB is one of the transcription factors regulating the synthesis of new proteins necessary for the formation of LTM. Insights into the biochemistry of CREB activity allow for the design of experiments to examine the mnemonic effects of both increases and decreases of CREB levels and function. The specificity and consistency of the results obtained by many laboratories using diverse model systems establish that CREB may be a key molecular player in the cellular events underlying memory formation.

2.2. MASSED AND SPACED TRAINING SCHEDULES AND MEMORY

In general, memory retention is stronger following training with multiple trials, than with one single trial (Bugelski, 1962; Cooper and Pantle, 1967; Zacks, 1969). Not only is the number of training trials an important determinant of memory retention but so too is the distribution of training trials over time. Spaced training (training in which the trials are presented with intervening rest intervals) is generally more effective than massed training (the same number of training trials presented with no or short intervening rest intervals) in producing strong LTM for a variety of memory tasks (Ewing et al., 1985; Rescorla, 1988). In species ranging from *Aplysia* (Carew et al., 1972; Pinsker et al., 1973; Frost et al., 1985; Cleary et al., 1998; Mauelshagen et al., 1998), *Drosophila* (Tully et al., 1994), *Chasmagnathus* crab (Freudenthal et al., 1998), mouse (Kogan et al., 1996), rat (Fanselow and Tighe, 1988; Barela, 1999; Josselyn et al., 2001) to human (Ebbinghaus, 1885), spaced training is needed to produce maximal LTM. Thus, spaced training is more likely to induce LTM than massed training. Extensive evidence from different species and behavioral tasks indicate that the levels of CREB play a key role in this trial-spacing effect.

3. MEMORY: THE ROLE OF CREB

One of the fundamental results that emerged from studies with protein synthesis inhibitors is that LTM, but not STM, requires protein synthesis following training. STM is thought to involve transient changes in synaptic strength, perhaps mediated by covalent modifications of preexisting proteins. By contrast, LTM requires both transcription and translation of genes, and is proposed to involve growth and restructuring of new synapses. CREB is thought to be one of the factors necessary for initiating the transcription of proteins required for LTM in a variety of species.

3.1. CREB AND ELECTROPHYSIOLOGICAL STUDIES OF LONG-TERM PLASTICITY IN *APLYSIA*

Tactile or electrical stimulation of the siphon of the marine mollusc *Aplysia* produces a defensive reflex in which the siphon and gill are withdrawn. If an electric shock is applied to the tail of *Aplysia* the subsequent reaction to siphon stimulation is increased, or sensitized. A single tail shock produces short-term sensitization of the withdrawal reflex that lasts several minutes, and does not require protein synthesis (Pinkser et al., 1970; Carew et al., 1971; Kandel et al., 1981). However, repeated intermittent (spaced) shocks to the tail produce long-term sensitization of the withdrawal reflex that lasts many hours and requires the synthesis of new proteins (Pinsker et al., 1973; Kandel and Schwartz, 1982; Frost et al., 1985; Goelet et al., 1986; Bailey et al., 1992; Cleary et al., 1998). Behavioral sensitization in *Aplysia* is mediated by facilitation of synaptic transmission between sensory (responding to the sensitizing stimulus) and motor neurons (mediating the withdrawal response) (Frost et al., 1985; Byrne, 1987). This monosynaptic connection is enhanced for a period of minutes following a single tail shock and for longer than 24 h following repeated spaced shocks (Walters et al., 1983; Buonomano and Byrne, 1990; Mercer et al., 1991; Cleary et al., 1998).

Long-term facilitation (LTF), a stable enhancement of synaptic function with properties that closely mirror behavioral long-term sensitization, may be observed in a co-culture preparation of sensory and motor neurons. One pulse of serotonin, a transmitter released by the sensitizing tail stimulus (Glanzman et al., 1989), produces short-term facilitation (STF) lasting only minutes, whereas five spaced pulses of serotonin produces LTF lasting longer than 24 h (Carew and Kandel, 1973; Walters et al., 1983; Montarolo et al., 1986; Rayport and Schacher, 1986; Sweatt and Kandel, 1989; Mercer et al., 1991; Emptage and Carew, 1993; Mauelshagen et al., 1996; Zhang et al., 1997). Similar to long-term behavioral sensitization, LTF is dependent on protein synthesis (Montarolo et al., 1986) whereas STF is not (Montarolo et al., 1986; Dash et al., 1990).

Thus, both STF and LTF may be triggered by serotonin, suggesting that this neurotransmitter may have distinct downstream actions that differentially mediate the formation of STF or LTF. Serotonin activates G protein receptors that are positively coupled to adenylyl cyclase. This increase in intracellular cAMP levels transiently activates cytoplasmic protein kinases, including PKA and the diacylglycerol–protein kinase C (DAG–PKC) system. This kinase activation, in turn, is thought to covalently modify (phosphorylate) a number of target proteins. One possibility is that this cascade culminates in the closing of K^+ channels, prolonging the action potential and increasing the influx of Ca^{2+}, leading to the augmentation of transmitter release, and culminating in STF (Castellucci et al., 1980; Sugita et al., 1992; Byrne et al., 1993).

A similar mechanism, however, cannot account for LTF. Although intracellular cAMP levels rise dramatically shortly after serotonin treatment, this increase is not present 24 h later, at a time when LTF is observed (Bernier et al., 1982). Instead, it may be that initial increases in cAMP activate PKA and MAP kinase cascades, triggering CREB-dependent transcription of genes whose products are required for LTF (Bacskai et al., 1993; Martin et al., 1997).

The first study to suggest that CREB is required for memory formation or plasticity was performed in *Aplysia* cultured neurons (Dash et al., 1990). LTF, but not STF, was blocked by injection of oligonucleotides with CRE sequences into cultured sensory neurons (Dash et al., 1990). Presumably, the CRE–oligonucleotides trap the CREB proteins needed for the transcriptional activation of genes that ultimately mediate LTF (Kaang et al., 1993; Alberini et al., 1994). Moreover, a similar injection of a reporter gene driven by a CRE-containing promoter shows that repeated pulses of serotonin that produce LTF also trigger CREB activation, while a single pulse of serotonin that does not produce LTF similarly does not trigger CREB activation (Kaang et al., 1993).

Cloning studies identified several CREB-like proteins in *Aplysia* (Bartsch et al., 1995). The *Aplysia* CREB1 gene encodes three proteins (ApCREB1a, ApCREB1b and ApCREB1c) by alternative splicing. CREB1a shares structural and functional homology with CREB transactivators in mammals. Injection of antibodies or antisense against CREB1a blocks the LTF normally induced by five pulses of serotonin. These treatments, however, have no effect on STF (Bartsch et al., 1995). Conversely, injection of phosphorylated recombinant ApCREB1a protein alone induces LTF and prevents further facilitation normally induced by additional pulses of serotonin (Bartsch et al., 1995). Thus, ApCREB1a seems both necessary and sufficient to induce LTF in cultured neurons (Bartsch et al., 1995).

On the other hand, ApCREB1b lacks a P-box and, therefore, cannot be phosphorylated by PKA, CaMK or PKC. Thus, ApCREB1b resembles mammalian ICER both structurally and functionally. CREB1b forms homodimers as well as heterodimers with CREB1a and represses CREB1a-mediated transactivation and LTF (Bartsch et al., 1995). Injection of antisense oligonucleotides specifically targeted to CREB1b lowers the threshold for producing LTF, such that a single pulse of serotonin (rather than five spaced pulses) now induces LTF. Thus, removing the inhibition presumably produced by CREB1b enhances formation of LTF (Bartsch et al., 1995).

Aplysia CREB1c is a cytoplasmic protein that does not contain a nuclear localization signal and is unable to bind DNA or form dimers. Injection of unphosphorylated CREB1c followed by a single pulse of serotonin enhances STF and induces LTF. Therefore, this cytoplasmic form of CREB may play an important role not only in the modulation of CREB-mediated transcription necessary for LTF but also in STF.

Aplysia CREB2 is structurally unrelated to *Aplysia* CREB1 but shares some homology with mouse ATF-4 (Hai et al., 1989). Pairing a single pulse of serotonin that normally induces STF (but not LTF) with antibodies against ApCREB2 produces LTF (Bartsch et al., 1995). Transfection studies in F9 cells show that ApCREB2 represses the function of ApCREB1 (a CREB activator). Thus, it may be that ApCREB2 inhibits LTF by forming heterodimers with ApCREB1, thereby masking the activation domains of ApCREB1 and inhibiting its function. However, other experiments also show that, under certain circumstances, ApCREB2 may function as an activator in F9 cells (Bartsch et al., 1995). Thus, when co-transfected with a PKA construct, ApCREB2 (the nominal repressor) and ApCREB1 (the nominal activator) produce similar increases in the levels of a LacZ reporter gene under the regulation of a promoter with five CRE sites. ATF-4, which shares homology with ApCREB2 (Hai et al., 1989; Bartsch et al., 1995), may also function as a transcriptional repressor (Karpinski et al., 1992) or an activator (Bartsch et al., 1995) depending on the conditions used. Thus, the precise mechanism underlying the effects that ApCREB2 exerts on LTF is unclear.

3.2. CREB AND MEMORY IN *DROSOPHILA*

Learning and memory in *Drosophila* may be assessed using a Pavlovian olfactory test in which flies learn to avoid a previously neutral odor that was paired with shock (conditioning stimulus +; CS+) in favor of another odor that was not paired with shock (CS−) in a T-maze (Tully, 1991). Several temporally distinct phases of memory are identified using this test, including STM and LTM. Similar to behavioral sensitization and facilitation in *Aplysia*, spaced training in *Drosophila* induces LTM that is dependent on protein synthesis (Tully et al., 1994). Massed training, on the other hand, produces (protein synthesis independent) STM but no LTM (Tully et al., 1994). For example, 10 spaced training trials, but not 48 massed training trials conducted over the same span of time, produce robust LTM (Yin et al., 1995).

Memory has been studied using both forward and reverse genetics in *Drosophila* (Tully, 1991). Using a forward genetic approach, the progeny of flies that were treated with a mutagen were screened for learning and memory impairments. Two mutants identified by this screen were subsequently determined to have disruptions in Ca^{2+}/CaM-stimulated adenylate cyclase (*rutabaga*) and in cAMP-specific phosphodiesterase (*dunce*), both key enzymes in the regulation of intracellular levels of cAMP (Byers et al., 1981; Tully, 1991; Levin et al., 1992).

Using a reverse genetic approach, CREB function was disrupted in *Drosophila* by the transgenic expression of a CREB transcriptional repressor (Yin et al., 1994). The *Drosophila* dCREB2 gene encodes two isoforms, dCREB2a, a transcriptional activator, and dCREB2b, a transcriptional repressor. dCREB2b binds CRE sites, but lacks the two exons required for transcriptional activation, and represses CREB-mediated transcription in cell culture. To control the onset of CREB repression in the transgenic fly, dCREB2b was placed under the control of a promoter that is activated by temperature increases (heat-shock promoter). Inducing dCREB2b prior to spaced training in the olfactory task completely disrupts LTM without affecting STM. The finding that STM is intact indicates that the overexpression of this CREB repressor did not disrupt learning or the perceptual responses to the odors used, shock reactivity, or motor performance necessary for this task. Mutating two amino acids that disrupt the dimerization domain of dCREB2b results in normal LTM, thus showing the specificity of the inhibition produced by the dominant negative CREB protein (Yin et al., 1994).

Multiple spaced training is required to produce maximal LTM in the olfactory task in normal *Drosophila* while massed training produces strong STM but no LTM. However, massed training alone is sufficient to produce maximal LTM if a CREB activator (dCREB2a) is overexpressed in transgenic flies prior to training. Furthermore, overexpression of this CREB activator produces robust LTM following only one training trial (Yin et al., 1995). Transgenic flies overexpressing a mutant activator, where Ser231 (similar to Ser133 of the mammalian CREB gene) was replaced by an Ala, do not show LTM after one training trial, indicating that phosphorylation of CREB is required for the enhancement of LTM (Yin et al., 1995). Together, these results show the importance of CREB in LTM formation in *Drosophila* and, furthermore, suggest that CREB may be a limiting component of this process.

3.3. PKA, CREB AND MEMORY IN HONEYBEES

The findings of the role of CREB in memory using *Drosophila* are in agreement with similar results from an associative olfactory conditioning task in honeybees (*Apis mellifera*). In this task, the proboscis extension reflex is conditioned by pairing an odor (the CS) with a sucrose reward (the US) (Bitterman et al., 1983; Menzel and Müller, 1996). A single trial produces activation of PKA in the antennal lobes and memory for the association, both of which decay over several hours (Hammer and Menzel, 1995; Menzel et al., 1996; Grunbaum and Müller, 1998). Multiple spaced training trials, on the other hand, produces prolonged activation of PKA in the antennal lobes and a prolonged memory (LTM that is protein-synthesis-dependent) (Müller, 1996, 2000; Grunbaum and Müller, 1998; Wüstenberg et al., 1998; Fiala et al., 1999).

Injection of antisense against the catalytic subunit of PKA into the brain of a bee or systemic injection of the PKA antagonist, RpBrcAMP, blocks LTM, but not STM (Fiala et al., 1999; Müller, 2000). These results suggest that PKA activation at the time of training is critical to LTM formation. Moreover, artificially prolonging PKA activation in the antennal lobes (by photorelease of cAMP) in combination with a single training trial is sufficient to induce LTM (Müller, 2000). Thus, similar to *Drosophila*, training that normally produces STM may produce LTM if CREB, or a pathway upstream of CREB, is artificially increased or activated.

3.4. CREB AND MEMORY IN SONG BIRDS

Young male zebra finch birds (*Taeniopygia guttata*) learn to sing from conspecifics during a song-sensitive period in development (Sakaguchi et al., 1999). This form of learning is sensitive to protein synthesis inhibitors (Chew et al., 1995) and triggers an increase in CREB phosphorylation in the Higher Vocal Center (HVC), a brain area that is critical for song learning (Sakaguchi et al., 1999). Neither exposure to songs from other species nor white noise increase the levels of phosphorylated CREB in the HVC, thus showing the specificity of this effect (Sakaguchi et al., 1999). Similar to mammalian CREB, zebra finch CREB (zCREB) contains a putative PKA phosphoacceptor site at Ser119 (homologous to Ser133 in mammals) and zCREB is readily phosphorylated at Ser119 by PKA in vitro. These results suggest that CREB activation may play a role in the production of long-lasting memory traces in the song system of zebra finches.

3.5. CREB AND MEMORY IN MAMMALS

The study of the role of CREB in mammalian memory began with the generation of a mouse in which the CREB gene was disrupted. A neomycin resistance (neo) gene was inserted into exon 2 of the CREB gene, which was thought to contain the translation initiation site for all CREB isoforms (Hummler et al., 1994). This neo insertion resulted in the loss of the two main CREB isoforms (α and δ) in the CREB$^{\alpha\delta-}$ mice (Hummler et al., 1994). However, the translation of a previously unknown CREB isoform (CREBβ) starts from exon 4, and consequently the insertion of neo gene into exon 2 did not disrupt this isoform. Instead, the usually low expression level of CREBβ, is upregulated in the CREB$^{\alpha\delta-}$ mutants (Blendy et al., 1996). The levels of CREM activator (τ) and repressor isoforms (α and β) are also increased in these mutants (Hummler et al., 1994). Despite this upregulation, the CREB$^{\alpha\delta-}$ mutation decreases CREB-dependent transcription in these mutants (Hummler et al., 1994; Blendy et al., 1996).

3.5.1. Fear conditioning

To determine whether the CREB$^{\alpha\delta-}$ mutation affects memory, mutant mice and wild-type (WT) littermate control mice were tested in a fear conditioning paradigm (Bourtchuladze et al., 1994; Kogan et al., 1996). Fear conditioning is a form of Pavlovian learning in which animals learn to fear a stimulus (CS; typically a context or discrete cue such as a tone) that previously has been paired with an aversive stimulus such as footshock (US) (Frankland et al., 1998; Fanselow and LeDoux, 1999; LeDoux, 2000). LTM, but not STM, for fear conditioning depends on the synthesis of new proteins (Abel et al., 1997; Schafe et al., 1999). Similarly, the CREB$^{\alpha\delta-}$ mutation disrupts LTM, but not STM, for both discrete cue and contextual fear conditioning.

A deficit in LTM is also found using a different paradigm to measure discrete cue fear conditioning, the fear-potentiated startle task (Falls et al., 2000). In the fear-potentiated startle paradigm, LTM is inferred from an increase in the amplitude of the acoustic startle response of mice when the startle reflex is elicited in the presence of a light (CS) that was previously paired with footshock (US) (Davis, 1992; Josselyn et al., 2001). Protein synthesis within the basolateral complex of the amygdala is critical for LTM of fear-potentiated startle (Gewirtz et al., 1998). CREB$^{\alpha\delta-}$ mutants show LTM deficits in this task (Falls et al., 2000).

Interestingly, overexpression of a CREB repressor (Ser133Ala mutation) in the forebrain does not affect memory to the same extent as the CREB$^{\alpha\delta-}$ mutation (Rammes et al., 2000). In one of the transgenic lines studied, a LTM deficit was observed following cued fear conditioning, even though synaptic plasticity in the amygdala (as assessed by long-term potentiation (LTP)) appeared unaffected. It could be that the milder phenotype observed in these transgenic mice is due to a milder decrease in CREB function, and/or to upregulation or compensation by other transcription factors. In contrast, recent studies with a transgenic mouse expressing an inducible CREB repressor (Ser133Ala), show that the induction of the CREB repressor produces profound LTM deficits in both contextual and discrete cue fear conditioning (Kida et al., in press).

Also important to note are the findings that CREB-mediated transcription is activated during fear-conditioning training that results in LTM. For example, transgenic mice with a β-galactosidase reporter construct under the regulation of a CRE-containing promoter (CRE-LacZ), show that fear-conditioning training induces CRE-mediated transcription in several brain regions. Contextual, but not discrete cue (tone), fear training induces LacZ expression in the hippocampus, while both forms of fear training induce LacZ expression in the amygdala (Impey et al., 1998b). Together, these findings suggest that although CREB seems to be critical for LTM, not all genetic lesions of CREB produce pronounced effects on LTM. Understanding why and how these different genetic manipulations of CREB affect memory will help to reveal the intricate relation between the regulation of CREB-mediated transcription and memory formation.

3.5.2. Social behavior

Besides a role in LTM for fear conditioning, CREB has also been implicated in other forms of memory, such as social memory. The response demands of the tasks used to assess social memory are different from those required by fear conditioning. Nevertheless, both social memory and fear conditioning memory may involve some overlapping brain regions (such as the hippocampus) (Winocur, 1990; Bunsey and Eichenbaum, 1995). In the social transmission of food preference task rodents develop a preference for foods recently smelled on the breath of other rodents (Galef and Wigmore, 1983; Strupp and Levitsky, 1984; Galef et al., 1988). CREB$^{\alpha\delta-}$ mutant mice show normal immediate memory but deficient LTM in this task (Kogan et al., 1996).

In another type of social memory, social recognition memory, the ability of rodents to remember conspecifics is evaluated (Thor and Holloway, 1982; Kogan et al., 2000). Social recognition is defined by a decrease in spontaneous investigation behaviors observed in a mouse re-exposed to a familiar conspecific. Lesions of the hippocampus disrupt longer-term social memory, but not memory assessed immediately following training (Kogan et al., 2000). Furthermore, protein synthesis inhibitors block LTM but not STM for social recognition in WT mice (Kogan et al., 2000). Similarly, CREB$^{\alpha\delta-}$ mice show intact STM but impaired LTM for social recognition (Kogan et al., 2000). Thus, in two types of social memory, CREB$^{\alpha\delta-}$ mice show LTM impairments but normal immediate or STM. The findings suggest that the CREB$^{\alpha\delta-}$ mutation specifically disrupts those processes required for the long-term retention of information, rather than its acquisition.

3.5.3. Spatial learning

In the hidden platform version of the Morris water maze (Morris, 1981), rodents learn to find a platform submerged in a pool of opaque water. In the visible platform version of this task, mice learn to find a marked escape platform. Hippocampal lesions disrupt performance on the hidden platform, but not the visible platform, version of the Morris water maze (Morris et al., 1982; Sutherland et al., 1982; Cho et al., 1999). CREB$^{\alpha\delta-}$ mice show a profound impairment in spatial learning and/or memory in the hidden version of this task (Bourtchuladze et al., 1994; Kogan et al., 1996). In contrast, learning in the visible platform version of the water maze is intact in the CREB$^{\alpha\delta-}$ mice, again showing the behavioral specificity of this mutation.

In agreement with these findings, injections directly into the dorsal hippocampus of antisense against CREB mRNA disrupt learning and/or memory of rats in the hidden platform version of the water maze (Guzowski and McGaugh, 1997). In contrast, injection of antisense 2 days following the completion of training does not affect subsequent performance in the water maze, indicating that decreasing CREB function does not disrupt expression of the memory. This finding is consistent with results from *Drosophila* and *Aplysia* suggesting that the critical period for CREB function in LTM is during, or shortly after, training.

Injection of CREB antisense oligonucleotides into the hippocampus first decreased the levels of α and δ CREB within 6 h of injection, but increased the levels of these isoforms 14 h later. This rebound effect may reflect mechanisms similar to those responsible for increases in the levels of CREBβ and of CREM isoforms in CREB$^{\alpha\delta-}$ mutants. Surprisingly, training conducted at a time when CREB levels were lower (within 6 h of injection) or higher (within 20 h of injection) relative to normal levels produced similar deficits. It could be, therefore, that unlike Pavlovian conditioning in *Drosophila*, increased levels of CREB do not facilitate the storage of complex information, such as that accumulated in spatial tasks.

3.5.4. Conditioned taste aversion

CREB is also required for the development of a conditioned taste aversion (CTA). Pairing malaise or sickness (for example, induced by lithium chloride, LiCl) with exposure to a novel taste (for example, saccharin) may produce a strong and long-lasting aversion to the novel taste. Infusion of the protein synthesis inhibitor, anisomycin, directly into the amygdala during training blocks LTM for this aversion (Lamprecht and Dudai, 1996). Similar infusion of oligoncucleotides against CREB into the amygdala decreases CTA memory measured 3–5 days, but not 2 h, following training (Lamprecht et al., 1997). Infusions of CREB sense into the amygdala or CREB antisense oligonucleotides into the basal ganglia do not disrupt LTM for the taste aversion. Furthermore, CREB antisense oligonucleotides produce no effect on memory retrieval if infused into the amygdala before the memory test, rather than during training. Interesting, the CREB$^{\alpha\delta-}$ mice also show impaired LTM for CTA (Josselyn et al., 1999).

The finding that CREB in the amygdala is important for the development of CTA memory is supported by the observation of an increase in the levels of phosphorylated CREB (pCREB) in the lateral nucleus of the amygdala specifically following CTA training. Importantly, this CTA training does not induce increases in pCREB in other brain regions such as the insular or gustatory cortex. Furthermore, increases in pCREB are not observed following administration of either the CS (e.g., saccharin) or the US (e.g., LiCl) alone (Swank, 2000), attesting to the associative nature of the observed increase in pCREB. In addition, training in a similar appetitive task that leads to LTM also leads to increases in pCREB. For example, in the olfactory preference task an odor is paired with stroking in neonate rats. This pairing produces a reliable preference for the odor (tested 24 h later) and triggers an increase in pCREB in the olfactory bulb. Similar to the findings with CTA, presentation of the odor alone or stroking alone does not induce LTM for the olfactory preference and does not induce an increase in pCREB levels in the olfactory bulbs (McLean et al., 1999).

3.6. TRIAL SPACING EFFECT

Robust LTM for contextual fear conditioning may be produced by a single training trial in WT mice. Despite normal sensory perception, the same single trial produces only a transient (<60 min) memory in CREB$^{\alpha\delta-}$ mice (Bourtchuladze et al., 1994). Not even training that produces maximal LTM in WT mice (5 trials with 1 min intervals), compensates for the profound contextual fear amnesia of these mutants. However, two spaced trials with a 1-h intertrial interval (ITI), that in WT control mice does not produce higher levels of freezing than a single trial, nevertheless induces robust LTM in CREB$^{\alpha\delta-}$ mutants (Kogan et al., 1996). Similarly, a single 5-min interaction with another mouse is sufficient to trigger LTM for social transmission of food preference in normal mice, but not in the CREB$^{\alpha\delta-}$ mutants. In CREB$^{\alpha\delta-}$ mice, LTM for socially transmitted food preferences requires spaced training (two trials with a 1-h ITI) (Kogan et al., 1996).

Learning to find the hidden platform in the Morris water maze is gradual, and, to master the task mice must remember what is learned over days. As with the fear conditioning and social transmission of food preference tasks, increasing the ITI (more than 10 min and as long as 24 h) and doubling the amount of training (20 trials instead of 10) overcomes the LTM deficit of the CREB$^{\alpha\delta-}$ mutants in the water maze (Kogan et al., 1996). The finding that spaced training rescues the LTM deficits in CREB$^{\alpha\delta-}$ mutants in three different tasks indicates that the profound deficits in LTM observed following relatively more massed training may not be attributed to deficits in sensory, motor or motivational processes.

Together these results indicate that massed training induces STM, but less robust LTM in WT animals, and spaced training rescues the LTM deficit in CREB$^{\alpha\delta-}$ mutant mice. However, does the reverse hold true? Does increasing CREB levels rescue or facilitate LTM following massed training? A recent experiment addressed these questions. WT rats given massed fear conditioning training (4 CS (light)–US (shock) pairings with ITIs of 3, 5 or 10 s) show no or weak LTM, as measured by fear-potentiated startle, compared to rats given the same number of training trials presented in a spaced fashion (ITIs of 8 min) (Josselyn et al., 2001). However, increasing CREB levels specifically in the basolateral amygdala of these rats via viral vector-mediated gene transfer increases LTM following massed fear training. Thus, training that normally induces STM (but little LTM) produces robust LTM if CREB levels are increased in the amygdala. The enhancing effect of CREB overexpression on LTM formation depends on phosphorylation of the viral encoded CREB gene as similar overexpression of mCREB (in which Ala was substituted for Ser133) does not enhance LTM following massed training. Consistent with the key role that the amygdala occupies in fear conditioning, overexpression of CREB in regions surrounding the amygdala or directly into the caudate nucleus does not facilitate the formation of LTM following massed training. Furthermore, increasing CREB levels at the time of training (memory formation), rather than testing (retrieval), seems critical for the facilitatory effects of CREB on LTM.

There are striking parallels among the studies showing that CREB affects the training schedules required to produce LTM. First, multiple spaced applications of serotonin are needed to trigger LTF between *Aplysia* synapses, while a single application of serotonin induces STF. However, a single application of serotonin given together with the CREB activator, ApCREB1a, produces LTF. Second, *Drosophila* given massed training for an olfactory avoidance task show no LTM. However, massed training, or even a single training trial, produces robust LTM if a CREB activator is overexpressed in flies before training. Third, rats given massed fear training show no or weak LTM as assessed by fear-potentiated startle. However, massed training induces robust LTM in rats if CREB is overexpressed in the basolateral amygdala. Finally, the LTM deficits for a variety of memory tasks observed in the CREB$^{\alpha\delta-}$ mutant mice (with decreased levels of CREB) are overcome with spaced training. Together, these results from a range of species and memory tests suggest that animals with relatively lower levels of activator CREB require longer rest intervals between trials to produce LTM whereas animals with higher levels of activator CREB may exhibit LTM following massed, as well as spaced, trials. Flies, mice and rats with presumably more activator CREB require less training than flies and mice with less activator CREB (Yin et al., 1994, 1995; Josselyn et al., 2001). Therefore, in species ranging from *Aplysia*, mice, rats and flies, manipulations of CREB function seem to directly affect the amount and distribution of training required for the induction of LTM. The ability of CREB to affect the schedules of training required for memory formation seems to be highly conserved across species.

A 10-min ITI is optimal for olfactory learning in WT *Drosophila* (Yin et al., 1995) and an 8-min ITI seems best for fear conditioning in WT rats using fear-potentiated startle (Josselyn et al., 2001). Nevertheless, CREB$^{\alpha\delta-}$ mutants trained with 1-h ITIs perform better than those trained with 10-min ITIs, suggesting that 10 min may be just below the threshold for optimal memory induction in these mutant mice (Kogan et al., 1996). The timing of these events is important because it may provide hints regarding the nature of underlying mechanisms. It may take 3–8 min for synaptic activation to trigger maximal CREB activation (phosphorylation) (Moore et al., 1996). After committing the general transcriptional machinery to genes with CRE promoters, it may take CREB a few more minutes before another round of transcription may be initiated (akin to a refractory period). Additionally, it is possible that the longer intervals result in optimal inactivation of phosphatases (Bito et al., 1996; Liu and Graybiel, 1996) that control the phosphorylation and activation of CREB transcription factors.

3.7. DISSECTING THE ROLE OF CREB IN MEMORY WITH AN INDUCIBLE, BRAIN-SPECIFIC TRANSGENE

The findings obtained using the CREB$^{\alpha\delta-}$ mice indicate that CREB function is required for LTM. However, it is important to test the role of CREB in memory with alternative tools, especially those that allow temporal control over CREB function. Thus, our laboratory developed a transgenic mouse expressing a brain-specific and inducible CREB repressor (Kida et al., in press).

This inducible CREB repressor (CREB^{IR}) mouse overexpresses the αCREB isoform with a mutation of Ser133 to Ala (αCREB^{S133A}) that represses endogenous CREB function (Gonzalez and Montminy, 1989; Brindle and Montminy, 1992). The inducibility of the system is provided by fusing the mutant CREB to a ligand-binding domain (LBD) of a human estrogen receptor with a G521R mutation (LBD^{G521R}), the activity of which is regulated not by estrogen but by the synthetic ligand, tamoxifen (TAM) (Danielian et al., 1993; Logie and Stewart, 1995; Feil et al., 1996). Therefore, in the absence of the inducer TAM, the LBD^{G521R}–CREB^{S133A} fusion protein is inactive (Feil et al., 1996). However, administration of TAM activates this inducible CREB repressor fusion protein, allowing it to compete with endogenous CREB and disrupt CRE-mediated transcription. The CREB^{IR} construct represses

CRE-mediated transcription of a CRE-luciferase reporter in a model system (COS cells) in a TAM-dependent manner. In the transgenic mice, the inducible CREB repressor is under the control of the αCaMKII promoter, which is active in excitatory neurons of forebrain areas including the hippocampus, amygdala, neocortex and striatum (Mayford et al., 1996). Thus, administration of the inducer TAM decreases in CREB function in forebrain areas.

Parametric studies using this transgenic mouse show that administration of the inducer, TAM, 6 or 12 h but not 30 min prior to contextual fear training produces a deficit in LTM measured 24 h following training. Importantly, similar administration of TAM to WT littermate mice, regardless of time of administration, produces no effect. These results suggest that the $CREB^{IR}$ system has good temporal and inducible control. Moreover, the finding that $CREB^{IR}$ mice administered TAM 24 h before training show no subsequent deficit in LTM suggests that the disruption of CREB function using this system is reversible within 24 h. STM for contextual fear conditioning is not disrupted in these transgenic mice administered TAM prior to training. Furthermore, acutely disrupting CREB function (by administering TAM to $CREB^{IR}$ mice) before a retrieval test produces no effect, indicating that CREB is not critically involved in the retrieval of fear memories.

4. PKA, CREB AND LONG-TERM POTENTIATION

Long-term potentiation (LTP) is the most extensively studied candidate cellular plasticity mechanism thought to underlie memory (Bliss and Collingridge, 1993). LTP refers to a class of long-lasting enhancements in synaptic efficacy with properties expected of a memory mechanism (long-lasting, associative, reversible, etc.). A variety of different studies suggest that an LTP-like mechanism may be involved in memory formation (Barnes, 1995; Maren and Baudry, 1995; Mayford et al., 1996). LTP is not a single phenomenon; rather, there are various forms of LTP with distinct time courses and underlying biochemical mechanisms (Huang et al., 1996).

The best studied LTP occurs between CA3 and CA1 pyramidal neurons of the hippocampus and is sensitive to blockers of CaMKs (Bliss and Collingridge, 1993; Chapman, 2001). One stimulus train typically produces LTP that dissipates within 1 to 2 h (early LTP or E-LTP), and is insensitive to inhibitors of protein synthesis (Frey et al., 1988, 1993; Huang and Kandel, 1994). Late-LTP (L-LTP) lasts much longer than E-LTP (>7 h), generally requires more stimulus trains to induce, and is blocked by protein synthesis inhibitors (Frey et al., 1993; Huang and Kandel, 1994).

In addition, L-LTP is blocked by pharmacological agents that inhibit PKA (Frey et al., 1993; Huang and Kandel, 1994; Nguyen et al., 1994; Woo et al., 2000). Moreover, transgenic mice expressing an inhibitory form of the regulatory subunit of PKA (R(AB)), which have significantly reduced levels of hippocampal PKA activity (approximately 40–50% of basal activity), show deficits in the late phase of L-LTP, even though synaptic transmission and the early phase of LTP are normal (Abel et al., 1997). These findings are consistent with those showing that the R(AB) mice, as well as systemic (Abel et al., 1997; Bourtchouladze et al., 1998), icv (Bourtchouladze et al., 1998; Schafe et al., 1999) or intra-amygdala (Ding et al., 1998; Schafe et al., 2000) administration of pharmacological agents that inhibit PKA to WT animals impair LTM for fear conditioning. Together, these findings are consistent with a model that proposes that PKA is required for both L-LTP and LTM formation.

In contrast, administration of low levels of rolipram, a phosphodiesterase type IV inhibitor that increases cAMP levels, may selectively enhance the formation of LTP and LTM (Barad et al., 1998). In WT hippocampal slices E-LTP decays to baseline levels several hours after tetanus. However, with the addition of rolipram, this stimulation protocol produces L-LTP that persists for several hours. Similarly, rolipram enhances LTM in contextual fear conditioning without affecting STM. Thus it could be that increasing cAMP, which may increase PKA activation and ultimately CREB phosphorylation, may facilitate both LTP and LTM.

Not only is PKA critically involved in L-LTP, but CREB may also play an important role. Hippocampal slices from the CA1 region of CREB$^{\alpha\delta-}$ mutant mice show normal E-LTP but impaired L-LTP (A.J.S., in prep). A single bout of high-frequency stimulation (100 Hz for 1 s with a 250-μs stimulus) induces a potentiation that lasts several hours in WT mice. In contrast, the same stimulus triggers a smaller potentiation that dissipates within 90 min in CREB$^{\alpha\delta-}$ mutants. Thirty minutes following this stimulation, however, slices from CREB$^{\alpha\delta-}$ mutants are potentiated, consistent with normal STM in these mice (Bourtchuladze et al., 1994). Thus, two genetic (CREB$^{\alpha\delta-}$ mutation and R(AB) transgenics) and several pharmacological manipulations that affect the PKA/CREB pathway produce comparable effects on LTP and LTM, suggesting that this second messenger pathway is required for L-LTP and LTM formation.

The hypothesis that CREB is important for L-LTP is also supported by the observation that the levels of pCREB increase in response to L-LTP-inducing synaptic stimuli in hippocampal slices (Matthies et al., 1997; Lu et al., 1999), amygdala slices (Huang et al., 2000), dissociated hippocampal neurons in vitro (Bito et al., 1996) and in the hippocampus in vivo (Schulz et al., 1999; Davis et al., 2000). Increases in CRE-mediated gene expression are produced by L-LTP inducing stimuli in CRE-reporter mice (Impey et al., 1996). Importantly, the increases in pCREB levels are stimulus-specific. That is, increases in pCREB are not observed following stimuli that do not result in L-LTP (such as low-frequency stimulation of the perforant pathway) but are observed following stimuli that result in L-LTP (such as high-frequency stimulation of the perforant pathway) (Schulz et al., 1999).

Long-term depression (LTD), a use-dependent depression between synapses (Linden and Connor, 1995), may be observed in the cerebellum and is thought to mediate some forms of motor learning (Thompson, 1986). Together with LTP-like phenomena, LTD-like phenomena may modulate the storage capacity of neuronal networks, by fine-tuning synaptic weights (Malenka, 1994). Similar to LTP, a late-phase form of LTD is blocked by protein synthesis inhibitors (Ahn et al., 1999). Moreover, inhibition of CREB function with a dominant negative isoform (A-CREB) also inhibits L-LTD in cerebellar slices (Ahn et al., 1999).

In addition to PKA, CaM kinases are located upstream from CREB and may be critically involved synaptic plasticity such as LTP. Antisense oligonucleotides against the α and β splice variants of CaMKIV disrupt CREB phosphorylation and mutant mice with a targeted disruption of the CaMKIV gene show impaired CREB phosphorylation and CREB-dependent transcription of c-Fos (Ho et al., 2000). Furthermore, these mutant mice show impaired LTP (measured in hippocampal CA1 neurons) and LTD (measured in cerebellar neurons). Interestingly, these CaMKIV mutant mice show no LTM deficits, suggesting that other signaling pathways may compensate for the impairment of CaMKIV-dependent pathway in these mutant mice.

4.1. MASSED VS. SPACED STIMULATION OF SYNAPSES AND LTP

Just as the interval between training trials is an important determinant of LTM in behavioral studies, the interval between tetani is important for the induction of L-LTP in hippocampal slices. Under certain conditions, 1-min intervals between periods of high-frequency stimulation (e.g., 100 Hz) produces unstable LTP that lasts only a few hours. The same tetani, however, delivered at 10-min intervals is more likely to trigger a stable, longer-lasting LTP (>4 h) that requires protein synthesis (Huang et al., 1996). Results from our laboratory suggest that increasing the interval between tetani may overcome the L-LTP deficits observed in the CREB$^{\alpha\delta-}$ mutants (A.J.S. and J.K., in preparation). Thus, the loss of the α and δ CREB activators does not block either L-LTP or LTM under all conditions. Instead, the mutants require spaced training to show normal LTM (Kogan et al., 1996) and spaced tetanization to show normal L-LTP.

Consistent with the findings in hippocampal cultures (Deisseroth et al., 1996), a brief stimulus (100 Hz) fails to activate CREB-dependent transcription in hippocampal slices from the CRE-LacZ (reporter) mice, and fails to induce L-LTP. In contrast, multiple-spaced tetani (three tetanic trains given with a 5-min ITI), that induce protein-synthesis-dependent L-LTP, also trigger an increase in LacZ expression in the CRE-LacZ hippocampal slices (Impey et al., 1996). This increase is detectable 2 h after stimulation and reaches a peak within 4–6 h. Both types of LTP stimuli used (a single tetanus or three tetani) increase the levels of CREB phosphorylation, although only the repeated tetanic stimulation increases LacZ expression (Impey et al., 1996).

It is important to note that even though the levels of CREB may help determine how tetanic patterns affect LTP, CREB activation itself may not be the only biological sensor that reads out these patterns. A recent study shows that the pattern of electrophysiological stimulation affects the duration of a kinase upstream of CREB. Thus, spaced membrane depolarizations, but not a single prolonged stimulus, produces a persistent activation of the MAPK pathway (Wu et al., 2001b; but see also Fields et al., 1997). Importantly, these authors also show that MAPK activation triggers a slow-onset activation (phosphorylation) of CREB (Wu et al., 2001a). Thus, in a treatment that may be akin to spaced training in behavioral experiments, spaced electrophysiological 'training' may activate a kinase upstream from CREB.

4.2. CREB AS A GAIN CONTROL DEVICE FOR MEMORY

Together, the data reviewed thus far suggest that CREB may have a specific computational role in memory formation. CREB may assist in 'setting the height of the bar' that must be attained in order for LTM and LTP formation to occur. High levels of CREB activity may allow neuronal circuits to acquire memory relatively quickly, while circuits with low CREB levels would acquire memory relatively slower. This gain control mechanism could be critical in determining the number and distribution of trials required to lay down memories in a given circuit.

It is important to note that CREB-like proteins function with many other transcription factors. There is evidence that the results obtained by manipulating CREB are also found in studies using other transcription factors (Alberini et al., 1994). Therefore, CREB may be simply one of several related gain control mechanisms that contribute to the regulation of memory acquisition in neural circuits.

5. CREB AND SYNAPTIC REMODELING

Behavioral long-term sensitization in *Aplysia* results not only in a stable increase in neurotransmitter release, but also in structural changes, such as the growth of new synapses. These structural changes include alterations in sensory neuron active zones and an increase in the number of presynaptic varicosities (Bailey and Chen, 1988). There are many parallels between agents that induce (or block) LTF and those that induce (or block) these structural changes, a finding that confirms the relationship between synaptic function, structural changes and memory formation in *Aplysia* (Bailey and Chen, 1988). Intracellular injection of cAMP into the intact ganglion (Nazif et al., 1991) and repeated pulses of serotonin that induce LTF in culture (Glanzman et al., 1990; Bailey et al., 1992) also induce the growth of new synapses. A single pulse of serotonin paired with an injection of an antibody against a CREB repressor (ApCREB2) is sufficient to induce the growth of new synapses, just as it is sufficient to induce LTF (Bartsch et al., 1995). These results indicate that CREB-dependent proteins are involved in the growth of new synapses during LTM formation in *Aplysia*.

However, CREB-mediated transcription seems to be necessary but not sufficient for the growth of new synaptic connections. Experiments using a single sensory neuron composed of two branches that contact two spatially separated motor neurons shows that local application of serotonin onto a single synapse induces LTF that is branch-specific (Martin et al., 1997; Casadio et al., 1999). This branch-specific LTF requires local protein synthesis and CREB activation in the nucleus of the presynaptic neuron. Repeated application of serotonin onto the cell body of the sensory neuron induces a cell-wide transient LTF (not beyond 48 h) that is CREB-dependent, but is not accompanied by synaptic growth. A similar pattern (transient LTF and no synaptic growth) is produced by injection of phospho-CREB1 into the sensory neuron. In order for this transient LTF to become stable and for growth to appear, a single pulse of serotonin at either synapse is required. Thus, CREB-mediated transcription cooperatively induces synaptic plastic changes in concert with local stimulation by serotonin.

In addition to a proposed role for CREB in synaptic remodeling in *Aplysia*, there may also be a role for CREB in structural plasticity in mammals. A recurrent theme in the study of development and learning is that these two forms of plasticity may share some common mechanistic features (Hebb, 1949; Carew et al., 1998). Thus, CREB may be important to both to the formation of memory and to the synaptic restructuring that takes place during development.

5.1. DEVELOPMENTAL PLASTICITY OF VISUAL CORTEX

During a critical period in development, manipulating the input to the visual system may dramatically change neuronal connections in the visual cortex (Hubel and Wiesel, 1998). For example, depriving visual input from one eye during this critical period shifts the responses of cortical neurons towards the non-deprived eye (Wiesel and Hubel, 1963; Gordon et al., 1996). This process may require new protein synthesis, and perhaps CREB-dependent transcription.

Monocular deprivation during this critical period produces an increase in the transcription of a CRE-reporter (CRE-LacZ) construct in the visual cortex of transgenic mice (Pham et al., 1999). Specifically, the primary visual cortex (V1) receiving the input from the non-deprived eye shows an increase in CRE-LacZ expression compared to the area of primary visual cortex receiving input from the deprived eye. Importantly, both the shift in visual cortical responses and the increase in CRE-mediated transcription occur only during the critical period. Moreover, binocular deprivation, which does not induce cortical plasticity (Wiesel and Hubel, 1965; Gordon et al., 1996), similarly does not increase CRE-mediated transcription in the visual cortex of the transgenic mice. These results suggest that CREB is important in developmental visual cortical plasticity.

5.2. CREB AND PLASTICITY OF THE BARREL CORTEX

Unlike the visual cortex, plasticity in the somatosensory cortex does not seem to be restricted to a brief developmental critical period. Under certain conditions, there is a preferential re-allocation of cortical resources to areas representing the most-used sensory inputs. That is, there may be a use-dependent shift in cortical resources (for review see Buonomano and Merzenich, 1998). In rodents, experience-dependent plasticity has been extensively studied in the barrel cortex, a region of the somatosensory cortex that represents the long facial vibrissa. The vibrissa are topographically represented in barrel-like structures in the cortex. Cells within a barrel respond primarily to stimulation of their corresponding (principal) whisker but also show weaker responses to stimulation of whiskers neighboring the principal (Woolsey, 1967; Woolsey and Van der Loos, 1970; Welker, 1971).

The anatomical map of the barrels shows plasticity during a developmental critical period such that removal of the vibrissa follicles shortly after birth only leads to abnormal anatomical development of the barrels (Van der Loos and Woolsey, 1973). In contrast, the functional map of the barrel cortex exhibits plasticity outside the critical window, even in adulthood. Removing or even trimming some of the whiskers in adulthood, no longer alters the size or morphology, but does change the functional properties of the barrels (Hand, 1982; Fox, 1992). The receptive field properties of neurons in barrel-columns corresponding to the spared whisker typically expand in size and exhibit potentiation (Hand, 1982; Fox, 1992; Glazewski and Fox, 1996). A corresponding depression of cortical responses to the areas representing the deprived whiskers is observed. Interestingly, the CREB$^{\alpha\delta-}$ mutation, which disrupts LTP and LTM (see above), also disrupts this use-dependent plasticity in the barrel cortex (Glazewski et al., 1999). Similar to the results summarized above for the visual cortex, manipulations (whisker deprivation) that trigger plasticity in the barrel cortex also increase the expression of a transgenic CRE-LacZ reporter in the spared barrel (Barth et al., 2000). Importantly, manipulations that do not lead to plasticity, also do not produce increases in LacZ expression. Thus, normal undeprived animals and animals in which all the whiskers have been removed show low levels of LacZ expression and do not exhibit potentiated barrel responses (Barth et al., 2000). Therefore, CREB may be important in structural plasticity that takes place during development and the use-dependent functional plasticity that takes place during adulthood. These findings are in agreement with the idea that learning and development may share common mechanistic features.

6. TARGET GENES OF CREB

6.1. POSSIBLE DOWNSTREAM TARGETS OF CREB IN *APLYSIA*

The long-lasting plastic changes induced by CREB may be orchestrated by one or a choir of CREB target proteins. Multiple pulses of serotonin activate CREB in *Aplysia* and induce a set of genes, any one of which may regulate synaptic plasticity. Several synaptic plasticity candidate genes have been identified and will be briefly discussed. The CCAAT/enhancer binding protein, ApC/EBP is a candidate gene as it contains CRE sites in the promoter region, is rapidly induced by cAMP and exhibits properties consistent with an immediate-early gene. Low levels of ApC/EBP are present in unstimulated sensory neurons but higher levels are observed during the initial phase of LTF. Decreasing the expression of ApC/EBP selectively blocks the formation of LTF but not STF (Alberini et al., 1994). Thus, ApC/EBP may be a transcriptional regulator downstream of CREB that induces the transcription of target genes necessary for LTF. Interestingly, inhibitory avoidance training in rats induces two homologs of ApC/EBP (C/EBP β and C/EBP α), suggesting that the function of ApC/EBP may be evolutionarily conserved (Taubenfeld et al., 2001).

Using a yeast two-hybrid screen with ApC/EBP as bait, another synaptic plasticity candidate gene was identified, the *Aplysia* Activating Factor, ApAF (Bartsch et al., 2000). Unlike ApC/EBP, ApAF is constitutively expressed in sensory neurons and is not up-regulated by serotonin application. ApAF shares homology with the mammalian PAR family of transcription factors, including TEF, E4BP4 and DBP. ApAF forms dimers with ApC/EBP and ApCREB2, but not with ApCREB1. PKA, CaMKII and PKC phosphorylate ApAF in vitro. Injection of recombinant ApAF into a co-cultured neuronal preparation converts the STF normally produced by one pulse of serotonin into LTF. However, unlike injection of ApCREB1a, injection of ApAF alone is not sufficient to induce LTF. Injection of antibodies against ApAF, or injection of a dominant negative form of ApAF, that contains only the bZIP domain, blocks the LTF produced by either five pulses of serotonin or injection of ApCREB1a itself (Bartsch et al., 2000). These manipulations, however, have no effect on STF. Thus, although ApAF does not directly interact with CREB1a, it may act downstream from, or somehow parallel to, CREB1a. It could be that ApAF heterodimerizes with ApCREB2, thereby removing the repression that ApCREB2 normally exerts on CRE-mediated transcription.

Another gene that is rapidly induced by LTF in *Aplysia* codes for ubiquitin C-terminal hydrolase (ApUch), an enzyme associated with proteosome-dependent proteolysis (Hegde et al., 1997). The ubiquitin–proteosome pathway degrades target proteins and ApUch enhances the degradation of substrates in proteosomes by removing ubiquitin. Injection of an antibody or antisense against ApUch blocks LTF, but not STF. Therefore, it seems that ubiquitin-dependent proteolysis is important for the induction of LTF, perhaps by removing or degrading inhibitory proteins. One possibility is that ubiquitin-mediated proteolysis cleaves the regulatory (R) subunit of PKA, freeing the catalytic (C) subunit, thereby producing a persistent activation of PKA (Bergold et al., 1992). Support for this comes from the finding that addition of C subunits of PKA may rescue the block of LTF produced by proteosome inhibitors. Thus, it could be that the activation of PKA C subunits (by ubiquitin-dependent degradation of R subunits of PKA) may be a key step in the phosphorylation of CREB necessary to produce LTF. Interestingly, a mutation of a related Angelman ubiquitin ligase in mice produces deficits in LTM for contextual fear conditioning and LTP (Jiang et al., 1998).

Another gene that may interact with CREB in mediating synaptic plasticity in *Aplysia* is *Aplysia* tolloid/BMP-1 (bone morphogenic protein)-like protein (ApTBL-1) (Liu et al., 1997). ApTBL-1 may function as a protease that activates growth factors such as transforming growth factor-β (TGF-β). Inhibitors of TGF-β block LTF produced by electrical stimulation, while treatment with TGF-β induces LTF (Zhang et al., 1997). Hence, LTF may involve the expression of ApTBL-1 (CREB-dependent?) that activates TGF-β, that, in turn, triggers a cascade of events that result in increased neurotransmitter release. This cascade may resemble that triggered by neurotrophins in hippocampal neurons (Kang and Schuman, 1996). It is interesting to note that TGF-β triggers LTF without inducing long-term increases in neuronal excitability or STF (Zhang et al., 1997) thus confirming the previous finding that these two phases of sensitization are independent (Emptage and Carew, 1993; Mauelshagen et al., 1996).

6.2. POSSIBLE DOWNSTREAM TARGETS OF CREB IN MAMMALS

Many genes have been identified as potential target genes for CREB in mammals based on the presence of CRE sites within their 5′-flanking region. CREB-mediated plasticity may be ultimately mediated by any one, or a chorus, of these target genes. Although the precise downstream targets of CREB that are essential for mediating synaptic plasticity remain unknown, some studies have begun to evaluate the potential target genes. Two criteria should be met in order for a gene to be a potential mediator of CREB-induced plasticity. First, the gene should be regulated by CREB and second, the gene should be involved in synaptic plasticity.

One set of potential downstream targets of CREB is the immediate-early genes. The immediate-early genes, including Fos, Zif268 (also known as EGR1 and NGFI-A) and HZF-3 (also known as NURR1), contain CRE sites in their promoter regions and are transcriptionally regulated by CREB (Berkowitz et al., 1989; Changelian et al., 1989; Sakamoto et al., 1991; Saucedo-Cardenas et al., 1997). Furthermore, these genes are induced by neural stimulation and behavioral experience (Nikolaev et al., 1992; Robertson, 1992; Dragunow, 1996; see also Chapter VIII by Kaczmarek as well as Chapter XII by Leah and Wilce, this volume). The immediate-early genes are thought to function as nuclear third messengers that mediate a secondary cascade of transcription that leads to the delayed expression of effector proteins.

Fos is perhaps the most studied immediate-early gene in the central nervous system. Expression of the c-*fos* gene is increased by manipulations that also induce LTM or hippocampal LTP (Worley et al., 1991; Hess et al., 1995; Lanahan et al., 1997; Guzowski et al., 2001). Similar to CREB, antisense against Fos blocks LTM, but STM, in several behavioral tasks (Lamprecht and Dudai, 1996; Grimm et al., 1997; Morrow et al., 1999; Tolliver et al., 2000).

A second well-studied immediate-early gene that may be a downstream target mediating at least some of the synaptic plasticity effects of CREB is zif268. The levels of zif268 are upregulated following LTP-inducing stimulation (Cole et al., 1989; Abraham et al., 1991; Richardson et al., 1992). Furthermore, mutant mice lacking the zif268 gene show deficits in L-LTP and LTM for a variety of learning tasks (Jones et al., 2001). Of particular interest is the finding that, similar to the CREB$^{\alpha\delta-}$ mutant mice (Kogan et al., 1996), the spatial memory deficits of the Zif268 mutant mice are rescued by spaced training (Jones et al., 2001).

Less is known about the role of the immediate-early gene, HZF-3, in plasticity. HZF-3 encodes a transcription factor that is a member of the nuclear hormone receptor superfamily (Law et al., 1992; Mages et al., 1995; Peña de Ortiz and Jamieson, 1996). *hzf-3* mRNA is induced in the hippocampus during acquisition of a spatial discrimination task or induction of mossy fiber-CA3 LTP (Onton et al., 1996; Peña de Ortiz et al., 2000). Furthermore, antisense against hzf-3 infused into CA1 or CA3 regions of the hippocampus impairs retention of a previously acquired spatial discrimination (Colón et al., 2000). Additional studies, perhaps using inducible transgenic or knockout approaches, are required to further assess the potential role of hzf-3 in CREB-dependent plasticity.

Recent studies show that CREB mediates the Ca^{2+}-dependent transcriptional activation of brain-derived neurotrophic factor (BDNF), a gene with CRE sites in its promoter region (Shieh et al., 1998; Tao et al., 1998). BDNF plays a critical role not only in neuronal development but also neuronal plasticity including synaptic transmission and LTP. Therefore, BDNF is a strong candidate for a CREB target that mediates some of the synaptic functions of CREB.

Additional candidates for CREB-regulated effector proteins await identification and characterization. Studies combining genetically modified mice that are impaired in CREB function and gene profiling techniques such as microarrays may be useful techniques to this end.

7. CONCLUSION

The findings reviewed here represent a significant advance in our understanding of memory formation and plasticity. In a dizzying range of species, including *Aplysia*, *Drosophila*, song birds, honey bees, mice and rats, CREB-dependent transcription has been shown to be crucial for the formation of LTM. The studies reviewed above show that altering CREB function (with pharmacology, antibodies, antisense, viral vectors, targeted and inducible genetic mutations) effect LTM. The convergence of these results argues that CREB function is required for memory formation. This extensive body of data also suggests that the levels of active CREB are an important determinant of the amount and schedule of training required for the formation of LTM; in general increases in CREB obviate spaced training, while decreases in the levels of this transcription factor may be overcome by extended spaced training. Additionally, these studies show that CREB may have a universal impact on memory formation. Tests as diverse as olfactory conditioning in flies, fear conditioning, spatial memory, conditioned taste aversion, social recognition and social transmission of food preferences in rodents demonstrate the involvement of CREB in memory formation. Furthermore, the data reviewed here also indicate that CREB plays a key role in synaptic plasticity such as LTP and LTD. In addition, evidence was reviewed suggesting that CREB is important in structural and functional plasticity in the cortex. This finding adds credence to the suggestion that mechanisms important for development are recapitulated during learning. Overall, these findings suggest that CREB has a highly conserved role in various types of plasticity.

8. ABBREVIATIONS

ApAF	*Aplysia* activating factor
ApUch	*Aplysia* ubiquitin C-terminal hydrolase
ATF	activating transcription factor
BDNF	brain-derived neurotrophic factor
C/EBP	CAAT/enhancer binding protein
CaM	calmodulin
CaMK	calmodulin kinase
cAMP	cyclic adenosine monophosphate
CBP	CREB binding protein
CK	casein kinase
CRE	cAMP responsive elements
CREB	cAMP responsive element binding protein
$CREB^{IR}$	CREB-inducible repressor
CREM	cAMP responsive element modulator
CS	conditioned stimulus
CTA	conditioned taste aversion
E-LTP	early LTP
ERK	extracellular signal-related kinase
GSK	glycogen synthase kinase

HAT	histone acetyltransferase
ICER	inducible cAMP early repressor
ITI	intertrial interval
KID	kinase-inducible domain
LiCl	lithium chloride
L-LTP	late LTP
LTD	long-term depression
LTF	long-term facilitation
LTM	long-term memory
LTP	long-term potentiation
NGF	nerve growth factor
NMDAR	*N*-methyl-D-aspartate receptor
PKA	protein kinase A
PKC	protein kinase C
PP	protein phosphatase
STF	short-term facilitation
STM	short-term memory
TAM	tamoxifen
TGF-β	transforming growth factor
Trk	tyrosine kinase
US	unconditioned stimulus
WT	wild type

9. ACKNOWLEDGEMENTS

The authors wish to thank S. Kushner and P.W. Frankland for helpful comments on the manuscript.

10. REFERENCES

Abel T, Nguyen PV, Barad M, Deuel TA, Kandel ER, Bourtchouladze R (1997): Genetic demonstration of a role for PKA in the late phase of LTP and in hippocampus-based long-term memory. *Cell 88*:615–626.

Abraham WC, Dragunow M, Tate WP (1991): The role of immediate early genes in the stabilization of long-term potentiation. *Mol Neurobiol 5*:297–314.

Ahn S, Ginty DD, Linden DJ (1999): A late phase of cerebellar long-term depression requires activation of CaMKIV and CREB. *Neuron 23*:559–568.

Alberini CM, Ghirardi M, Metz R, Kandel ER (1994): C/EBP is an immediate-early gene required for the consolidation of long-term facilitation in Aplysia. *Cell 76*:1099–1114.

Arias J, Alberts A, Brindle P, Claret F, Smeal T, Karin M, Feramisco J, Montminy M (1994): Activation of cAMP and mitogen responsive genes relies on a common nuclear factor. *Nature 370*:226–229.

Bacskai BJ, Hochner B, Mahaut SM, Adams SR, Kaang BK, Kandel ER, Tsien RY (1993): Spatially resolved dynamics of cAMP and protein kinase A subunits in Aplysia sensory neurons. *Science 260*:222–226.

Bailey CH, Chen M (1988): Long-term memory in Aplysia modulates the total number of varicosities of single identified sensory neurons. *Proc Natl Acad Sci USA 85*:2373–2377.

Bailey CH, Montarolo P, Chen M, Kandel ER, Schacher S (1992): Inhibitors of protein and RNA synthesis block structural changes that accompany long-term heterosynaptic plasticity in *Aplysia*. *Neuron 9*:749–758.

Bannister AJ, Kouzarides T (1995): CBP-induced stimulation of c-Fos activity is abrogated by E1A. *EMBO J 14*:4758–4762.

Bannister AJ, Kouzarides T (1996): The CBP co-activator is a histone acetyltransferase. *Nature 384*:641–643.

Barad M, Bourtchouladze R, Winder DG, Golan H, Kandel E (1998): Rolipram, a type IV-specific phosphodiesterase inhibitor, facilitates the establishment of long-lasting long-term potentiation and improves memory. *Proc Natl Acad Sci USA 95*:15020–15025.

Barela PB (1999): Theoretical mechanisms underlying the trial-spacing effect in Pavlovian fear conditioning. *J Exp Psychol: Anim Behav Process 25*:177–193.

Barnes CA (1995): Involvement of LTP in memory: are we searching under the street light?. *Neuron 15*:751–754.

Barth AL, McKenna M, Glazewski S, Hill P, Impey S, Storm D, Fox K (2000): Upregulation of cAMP response element-mediated gene expression during experience-dependent plasticity in adult neocortex. *J Neurosci 20*:4206–4216.

Bartsch D, Ghirardi M, Skehel PA, Karl KA, Herder SP, Chen M, Bailey CH, Kandel ER (1995): *Aplysia* CREB2 represses long-term facilitation: relief of repression converts transient facilitation into long-term functional and structural change. *Cell 83*:979–992.

Bartsch D, Ghirardi M, Casadio A, Giustetto M, Karl KA, Zhu H, Kandel ER (2000): Enhancement of memory-related long-term facilitation by ApAF, a novel transcription factor that acts downstream from both CREB1 and CREB2. *Cell 103*:595–608.

Bergold PJ, Beushausen SA, Sacktor TC, Cheley S, Bayley H, Schwartz JH (1992): A regulatory subunit of the cAMP-dependent protein kinase down-regulated in aplysia sensory neurons during long-term sensitization. *Neuron 8*:387–397.

Berkowitz LA, Riabowol KT, Gilman MZ (1989): Multiple sequence elements of a single functional class are required for cyclic AMP responsiveness of the mouse c-fos promoter. *Mol Cell Biol 9*:4272–4281.

Bernier L, Castellucci VF, Kandel ER, Schwartz JH (1982): Facilitatory transmitter causes a selective and prolonged increase in adenosine 3′:5′-monophosphate in sensory neurons mediating the gill and siphon withdrawal reflex in *Aplysia. J Neurosci 2*:1682–1691.

Bito H, Deisseroth K, Tsien RW (1996): CREB phosphorylation and dephosphorylation: a Ca^{2+}- and stimulus duration-dependent switch for hippocampal gene expression. *Cell 87*:1203–1214.

Bitterman ME, Menzel R, Fietz A, Schafer S (1983): Classical conditioning of proboscis extension in honeybees (*Apis mellifera*). *J Comp Psychol 97*:107–119.

Blendy JA, Kaestner KH, Schmid W, Gass P, Schutz G (1996): Targeting of the CREB gene leads to up-regulation of a novel CREB mRNA isoform. *EMBO J 15*:1098–1106.

Blenis J, Chung J, Erikson E, Alcorta DA, Erikson RL (1991): Distinct mechanisms for the activation of the RSK kinases/MAP2 kinase/pp90rsk and pp70-S6 kinase signaling systems are indicated by inhibition of protein synthesis. *Cell Growth Differ 2*:279–285.

Bliss TV, Collingridge GL (1993): A synaptic model of memory: long-term potentiation in the hippocampus. *Nature 361*:31–39.

Bonni A, Ginty DD, Dudek H, Greenberg ME (1995): Serine 133-phosphorylated CREB induces transcription via a cooperative mechanism that may confer specificity to neurotrophin signals. *Mol Cell Neurosci 6*:168–183.

Bourtchuladze R, Frenguelli B, Blendy J, Cioffi D, Schutz G, Silva AJ (1994): Deficient long-term memory in mice with a targeted mutation of the cAMP-responsive element-binding protein. *Cell 79*:59–68.

Bourtchouladze R, Abel T, Berman N, Gordon R, Lapidus K, Kandel ER (1998): Different training procedures recruit either one or two critical periods for contextual memory consolidation, each of which requires protein synthesis and PKA. *Learn Mem 5*:365–374.

Brindle PK, Montminy MR (1992): The CREB family of transcription activators. *Curr Opin Genet Dev 2*:199–204.

Bugelski BR (1962): Presentation time, total time, and mediation in paired associate learning. *J Exp Psychol 63*:25–29.

Bullock BP, Habener JF (1998): Phosphorylation of the cAMP response element binding protein CREB by cAMP-dependent protein kinase A and glycogen synthase kinase-3 alters DNA-binding affinity, conformation, and increases net charge. *Biochemistry 37*:3795–3809.

Bunsey M, Eichenbaum H (1995): Selective damage to the hippocampal region blocks long-term retention of a natural and nonspatial stimulus–stimulus association. *Hippocampus 5*:546–556.

Buonomano DV, Byrne JH (1990): Long-term synaptic changes produced by a cellular analog of classical conditioning in *Aplysia. Science 249*:420–423.

Buonomano DV, Merzenich MM (1998): Cortical plasticity: from synapses to maps. *Annu Rev Neurosci 21*:149–186.

Byers D, Davis RL, Kiger Jr. JA (1981): Defect in cyclic AMP phosphodiesterase due to the dunce mutation of learning in *Drosophila melanogaster. Nature 289*:79–81.

Byrne JH (1987): Cellular analysis of associative learning. *Physiol Rev 67*:329–439.

Byrne JH, Zwartjes R, Homayouni R, Critz SD, Eskin, A (1993): Roles of second messenger pathways in neuronal plasticity and in learning and memory. Insights gained from *Aplysia*. In: Nairn S (Ed), *Advances in Second Messenger and Phosphoprotein Research*. New York: Raven Press Ltd, Vol 27, pp 47–108.

Carew TJ, Kandel ER (1973): Acquisition and retention of long-term habituation in *Aplysia*: correlation of behavioral and cellular processes. *Science 182*:1158–1160.

Carew TJ, Castellucci VF, Kandel ER (1971): An analysis of dishabituation and sensitization of the gill-withdrawal reflex in *Aplysia*. *Int J Neurosci 2*:79–98.

Carew TJ, Pinsker HM, Kandel ER (1972): Long-term habituation of a defensive withdrawal reflex in *Aplysia*. *Science 175*:451–454.

Carew TJ, Menzel R, Shatz C. (1998): Points of contact between development and learning. In: Carew TJ, Menze R, Shatz C (Eds), *Mechanistic Relationship between Development and Learning*. Chichester: Wiley, pp 1–14.

Casadio A, Martin KC, Giustetto M, Zhu H, Chen M, Bartsch D, Bailey CH, Kandel ER (1999): A transient, neuron-wide form of CREB-mediated long-term facilitation can be stabilized at specific synapses by local protein synthesis. *Cell 99*:221–237.

Castellucci VF, Kandel ER, Schwartz JH, Wilson FD, Nairn AC, Greengard P (1980): Intracellular injection of the catalytic subunit of cyclic AMP-dependent protein kinase simulates facilitation of transmitter release underlying behavioral sensitization in *Aplysia*. *Proc Natl Acad Sci USA 77*:7492–7496.

Changelian PS, Feng P, King TC, Milbrandt J (1989): Structure of the NGFI-A gene and detection of upstream sequences responsible for its transcriptional induction by nerve growth factor. *Proc Natl Acad Sci USA 86*:377–381.

Chapman PF (2001): The diversity of synaptic plasticity. *Nat Neurosci 4*:556–558.

Chawla S, Hardingham GE, Quinn DR, Bading H (1998): CBP: a signal-regulated transcriptional coactivator controlled by nuclear calcium and CaM kinase IV. *Science 281*:1505–1509.

Chen RH, Sarnecki C, Blenis J (1992): Nuclear localization and regulation of erk- and rsk-encoded protein kinases. *Mol Cell Biol 12*:915–927.

Chew SJ, Mello C, Nottebohm F, Jarvis E, Vicario DS (1995): Decrements in auditory responses to a repeated conspecific song are long-lasting and require two periods of protein synthesis in the songbird forebrain. *Proc Natl Acad Sci USA 92*:3406–3410.

Cho Y, Friedman E, Silva A (1999): Ibotenate lesions of the hippocampus impair spatial learning but not contextual fear conditioning in mice. *Behav Brain Res 98*:77–87.

Chrivia JC, Kwok RPS, Lamb N, Hagiwara M, Montminy MR, Goodman RH (1993): Phosphorylated CREB binds specifically to the nuclear protein CBP. *Nature 365*:855–859.

Cleary LJ, Lee WL, Byrne JH (1998): Cellular correlates of long-term sensitization in Aplysia. *J Neurosci 18*:5988–5998.

Cole AJ, Saffen DW, Baraban JM, Worley PF (1989): Rapid increase of an immediate early gene messenger RNA in hippocampal neurons by synaptic NMDA receptor activation. *Nature 340*:474–476.

Cole TJ, Copeland NG, Gilbert DJ, Jenkins NA, Schutz G, Ruppert S (1992): The mouse CREB (cAMP responsive element binding protein) gene: structure, promoter analysis, and chromosomal localization. *Genomics 13*:974–982.

Colón W, Colón M, Ren K, Carrasquillo Y, Santos I, Peña de Ortiz S (2000): Transcription and recombination mechanisms for spatial learning. *Soc Neurosci Abstr 26*:1747.

Comb M, Birnberg NC, Seasholtz A, Herbert E, Goodman HM (1986): A cyclic AMP- and phorbol ester-inducible DNA element. *Nature 323*:353–356.

Cooper EH, Pantle AJ (1967): The Total-Time Hypothesis in Verbal Learning. *Psychol Bull 68*:25–29.

Danielian PS, White R, Hoare SA, Fawell SE, Parker MG (1993): Identification of residues in the estrogen receptor that confer differential sensitivity to estrogen and hydroxytamoxifen. *Mol Endocrinol 7*:232–240.

Dash PK, Hochner B, Kandel ER (1990): Injection of the cAMP-responsive element into the nucleus of *Aplysia* sensory neurons blocks long-term facilitation. *Nature 345*:718–721.

Dash PK, Karl KA, Colicos MA, Prywes R, Kandel ER (1991): cAMP response element-binding protein is activated by Ca^{2+}/calmodulin- as well as cAMP-dependent protein kinase. *Proc Natl Acad Sci USA 88*:5061–5065.

Davis HP, Squire LR (1984): Protein synthesis and memory. *Psychol Bull 96*:518–559.

Davis M (1992): The role of the amygdala in fear-potentiated startle: implications for animal models of anxiety. *Trends Pharmacol Sci 13*:35–41.

Davis S, Vanhoutte P, Pages C, Caboche J, Laroche S (2000): The MAPK/ERK cascade targets both Elk-1 and cAMP response element-binding protein to control long-term potentiation-dependent gene expression in the dentate gyrus in vivo. *J Neurosci 20*:4563–4572.

De Cesare D, Jacquot S, Hanauer A, Sassone-Corsi P (1998): Rsk-2 activity is necessary for epidermal growth factor-induced phosphorylation of CREB protein and transcription of c-fos gene. *Proc Natl Acad Sci USA 95*:12202–12207.

Deak M, Clifton AD, Lucocq LM, Alessi DR (1998): Mitogen- and stress-activated protein kinase-1 (MSK1) is directly activated by MAPK and SAPK2/p38, and may mediate activation of CREB. *EMBO J 17*:4426–4441.

Deisseroth K, Bito H, Tsien RW (1996): Signaling from synapse to nucleus: postsynaptic CREB phosphorylation during multiple forms of hippocampal synaptic plasticity. *Neuron 16*:89–101.

Deisseroth K, Heist EK, Tsien RW (1998): Translocation of calmodulin to the nucleus supports CREB phosphorylation in hippocampal neurons. *Nature 392*:198–202.

Ding C, Lee YL, Davis M (1998): Role of PKA and CaM kinases in fear conditioning assessed with fear-potentiated startle using local infusion of Rp-8-Br-cAMP, KN-62, or KN-93 into the amygdala. *Soc Neurosci Abstr 24*:926.

Dragunow M (1996): A role for immediate-early transcription factors in learning and memory. *Behav Genet 26*:293–299.

Ebbinghaus, H. (1885): *Uber das Gedachtnis*. New York: Dover.

Emptage NJ, Carew TJ (1993): Long-term synaptic facilitation in the absence of short-term facilitation in *Aplysia* neurons. *Science 262*:253–256.

Enslen H, Sun P, Brickey D, Soderling SH, Klamo E, Soderling TR (1994): Characterization of Ca^{2+}/calmodulin-dependent protein kinase IV. Role in transcriptional regulation. *J Biol Chem 269*:15520–15527.

Ewing MF, Larew MB, Wagner AR (1985): Distribution-of-trials effects in Pavlovian conditioning: An apparent involvement of inhibitory backward conditioning with short intertrial intervals. *J Exp Psychol: Anim Behav Process 11*:38–58.

Falls WA, Kogan JH, Silva AJ, Willott JF, Carlson S, Turner JG (2000): Fear-potentiated startle, but not prepulse inhibition of startle, is impaired in $CREB^{\alpha\delta-/-}$ mutant mice. *Behav Neurosci 114*:998–1004.

Fanselow MS, Kim JJ (1994): Acquisition of contextual Pavlovian fear conditioning is blocked by application of an NMDA receptor antagonist D,L-2-amino-5-phosphonovaleric acid to the basolateral amygdala. *Behav Neurosci 108*:210–212.

Fanselow MS, LeDoux JE (1999): Why we think plasticity underlying Pavlovian fear conditioning occurs in the basolateral amygdala. *Neuron 23*:229–232.

Fanselow MS, Tighe TJ (1988): Contextual conditioning with massed versus distributed unconditional stimuli in the absence of explicit conditional stimuli. *J Exp Psychol: Anim Behav Process 14*:187–199.

Feil R, Brocard J, Mascrez B, LeMeur M, Metzger D, Chambon P (1996): Ligand-activated site specific recombination in mice. *Proc Natl Acad Sci USA 93*:10887–10890.

Fiala A, Müller U, Menzel R (1999): Reversible downregulation of protein kinase A during olfactory learning using antisense technique impairs long-term memory formation in the honeybee, *Apis mellifera*. *J Neurosci 19*:10125–10134.

Fields RD, Eshete F, Stevens B, Itoh K (1997): Action potential-dependent regulation of gene expression: temporal specificity in Ca^{2+}, cAMP-responsive element binding proteins, and mitogen-activated protein kinase signaling. *J Neurosci 17*:7252–7266.

Fimia GM, De Cesare D, Sassone-Corsi P (1999): CBP-independent activation of CREM and CREB by the LIM-only protein ACT. *Nature 398*:165–169.

Finkbeiner S, Tavazoie SF, Maloratsky A, Jacobs KM, Harris KM, Greenberg ME (1997): CREB: a major mediator of neuronal neurotrophin responses. *Neuron 19*:1031–1047.

Fiol CJ, Williams JS, Chou CH, Wang QM, Roach PJ, Andrisani OM (1994): A secondary phosphorylation of CREB341 at Ser129 is required for the cAMP-mediated control of gene expression. A role for glycogen synthase kinase-3 in the control of gene expression. *J Biol Chem 269*:32187–32193.

Foulkes NS, Borrelli E, Sassone-Corsi P (1991): CREM gene: use of alternative DNA-binding domains generates multiple antagonists of cAMP-induced transcription. *Cell 64*:739–749.

Foulkes NS, Mellstrom B, Benusiglio E, Sassone-Corsi P (1992): Developmental switch of CREM function during spermatogenesis: from antagonist to activator. *Nature 355*:80–84.

Fox K (1992): A critical period for experience-dependent synaptic plasticity in rat barrel cortex. *J Neurosci 12*:1826–1838.

Frankland PW, Cestari V, Filipkowski RW, McDonald RJ, Silva AJ (1998): the dorsal hippocampus is essential for complex discrimination but not for contextual conditioning. *Behav Neurosci 112*:863–874.

Freudenthal R, Locatelli F, Hermitte G, Maldonado H, Lafourcade C, Delorenzi A, Romano A (1998): Kappa-B like DNA-binding activity is enhanced after spaced training that induces long-term memory in the crab Chasmagnathus. *Neurosci Lett 242*:143–146.

Frey U, Krug M, Reymann KG, Matthies H (1988): Anisomycin, an inhibitor of protein synthesis, blocks late phases of LTP phenomena in the hippocampal CA1 region in vitro. *Brain Res 452*:57–65.
Frey U, Huang YY, Kandel ER (1993): Effects of cAMP simulate a late stage of LTP in hippocampal CA1 neurons. *Science 260*:1661–1664.
Frost WN, Castellucci VF, Hawkins RD, Kandel ER (1985): Monosynaptic connections made by the sensory neurons of the gill- and siphon-withdrawal reflex in *Aplysia* participate in the storage of long-term memory for sensitization. *Proc Natl Acad Sci USA 82*:8266–8269.
Galef Jr. BG, Wigmore SW (1983): Transfer of information concerning distant foods: a laboratory investigation of the 'information-centre' hypothesis. *Anim Behav 31*:748–758.
Galef Jr. BG, Mason JR, Preti G, Bean NJ (1988): Carbon disulfide: a semiochemical mediating socially induced diet choice in rats. *Physiol Behav 42*:119–124.
Gewirtz JC, Josselyn SA, Walker DL, Davis M (1998): Inhibition of protein synthesis in the amygdala blocks acquisition of fear potentiated startle, even after pretraining with a different cue: A test of the 'synaptic tagging' hypothesis. *Soc Neurosci Abstr 24*:926.
Glanzman DL, Mackey SL, Hawkins RD, Dyke AM, Lloyd PE, Kandel ER (1989): Depletion of serotonin in the nervous system of Aplysia reduces the behavioral enhancement of gill withdrawal as well as the heterosynaptic facilitation produced by tail shock. *J Neurosci 9*:4200–4213.
Glanzman DL, Kandel ER, Schacher S (1990): Target-dependent structural changes accompanying long-term synaptic facilitation in *Aplysia* neurons. *Science 249*:799–802.
Glazewski S, Fox K (1996): Time course of experience-dependent synaptic potentiation and depression in barrel cortex of adolescent rats. *J Neurophysiol 75*:1714–1729.
Glazewski S, Barth AL, Wallace H, McKenna M, Silva A, Fox K (1999): Impaired experience-dependent plasticity in barrel cortex of mice lacking the alpha and delta isoforms of CREB. *Cereb Cortex 9*:249–256.
Goelet P, Castellucci VF, Schacher S, Kandel E (1986): The long and short of long-term memory — a molecular framework. *Nature 322*:419–422.
Gonzalez GA, Montminy MR (1989): Cyclic AMP stimulates somatostatin gene transcription by phosphorylation of CREB at serine 133. *Cell 59*:675–680.
Gonzalez GA, Yamamoto KK, Fischer WH, Karr D, Menzel P, Biggs WD, Vale WW, Montminy MR (1989): A cluster of phosphorylation sites on the cyclic AMP-regulated nuclear factor CREB predicted by its sequence. *Nature 337*:749–752.
Gordon JA, Cioffi D, Silva AJ, Stryker MP (1996): Deficient plasticity in the primary visual cortex of alpha-calcium/calmodulin-dependent protein kinase II mutant mice. *Neuron 17*:491–499.
Grimm R, Schicknick H, Riede I, Gundelfinger ED, Herdegen T, Zuschratter W, Tischmeyer W (1997): Suppression of c-fos induction in rat brain impairs retention of a brightness discrimination reaction. *Learn Mem 3*:402–413.
Grunbaum L, Müller U (1998): Induction of a specific olfactory memory leads to a long-lasting activation of protein kinase C in the antennal lobe of the honeybee. *J Neurosci 18*:4384–4392.
Guzowski JF, McGaugh JL (1997): Antisense oligodeoxynucleotide-mediated disruption of hippocampal cAMP response element binding protein levels impairs consolidation of memory for water maze training. *Proc Natl Acad Sci USA 94*:2693–2698.
Guzowski JF, Setlow B, Wagner EK, McGaugh JL (2001): Experience-dependent gene expression in the rat hippocampus after spatial learning: a comparison of the immediate-early genes arc, c-fos, and zif268. *J Neurosci 21*:5089–5098.
Hagiwara M, Alberts A, Brindle P, Meinkoth J, Feramisco J, Deng T, Karin M, Shenolikar S, Montminy M (1992): Transcriptional attenuation following cAMP induction requires PP-1-mediated dephosphorylation of CREB. *Cell 70*:105–113.
Hagiwara M, Brindle P, Harootunian A, Armstrong R, Rivier J, Vale W, Tsien R, Montminy MR (1993): Coupling of hormonal stimulation and transcription via the cyclic AMP-responsive factor CREB is rate limited by nuclear entry of protein kinase A. *Mol Cell Biol 13*:4852–4859.
Hai T, Curran T (1991): Cross-family dimerization of transcription factors Fos/Jun and ATF/CREB alters DNA binding specificity. *Proc Natl Acad Sci USA 88*:3720–3724.
Hai T, Liu F, Coukos W, Green M (1989): Transcription factor ATF cDNA clones: an extensive family of leucine zipper proteins able to selectively form DNA-binding heterodimers. *Genes Dev 3*:2083–2090.
Hammer M, Menzel R (1995): Learning and memory in the honeybee. *J Neurosci 15*:1617–1630.
Hand PJ (1982): Plasticity of the rat cortical barrel system. In: Strick P, Morrison AD (Ed), *Changing Concepts of the Nervous System*. New York: Academic Press, pp 49–75.
Hardingham GE, Chawla S, Johnson CM, Bading H (1997): Distinct functions of nuclear and cytoplasmic calcium in the control of gene expression. *Nature 385*:260–265.

Hardingham GE, Chawla S, Cruzalegui FH, Bading H (1999): Control of recruitment and transcription-activating function of CBP determines gene regulation by NMDA receptors and L-type calcium channels. *Neuron 22*:789–798.

Hardingham GE, Arnold FJ, Bading H (2001): Nuclear calcium signaling controls CREB-mediated gene expression triggered by synaptic activity. *Nat Neurosci 4*:261–267.

Hebb DO (1949): *Organization of Behavior*. New York: Wiley.

Hegde AN, Inokuchi K, Pei W, Casadio A, Ghirardi M, Chain DG, Martin KC, Kandel ER, Schwartz JH (1997): Ubiquitin C-terminal hydrolase is an immediate-early gene essential for long-term facilitation in Aplysia. *Cell 89*:115–126.

Hess US, Lynch G, Gall CM (1995): Changes in c-*fos* mRNA expression in rat brain during odor discrimination learning: differential involvement of hippocampal subfields CA1 and CA3. *J Neurosci 15*:4786–4795.

Ho N, Liauw JA, Blaeser F, Wei F, Hanissian S, Muglia LM, Wozniak DF, Nardi A, Arvin KL, Holtzman DM, Linden DJ, Zhuo M, Muglia LJ, Chatila TA (2000): Impaired synaptic plasticity and cAMP response element-binding protein activation in Ca^{2+}/calmodulin-dependent protein kinase type IV/Gr-deficient mice. *J Neurosci 20*:6459–6472.

Hoeffler JP, Meyer TE, Yun Y, Jameson JL, Habener JF (1988): Cyclic AMP-responsive DNA-binding protein: structure based on a cloned placental cDNA. *Science 242*:1430–1433.

Hoeffler JP, Meyer TE, Waeber G, Habener JF (1990): Multiple adenosine 3′,5′-cyclic monophosphate response element DNA-binding proteins generated by gene diversification and alternative exon splicing. *Mol Endocrinol 4*:920–930.

Hoeffler JP, Lustbader JW, Chen CY (1991): Identification of multiple nuclear factors that interact with cyclic adenosine 3′,5′-monophosphate response element-binding protein and activating transcription factor-2 by protein–protein interactions. *Mol Endocrinol 5*:256–266.

Hu SC, Chrivia J, Ghosh A (1999): Regulation of CBP-mediated transcription by neuronal calcium signaling. *Neuron 22*:799–808.

Huang YY, Kandel ER (1994): Recruitment of long-lasting and protein kinase A-dependent long-term potentiation in the CA1 region of hippocampus requires repeated tetanization. *Learn Mem 1*:74–82.

Huang YY, Nguyen PV, Abel T, Kandel ER (1996): Long lasting forms of synaptic potentiation in the mammalian hippocampus. *Learn Mem 3*:74–85.

Huang YY, Martin KC, Kandel ER (2000): Both protein kinase A and mitogen-activated protein kinase are required in the amygdala for the macromolecular synthesis-dependent late phase of long-term potentiation. *J Neurosci 20*:6317–6325.

Hubel DH, Wiesel TN (1998): Early exploration of the visual cortex. *Neuron 20*:401–412.

Hummler E, Cole TJ, Blendy JA, Ganss R, Aguzzi A, Schmid W, Beermann F, Schutz G (1994): Targeted mutation of the cAMP response element binding protein (CREB) gene: compensation within the CREB/ATF family of transcription factors. *Proc Natl Acad Sci USA 91*:5647–5651.

Hurst HC, Totty NF, Jones NC (1991): Identification and functional characterisation of the cellular activating transcription factor 43 (ATF-43) protein. *Nucleic Acids Res 19*:4601–4609.

Impey S, Mark M, Villacres EC, Poser S, Chavkin C, Storm DR (1996): Induction of CRE-mediated gene expression by stimuli that generate long-lasting LTP in area CA1 of the hippocampus. *Neuron 16*:973–982.

Impey S, Obrietan K, Wong ST, Poser S, Yano S, Wayman G, Deloulme JC, Chan G, Storm DR (1998a): Cross talk between ERK and PKA is required for Ca^{2+} stimulation of CREB-dependent transcription and ERK nuclear translocation. *Neuron 21*:869–883.

Impey S, Smith DM, Obrietan K, Donahue R, Wade C, Storm DR (1998b): Stimulation of cAMP response element (CRE)-mediated transcription during contextual learning. *Nat Neurosci 1*:595–601.

Jiang YH, Armstrong D, Albrecht U, Atkins CM, Noebels JL, Eichele G, Sweatt JD, Beaudet AL (1998): Mutation of the Angelman ubiquitin ligase in mice causes increased cytoplasmic p53 and deficits of contextual learning and long-term potentiation. *Neuron 21*:799–811.

Jones MW, Errington ML, French PJ, Fine A, Bliss TV, Garel S, Charnay P, Bozon B, Laroche S, Davis S (2001): A requirement for the immediate early gene Zif268 in the expression of late LTP and long-term memories. *Nat Neurosci 4*:289–296.

Josselyn SA, Kida S, Silva, AJ (1999): CREB and long-term memory for conditioned taste aversion. *Soc Neurosci Abs 25*:645.

Josselyn SA, Shi C, Carlezon Jr. WA, Neve RL, Nestler EJ, Davis M (2001): Long-term memory is facilitated by cAMP response element-binding protein overexpression in the amygdala. *J Neurosci 21*:2404–2412.

Kaang BK, Kandel ER, Grant SG (1993): Activation of cAMP-responsive genes by stimuli that produce long-term facilitation in Aplysia sensory neurons. *Neuron 10*:427–435.

Kamei Y, Xu L, Heinzel T, Torchia J, Kurokawa R, Gloss B, Lin SC, Heyman RA, Rose DW, Glass CK, Rosenfeld MG (1996): A CBP integrator complex mediates transcriptional activation and AP-1 inhibition by nuclear receptors. *Cell 85*:403–414.

Kandel ER, Schwartz JH (1982): Molecular biology of learning: Modulation of transmitter release. *Science 218*:433–443.

Kandel ER, Klein M, Bailey CH, Hawkins RD, Castellucci VF, Lubit BW, Schwartz JH (1981): Serotonin, cyclic AMP, and the modulation of the calcium current during behavioral arousal. In: Jacobs BL, Gelperin A (Eds), *Serotonin Neurotransmission and Behavior*. Cambridge, MA: MIT Press, pp 211–254.

Kang H, Schuman EM (1996): A requirement for local protein synthesis in neurotrophin-induced hippocampal synaptic plasticity. *Science 273*:1402–1406.

Karpinski BA, Morle GD, Huggenvik J, Uhler MD, Leiden JM (1992): Molecular cloning of human CREB-2: an ATF/CREB transcription factor that can negatively regulate transcription from the cAMP response element. *Proc Natl Acad Sci USA 89*:4820–4824.

Kerppola T, Curran T (1995): Transcription. Zen and the art of Fos and Jun. *Nature 373*:199–200.

Kogan JH, Frankland PW, Blendy JA, Coblentz J, Marowitz Z, Schutz G, Silva AJ (1996): Spaced training induces normal long-term memory in CREB mutant mice. *Curr Biol 7*:1–11.

Kogan JH, Frankland PW, Silva AJ (2000): Long-term memory underlying hippocampus-dependent social recognition in mice. *Hippocampus 10*:47–56.

Kouzarides T (1999): Histone acetylases and deacetylases in cell proliferation. *Curr Opin Genet Dev 9*:40–48.

Kwok RPS, Lundblad J, Chrivia J, Richards J, Bachinger H, Brennan R, Roberts S, Green M, Goodman R (1994): Nuclear protein CBP is a coactivator for the transcription factor CREB. *Nature 370*:223–226.

Lamprecht R, Dudai Y (1996): Transient expression of c-fos in rat amygdala during training is required for encoding conditioned taste aversion memory. *Learn Mem 3*:31–41.

Lamprecht R, Hazvi S, Dudai Y (1997): cAMP response element-binding protein in the amygdala is required for long- but not short-term conditioned taste aversion memory. *J Neurosci 17*:8443–8450.

Lanahan A, Lyford G, Stevenson GS, Worley PF, Barnes CA (1997): Selective alteration of long-term potentiation-induced transcriptional response in hippocampus of aged, memory-impaired rats. *J Neurosci 17*:2876–2885.

Laoide BM, Foulkes NS, Schlotter F, Sassone-Corsi P (1993): The functional versatility of CREM is determined by its modular structure. *EMBO J 12*:1179–1191.

Law SW, Conneely OM, DeMayo FJ, O'Malley BW (1992): Identification of a new brain-specific transcription factor, Nurr1. *Mol Endocrinol 6*:2129–2135.

LeDoux JE (1992): Brain mechanisms of emotion and emotional learning. *Curr Biol 2*:191–197.

LeDoux JE (2000): Emotion circuits in the brain. *Annu Rev Neurosci 23*:155–184.

Lee EHY, Lee CP, Wang HI, Lin WR (1993): Hippocampal CRF, NE, and NMDA system interactions in memory processing in the rat. *Synapse 14*:144–153.

Levin LR, Han PL, Hwang PM, Feinstein PG, Davis RL, Reed RR (1992): The Drosophila learning and memory gene rutabaga encodes a Ca^{2+}/Calmodulin-responsive adenylyl cyclase. *Cell 68*:479–489.

Linden DJ, Connor JA (1995): Long-term synaptic depression. *Annu Rev Neurosci 18*:319–357.

Liu FC, Graybiel AM (1996): Spatiotemporal dynamics of CREB phosphorylation: transient versus sustained phosphorylation in the developing striatum. *Neuron 17*:1133–1144.

Liu QR, Hattar S, Endo S, MacPhee K, Zhang H, Cleary LJ, Byrne JH, Eskin A (1997): A developmental gene (tolloid/BMP-1) is regulated in Aplysia neurons by treatments that induce long term sensitization. *J Neurosci 17*:755–765.

Logie C, Stewart AF (1995): Ligand-regulated site-specific recombination. *Proc Natl Acad Sci USA 92*:5940–5964.

Loriaux MM, Brennan RG, Goodman RH (1994a): Modulatory function of CREB. CREM alpha heterodimers depends upon CREM alpha phosphorylation. *J Biol Chem 69*:28839–28843.

Loriaux MM, Rehfuss RP, Brennan RG, Goodman RH (1994b): Engineered leucine zippers show that hemiphosphorylated CREB complexes are transcriptionally active. *Proc Natl Acad Sci USA 90*:9046–9050.

Lu YF, Kandel ER, Hawkins RD (1999): Nitric oxide signaling contributes to late-phase LTP and CREB phosphorylation in the hippocampus. *J Neurosci 19*:10250–10261.

Mages HW, Rilke O, Bravo R, Senger G, Kroczek RA (1995): NOT, a human immediate-early response gene closely related to the steroid/thyroid hormone receptor NAK1/TR3. *Mol Endocrinol 8*:1583–1591.

Malenka RC (1994): Synaptic plasticity in the hippocampus: LTP and LTD. *Cell 78*:535–538.

Maren S, Baudry M (1995): Properties and mechanisms of long-term synaptic plasticity in the mammalian brain: relationships to learning and memory. *Neurobiol Learn Mem 63*:1–18.

Martin KC, Casadio A, Zhu HEY, Rose JC, Chen M, Bailey CH, Kandel ER (1997): Synapse-specific, long-term facilitation of aplysia sensory to motor synapses: a function for local protein synthesis in memory storage. *Cell 91*:927–938.

Matthews RP, Guthrie CR, Wailes LM, Zhao X, Means AR, McKnight GS (1994): Calcium/calmodulin-dependent protein kinase types II and IV differentially regulate CREB-dependent gene expression. *Mol Cell Biol 14*:6107–6116.

Matthies H (1989): In search of cellular mechanisms of memory. *Prog Neurobiol 32*:277–349.

Matthies H, Schulz S, Thiemann W, Siemer H, Schmidt H, Krug M, Hollt V (1997): Design of a multiple slice interface chamber and application for resolving the temporal pattern of CREB phosphorylation in hippocampal long-term potentiation. *J Neurosci Methods 78*:173–179.

Mauelshagen J, Parker GR, Carew TJ (1996): Dynamics of induction and expression of long-term synaptic facilitation in Aplysia. *J Neurosci 16*:7099–7108.

Mauelshagen J, Sherff CM, Carew TJ (1998): Differential induction of long-term synaptic facilitation by spaced and massed applications of serotonin at sensory neuron synapses of *Aplysia californica. Learn Mem 5*:49–52.

Mayford M, Baranes D, Podsypanina K, Kandel ER (1996): The 3′-untranslated region of CaMKII alpha is a cis-acting signal for the localization and translation of mRNA in dendrites. *Proc Natl Acad Sci USA 93*:13250–13255.

McLean JH, Harley CW, Darby-King A, Yuan Q (1999): pCREB in the neonate rat olfactory bulb is selectively and transiently increased by odor preference-conditioned training. *Learn Mem 6*:608–618.

Menzel R, Müller U (1996): Learning and memory in honeybees: from behavior to neural substrates. *Annu Rev Neurosci 19*:379–404.

Menzel R, Hammer M, Müller U, Rosenboom H (1996): Behavioral, neural and cellular components underlying olfactory learning in the honeybee. *J Physiol, Paris 90*:395–398.

Mercer AR, Emptage NJ, Carew TJ (1991): Pharmacological dissociation of modulatory effects of serotonin in Aplysia sensory neurons. *Science 54*:1811–1813.

Molina CA, Foulkes NS, Lalli E, Sassone-Corsi P (1993): Inducibility and negative autoregulation of CREM: an alternative promoter directs the expression of ICER, an early response repressor. *Cell 75*:875–886.

Montarolo PG, Goelet P, Castelucci VF, Morgan J, Kandel ER, Schacher S (1986): A critical period for macromolecular synthesis in long-term heterosynaptic facilitation in *Aplysia. Science 234*:1249–1254.

Montminy MR, Bilezikjian LM (1987): Binding of a nuclear protein to the cyclic-AMP response element of the somatostatin gene. *Nature 328*:175–178.

Montminy MR, Sevarino KA, Wagner JA, Mandel G, Goodman RH (1986): Identification of a cyclic-AMP-responsive element within the rat somatostatin gene. *Proc Natl Acad Sci USA 83*:6682–6686.

Moore AN, Waxham MN, Dash PK (1996): Neuronal activity increases the phosphorylation of the transcription factor cAMP response element-binding protein (CREB) in rat hippocampus and cortex. *J Biol Chem 271*:14214–14220.

Morris RGM (1981): Spatial localization does not require the presence of local cues. *Learn Motiv 12*:239–260.

Morris RGM, Garrud P, Rawlins JNP, O'Keefe J (1982): Place navigation impaired in rats with hippocampal lesions. *Nature 297*:681–683.

Morrow BA, Elsworth JD, Inglis FM, Roth RH (1999): An antisense oligonucleotide reverses the footshock-induced expression of fos in the rat medial prefrontal cortex and the subsequent expression of conditioned fear-induced immobility. *J Neurosci 19*:5666–5673.

Müller U (1996): Inhibition of nitric oxide synthase impairs a distinct form of long-term memory in the honeybee, *Apis mellifera. Neuron 16*:541–549.

Müller U (2000): Prolonged activation of cAMP-dependent protein kinase during conditioning induces long-term memory in honeybees. *Neuron 27*:159–168.

Nakajima T, Uchida C, Anderson SF, Lee CG, Hurwitz J, Parvin JD, Montminy M (1997a): RNA helicase A mediates association of CBP with RNA polymerase II. *Cell 90*:1107–1112.

Nakajima T, Uchida C, Anderson SF, Parvin JD, Montminy M (1997b): Analysis of a cAMP-responsive activator reveals a two-component mechanism for transcriptional induction via signal-dependent factors. *Genes Dev 11*:738–747.

Nazif FA, Byrne JH, Cleary LJ (1991): cAMP induces long-term morphological changes in sensory neurons of *Aplysia. Brain Res 539*:324–327.

Nguyen PV, Abel T, Kandel ER (1994): Requirement of a critical period of transcription for induction of a late phase of LTP. *Science 265*:1104–1107.

Nikolaev E, Kaminska B, Tischmeyer W, Matthies H, Kaczmarek L (1992): Induction of expression of genes encoding transcription factors in the rat brain elicited by behavioral training. *Brain Res Bull 28*:479–484.

Nordheim A (1994): Transcription factors. CREB takes CBP to tango. *Nature 370*:177–178.

Ogryzko VV, Schiltz RL, Russanova V, Howard BH, Nakatani Y (1996): The transcriptional coactivators p300 and CBP are histone acetyltransferases. *Cell 87*:953–959.

Onton JA, Peña de Ortiz S, Reyes JA, Vallejos CR, Barea-Rodriguez EJ, Martinez Jr. JL (1996): HZF-3 is an immediate-early gene in mossy fiber LTP and is induced by training in the Morris water maze. *Soc Neurosci Abstr 22*:1515.

Parker D, Jhala US, Radhakrishnan I, Yaffe MB, Reyes C, Shulman AI, Cantley LC, Wright PE, Montminy M (1998): Analysis of an activator:coactivator complex reveals an essential role for secondary structure in transcriptional activation. *Mol Cell 2*:353–359.

Peña de Ortiz S, Jamieson GA (1996): HZF-3, an immediate-early orphan receptor homologous to NURR1/NOT: induction upon membrane depolarization and seizures. *Mol Brain Res 38*:1–13.

Peña de Ortiz S, Maldonado-Vlaar CS, Carrasquillo Y (2000): Hippocampal expression of the orphan nuclear receptor gene *hzf-3/nurr1* during spatial discrimination learning. *Neurobiol Learn Mem 74*:161–171.

Pham TA, Impey S, Storm DR, Stryker MP (1999): CRE-mediated gene transcription in neocortical neuronal plasticity during the developmental critical period. *Neuron 22*:63–72.

Pinkser H, Castellucci VF, Kupfermann I, Kandel ER (1970): Habituation and dishabituation of the gill-withdrawal reflex in *Aplysia*. *Science 167*:1740–1742.

Pinsker HM, Hening WA, Carew TJ, Kandel ER (1973): Long-term sensitization of a defensive withdrawal reflex in *Aplysia*. *Science 182*:1039–1042.

Rammes G, Steckler T, Kresse A, Schutz G, Zieglgansberger W, Lutz B (2000): Synaptic plasticity in the basolateral amygdala in transgenic mice expressing dominant-negative cAMP response element-binding protein (CREB) in forebrain. *Eur J Neurosci 12*:2534–2546.

Rayport SG, Schacher S (1986): Synaptic plasticity in vitro: cell culture of identified *Aplysia* neurons mediating short-term habituation and sensitization. *J Neurosci 6*:759–763.

Rehfuss RP, Walton KM, Loriaux MM, Goodman RH (1991): The cAMP-regulated enhancer-binding protein ATF-1 activates transcription in response to cAMP-dependent protein kinase A. *J Biol Chem 266*:18431–18434.

Rescorla RA (1988): Behavioral studies of Pavlovian conditioning. *Annu Rev Neurosci 11*:111–152.

Richards JP, Bachinger HP, Goodman RH, Brennan RG (1996): Analysis of the structural properties of cAMP-responsive element-binding protein (CREB) and phosphorylated CREB. *J Biol Chem 271*:13716–13723.

Richardson CL, Tate WP, Mason SE, Lawlor PA, Dragunow M, Abraham WC (1992): Correlation between the induction of an immediate early gene, Zif/268, and long-term potentiation in the dentate gyrus. *Brain Res 580*:147–154.

Robertson HA (1992): Immediate-early genes, neuronal plasticity, and memory. *Biochem Cell Biol 70*:729–737.

Sakaguchi H, Wada K, Maekawa M, Watsuji T, Hagiwara M (1999): Song-induced phosphorylation of cAMP response element-binding protein in the songbird brain. *J Neurosci 19*:3973–3981.

Sakamoto KM, Bardeleben C, Yates KE, Raines MA, Golde DW, Gasson JC (1991): 5′ upstream sequence and genomic structure of the human primary response gene, EGR-1/TIS8. *Oncogene 6*:867–871.

Sassone-Corsi P (1995): Transcription factors responsive to cAMP. *Annu Rev Cell Dev Biol 11*:355–377.

Saucedo-Cardenas O, Kardon R, Ediger TR, Lydon JP, Coneely OM (1997): Cloning and structural organization of the gene encoding the murine nuclear receptor transcription factor, NURR1. *Gene 187*:135–139.

Schafe GE, LeDoux JE (2000): Memory consolidation of auditory Pavlovian fear conditioning requires protein synthesis and protein kinase A in the amygdala. *J Neurosci 20*:RC96.

Schafe GE, Nadel NV, Sullivan GM, Harris A, LeDoux JE (1999): Memory consolidation for contextual and auditory fear conditioning is dependent on protein synthesis, PKA, and MAP kinase. *Learn Mem 6*:97–110.

Schafe GE, Atkins CM, Swank MW, Bauer EP, Sweatt JD, LeDoux JE (2000): Activation of ERK/MAP kinase in the amygdala is required for memory consolidation of pavlovian fear conditioning. *J Neurosci 20*:8177–8187.

Schulz S, Siemer H, Krug M, Hollt V (1999): Direct evidence for biphasic cAMP responsive element-binding protein phosphorylation during long-term potentiation in the rat dentate gyrus in vivo. *J Neurosci 19*:5683–5692.

Shaywitz AJ, Greenberg ME (1999): CREB: a stimulus-induced transcription factor activated by a diverse array of extracellular signals. *Annu Rev Biochem 68*:821–861.

Sheng M, Greenberg ME (1990): The regulation and function of c-fos and other immediate early genes in the nervous system. *Neuron 4*:477–485.

Sheng M, Dougan ST, McFadden G, Greenberg ME (1988): Calcium and growth factor pathways of c-fos transcriptional activation require distinct upstream regulatory sequences. *Mol Cell Biol 8*:2787–2796.

Sheng M, McFadden G, Greenberg ME (1990): Membrane depolarization and calcium induce c-fos transcription via phosphorylation of transcription factor CREB. *Neuron 4*:571–582.

Sheng M, Thompson MA, Greenberg ME (1991): CREB: a $Ca^{(2+)}$-regulated transcription factor phosphorylated by calmodulin-dependent kinases. *Science 252*:1427–1430.

Shieh PB, Hu SC, Bobb K, Timmusk T, Ghosh A (1998): Identification of a signaling pathway involved in calcium regulation of BDNF expression. *Neuron 20*:727–740.

Short JM, Wynshaw-Boris A, Short HP, Hanson RW (1986): Characterization of the phosphoenolpyruvate carboxykinase (GTP) promoter-regulatory region. II. Identification of cAMP and glucocorticoid regulatory domains. *J Biol Chem 261*:9721–9726.

Strupp BJ, Levitsky DA (1984): Social transmission of food preferences in adult Hooded rats (Rattus norvegicus). *J Comp Psychol 98*:257–266.

Sugita S, Goldsmith JR, Baxter DA, Byrne JH (1992): Involvement of protein kinase C in serotonin-induced spike broadening and synaptic facilitation in sensorimotor connections of *Aplysia. J Neurophysiol 68*:643–651.

Sun P, Enslen H, Myung PS, Maurer RA (1994): Differential activation of CREB by Ca^{2+}/calmodulin-dependent protein kinases type II and type IV involves phosphorylation of a site that negatively regulates activity. *Genes Dev 8*:2527–2539.

Sun P, Lou L, Maurer RA (1996): Regulation of activating transcription factor-1 and the cAMP response element-binding protein by Ca^{2+}/calmodulin-dependent protein kinases type I, II, and IV. *J Biol Chem 271*:3066–3073.

Sutherland RJ, Kolb B, Whishaw IQ (1982): Spatial mapping: definitive disruption by hippocampal or medial frontal cortical damage in the rat. *Neurosci Lett 31*:271–276.

Swank MW (2000): Phosphorylation of MAP kinase and CREB in mouse cortex and amygdala during taste aversion learning. *Neuroreport 11*:1625–1630.

Sweatt JD, Kandel ER (1989): Persistent and transcriptionally-dependent increase in protein phosphorylation in long-term facilitation of *Aplysia* neurons. *Nature 339*:51–54.

Swope DL, Mueller CL, Chrivia JC (1996): CREB-binding protein activates transcription through multiple domains. *J Biol Chem 271*:28138–28145.

Tan Y, Rouse J, Zhang A, Cariati S, Cohen P, Comb MJ (1996): FGF and stress regulate CREB and ATF-1 via a pathway involving p38 MAP kinase and MAPKAP kinase-2. *EMBO J 15*:4629–4642.

Tao X, Finkbeiner S, Arnold DB, Shaywitz AJ, Greenberg ME (1998): Ca^{2+} influx regulates BDNF transcription by a CREB family transcription factor-dependent mechanism. *Neuron 20*:709–726.

Taubenfeld SM, Wiig KA, Monti B, Dolan B, Pollonini G, Alberini CM (2001): Fornix-dependent induction of hippocampal CCAAT enhancer-binding protein [beta] and [delta] Co-localizes with phosphorylated cAMP response element-binding protein and accompanies long-term memory consolidation. *J Neurosci 21*:84–91.

Thompson RF (1986): The neurobiology of learning and memory. *Science 233*:941–947.

Thor DH, Holloway WR (1982): Social memory of the male laboratory rat. *J Comp Physiol Psychol 96*:1000–1006.

Tolliver BK, Sganga MW, Sharp FR (2000): Suppression of c-fos induction in the nucleus accumbens prevents acquisition but not expression of morphine-conditioned place preference. *Eur J Neurosci 12*:3399–3406.

Tully T (1991): Genetic dissection of learning and memory in *Drosophila melanogaster*. In: Madden J (Ed), *Neurobiology of Learning, Emotion and Affect*. New York: Raven Press, pp 30–66.

Tully T, Preat T, Boynton SC, Del Vecchio M (1994): Genetic dissection of consolidated memory in *Drosophila. Cell 79*:35–47.

Van der Loos H, Woolsey TA (1973): Somatosensory cortex: structural alterations following early injury to sense organs. *Science 179*:395–398.

Waeber G, Meyer TE, LeSieur M, Hermann HL, Gerard N, Habener JF (1991): Developmental stage-specific expression of cyclic adenosine 3′,5′-monophosphate response element-binding protein CREB during spermatogenesis involves alternative exon splicing. *Mol Endocrinol 5*:1418–1430.

Walters ET, Byrne JH, Carew TJ, Kandel ER (1983): Mechanoafferent neurons innervating tail of *Aplysia*. II. Modulation by sensitizing stimuli. *J Neurophysiol 50*:1543–1559.

Welker C (1971): Microelectrode delineation of fine grain somatotopic organization of (SmI) cerebral neocortex in albino rat. *Brain Res 26*:259–275.

Wiesel TN, Hubel DH (1963): Single-cell responses in striate cortex of kittens deprived of vision in one eye. *J Neurophysiol 26*:111–115.

Wiesel TN, Hubel DH (1965): Comparison of the effects of unilateral and bilateral eye closure on cortical unit responses in kittens. *J Neurophysiol 28*:1029–1040.

Winocur G (1990): Anterograde and retrograde amnesia in rats with dorsal hippocampal or dorsomedial thalamic lesions. *Behav Brain Res 38*:145–154.

Woo NH, Duffy SN, Abel T, Nguyen PV (2000): Genetic and pharmacological demonstration of differential recruitment of cAMP-dependent protein kinases by synaptic activity. *J Neurophysiol 84*:2739–2745.

Woolsey TA (1967): Somatosensory, auditory and visual cortical areas of the mouse. *Johns Hopkins Med J 121*:91–112.

Woolsey TA, Van der Loos H (1970): The structural organization of layer IV in the somatosensory region (SI) of mouse cerebral cortex. The description of a cortical field composed of discrete cytoarchitectonic units. *Brain Res 17*:205–242.

Worley PF, Bhat RV, Baraban JM, Erickson CA, McNaughton BL, Barnes CA (1991): Thresholds for synaptic activation of transcription factors in hippocampus: correlation with long-term enhancement. *J Neurosci 11*:4776–4786.

Wu GY, Deisseroth K, Tsien RW (2001a): Activity-dependent CREB phosphorylation: Convergence of a fast, sensitive calmodulin kinase pathway and a slow, less sensitive mitogen-activated protein kinase pathway. *Proc Natl Acad Sci USA 98*:2808–2813.

Wu GY, Deisseroth K, Tsien RW (2001b): Spaced stimuli stabilize MAPK pathway activation and its effects on dendritic morphology. *Nat Neurosci 4*:151–158.

Wu X, McMurray CT (2001): Calmodulin kinase II attenuation of gene transcription by preventing cAMP response element-binding protein (CREB) dimerization and binding of the CREB-binding protein. *J Biol Chem 276*:1735–1741.

Wüstenberg D, Gerber B, Menzel R (1998): Short communication: long- but not medium-term retention of olfactory memories in honeybees is impaired by actinomycin D and anisomycin. *Eur J Neurosci 10*:2742–2745.

Xing J, Ginty DD, Greenberg ME (1996): Coupling of the RAS–MAPK pathway to gene activation by RSK2, a growth factor-regulated CREB kinase. *Science 273*:959–963.

Yamamoto KK, Gonzalez GA, Menzel P, Rivier J, Montminy MR (1990): Characterization of a bipartite activator domain in transcription factor CREB. *Cell 60*:611–617.

Yin J, Wallach JS, Vecchio MD, Wilder EL, Zhou H, Quinn WG, Tully T (1994): Induction of a dominant-negative CREB transgene specifically blocks long-term memory in Drosophila melanogaster. *Cell 79*:49–58.

Yin J, Del Vecchio M, Zhou H, Tully T (1995): CREB as a memory modulator: Induced expression of a dCREB2 activator isoform enhances long-term memory in Drosophila. *Cell 81*:107–115.

Yun YD, Dumoulin M, Habener JF (1990): DNA-binding and dimerization domains of adenosine 3′,5′-cyclic monophosphate-responsive protein CREB reside in the carboxyl-terminal 66 amino acids. *Mol Endocrinol 4*:931–939.

Zacks RT (1969): Invariance of total learning time under different conditions of practice. *J Exp Psychol 82*:289–296.

Zhang F, Endo S, Cleary LJ, Eskin A, Byrne JH (1997): Role of transforming growth factor-β in long term synaptic facilitation in *Aplysia*. *Science 275*:1318–1320.

Subject Index